VOLUME SIXTY EIGHT

ADVANCES IN BOTANICAL RESEARCH

New Light on Alkaloid Biosynthesis and Future Prospects

ADVANCES IN BOTANICAL RESEARCH

Series Editors

VOLUME SIXTY EIGHT

ADVANCES IN BOTANICAL RESEARCH

New Light on Alkaloid Biosynthesis and Future Prospects

Volume Editor

NATHALIE GIGLIOLI-GUIVARC'H

Equipe d'Accueil EA 2106 'Biomolécules et Biotechnologies Végétales', Université François-Rabelais de Tours, Tours, France

ELSEVIER

AMSTERDAM • BOSTON • HEIDELBERG • LONDON
NEW YORK • OXFORD • PARIS • SAN DIEGO
SAN FRANCISCO • SINGAPORE • SYDNEY • TOKYO
Academic Press is an imprint of Elsevier

Academic Press is an imprint of Elsevier
32 Jamestown Road, London NW1 7BY, UK
Radarweg 29, PO Box 211, 1000 AE Amsterdam, The Netherlands
The Boulevard, Langford Lane, Kidlington, Oxford, OX5 1GB, UK
225 Wyman Street, Waltham, MA 02451, USA
525 B Street, Suite 1800, San Diego, CA 92101-4495, USA

First edition 2013

ISBN: 978-0-12-408061-4
ISSN: 0065-2296

Printed and bound by CPI Group (UK) Ltd, Croydon, CR0 4YY

Transferred to digital print 2012

CONTENTS

Contributors *ix*
Preface *xiii*

1. Towards Complete Elucidation of Monoterpene Indole Alkaloid Biosynthesis Pathway: *Catharanthus roseus* as a Pioneer System **1**
Vonny Salim and Vincenzo De Luca

1. Introduction 2
2. Division of MIA Biosynthesis Pathway 5
3. Organisation and Spatial Separation of MIA Biosynthesis 18
4. Large-Scale Genomic Approaches in Functional Characterisation of Genes Involved in MIA Biosynthesis 21
5. Metabolic Engineering of the MIA Biosynthesis Pathway 24
6. Conclusions and Perspectives 27

Acknowledgements 29
References 29

2. Increasing the Pace of New Discoveries in Tropane Alkaloid Biosynthesis **39**
Jan Jirschitzka, Franziska Dolke, and John C. D'Auria

1. Introduction 40
2. Tropane Alkaloids in Plants: From Herbs To Trees 44
3. Biosynthesis of Tropane Alkaloids 49
4. Metabolic Engineering of Tropane Alkaloids 59
5. Conclusions 63

Acknowledgements 63
References 63

3. Deciphering the Evolution, Cell Biology and Regulation of Monoterpene Indole Alkaloids **73**
Benoit St-Pierre, Sébastien Besseau, Marc Clastre, Vincent Courdavault, Martine Courtois, Joel Crèche, Eric Ducos, Thomas Dugé de Bernonville, Christelle Dutilleul, Gaëlle Glévarec, Nadine Imbault, Arnaud Lanoue, Audrey Oudin, Nicolas Papon, Olivier Pichon, and Nathalie Giglioli-Guivarc'h

1. Introduction 74
2. Regulation of MIA Biosynthesis Pathway 86

3. Spatial Organization of MIA Biosynthesis 93
4. Conclusions 101
Acknowledgements 102
References 102

4. Biosynthesis and Catabolism of Purine Alkaloids 111
Hiroshi Ashihara, Takao Yokota, and Alan Crozier

1. Introduction 112
2. Occurrence of Purine Alkaloids in Plant Kingdom 113
3. Biosynthesis of Purine Alkaloids 115
4. Catabolism of Caffeine 126
5. Distribution of Purine Alkaloids in Tissues and Cells 128
6. Biotechnology of Purine Alkaloids 129
7. *In Planta* Function of Purine Alkaloids 131
8. Conclusions and Perspectives 133
References 133

5. Camptothecin: Biosynthesis, Biotechnological Production and Resistance Mechanism(s) 139
Supaart Sirikantaramas, Mami Yamazaki, and Kazuki Saito

1. Introduction 140
2. Camptothecin Biosynthetic Pathway 143
3. Biotechnological Production of Camptothecin and Its Derivatives 150
4. Resistance Mechanism(s) in Camptothecin-Producing Plants 154
5. Conclusions 155
Acknowledgements 156
References 156

6. Improved Production of Plant Isoquinoline Alkaloids by Metabolic Engineering 163
Fumihiko Sato

1. Introduction 164
2. Isoquinoline Alkaloid Biosynthesis and Pathway Characterization 165
3. Metabolic Pathway Engineering 171
4. Perspectives 176
5. Conclusions 177
Acknowledgement 177
References 177

7. Bioengineering of Isoquinoline Alkaloid Production in Microbial Systems **183**

Akira Nakagawa, Eitaro Matsumura, Fumihiko Sato, and Hiromichi Minami

1. Introduction 184
2. Design of IQA Biosynthetic Pathways and Their Enzymes in Bacteria 186
3. Perspectives 198
4. Conclusions 199

Acknowledgements 199
References 199

8. From Bioreactor to Entire Plants: Development of Production Systems for Secondary Metabolites **205**

Thi Khieu Oanh Nguyen, Rebecca Dauwe, Frédéric Bourgaud, and Eric Gontier

1. Introduction 206
2. General Strategies to Produce Secondary Metabolites from Plant Material 208
3. Tropane Alkaloids Produced by Solanaceae 210
4. Furanocoumarins Produced by Rutaceae 211
5. Evaluating Plant Bioproduction and Selecting High-Producing Genotypes 212
6. First Steps in the Development of Agronomical Bioproduction 216
7. Biotechnological Approaches: From Callus to Organ *In Vitro* Culture 217
8. Process Engineering and Low-Cost 'Bioreactor' Design 218
9. Comparison Between Biotech (*In Vitro*) and Field Production 222
10. Hydroponics and Aeroponics as Open Bioreactors for the Production and Exploitation of High-Quality Biomass 223
11. Future Perspectives and Limitations; Concluding Remarks 226

Acknowledgements 228
References 228

9. New Methods of Analysis and Investigation of Terpenoid Indole Alkaloids **233**

Guitele Dalia Goldhaber-Pasillas, Young Hae Choi, and Robert Verpoorte

1. Introduction 235
2. Extraction and Purification Methods 236
3. Analytical Methods 242
4. Applications in Fingerprint Analysis 260
5. Conclusions 262

References 263

10. Like Cures Like: Caffeine Immunizes Plants Against Biotic Stresses **273**

Hiroshi Sano, Yun-Soo Kim, and Yong-Eui Choi

1. Introduction 274
2. Caffeine Biosynthesis 276
3. Caffeine-Producing Transgenic Plants 278
4. Biotic Stress Tolerance 280
5. Activation of Defence System 283
6. Caffeine Targets 284
7. Caffeine Signal Cascade 287
8. Plant Immunization 291
9. Concluding Remarks 293

Acknowledgements 295
References 295

11. Alkaloids from Marine Bacteria **301**

Sergey B. Zotchev

1. Introduction 302
2. Occurrence and Biological Activities of Alkaloids from Marine Bacteria 302
3. Biosynthesis of Alkaloids in Marine Bacteria 319
4. Conclusions 327

Acknowledgements 327
References 327

Author Index *335*
Subject Index *367*

CONTRIBUTORS

Hiroshi Ashihara
Department of Biological Sciences, Ochanomizu University, Otsuka, Bunkyo-ku, Tokyo, Japan

Sébastien Besseau
EA 2106 'Biomolécules et Biotechnologies Végétales', Université François-Rabelais de Tours, Tours, France

Frédéric Bourgaud
Université de Lorraine; INRA, Laboratoire Agronomie et Environnement, UMR1121, ENSAIA, Vandœuvre, and Plant Advanced Technologies SA, Vandœuvre, France

Yong-Eui Choi
Department of Forest Resources, Kangwon National University, Chuncheon, Republic of Korea

Young Hae Choi
Natural Products Laboratory, Institute of Biology Leiden, Leiden University, Leiden, The Netherlands

Marc Clastre
EA 2106 'Biomolécules et Biotechnologies Végétales', Université François-Rabelais de Tours, Tours, France

Vincent Courdavault
EA 2106 'Biomolécules et Biotechnologies Végétales', Université François-Rabelais de Tours, Tours, France

Martine Courtois
EA 2106 'Biomolécules et Biotechnologies Végétales', Université François-Rabelais de Tours, Tours, France

Joel Crèche
EA 2106 'Biomolécules et Biotechnologies Végétales', Université François-Rabelais de Tours, Tours, France

Alan Crozier
School of Medicine, College of Medical, Veterinary and Life Sciences, University of Glasgow, Glasgow, United Kingdom

John C. D'Auria
Department of Biochemistry, Max Planck Institute for Chemical Ecology, Jena, Germany

Rebecca Dauwe
Plant Biology & Innovation research Unit EA3900-UPJV, Université de Picardie Jules Verne, PRES UFECAP, Faculty of Sciences, Ilot des poulies, Amiens, France

Thomas Dugé de Bernonville
EA 2106 'Biomolécules et Biotechnologies Végétales', Université François-Rabelais de Tours, Tours, France

Vincenzo De Luca
Department of Biological Sciences, Brock University, St. Catharines, Ontario, Canada

Franziska Dolke
Department of Bioorganic Chemistry, Max Planck Institute for Chemical Ecology, Jena, Germany

Eric Ducos
EA 2106 'Biomolécules et Biotechnologies Végétales', Université François-Rabelais de Tours, Tours, France

Christelle Dutilleul
EA 2106 'Biomolécules et Biotechnologies Végétales', Université François-Rabelais de Tours, Tours, France

Nathalie Giglioli-Guivarc'h
EA 2106 'Biomolécules et Biotechnologies Végétales', Université François-Rabelais de Tours, Tours, France

Gaëlle Glévarec
EA 2106 'Biomolécules et Biotechnologies Végétales', Université François-Rabelais de Tours, Tours, France

Guitele Dalia Goldhaber-Pasillas
Natural Products Laboratory, Institute of Biology Leiden, Leiden University, Leiden, The Netherlands

Eric Gontier
Plant Biology & Innovation research Unit EA3900-UPJV, Université de Picardie Jules Verne, PRES UFECAP, Faculty of Sciences, Ilot des poulies, Amiens, and Plant Advanced Technologies SA, Vandœuvre, France

Nadine Imbault
EA 2106 'Biomolécules et Biotechnologies Végétales', Université François-Rabelais de Tours, Tours, France

Jan Jirschitzka
Department of Biochemistry, Max Planck Institute for Chemical Ecology, Jena, Germany

Yun-Soo Kim
Department of Forest Resources, Kangwon National University, Chuncheon, Republic of Korea

Arnaud Lanoue
EA 2106 'Biomolécules et Biotechnologies Végétales', Université François-Rabelais de Tours, Tours, France

Eitaro Matsumura
Research Institute for Bioresources and Biotechnology, Ishikawa Prefectural University, Suematsu, Nonoichi, Ishikawa, Japan

Hiromichi Minami
Research Institute for Bioresources and Biotechnology, Ishikawa Prefectural University, Suematsu, Nonoichi, Ishikawa, Japan

Akira Nakagawa
Research Institute for Bioresources and Biotechnology, Ishikawa Prefectural University, Suematsu, Nonoichi, Ishikawa, Japan

Thi Khieu Oanh Nguyen
Plant Biology & Innovation research Unit EA3900-UPJV, Université de Picardie Jules Verne, PRES UFECAP, Faculty of Sciences, Ilot des poulies, Amiens, France

Audrey Oudin
EA 2106 'Biomolécules et Biotechnologies Végétales', Université François-Rabelais de Tours, Tours, France

Nicolas Papon
EA 2106 'Biomolécules et Biotechnologies Végétales', Université François-Rabelais de Tours, Tours, France

Olivier Pichon
EA 2106 'Biomolécules et Biotechnologies Végétales', Université François-Rabelais de Tours, Tours, France

Kazuki Saito
Graduate School of Pharmaceutical Sciences, Chiba University, Chuo-ku, Chiba, and RIKEN Center for Sustainable Resource Science, Tsurumi-ku, Yokohama, Japan

Vonny Salim
Department of Biological Sciences, Brock University, St. Catharines, Ontario, Canada

Hiroshi Sano
Research and Education Center for Genetic Information, Nara Institute of Science and Technology, Nara, Japan, and Department of Forest Resources, Kangwon National University, Chuncheon, Republic of Korea

Fumihiko Sato
Division of Integrated Life Science, Graduate School of Biostudies, Kyoto University, Oiwake-cho, Kitashirakawa, Sakyo-ku, Kyoto, Japan

Supaart Sirikantaramas
Department of Biochemistry, Faculty of Science, Chulalongkorn University, Bangkok, Thailand

Benoit St-Pierre
EA 2106 'Biomolécules et Biotechnologies Végétales', Université François-Rabelais de Tours, Tours, France

Robert Verpoorte
Natural Products Laboratory, Institute of Biology Leiden, Leiden University, Leiden, The Netherlands

Mami Yamazaki
Graduate School of Pharmaceutical Sciences, Chiba University, Chuo-ku, Chiba, Japan

Takao Yokota
Department of Biosciences, Teikyo University, Utsunomiya, Japan

Sergey B. Zotchev
Department of Biotechnology, Norwegian University of Science and Technology, Trondheim, Norway

PREFACE

Over the past decade, applications of high-throughput technologies, such as expressed sequence tag databases, DNA microarrays, and proteome and metabolome analyses, have considerably accelerated the discovery of new components and mechanisms involved in the assembly of alkaloids in plants. Combined with an intensive worldwide research programme and using several technical breakthroughs in biochemical, molecular, cellular, and physiological research, they have contributed to impressive advancements in our understanding of alkaloid biosynthesis. In particular, many enzymes acting in alkaloid biosynthetic pathways have been identified and characterized at the molecular level and numerous regulation processes have been also deciphered, highlighting the specific roles of phytohormones in the regulation of alkaloid biosynthesis as well as in their trafficking and storage. Different approaches based on microscopy analysis have also contributed to the elucidation of the complex and original architectures of alkaloid biosynthetic pathways showing the distribution of the high number of enzymatic steps in different tissues but also in different subcellular compartments. Alkaloids classification, usually based on their chemical structure, is thus reassessed according to a new vision of their metabolism and better knowledge of their biological and ecological activities. This complex organization can explain, in part, the difficulties encountered in the attempts to improve alkaloid production *in planta* as well as the challenges to generate these products via chemical synthesis.

Alkaloids have important biological activities, many of which have medicinal properties and are used in the treatment of human ailments, explaining our great interest not only in the identification of new natural molecules but also in the development of alkaloid production processes. We still believe that these alkaloids, and drugs developed from them, could be part of our arsenal of medicines used to cure serious diseases such as cancer or AIDS. However, plants contain only low levels of alkaloids. Therefore, improving their production by chemical synthesis or by increasing natural synthesis in plants remains a challenge and motivates research in this field. One of the ultimate goals of current research is to transpose part of the recent discoveries to the development of metabolic engineering strategies to overcome the usually very low yield of alkaloid production *in planta*. In this context, we have witnessed, over the past 5 years, the emergence of new processes including development of yeast or bacterial platforms for the fermentative production of plant alkaloids.

The understanding of alkaloid biosynthetic pathways improves every year and there are now 12,000 natural compounds recognized as alkaloids, according to the discovery and the characterization of new natural molecules. Alkaloids present a relatively large prevalence in nature and are relatively common chemicals in all kingdoms of living organisms. More than 20% of identified plants are able to produce alkaloids of one form or another. These alkaloids show a very large degree of diversity that is at least equivalent to that observed between the plant species themselves. This structural diversity is probably the result of specific biochemical differentiation over the course of evolutionary time which reflects changing interactions of molecules with biological targets leading to adaption of plants to their changing environment. Thus, in our quest to identify new bioactive molecules after having largely explored the plant kingdom, we are now greatly interested in marine organisms. Indeed, if several plant-derived alkaloids are now classed as leading drugs in the treatment of different types of cancer, marine-derived alkaloids, isolated from aquatic fungi, cyanobacteria, sponges, algae, and tunicates, have been found to also exhibit various anti-cancer activities suggesting exciting perspectives.

Alkaloids have already been the subject of many books and academic works from various scientific fields. *New Light on Alkaloid Biosynthesis and Future Prospects* is intended to present different alkaloid families displaying the best characterized biosynthetic pathways and ecological role and the most advanced biotechnological developments. The book is divided into eleven chapters. Chapters 1 to 6 focus on plant alkaloid biosynthesis with a particular emphasis on monotepernoid indole alkaloids, tropane alkaloids, purine alkaloids and isoquinoline alkaloids. This presentation is completed by the chapter 9 which aims at presenting the technological advances leading to the identification and the characterization of new alkaloids and the chapter 10 which illustrates a distinct ecological role of alkaloids. Moreover the final chapter 11 gives an overview of marine alkaloids since very interesting discoveries have been recently made. Four chapters from 5 to 8 are more dedicated to the presentation of recent biotechnological developments leading to improvement in alkaloids production. Two of them present example(s) of the biotechnological production of camptothecin and the development of plant systems adapted to the production of specialized metabolites. The two others describe the ways used to improve the production of isoquinoline alkaloids by metabolic engineering in transgenic plants or by using an emerging concept of fermentative production of plant alkaloids by heterologous microbial systems.

In conclusion, I wanted here to acknowledge all the authors who agreed to share their knowledge on alkaloids and to contribute to this book and also the colleagues of my laboratory for their assistance in the preparation of this book. I hope it will be of interest not only for researchers but also for students or anyone who is interested in the field of specialized metabolite research.

NATHALIE GIGLIOLI-GUIVARC'H
March 2013

CHAPTER ONE

Towards Complete Elucidation of Monoterpene Indole Alkaloid Biosynthesis Pathway: *Catharanthus roseus* as a Pioneer System

Vonny Salim, Vincenzo De Luca[1]
Department of Biological Sciences, Brock University, St. Catharines, Ontario, Canada
[1]Corresponding author: e-mail address: vdeluca@brocku.ca

Contents

1. Introduction 2
2. Division of MIA Biosynthesis Pathway 5
   2.1 Early monoterpene biosynthesis 5
   2.2 Iridoid biosynthesis 10
   2.3 Early MIA biosynthesis 12
   2.4 The late MIA biosynthesis pathway 16
3. Organisation and Spatial Separation of MIA Biosynthesis 18
   3.1 Epidermis as an important biosynthetic site of MIAs and their precursors 19
   3.2 The use of epidermis-enriched transcriptomic resources for gene discovery 20
4. Large-Scale Genomic Approaches in Functional Characterisation of Genes Involved in MIA Biosynthesis 21
   4.1 The shared pathways among Apocynaceae family 21
   4.2 Tools for screening the candidate genes 22
5. Metabolic Engineering of the MIA Biosynthesis Pathway 24
6. Conclusions and Perspectives 27
Acknowledgements 29
References 29

Abstract

The development of various plant-based engineering efforts has been facilitated by recent large-scale transcriptomic resources. In consideration of the progress in the study of monoterpene indole alkaloid (MIA) metabolism achieved in the last decade, some strategies have been developed for metabolic engineering efforts. However, unidentified biosynthetic genes in the pathway limit this potential. *Catharanthus roseus* is the most well-studied medicinal plant owing to its production of valuable anticancer dimeric MIAs such as vinblastine. This chapter highlights the cell-, organ-, development- and

Advances in Botanical Research, Volume 68
ISSN 0065-2296
http://dx.doi.org/10.1016/B978-0-12-408061-4.00001-8

environment-specific organisation of MIA biosynthesis and describes the intra- and intercellular trafficking of MIAs required for their assembly within *C. roseus*. The combined use of cell- and organ-specific transcriptome databases of several MIA-accumulating plants is facilitating combined bioinformatic approaches to identify MIA candidate genes. Virus-induced gene silencing is being used to screen candidate genes for their involvement in MIA biosynthesis, and the function of selected genes can be identified by the expression and assay of recombinant proteins in bacterial or yeast systems. These new tools show great promise for a more rapid discovery of new genes involved in whole MIA pathways that enhance the potential of reconstituting them in heterologous microorganisms for the production of any valuable MIA.

ABBREVIATIONS

IPAP cells internal phloem associated parenchyma cells
LAMT loganic acid methyltransferase
MeJA methyl jasmonate
MEP methylerythritol 4-phosphate
MIA monoterpene indole alkaloid
MVA mevalonic acid
VIGS virus-induced gene silencing

1. INTRODUCTION

Monoterpene indole alkaloids (MIAs) are one of the most diverse groups of plant secondary metabolites of the Apocynaceae, Loganiaceae and Rubiaceae plant families. MIAs comprise approximately 3000 compounds that exhibit powerful biological activities. While the roles of some MIAs have been described in plant defense against herbivores and pathogens (Luijendijk, Vandermeijden, & Verpoorte, 1996), key derivatives have been exploited for therapeutic purposes. Several drugs have been developed including anticancer agents such as vinblastine and vincristine from *Catharanthus roseus* (Madagascar periwinkle) and camptothecin from *Camptotheca acuminata*, antiarrhythmic agents such as ajmaline, agents for the treatment of neurological disorders such as serpentine from *Rauwolfia serpentina* and antimalarial agents such as quinine from *Cinchona ledgeriana* (Fig. 1.1). Among the best-characterised plants investigated, the MIA biosynthesis pathways from *C. roseus* have been extensively studied at the biochemical and molecular levels (O'Connor & Maresh, 2006; Ziegler & Facchini, 2008). While the Madagascar periwinkle remains the only commercial source for vinblastine and vincristine, this plant also accumulates

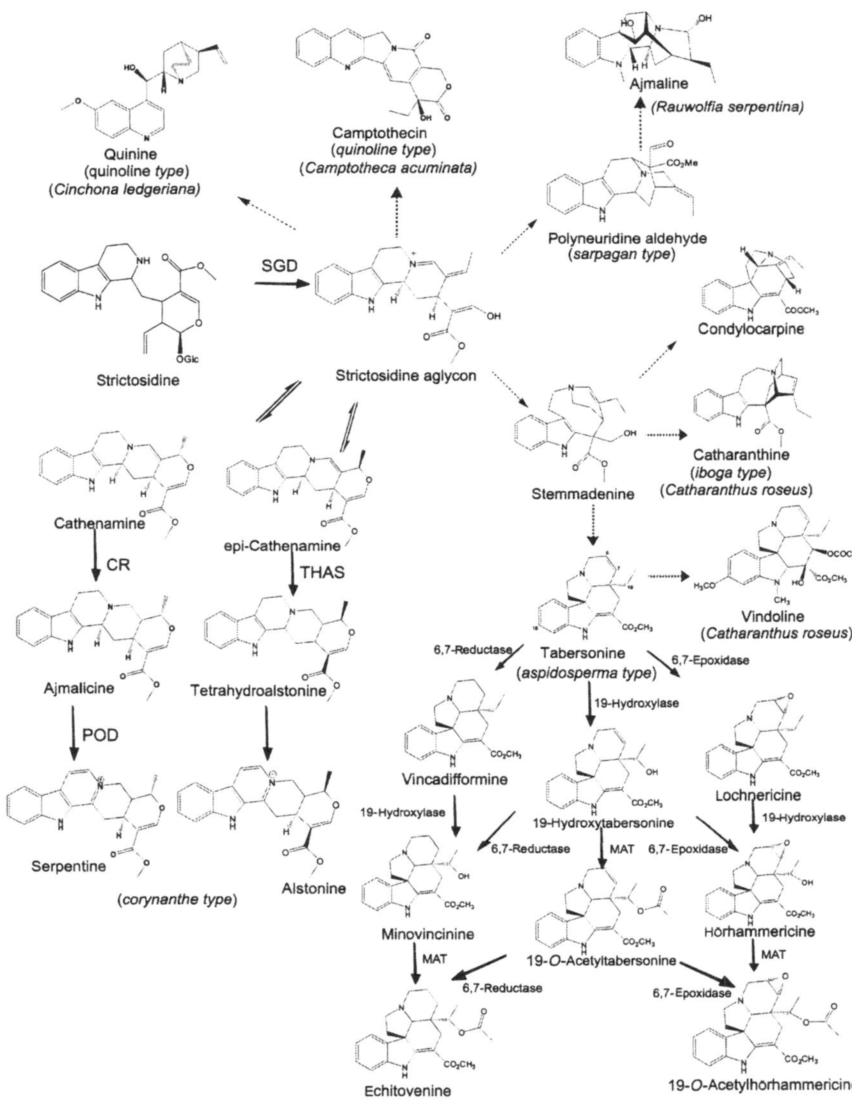

Figure 1.1 The representation of chemical structures of different types of alkaloids derived from strictosidine, namely iboga (catharanthine), aspidosperma (tabersonine that is elaborated into separate pathways to produce vindoline in *Catharanthus* leaves, or MIAs such as lochnericine and hörhammericine in roots) and corynanthe (ajmalicine and serpentine in *Catharanthus*), sarpagan types in *Rauwolfia serpentina*, quinoline types such as camptothecin in *Camptotheca acuminata* and quinine in *Cinchona ledgeriana*. Dashed arrows show the multiple steps in the biosynthetic pathway. Abbreviations: SGD, strictosidine β-D-glucosidase, CR, cathenamine reductase, THAS, tetrahydroalstonine synthase, POD, peroxidase; MAT, minovincinine 19-*O*-acetyltransferase. *Adapted from El-Sayed and Verpoorte (2007) and Giddings et al. (2011).*

many different types of MIAs (O'Connor & Maresh, 2006; van der Heijden, Jacobs, Snoeijer, Hallard, & Verpoorte, 2004) in addition to these potent inhibitors of microtubule formation that have been developed to treat leukaemia, Hodgkin's lymphoma and other types of cancer.

The early stages of MIA biosynthesis in *C. roseus* involve the formation of the iridoid-secologanin derived from isoprenoid biosynthesis and its condensation with tryptamine to yield the central intermediate, strictosidine, the common precursor for highly divergent MIAs (Fig. 1.1) that include catharanthine and vindoline. The two monomers are then coupled to form the anticancer dimeric MIA, vinblastine (Costa et al., 2008). Despite the efforts by synthetic chemists to produce these valuable secondary metabolites (Ishikawa, Colby, & Boger, 2008; Kuboyama, Yokoshima, Tokuyama, & Fukuyama, 2004), their industrial production still depends on extraction and purification from *C. roseus* leaves with low yields (Gueritte, Bac, Langlois, & Potier, 1980). Therefore, an alternative method to improve the production of these valuable molecules would be advantageous. Some efforts will include the transfer of biosynthetic pathways identified in *C. roseus* or other MIA-producing plants to microorganisms or to other plant species, which will require enormous technological breakthroughs to be realised. In order to contemplate such large scale pathway engineering, the identification and characterization of the remaining biosynthetic genes and other proteins involved in the target pathway is necessary. In addition, MIA biosynthesis in *C. roseus* has been demonstrated to occur in distinct cell types. The translocation of intermediates between cells requires the identification of the transporters involved and shows the additional complexities of MIA biosynthesis that may require understanding and characterization in order to facilitate metabolic engineering of the pathway.

Although a number of genes involved in the biosynthesis of catharanthine and vindoline in *C. roseus* have been identified by traditional forward genetic approaches (enzyme isolation from the plants, protein purification followed by sequencing and recombinant protein expression in appropriate hosts), many genes remain to be elucidated (De Luca, Salim, Levac, Atsumi, & Yu, 2012). Homology-based cloning approaches that use sequence similarity to help identify other gene family members, such as acetyltransferase and cytochrome P450 genes, involved in the pathway have been applied. However, this method relies on testing many candidate genes for functional analysis through tedious processes and is often limited by the availability of the substrate for the reaction being identified.

Alternatively, the plant can be stimulated with an elicitor to up-regulate the likely MIA genes involved that lead to increases in MIA production (Ziegler & Facchini, 2008). In the last decade, modern genomic and molecular biology methods have begun to accelerate the discovery of MIA biosynthesis steps (De Luca, Salim, Atsumi, & Yu, 2012 (Science Review); Facchini et al., 2012; Góngora-Castillo et al., 2012; O'Connor, 2012) through the application of high-throughput technologies, including expressed sequence tags (ESTs), DNA microarray analyses, proteomics and metabolomics. Large-scale medicinal plant genome projects such as Phytometasyn (Facchini et al., 2012) and the Medicinal Plant Genomics Consortium (http://www.medicinalplantgenomics.msu.edu/; Góngora-Castillo et al., 2012) have included studies with *C. roseus,* and candidate genes for the remaining steps in MIA biosynthesis in this species are likely to be revealed in the next few years.

This chapter focuses on the biochemistry of MIA biosynthesis, its cell- and organ-specific localization and its regulation by developmental and environmental cues, together with the intra- and inter-cellular trafficking of biosynthetic intermediates required for elaboration of MIAs in the *C. roseus* model system. The chapter also describes how database mining together with virus-induced gene silencing (VIGS) is being used to speed up the discovery of new MIA genes in *C. roseus*.

2. DIVISION OF MIA BIOSYNTHESIS PATHWAY

MIA biosynthetic pathways will be illustrated in four different stages: early steps in monoterpene biosynthesis, iridoid biosynthesis, early MIA biosynthesis and late MIA biosynthesis that emphasise the vindoline pathway in *C. roseus*.

2.1. Early monoterpene biosynthesis

2.1.1 Biosynthetic genes involved in the early monoterpene pathway

Terpenes are the largest class of secondary metabolites with over 30,000 compounds derived from C_5 isoprenoid units. These isoprene units condense to form $C_{5\times}$ moieties such as monoterpenes (C_{10}), sesquiterpene (C_{15}), diterpene (C_{20}), triterpene (C_{30}), tetraterpene (C_{40}) and polyterpenes ($C_{5\times}$). Terpenes are known to have many biological and physiological functions that affect the normal growth and development of plants as they are the precursors of chlorophyll and of hormones such as cytokinins, gibberellins, abscisic acid and brassinosteroids (Rodriguez-Conception & Boronat, 2002). The early steps in the isoprenoid pathway consist of the enzymatic steps involved in the synthesis of isopentenyl diphosphate (IPP). Within plants, the biosynthesis of IPP can occur via two metabolic pathways: the first known as the mevalonic acid (MVA) pathway was discovered in the 1950s and occurs in animals, plants, fungi and some bacteria; the second is the methylerythritol 4-phosphate (MEP) pathway discovered in the mid-1990s and found in most bacteria and plants, but absent in archaebacteria, fungi and animals (Rodriguez-Conception & Boronat, 2002; Rohmer, 1999; Fig. 1.2).

The mevalonate pathway starts with the coupling of two molecules of acetyl-CoA to form acetoacetyl-CoA catalysed by acetoacetyl-CoA thiolase. Condensation of acetoacetyl-CoA with another unit of acetyl-CoA to form 3-hydroxy-3-methyl-glutaryl-CoA (HMG-CoA) is then catalysed by HMG-CoA synthase. This intermediate is then reduced to form MVA by HMG-CoA reductase, which is subsequently phosphorylated to produce 5-diphosphomevalonate by mevalonate kinase (MVAK). 5-Diphosphomevalonate is then decarboxylated by 5-diphosphomevalonate decarboxylase to IPP. While the MVA pathway, found within the cytosol, provides isoprene units for the assembly of sesquiterpenes and triterpenes (Lange & Croteau, 1999; Newman & Chappell, 1999), the plastid-localised MEP pathway leads to the synthesis of monoterpenes, diterpenes and tetraterpenes as shown in Fig. 1.2. It has been suggested that the biosynthesis of both pathways is highly regulated with cross-talk between the two pathways across the cytosolic and plastid compartments (Eisenreich, Rohdich, & Bacher, 2001; El-Sayed & Verpoorte, 2007; Oudin, Courtois, Rideau, & Clastre, 2007).

The initial step of the MEP pathway involves glyceraldehyde 3-phosphate and pyruvate that condense to form 1-deoxy-D-xylulose 5-phosphate (DXP). cDNA encoding 1-deoxy-D-xylulose 5-phosphate synthase (DXS), part of a family of transketolases from *C. roseus*, has been isolated and characterised (Chahed et al., 2000). This metabolite is then reduced and isomerised to produce 2-*C*-methyl-D-erythritol-4-phosphate by DXP reductoisomerase (DXR). In the latter step, this intermediate is then condensed with CTP to generate 4-(cytidine 5′-diphospho)-2-*C*-methyl-D-erythritol by 4-cytidyl-diphospho-2-*C*-methyl-D-erythritol synthase (MECS). Both DXR and MECS have been isolated from *C. roseus* and shown to be up-regulated in MIA-producing cell cultures (Rohdich, Kis, Bacher, & Eisenreich, 2001; Veau et al., 2000). This intermediate is then phosphorylated by ATP to form 2-phospho-4-(cytidine

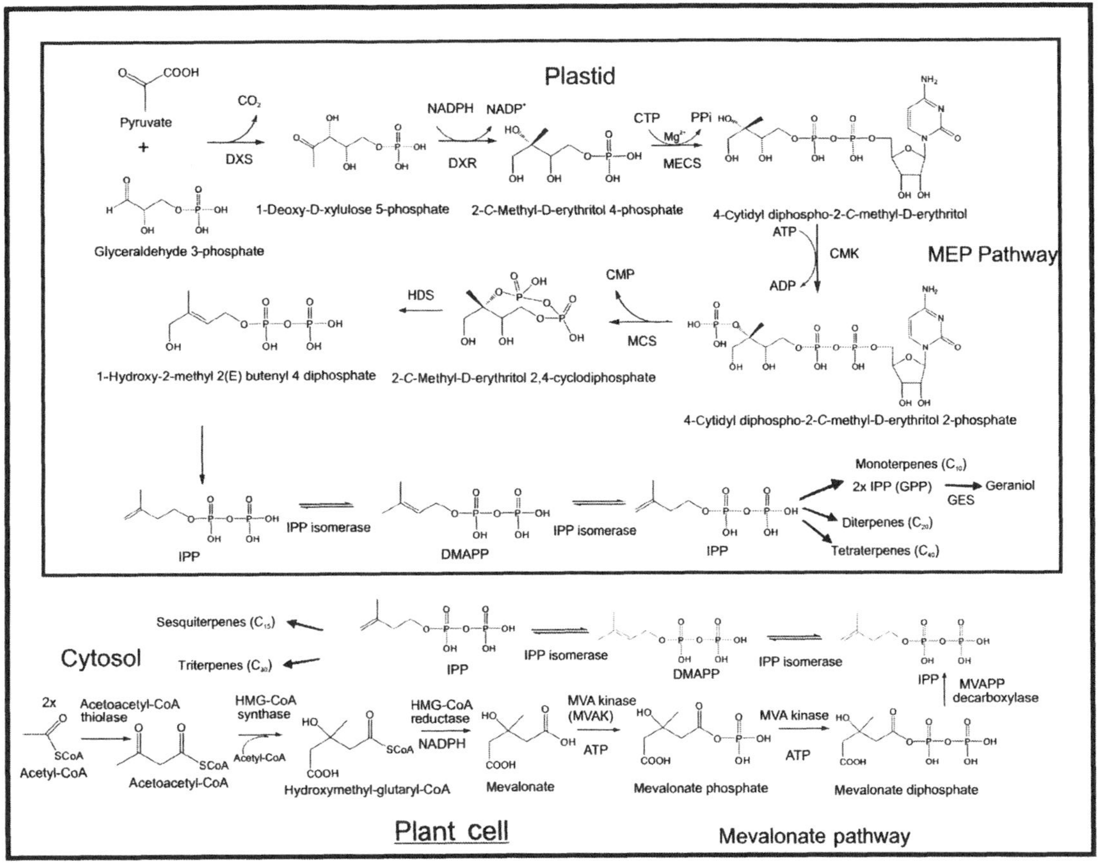

Figure 1.2 The biosynthetic steps of the early monoterpene pathway that involves the plastidic MEP pathway. Diagram illustrating the subcellular spatial separation and the source of the plant-derived terpenes. Abbreviations: MEP, 2-*C*-methyl-D-erythritol-4-phosphate, DXS, 1-deoxy-D-xylulose 5-phosphate-synthase, DXR, 1-deoxy-D-xylulose 5-phosphate reductase, MECS, 4-cytidyl-diphospho-2-*C*-methyl-D-erythritol synthase, CMK, 4-cytidyl-diphospho-2-*C*-methyl-D-erythritol kinase, MCS, 2-*C*-methyl-D-erythritol 2,4-cyclodiphosphate synthase, HDS, hydroxymethylbutenyl diphosphate synthase, IPP, Isopentenyl diphosphate, DMAPP, dimethylallyl diphosphate, GES, Geraniol synthase, GPP, Geraniol diphosphate, HMG-CoA—hydroxymethyl-glutaryl-CoA, MVA, mevalonate, MVAPP, mevalonate diphosphate. *Adapted from El-Sayed and Verpoorte (2007).*

5′-diphospho)-2-*C*-methyl-D-erythritol by a kinase (MCK), and the cytidine nucleotide is removed to form 2-*C*-methyl-D-erythritol-2,4-cyclodiphosphate by 2-*C*-methyl-D-erythritol 2,4-cyclodiphosphate synthase (MCS). This intermediate is further dehydrated and reduced to form IPP (El-Sayed & Verpoorte, 2007; Oudin, Mahroug, et al., 2007; Rodriguez-Conception & Boronat, 2002; Rohdich et al., 2001; Rohmer, 1999). This latter step has been functionally characterised in *C. roseus*, known as hydroxymethylbutenyl diphosphate synthase (HDS) (Oudin, Mahroug, et al., 2007).

Isomerisation of IPP to form dimethylallyl diphosphate (DMAPP) is catalysed by IPP isomerase as a key step in isoprenoid biosynthesis (Ramos-Valvidia, van der Heijden, & Verpoorte, 1997; Fig. 1.2). DMAPP is condensed with one IPP to form geranyl diphosphate (GPP), which is the precursor for the monoterpenes (Contin, van der Heijden, Lefeber, & Verpoorte, 1998). Interestingly, analysis of leaf-epidermis-enriched cDNA libraries has identified 4-diphosphocytidyl-2-*C*-methyl-D-erythritol synthase (MECS) along with four MVA pathway genes, 3-hydroxy-3-methyl-glutaryl-CoA reductase (HMGR), 3-ketoacyl CoA thiolase, acetoacetyl-CoA thiolase and HMG-CoA synthase, and three genes common to the MEP/MVA pathways, namely, IPP isomerase, farnesyl diphosphate synthase and geranyl diphosphate synthase (GPPS) (Murata, Roepke, Gordon, & De Luca, 2008). Recently, geraniol synthase (GES) that catalyses the conversion of GPP to geraniol has been cloned and characterised from *C. roseus* (Simkin et al., 2013).

Since the discovery of the MEP pathway in higher plants, the metabolic source of the terpene moiety has been re-established. Contin et al. (1998) performed feeding studies with ^{13}C glucose and showed that the terpenoid moiety of secologanin in cell suspension cultures of *C. roseus* is not MVA-derived, but is clearly formed from the MEP pathway. The consistent results of feeding studies with cultures of *Ophiorrhiza pumila* also suggest the utilisation of the MEP pathway in secologanin biosynthesis (Yamazaki, Sudo, Yamazaki, Aimi, & Saito, 2003). Furthermore, transcript analysis by both Veau et al. (2000) and Chahed et al. (2000) showed that the expression of the MEP pathway genes is more highly correlated with the accumulation of the MIA ajmalicine, rather than the expression of the MVA pathway genes. Based on these results, MEP pathway-derived secologanin is consistent with the known origins of monoterpenes (Eisenreich et al., 2001).

2.1.2 Localisation of the early monoterpene biosynthesis pathway

The biosynthesis of monoterpenes in plants has been shown to occur in plastids, where the MEP pathway is localised (Lange & Croteau, 1999; Rohmer, 1999). Furthermore, *in situ* hybridisation studies have suggested that three genes from the MEP pathway (*1-deoxy-D-xylulose 5-phosphate synthase* (*dxs*), *1-deoxy-D-xylulose 5-phosphate reductoisomerase* (*dxr*) and *2-C-methyl-D-erythritol 2,4-cyclodiphosphate synthase* (*mecs*)) are preferentially expressed in the internal phloem-associated parenchyma (IPAP) cells of the vasculature (Burlat, Oudin, Courtois, Rideau, & St-Pierre, 2004). Based on these data, IPAP cells seem to play important roles in providing the precursors for MIA biosynthesis. Immunogold-labelling studies of HDS (Oudin, Mahroug, et al., 2007) localised this MEP pathway protein to IPAP, mesophyll and epidermal cells. However, the signal found in IPAP cells was 100- to 250-fold greater than in epidermal cells. Recent studies to characterise geraniol synthase also localised its expression to IPAP cell plastids (Simkin et al., 2013). Together these and other studies strongly suggest that the entire MEP pathway is preferentially expressed in IPAP cells of *Catharanthus* leaves.

2.1.3 Gene regulation of early monoterpene biosynthesis

Studies with various tissues such as seedlings and plant organs at various developmental stages, hairy root cultures and organ cultures suggest that MIA biosynthesis is developmentally regulated (El-Sayed & Verpoorte, 2007; Memelink & Gantet, 2007). For example, a root-specific transcription factor CrWRKY1 has been studied. Over-expression of CrWRKY1 in roots resulted in the accumulation of serpentine in this specific organ (Suttipanta et al., 2011). Furthermore, coordinated expression of genes involved in the pathway is important for optimised metabolite production that is required for plant response to its environment. Methyl jasmonate (MeJA) is one of the signalling molecules that have been studied for triggering defense against pathogens and herbivores. MeJA appears to trigger a signal transduction cascade that activates the octadecanoid-derivative responsive *Catharanthus* APETALA-3 binding protein known as ORCA3. The MeJA-mediated transcription factor cascade includes involvement of the basic helix-loop-helix protein CrMYC2 (Zhang et al., 2011) that appears to regulate ORCA3. Cell suspension cultures of *C. roseus* that over-express ORCA3 up-regulated expression of *1-deoxy-D-xylulose 5-phosphate synthase* (*dxs*) gene (El-Sayed & Verpoorte, 2007; van der Fits & Memelink, 2000). *Catharanthus* gene-profiling data also showed that additional genes were up-regulated in response to MeJA treatment (*1-deoxy-D-xylulose 5-phosphate synthase* (*dxs*), *2-C-methyl-D-erythritol 2,4,-cyclodiphosphate synthase* (*mecs*), *HDS* and *geranyl pyrophosphate synthase* (*gpps*); Rischer et al., 2006). Although these data support the MeJA-mediated up-regulation of the MEP pathway, little information is available on the fate of the IPP produced as a result of this induction for the formation of various terpenes and MIAs.

2.2. Iridoid biosynthesis

2.2.1 Biosynthetic genes involved in the iridoid pathway

The first step to iridoid biosynthesis oxidises geraniol to generate 10-hydroxygeraniol by a cytochrome P450 monooxygenase (CYP76B6), which was purified, cloned and functionally characterised from *C. roseus* (Collu et al., 2001; Meehan & Coscia, 1973; Fig. 1.3). 10-Hydroxygeraniol is further oxidised into the dialdehyde 10-oxogeranial by an oxidoreductase (Ikeda et al., 1991). 10-Oxogeraniol is converted to iridodial by cyclisation (Sanchez-Iturbe, Galaz-Avalos, & Loyola-Vargas, 2005; Uesato, Ikeda, Fujita, Inouye, & Zenk, 1987). Recently, this NADPH-dependent cyclase, iridodial synthase (IRS), has been cloned and characterised from *C. roseus* (Geu-Flores et al., 2012; Fig. 1.3). The cyclised product is then further oxidised to form deoxyloganetic acid that involves a cytochrome P450 enzyme, followed by glucosylation (GT) to produce deoxyloganic acid and then further hydroxylated (DLH) to form loganic acid (Madyastha, Guarnaccia, Baxter, & Coscia, 1973; Fig. 1.3). Methylation of loganic acid to form loganin is catalysed by loganic acid methyltransferase (LAMT) that has been cloned and characterised from *C. roseus* (Murata et al., 2008). Finally, the cleavage of the cyclopentane ring of loganin to secologanin by secologanin synthase (SLS) is of particular interest as it is one of the unusual P450s involved in secondary metabolism (Mizutani & Sato, 2011). This enzyme, first detected in cell suspension culture extracts of *Lonicera japonica* (Yamamoto, Katano, Ooi, & Inoue, 2000), was suggested to be membrane associated and to be a P450 with requirements for NADPH and O_2. Molecular cloning of SLS (CYP72A1) from *C. roseus* was followed by its functional expression, and biochemical characterisation in *Escherichia coli* confirmed its ability to convert loganin into secologanin (Irmler et al., 2000; Fig. 1.3).

2.2.2 Localisation of the iridoid pathway

The enzymes of the iridoid pathway are localised to separate subcellular compartments and to involve more than one cell type. As geraniol synthase

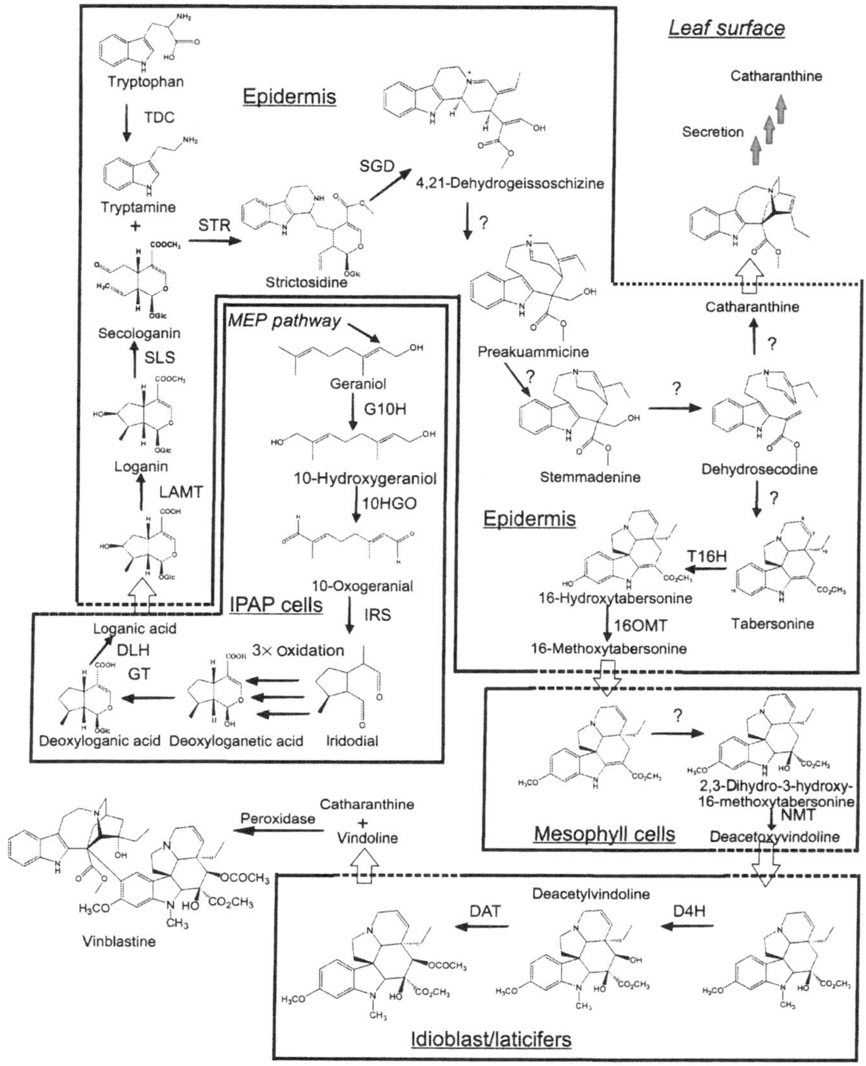

Figure 1.3 The MIA biosynthetic pathway is compartmentalised in different *Catharanthus* leaf cells. The early iridoid pathway is localised to the internal phloem associated parenchyma (IPAP) cells, while the late iridoid pathway and most of the MIA biosynthesis are associated with the leaf epidermis, and the late vindoline pathway is localised to special idioblast/laticifer cells. Open arrows show transport of metabolites. Abbreviations: TDC, tryptophan decarboxylase, STR, strictosidine synthase, SGD, strictosidine-β-D-glucosidase, G10H, geraniol 10-hydroxylase, 10HGO-10-hydroxygeraniol oxidoreductase, IRS, iridoid synthase, GT, glucosyltransferase, DLH, deoxyloganic acid hydroxylase, LAMT, loganic acid methyltransferase, SLS, secologanin synthase, T16H, tabersonine 16-hydroxylase, 16OMT, 16-hydroxytabersonine-*O*-methyltransferase, NMT, *N*-methyltransferase, D4H, deacetoxyvindoline 4-hydroxylase; DAT, deacetylvindoline acetyltransferase. *Adapted from Roepke et al. (2010).*

has been localised to the plastid, geraniol may then be exported to the cytosol where it would be converted to 10-OH geraniol by G10H associated first with vacuolar membranes and later with the endoplasmic reticulum (Guirimand et al., 2009; Madyastha, Ridgway, Dwyer, & Coscia, 1977) and is associated with membrane-bound flavin containing NADPH: cytochrome P450 reductase (CPR) normally required for cytochrome P450 reactions. In addition to localisation of three identified MEP pathway genes by *in situ* hybridisation, Burlat et al. (2004) also showed that G10H is preferentially expressed in the IPAP cells. Very recent studies with IRS also localised expression of this gene to IPAP cells (Geu-Flores et al., 2012), while the second to last (Roepke et al., 2010) and the last steps (Irmler et al., 2000) in secologanin biosynthesis appear to be localised to epidermal cells. Together, these data suggest that IPAP cells are specialised to supply the monoterpene geraniol for the iridoid pathway, at least up to the cyclisation step. While the cellular localisation of genes involved in the formation of the carboxyl group, glucosylation and hydroxylation to produce loganic acid has yet to be determined, it is likely that they may also occur in IPAP cells (Fig. 1.3). If this is correct, an undetermined loganic acid transporter may then be involved in its export from IPAP cells to the epidermis for final elaboration into secologanin.

2.2.3 Gene regulation of iridoid biosynthesis

While few data are available, there seem to be similarities in the regulation of the early steps of iridoid biosynthesis by a MeJA signalling pathway. Gene-profiling data showed that expression of G10H and 10HGO was significantly higher in cells treated with MeJA, while the terminal leaf epidermis-localised SLS expression was not affected (Rischer et al., 2006). While these studies may indicate that the early steps of iridoid biosynthesis are regulated differently than the later steps, it is not clear if the spatial separation of this pathway may also play a role in this process.

2.3. Early MIA biosynthesis

2.3.1 Biosynthetic genes involved in the early MIA pathway

The central precursor strictosidine is formed by a stereoselective Pictet–Spengler condensation mediated by strictosidine synthase (STR) of tryptamine derived from tryptophan by a pyridoxal-phosphate-dependent tryptophan decarboxylase (TDC) (De Luca, Marineau, & Brisson, 1989) and secologanin (Fig. 1.1; deWaal, Meijer, & Verpoorte, 1995; Maresh et al., 2008; Stöckigt & Zenk, 1977; Treimer & Zenk, 1979). TDC has been cloned and functionally characterised from different MIA-producing plants (De Luca et al., 1989; Lopez-Meyer & Nessler, 1997; Yamazaki et. al., 2003). STR was first cloned from *R. serpentina* and functionally expressed in *E. coli* (Kutchan, Hampp, Lottspeich, Beyreuther, & Zenk, 1988). Later, STR orthologs were isolated from *C. roseus* and *O. pumila* (Mcknight, Roessner, Devagupta, Scott, & Nessler, 1990; Pasquali et al., 1992; Yamazaki et al., 2003). The glucose moiety of strictosidine is subsequently removed by strictosidine β-D-glucosidase (SGD) (Fig. 1.1). Further purification of SGD from *C. roseus* cell cultures revealed that SGD has a strong affinity for strictosidine as substrate and a high molecular mass that exists as an aggregate of multiple 63-kDa subunits (Luijendijk, Stevens, & Verpoorte, 1998). While trypsin digestion was performed to solubilise the enzyme without the loss of activity, this stability to proteases has been used to suggest a putative but uncharacterised role for this enzyme in plant defense. SGD was later functionally characterised from *C. roseus* and *R. serpentina* (Geerlings, Ibanez, Memelink, van der Heijden, & Verpoorte, 2000; Gerasimenko, Sheludko, Ma, & Stöckigt, 2002). Encoded by a single gene in *C. roseus*, SGD shares about 60% homology at the amino acid level with other plant glucosidases (Geerlings et al., 2000). The three-dimensional structure of the SGD enzyme was also studied to reveal its catalytic mechanism (Barleben, Panjikar, Ruppert, Koepke, & Stöckigt, 2007). Interestingly, the production of a versatile strictosidine aglycone is the driving force combined with a number of different uncharacterised enzyme-mediated reactions for the remarkable structural diversity of MIAs found in Nature.

The ring arrangements after the formation of this reactive hemiacetal intermediate seem to be species-dependent (Szabo, 2008; Zhu, Guggisberg, Kalt-Hadamowsky, & Hesse, 1990). These diverse metabolites are produced only in certain plant families (Apocynaceae, Loganiaceae and Rubiaceae), and each member produces a subset of compounds that contributes to its varying biological function (Szabo, 2008). Within *C. roseus*, different arrangements of strictosidine aglycone yield three major (corynanthe, iboga and aspidosperma) classes of MIAs (Qureshi & Scott, 1968; Fig. 1.1).

Downstream steps of strictosidine aglycone formation leading to ajmalicine have been partially characterised. Following the conversion of strictosidine by *C. roseus* SGD, carbinolamine serves as an intermediate to produce cathenamine that is then reduced to form ajmalicine by cathenamine reductase (CR). Two different CRs with requirements for NADPH as cofactors have been identified and are detected at low levels in *C. roseus* cell cultures (El-Sayed & Verpoorte, 2007): one reduces cathenamine into ajmalicine and 19-epiajmalicine, while the other converts the iminium form of cathenamine into tetrahydroalstonine by tetrahydroalstonine synthase (THAS) (Hemscheidt & Zenk, 1985). Later, ajmalicine can be converted into serpentine by a vacuolar peroxidase (POD) from *C. roseus* (Blom et al., 1991; Fig. 1.1).

The preakaummicine, strychnos-type intermediate may be the common precursor for the aspidosperma, strychnos and iboga alkaloids (Fig. 1.1). Although mechanisms of preakuammicine formation from strictosidine aglycon have been proposed, preakuammicine has not been detected in plant extracts, mainly due to its lability. Moreover, preakuammicine can be reduced to stemmadenine, a more stable intermediate that can be rapidly consumed in the cell culture to yield catharanthine and tabersonine skeletons, as well as condylocarpine, but not as an intermediate (El-Sayed, Choi, Frédérich, Roytrakul, & Verpoorte, 2004; Fig. 1.1). Stemmadenine may be rearranged to generate dehydrosecodine that can lead to catharanthine-type alkaloids by a Diels–Alder reaction (O'Connor & Maresh, 2006; Qureshi & Scott, 1968).

Although genes involved in the pathways leading to catharanthine and tabersonine have not been fully isolated, some branch pathways through tabersonine in *C. roseus* and polyneuridine aldehyde have been studied most thoroughly. The six steps that catalyse the conversion of tabersonine to vindoline have been described in detail (Liscombe, Usera, & O'Connor, 2010; Ziegler & Facchini, 2008).

2.3.2 Localization of early MIA biosynthesis

While at least three cell types appear to be required to elaborate MIAs (Fig. 1.3), this biosynthetic pathway also requires the coordination of IPAP cell-based loganic acid biosynthesis with its transport to the leaf epidermis for conversion to secologanin and with the early MIA Pathway. *In situ* hybridisation and immunological studies have localised tryptophan decarboxylase (*tdc*) transcript and antigen as well as strictosidine synthase (*str*) transcript to the epidermis of stems, leaves, flower buds and most protoderm and cortical cells of the apical meristems in root tips (St-Pierre, Vázquez-Flota, & De Luca, 1999). Moreover, RT-PCR of laser-capture microdissected cells showed that *tdc*, *str* and strictosidine β-D-glucosidase (*sgd*) were preferentially expressed in the epidermis of *C. roseus* (Murata & De Luca, 2005). These data provide strong evidence that leaf epidermal cells are biosynthetically active sites for early MIA biosynthesis. Within epidermal cells, it seems that the first step of MIA biosynthesis catalysed by STR is localised to vacuoles (McKnight, Bergey, Burnett, & Nessler, 1991; Stevens, Blom, & Verpoorte, 1993), suggesting that tryptamine and secologanin are imported through the vacuole membrane from the cytosol where these metabolites are biosynthesised (De Luca & Cutler, 1987). Strictosidine must then be exported out of the vacuole into the cytosol for deglucosylation by SGD suggested to be associated with the endoplasmic reticulum as the SGD protein also has a putative targeting signal peptide (Geerlings et al., 2000). However, more recent studies have provided intriguing results reporting that SGD may mostly be associated with the nucleus (Guirimand et al., 2010). While this result is difficult to reconcile with the involvement of SGD with MIA biosynthesis, it suggests that elaboration of downstream MIAs from strictosidine is likely to be significantly more complex than originally anticipated. While early studies (Hemscheidt & Zenk, 1985) have suggested that separate tetrahydroalstonine, ajmalicine and epiajmalicine synthase enzymes are involved in the biosynthesis of these three MIAs, the correct recombinant reductases have yet to be isolated and functionally characterised. Studies with plant vacuoles have further suggested the existence of highly efficient transporters that mobilise newly synthesised ajmalicine for oxidation into serpentine that is then trapped in the vacuole by an ion-trap mechanism (Blom et al., 1991; El-Sayed & Verpoorte, 2007). The reactivity of the strictosidine aglycone intermediates required for developing enzyme assays together with other unknown constraints appears to make it very difficult to characterise these sections of the MIA biosynthesis pathway.

2.3.3 Gene regulation of early MIA biosynthesis

In addition to MeJA, a fungal elicitor such as *Phythium aphanidermatum* is also known to induce the expression of strictosidine synthase (*str*) and tryptophan decarboxylase (*tdc*) (Roewer, Cloutier, Nessler, & De Luca, 1992). Overexpression of ORCA3 transcription factor in cell cultures activated expression of *STR*, *TDC* and *SGD* (van der Fits & Memelink, 2000). This result was supported by gene-profiling data (Rischer et al., 2006), except for *STR* whose expression levels did not change. Another transcription factor associated with MIA biosynthesis is *ORCA2* that was identified after a yeast one-hybrid screen using MeJA or elicitor-responsive regulatory elements in the *STR* promoter (Menke, Parchmann, Mueller, Kijne, & Memelink, 1999). Further promoter analysis showed that it contains a G-Box element that is involved in binding the MeJA-responsive transcription factor CrMYC1 (Chatel et al., 2003). Other Zinc Cys_2/His_2-type, Zinc-finger transcription factors such as *ZCT1*, *ZCT2*, *ZCT3* have been shown to bind to the promoter of *TDC* and *STR* functioning as repressors of the MeJA response (Pauw et al., 2004). These data suggest that early MIA pathway genes (*TDC*, *STR* and *SGD*) respond to MeJA treatment in a well-coordinated manner. It is interesting that the accumulation of catharanthine, tabersonine and vindoline is also increased in response to MeJA treatment (Aerts, Schafer, Hesse, Baumann, & Slusarenko, 1996), but the molecular basis of this observation is not known.

2.4. The late MIA biosynthesis pathway

2.4.1 Biosynthetic genes involved in the late MIA/vindoline pathway in *C. roseus*

The late steps of MIA biosynthesis in *C. roseus* include the production of the well-studied tabersonine derivatives specific to different plant organs. In root tissues, tabersonine is oxidised into hörhammericine and lochnericine while it is converted into vindoline in the above-ground organs (Fig. 1.1; Laflamme, St-Pierre, & De Luca, 2001). Roots oxidise tabersonine to its epoxide via a microsomal cytochrome P450 monooxygenase requiring NADPH and molecular oxygen (tabersonine 6,7 epoxidase) in *C. roseus* hairy root cultures (Rodriguez, Compagnon, Crouch, St-Pierre, & De Luca, 2003). In addition, the root-tip-specific minovincinine 19-*O*-acetyltransferase (MAT) gene involved in the formation of 6,7-dehydroechitovenine and/or 19-*O*-acetylhorhammericine has been functionally characterised (Laflamme et al., 2001; Fig. 1.1). Another cytochrome P450 monooxygenase involved in this root-specific pathway is recently cloned and characterised CYP71BJ1 that hydroxylates at the C-19 of tabersonine and lochnericine (Giddings et al., 2011).

Vindoline, one of the monomers of the bisindole alkaloid vinblastine, is derived from tabersonine through six enzymatic steps. The first step involves hydroxylation of tabersonine by tabersonine 16-hydroxylase (T16H), another cytochrome P450 monooxygenase (Schröder et al., 1999; St-Pierre & De Luca, 1995). Expression of *T16H* is strongly induced by light with low activity found in etiolated seedlings. The hydroxyl group of 16-hydroxytabersonine is then O-methylated by an *S*-adenosyl-methionine (AdoMet) 16-hydroxytabersonine-16-*O*-methyltransferase (16OMT) that has been cloned and characterised to yield 16-methoxytabersonine (Fahn, Laussermair, Deus-Neumann, & Stöckigt, 1985; Levac, Murata, Kim, & De Luca, 2008). The next uncharacterised step converts 16-methoxytabersonine to 16-methoxy-2,3-dihydro-3-hydroxytabersonine, which is further N-methylated to produce desacetoxyvindoline. This *N*-methyltransferase has been cloned and functionally characterised (Liscombe et al., 2010) while its activity was localised previously within the thylakoid membrane of chloroplasts (De Luca & Cutler, 1987) and was detected in differentiated plants but not in plant cell cultures (Dethier & De Luca, 1993). Desacetoxyvindoline is hydroxylated by the oxoglutarate-dependent dioxygenase, desacetoxyvindoline 4-hydroxylase (D4H) (De Carolis, Chan, Balsevich, & De Luca, 1990; De Carolis & De Luca, 1993; Vazquez-Flota, De Carolis, Alarco, & De Luca, 1997). The last step of vindoline biosynthesis is catalysed by deacetylvindoline *O*-acetyltransferase (DAT), part of the BAHD (benzylalcohol-*O*-acetyl-, anthocyanin-*O*-hydroxycinnamoyl-, anthranilate-*N*-hydroxycinnamoyl/benzoyl- and deacetylvindoline 4-*O*-acetyltransferase) family that is responsible for acetylation of desacetoxyvindoline to yield vindoline (De Luca & Cutler, 1987; St-Pierre, Laflamme, Alarco, & De Luca, 1998). Interestingly, DAT shares 78% homology at the amino acid level with minovincinine 19-*O*-acetyltransferase MAT (Laflamme et al., 2001). While DAT accepts only its natural substrate, MAT shows a slight activity towards deacetylvindoline, the DAT substrate. The last two terminal steps in vindoline biosynthesis are light-regulated and found only in differentiated plant materials.

2.4.2 Localisation of vindoline biosynthesis

It is clear that MIA biosynthesis in *C. roseus* occurs in several different cell types (Guirimand et al., 2011; Murata et al., 2008) and is developmentally regulated (De Luca et al., 1986; Facchini & De Luca, 2008). Although some transcripts and activities specific to vindoline biosynthesis have been detected in hairy root and cell suspension cultures, these tissues do not produce vindoline. The conversion of tabersonine to vindoline occurs in various compartments: the first enzyme that acts on tabersonine is associated with the endoplasmic reticulum membrane, while the second enzyme, 16-OMT, is believed to be cytosolic (St-Pierre & De Luca, 1995). Furthermore, *N*-methyltransferase is believed to be associated with the thylakoid within the chloroplast suggesting that this enzyme is found within a chloroplast-rich part of leaf tissue, the mesophyll layer (De Luca & Cutler, 1987; Dethier & De Luca, 1993; Murata & De Luca, 2005). It has been proposed that 16-methoxytabersonine may be transported from the leaf epidermis where 16-OMT is found (Levac et al., 2008) to other cell types for N-methylation, passing through cell walls or via the plasmadesmata. At the subcellular level, the methylated intermediate is then transported to the cytosol for further hydroxylation and acetylation (De Carolis et al., 1990; De Luca & Cutler, 1987). RNA blot hybridisation studies indicated that the enzyme activity followed the levels of *d4h* transcripts occurring primarily in young leaves that decline with age and lower levels occurred in the stem and fruits. St-Pierre et al. (1999) reported that expression of D4H and DAT was localised to laticifer and idioblasts cells of leaves, stems and flower buds. These results explain the failure of vindoline production by cell culture technology as the late steps of vindoline biosynthesis take place in different cells and only in above-ground tissues.

2.4.3 Vindoline biosynthesis is modulated by MeJA and light

The vindoline pathway of *C. roseus* can be modulated by application of MeJA or by light to developing seedlings. MeJA treatment increased the expression level of D4H (Vazquez-Flota & De Luca, 1998a, 1998b) during seedling development. Furthermore, over-expression of the ORCA3 transcription factor (van der Fits & Memelink, 2000), but not MeJA treatment (Rischer et al., 2006), up-regulated *D4H* expression in cell cultures.

Exposure of *C. roseus* cell cultures to light increased the expression of T16H within 22–28 h of treatment (Schroder et al., 1999). Etiolated seedlings exposed to full or red light induced the expression of *D4H* together with enzyme activity and this could be reversed by far-red light treatment (Vazquez-Flota & De Luca, 1998a). Further studies suggested that the D4H activity could be post-translationally modified by an uncharacterised phytochrome-assisted mechanism. Similarly, DAT involved in the last step in vindoline biosynthesis appears to be activated by light through the involvement of phytochrome that mediates the reversible activation of *DAT* (Aerts & De Luca, 1992). Studies with purified DAT have shown it to be more strongly inhibited by tabersonine and Coenzyme A than by tryptamine, secologanin and vindoline (Power, Kurz, & De Luca, 1990), and raise questions whether such pathway precursors or co-substrates could modulate the *in vivo* activity of DAT.

3. ORGANISATION AND SPATIAL SEPARATION OF MIA BIOSYNTHESIS

While enzymes involved in MIA biosynthesis are associated with various subcellular compartments such as the cytosol, vacuole, endoplasmic reticulum and chloroplast, the division of MIA biosynthesis between multiple cells deserves particular attention. While little is known about the process, subcellular trafficking of intermediates plays a significant role in controlling MIA biosynthesis to effectively channel iridoids/MIA intermediates between cells and remove toxic MIA end products from the cytosol (Fig. 1.3). In this context, the low-level accumulation of dimeric anticancer MIA vinblastine could be due to the spatial separation between catharanthine that is secreted into the leaf exudates and vindoline that likely accumulates within idioblast and laticifer leaf cells (Roepke et al., 2010). This suggests that the MIA part of biosynthesis takes place in leaf epidermal cells in order to allow the secretion of catharanthine and vindoline pathway intermediates to their respective locations to prevent the accumulation of dimeric MIAs that are known to be toxic to *C. roseus* plants. In order to understand this complex compartmentalisation, the transporters involved need to be identified in relation with the supply of intermediates from one cell type to another and MIA product accumulation in a separate cell type or location.

This multi-cell-type coordination of MIA biosynthesis in leaves contrasts strongly with the occurrence of the MEP pathway, the secologanin pathway, tryptamine production and the complement of enzymes leading to root-based MIAs within protoderm and cortical cells around the root apical meristem. It is interesting that these root cells preferentially expressed *TDC*, *STR* and the terminal *MAT* involved in the assembly of 6,7-dehydroechitovenine and/or 19-*O*-acetylhörhammericine (Laflamme et al., 2001; St-Pierre et al., 1999), while D4H and DAT transcripts were not detected in the underground tissue. Remarkably, few studies have been performed to analyse the reasons for the same cell MIA biosynthesis occurring in roots in relation to the multi-cell MIA biosynthesis occurring in leaves of *Catharanthus*.

3.1. Epidermis as an important biosynthetic site of MIAs and their precursors

The involvement of at least three cell types, including IPAP, epidermis and idioblast/laticifers, indicates the translocation of at least one pathway intermediate in the MIA biosynthesis. Early studies have shown that TDC and STR are most abundant in *C. roseus* roots even though they are also detected in aerial organs (Pasquali et al., 1992). *In situ* hybridisation studies imply that the transcripts of the early MIA pathway, TDC, SLS and STR, are associated with the epidermis of leaves, stems and flower buds, while the later steps that involve D4H and DAT transcripts are localised to the specialised laticifer and idioblast cells of the same organs (De Luca & St-Pierre, 2000; Irmler et al.,

2000; Vazquez-Flota & De Luca, 1998a, 1998b). Furthermore, gene transcripts encoding enzymes in the MEP pathway and geraniol 10-hydroxylase (G10H) are associated with the IPAP cells of young *C. roseus* aerial organs (Burlat et al., 2004). These data suggest the potential translocation of vindoline biosynthetic intermediates from the IPAP cells to the leaf epidermis and from the epidermis to the latificifers and idioblasts (Fig. 1.3). As the epidermis seems to be the centralised site for translocation of the pathway intermediates, efforts towards RNA isolation from this particular cell type to identify more genes involved in the entire pathway can be beneficial. This approach has been applied in a number of cases such as the sequencing of specialised glandular trichome cells of *Mentha piperita* to dissect the menthol and polymethylated flavone biosynthesis pathways (Lange et al., 2000; Schilmiller, Last, & Pichersky, 2008). The carborundum abrasion (CA) technique was developed to extract RNA from the leaf epidermis, and cDNA libraries were generated to produce ESTs that can be exploited for gene discovery (Murata, Bienzle, Brandle, Sensen, & De Luca, 2006; Murata & De Luca, 2005; Murata et al., 2008). In addition, the CA technique can be used to isolate large-scale active epidermal protein from the leaf tissues such as in the purification of 16OMT or for enzyme assays of LAMT in which LAMT activity was enriched in protein extract from leaf epidermis compared with whole leaves. (Levac et al., 2008; Murata & De Luca, 2005; Murata et al., 2008).

3.2. The use of epidermis-enriched transcriptomic resources for gene discovery

While random sequencing of leaf epidermis-enriched cDNA libraries (Murata et al., 2008) has assisted in the identification of genes involved in MIA biosynthesis in *C. roseus*, such as characterisation of LAMT by homology-based cloning (Roepke et al., 2010), the availability of this type of database can be used in comparative bioinformatic studies with the large transcriptome *Catharanthus* database (http://www.phytometasyn.ca/) to identify other novel MIA genes. For example, an interesting candidate gene that is first pulled out from the whole leaf EST library can be verified for its preferential expression in the leaf epidermal cells (De Luca, Salim, Levac, et al., 2012). As biosynthetic genes in the catharanthine pathway have not been identified, the leaf epidermis-enriched cDNA library represents an excellent auxilliary source for the cloning and characterisation of members of this pathway.

4. LARGE-SCALE GENOMIC APPROACHES IN FUNCTIONAL CHARACTERISATION OF GENES INVOLVED IN MIA BIOSYNTHESIS

Large-scale transcriptomic projects have recently opened a new approach to revealing genes associated with MIA biosynthesis. Early EST-based approaches yielded more than 25,000 annotated ESTs based on partial sequencing of cDNA libraries from *C. roseus* organs and tissues (Murata et al., 2006; Rischer et al., 2006; Shukla, Shasany, Gupta, & Khanuja, 2006). The correlation networks that combine transcriptomic and metabolomics data from *C. roseus* treated with MeJA by using the cDNA/amplified fragment length polymorphism approach was performed to show a correlation between the biosynthetic gene expression and the subsequent metabolite product in the MIA biosynthesis pathway (Rischer et al., 2006). In general, the results suggest a coordinated increase in the expression of genes associated with MIA metabolism. Furthermore, the number of sequences related to alkaloid-accumulating plants, such as *C. roseus*, continues to increase as sequencing of various tissues such as stems, roots, flowers, elicited cell cultures and hairy roots including a TDC silenced line is announced (Góngora-Castillo et al., 2012; Runguphan, Maresh, & O'Connor, 2009). These large-scale transcriptome sequences were generated by the Medicinal Plant Genomics Consortium (http://medicinalplantgenomics.msu.edu) that has been initiated in an effort to utilise next-generation sequencing for transcriptomic analysis of 14 medicinal plants.

4.1. The shared pathways among Apocynaceae family

The recent large-scale sequencing project, PhytoMetaSyn, has produced transcriptomic profiles for over 70 medicinal plant species (http://www.phytometasyn.ca/) that produce secondary metabolites with high values (Facchini et al., 2012). In this project, ESTs from eight alkaloid-producing plants belonging to the order Gentianales with seven Apocynaceae family members that include three species of *Catharanthus* (*roseus*, *longifolius* and *ovalis*), *Vinca minor*, *R. serpentina*, *Tabernaemontana elegans*, *Amsonia hubrichtii* and one member of the Rubiaceae (*C. ledgeriana*) have been obtained. In addition, one species, *L. japonica* (Caprifoliaceae) that produces only secologanin (Kawai, Kuroyanagi, & Ueno, 1988), has been sequenced. While over 56,000 ESTs have been reported from leaf, leaf epidermis and roots of *C. roseus* as part of the NapGen (Natural Products Genomic Research) project(Murata et al., 2006), the PhytoMetaSyn project expands the availability of EST that would enable the transcriptome analysis within the same family. As Apocynaceae members share the same pathways for the biosynthesis of various MIAs, the orthologous genes that have been characterised from *C. roseus* to be involved in the MEP, iridoid and early MIA pathways can be found in the other seven species (Fig. 1.4).

Similarly, the putative iridoid biosynthetic genes from *L. japonica* can be represented in the database. The approach of sequencing several species from the same family provides the advantage of targeting the genes associated with a certain part of the pathway. As a result, this method will facilitate the gene discovery process.

4.2. Tools for screening the candidate genes

With the increase in the number of publicly available sequences from MIA-producing species, the need for an effective tool to screen a large number of candidate MIA genes has increased. RNA interference (RNAi) is one of the reverse genetics approach that has been used to silence genes in cell suspension

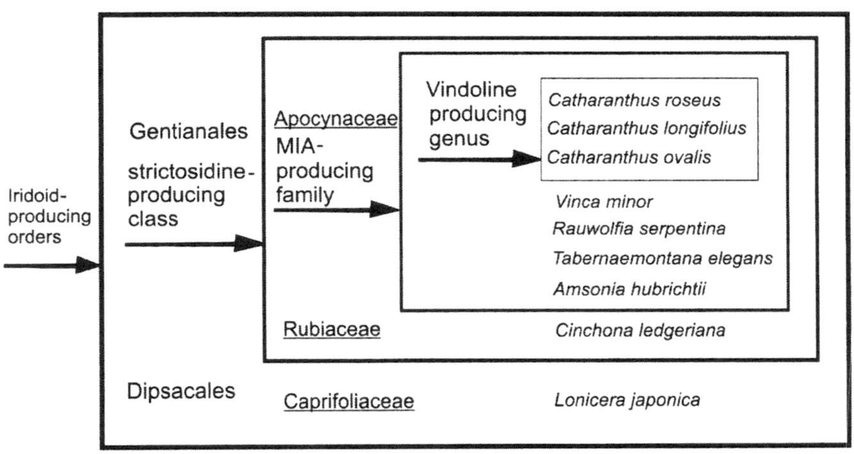

Figure 1.4 Diagram describing an approach for gene discovery by orthology-based comparison among MIA- and iridoid-producing plants. The target genes in MEP and iridoid pathways should be found in the databases in commonly iridoid-producing orders, such as Gentianales and one iridoid-only-producing species, *Lonicera japonica* from the Caprifoliaceae family, part of the Dipsacales order. The target genes for the early MIA pathway can be pulled out from all Gentianales (*Apocynaceae* and *Rubiaceae*) plant databases and those for the late MIA pathway such as vindoline should be uniquely found in the *Catharanthus* databases.

cultures (Courdavault et al., 2005; Papon et al., 2004) and in hairy root cultures (Runguphan et al., 2009). RNAi involves the introduction of short double-stranded RNA during plant transformation to interfere with expression of the target gene. Although this technique has been successful in suppressing the production of metabolites in the MIA pathway, experiments that involve RNAi take months or even years before the results become apparent. In addition, this approach is not ideal for characterisation of genes in the pathway as it shows restricted metabolism in cell culture systems, such as vindoline biosynthesis in *C. roseus*. Furthermore, efficient plant transformation strategies are not always available for most medicinal plants, including *C. roseus*.

VIGS has become an effective tool for gene functional analysis. This method exploits the plant defense mechanism against virus infection that leads to degradation of the corresponding mRNA transcripts of a host gene that has been targeted for silencing (Burch-Smith, Anderson, Martin, & Dinesh-Kumar, 2004). The binary tobacco rattle virus (pTRV) vector system developed and tested in *Nicotiana benthamiana* (Dinesh-Kumar, Anandalakshmi, Marathe, Schiff, & Liu, 2003; Liu, Schiff, Marathe, & Dinesh-Kumar, 2002; Ratcliff, Martin-Hernandez, & Baulcombe, 2001) has also been used with success in other plant systems. The first advantage of VIGS is the rapidity and ease of use of this gene-silencing process. The experiments can be performed in *C. roseus* in as few as 8 weeks starting from seed germination to screening of individual silenced plants for reduced gene expression and changing metabolite profiles (De Luca, Salim, Levac, et al., 2012). Other convenient advantages include the non-requirement for stable plant transformation and the requirement of only partial sequence information to achieve adequate gene silencing (Senthil-Kumar & Mysore, 2011). Its disadvantages may include non-targeted gene silencing and possible toxic effects from the metabolites being accumulated. Several genes from some medicinal plants have been silenced using this technique such as *Papaver somniferum* (Hileman, Drea, Martino, Litt, & Irish, 2005; Wijekoon & Facchini, 2012), *Eschscholzia californica* (Wege, Scholz, Gleissberg, & Becker, 2007), *Aquilegia* (Gould & Kramer, 2007) and *Thalictrum* (Di Stilio et al., 2010). VIGS was also used to suppress the CYP80F1 family that may convert littorine to hyoscyamine in *Hyoscyamus niger*, tropane alkaloid-producing species (Li et al., 2006). The method using the pTRV vector system for silencing in *C. roseus* has been employed when three known steps in vindoline biosynthesis are silenced (Liscombe & O'Connor, 2011). The *N*-methyltransferase in *C. roseus* was successfully silenced and resulted in the accumulation of the NMT substrate of 16-methoxy-2,3-dihydro-3-hydroxytabersonine, and a decreased level of vindoline

(Liscombe & O'Connor, 2011). An alternative method for pTRV vector delivery into *Catharanthus* plants has also been reported (De Luca, Salim, Levac, et al., 2012). Furthermore, the effectiveness of VIGS in silencing the target genes can be observed by suppressing the phytoene desaturase gene that is involved in chlorophyll biosynthesis. The silencing of this gene produces visible white sectors on the affected leaf and serves as a marker that VIGS works effectively in the species being examined (De Luca, Salim, Levac, et al., 2012).

VIGS technology has been applied in the functional characterisation of genes involved in O-demethylation of thebaine to codeine and of codeine to morphine in benzylisoquinoline alkaloid (BIA) biosynthesis of opium poppy (Hagel & Facchini, 2010). As VIGS experiments result in the accumulation of the substrates when the target gene is silenced, VIGS may provide the substrates required for the enzyme assays in order to confirm the gene function by expressing the proteins in recombinant systems. In the case of silencing this opium poppy demethylase, the accumulated BIA intermediates can be purified by thin-layer chromatography and the substrates can be tested using recombinant proteins. Hence, VIGS offers another advantage by providing difficult to synthesise or commercially unavailable substrates for confirming the biochemical function of the candidate gene.

VIGS is an efficient tool for gene function analysis in terms of narrowing down the large number of candidate genes that may be involved in the MIA biosynthesis pathway before performing detailed studies of enzyme characterisation. As the number of available candidate genes continues to increase with the rise of large-scale genomics technologies, VIGS offers a relatively rapid and efficient first screening step for MIA pathway gene identification. With the availability of this screening technique, *C. roseus* should continue to be a leading model system for studying MIA biosynthesis. However, VIGS remains restricted to certain species and its efficiency has been mostly shown in the characterisation of biosynthetic pathway genes, while genomics approaches lead to the identification of candidate genes from certain steps of the reactions in the pathway.

5. METABOLIC ENGINEERING OF THE MIA BIOSYNTHESIS PATHWAY

Identification of biosynthetic genes in the MIA biosynthesis pathway has permitted various efforts towards metabolic engineering. The goal of this type of technology is primarily to enhance the synthetic capacity of preferred products by the over-expression of pathway genes or by diverting pathways for the production of desired novel metabolites. This attempt requires insights into each reaction pathway step, its organisation and its regulation. Some complications include those imposed by enzymes that may be key rate-limiting steps in the biosynthetic pathway. Proper compartmentalisation may also be required to maintain suitable trafficking patterns. The first trial at metabolic engineering in MIA biosynthesis was the expression of constitutive TDC and STR transgenes using *C. roseus* cell cultures (Canel et al., 1998; Facchini, 2001). While over-expression of enzymes involved in the formation of tryptophan indole moiety, anthranilate synthase, caused an increase in the level of tryptophan without enhanced production of MIAs (Hong, Peebles, Shanks, San, & Gibson, 2006; Hughes, Hong, Gibson, Shanks, & San, 2004), the monoterpenoid pathway has been considered the rate-limiting step in MIA production. When the MEP pathway gene HMGR was over-expressed, alkaloid accumulated significantly, especially with one line that accumulated serpentine up to sevenfold compared to controls, while no effect on catharanthine was observed (Ayora-Talavera, Chappell, Lozoya-Gloria, & Loyola-Vargas, 2002). Moreover, studies with hairy root cultures over-expressing DAT caused accumulation of hörhammericine compared with control roots (Magnotta, Murata, Chen, & De Luca, 2007). This study shows that the potential interactions between DAT and the root-specific MAT might affect the conversion of hörhammericine to 19-*O*-acetyl-hörhammericine (Fig. 1.1). In terms of regulatory control of the MIA pathway, ORCA3 was suggested to activate the expression of anthranilate synthase, TDC, STR, D4H, CPR and *D*-1-deoxyxylulose 5-phosphate synthase, but not G10H, SGD and DAT (van der Fits & Memelink, 2000). Interestingly, in this constitutive expression of the ORCA3 system, addition of exogenous loganin was required to increase MIA levels. Therefore, coordinated over-expression of transcriptional activators ORCA2 and ORCA3 with the silencing of repressors such as G-box binding factors 1 and 2 and the zinc finger protein family might be a solution to activate the MIA pathway (Menke et al., 1999; Pauw et al., 2004; Siberil et al., 2001; van der Fits & Memelink, 2000). Although various approaches to metabolic engineering in cell culture systems have been reported, whole-plant metabolic engineering remains interesting for improving the productivity of MIAs, especially the potential of mutation breeding by screening for *C. roseus* germplasm that shows interesting genotypes, such as a high level of vinblastine and low vindoline production. In this case, a low vindoline line (Magnotta, Murata,

Chen, & De Luca, 2006) can be useful for further investigation of the later MIA biosynthesis pathway.

Once the genes in certain parts of a pathway are characterised, these multiple steps can be introduced in a suitable system such as yeast to improve the productivity of desired metabolites. Alternatively, plants can be engineered to produce non-natural MIAs that can be useful to produce new drugs with MIA backbones while reducing the side effects of the drugs. For example, TDC expressed in transformed roots with chlorinated tryptophan was able to produce chlorotryptamine derivatives and eventually chlorinated MIAs such as 10-chlorocatharanthine and 15-chlorotabersonine (Runguphan, Qu, & O'Connor, 2010). This example suggested the ability of plants to serve as factories for producing useful or even rare alkaloids, although their efficiency in plant systems needs to be further tested.

Several biosynthetic as well as regulatory genes have been used to genetically modify the production of MIAs in *C. roseus*. However, the most stable transformation methods in this species have been developed for roots and cell cultures (Pasquali, Porto, & Fett-Neto, 2006; Zárate & Verpoorte, 2007; Zhao & Verpoorte, 2007). While some success in plant cell culture technology for large-scale production of secondary metabolites has been reported, some drawbacks associated with this effort are low or unstable production of certain targeted metabolites, for example, the incapabilities of *C. roseus* cell culture to produce vindoline (Zhao & Verpoorte, 2007).

Another type of metabolic engineering study involves the substrate specificity of the enzyme (Ziegler & Facchini, 2008). Some studies that focused on the enzyme molecular structures used the crystal structure to generate the rational site-directed mutation. Mutant STR can also be coupled to SGD and has resulted in a large alkaloid library that can be screened for new drugs. For instance, the crystal structure of STR complexed with strictosidine was analysed to produce rational site-directed mutations. Of several STR-mutants tested for their substrate specificity, one showed a better turnover of some secologanin analogues compared with wild-type STR (Chen, Galan, Coltharp, & O'Connor, 2006; McCoy & O'Connor, 2006). Remarkably, two site-directed STR mutants were able to generate several strictosidine analogs with various substitutions that include halogenated derivatives (Bernhardt, McCoy, & O'Connor, 2007; Ma, Panjikar, Koepke, Loris, & Stöckigt, 2006). These analogues could eventually be fed to *C. roseus* hairy root cultures to produce modified MIAs.

Considering that certain parts of MIA biosynthesis genes have been elucidated, the current efforts are focused on transferring known parts of pathways to organisms such as *Saccharomyces cerevisieae* (Facchini et al., 2012). Some examples of the ability of yeast to carry out multiple-step plant pathways have been reported (Facchini et al., 2012) such as the production of the antimalarial drug artemisinic acid in engineered yeast (Ro, Paradise, Ouellet, Fisher, & Newman, 2006). While the major advantages of yeast as a production system have been well advertised in the literature, the problems solved by initial efforts to produce artemisinic acid will help to facilitate expression of large and more complex pathways such as those for MIAs. Metabolic engineering of MIA biosynthesis is likely to reveal new problems that will have to be addressed including the challenge of functional expression of dozens of genes within the heterologous system and the expression of steps that may be regulated by light. The importance of compartmentalisation, metabolite trafficking, and regulation of the MIA pathway in *planta* also need to be considered for the successful integration of this pathway in yeast.

6. CONCLUSIONS AND PERSPECTIVES

It is apparent from this review that MIA biosynthesis is highly organised. In *C. roseus*, the pathway takes place in at least three different cell types, namely the IPAP, the epidermis and the laticifer/idioblast cells. This separation demonstrates that mobile metabolites are shuttled between cell types. However, the mechanism of the transport and what intermediates are translocated need to be further investigated. So far, the major site of MEP biosynthesis as well as the early steps of iridoid biosynthesis is within the IPAP cells. An unknown iridoid likely to be loganic acid is then transported to the epidermis where the last two steps of iridoid biosynthesis occur. While early MIA biosynthesis and the first two steps in vindoline biosynthesis are localised to the epidermis, elaboration of strictosidine aglycone to catharanthine or tabersonine needs to be addressed to confirm that these steps also occur within the epidermal cells. An unidentified intermediate in vindoline biosynthesis is then transported from the epidermal cells to the mesophyll cells and eventually to the idioblast/laticifer cells, which explains vindoline accumulation within the leaf. Furthermore, localisation of catharanthine biosynthesis can be evidenced by the accumulation of this metabolite on the leaf surface. This spatial separation of catharanthine and vindoline explains the low-level production of the dimeric

MIA vinblastine. The different situation in root systems where the entire MIA pathway from early precursors to final products occurs in the same cell type needs to be more fully studied for comparative purposes. In addition, the MIA pathway can be induced by signalling molecules such as MeJA that leads to the regulation of the pathway by ORCA3 transcription factors, as it is evident that many genes in the MEP, iridoid and MIA pathways are up-regulated in response to MeJA.

Since the 1960s, the MIA biosynthesis pathways have been investigated by chemical approaches, including isotopic-labelling experiments. The target of this approach was mainly to determine the intermediates in the pathway. In the last few years, the majority of studies have been on developing tools for expanding the functional characterisation of genes involved in the MIA biosynthesis, especially by producing large-scale transcriptomes of different MIA-producing plant species and improved comparative bioinformatic approaches for identifying candidate genes. Although numerous MIA biosynthetic genes in *C. roseus* have been elucidated, the intermediate steps from strictosidine aglycone to ajmalicine, catharanthine and tabersonine remain uncharacterised at the biochemical and molecular level. The growing transcriptomic information from various MIA-producing species complemented by organ/tissue-specific transcriptomic data from certain species (such as the epidermal-enriched cDNA library in *C. roseus*) is being used to create a list of candidate genes for functional characterisation. The development of the VIGS method has expedited the gene discovery process as this tool can be used to screen the candidate genes by analysing the changes in MIA metabolite profiles. The accumulation of the MIA intermediates in VIGS-treated plants may also supply the potential substrates that can then be purified to complete the biochemical characterisation of the candidate gene in recombinant systems such as microorganisms or transgenic plants.

Future research could include the investigation of spatial localisation of the MIA pathway within other Apocynaceae members if there are similar patterns of compartmentalisation as *in C. roseus* observed in other species. A detailed understanding of metabolic regulation that is also influenced by complex organisation with intra- and inter-cellular translocation of intermediates is essential to achieve the goal of metabolic engineering of MIA metabolism that may involve the assembly of the entire pathway in microorganisms for large-scale production of valuable MIAs. The expansion of our knowledge of genomic studies, biochemistry and molecular biology of MIA biosynthesis will facilitate the discovery of novel genes that can be utilised in engineered systems for manufacturing biologically active MIAs as well as for promoting the drug discovery process.

ACKNOWLEDGEMENTS

This work was supported by a Natural Sciences and Engineering Research Council of Canada (NSERC) Discovery Grant (V. D. L.), NSERC/BARD/Agriculture Canada team grant, Canada Research Chairs (V. D. L.), Genome Canada, Genome Alberta, Genome Prairie, Genome British Columbia, the Canada Foundation for Innovation, the Ontario Ministry of Research and Innovation, the National Research Council of Canada and other government and private sector partners.

REFERENCES

Aerts, R., & De Luca, V. (1992). Phytochrome is involved in light-regulation of vindoline biosynthesis in *Catharanthus*. *Plant Physiology*, *100*, 1029–1032.

Aerts, R., Schafer, A., Hesse, M., Baumann, T., & Slusarenko, A. (1996). Signaling molecules and the synthesis of alkaloids in *Catharanthus roseus* seedlings. *Phytochemistry*, *42*, 417–422.

Ayora-Talavera, T., Chappell, J., Lozoya-Gloria, E., & Loyola-Vargas, V. M. (2002). Overexpression in *Catharanthus roseus* hairy roots of a truncated hamster 3-hydroxy-3-methylglutaryl-CoA reductase gene. *Applied Biochemistry and Biotechnology*, *97*, 135–145.

Barleben, L., Panjikar, S., Ruppert, M., Koepke, J., & Stöckigt, J. (2007). Molecular architecture of strictosidine glucosidase: The gateway to the biosynthesis of the monoterpenoid indole alkaloid family. *The Plant Cell*, *19*, 2886–2897.

Bernhardt, P., McCoy, E., & O'Connor, S. E. (2007). Rapid identification of enzyme variants for reengineered alkaloid biosynthesis in periwinkle. *Chemistry and Biology*, *14*, 888–897.

Blom, T. J. M., Sierra, M., van Vliet, T. B., Franke-van Dijk, M. E. I., de Koning, P., van Iren, F., et al. (1991). Uptake and accumulation of ajmalicine into isolated vacuoles of cultured cells of *Catharanthus roseus* (L.) G. Don. and its conversion into serpentine. *Planta*, *183*, 170–177.

Burch-Smith, T. M., Anderson, J. C., Martin, G. B., & Dinesh-Kumar, S. P. (2004). Applications and advantages of virus-induced gene silencing for gene function studies in plants. *The Plant Journal*, *39*, 734–746.

Burlat, V., Oudin, A., Courtois, M., Rideau, M., & St-Pierre, B. (2004). Co-expression of three MEP pathway genes and geraniol 10-hydroxylase in internal phloem parenchyma of *Catharanthus roseus* implicates multicellular translocation of intermediates during the biosynthesis of monoterpene indole alkaloids and isoprenoid-drived primary metabolites. *The Plant Journal*, *38*, 131–141.

Canel, C., Lopes-Cardoso, M. I., Whitmer, S., van der Fits, L., Pasquali, G., van der Heijden, R., et al. (1998). Effects of overexpression of strictosidine synthase and tryptophan decarboxylase on alkaloid production by cell cultures of *Catharanthus roseus*. *Planta*, *205*, 414–419.

Chahed, K., Oudin, A., Guivarc'h, N., Hamdi, S., Chenieux, J., Rideau, M., et al. (2000). 1-Deoxy-D-xylulose 5-phosphate from periwinkle: cDNA identification and induced gene expression in terpenoid indole alkaloid-producing cells. *Plant Physiology and Biochemistry*, *38*, 559–566.

Chatel, G., Montiel, G., Pré, M., Memelink, J., Thiersault, M., St-Pierre, B., et al. (2003). CrMYC1, a *Catharanthus roseus* elicitor- and jasmonate-responsive bHLH transcription factor that binds the G-box element of the strictosidine synthase gene promoter. *Journal of Experimental Botany*, *54*, 2587–2588.

Chen, S., Galan, M. C., Coltharp, C., & O'Connor, S. E. (2006). Redesign of a central enzyme in alkaloid biosynthesis. *Chemistry and Biology*, *13*, 1137–1141.

Collu, G., Unver, N., Peltenburg-Looman, A. M., van der Heijden, R., Verpoorte, R., & Memelink, J. (2001). Geraniol 10-hydroxylase, a cytochrome P450 enzyme involved in terpenoid indole alkaloid biosynthesis. *FEBS Letters, 508*, 215–220.
Contin, A., van der Heijden, R., Lefeber, A. W. M., & Verpoorte, R. (1998). The iridoid glucoside secologanin is derived from the novel triose phosphate/pyruvate pathway in a *Catharanthus roseus* cell culture. *FEBS Letters, 434*, 413–416.
Costa, M. M., Hilliou, F., Duarte, P., Pereira, L. G., Almeida, L., Leech, M., et al. (2008). Molecular cloning and characterization of a vacuolar class III peroxidase involved in the metabolism of anticancer alkaloids in *Catharanthus roseus*. *Plant Physiology, 146*, 403–417.
Courdavault, V., Thiersault, M., Courtois, M., Gantet, P., Oudin, A., Doireau, P., et al. (2005). CaaX-prenyl transferases are essential for expression of genes involved in the early stages of monoterpenoid biosynthetic pathway in *Catharanthus roseus* cells. *Plant Molecular Biology, 57*, 855–870.
De Carolis, E., Chan, F., Balsevich, J., & De Luca, V. (1990). Isolation and characterization of a 2-oxoglutarate-dependent dioxygenase involved in the second-to-last step in vindoline biosynthesis. *Plant Physiology, 94*, 1323–1329.
De Carolis, E., & De Luca, V. (1993). Purification, characterization, and kinetic analysis of a 2-oxoglutarate-dependent dioxygenase involved in vindoline biosynthesis from Catharanthus roseus. *Journal of Biological Chemistry, 268*, 5504–5511.
De Luca, V., Balsevich, J., Tyler, R. T., Eilert, U., Panchuk, B. D., & Kurz, W. G. W. (1986). Biosynthesis of indole alkaloids: Developmental regulation of the biosynthetic pathway from tabersonine to vindoline in Catharanthus roseus. *Journal of Plant Physiology, 125*, 147–156.
De Luca, V., & Cutler, A. J. (1987). Subcellular localization of enzymes involved in indole alkaloid biosynthesis in *Catharanthus roseus*. *Plant Physiology, 85*, 1099–1102.
De Luca, V., Marineau, C., & Brisson, N. (1989). Molecular cloning and analysis of cDNA encoding a plant tryptophan decarboxylase: Comparison with animal dopa decarboxylases. *Proceedings of the National Academy of Sciences of the United States of America, 86*, 2582–2586.
De Luca, V., Salim, V., Atsumi, S. M., & Yu, F. (2012). Mining the biodiversity of plants: A revolution in the making. *Science, 336*, 1658–1661.
De Luca, V., Salim, V., Levac, D., Atsumi, S. M., & Yu, F. (2012). Discovery and functional analysis of monoterpenoid indole alkaloid pathways in plants. *Methods in Enzymology, 515*, 207–229.
De Luca, V., & St-Pierre, B. (2000). The cell and developmental biology of alkaloid biosynthesis. *Trends in Plant Science, 4*, 168–173.
Dethier, M., & De Luca, V. (1993). Partial purification of an *N*-methyltransferase involved in vindoline biosynthesis in *Catharanthus roseus*. *Phytochemistry, 32*, 673–678.
deWaal, A., Meijer, A., & Verpoorte, R. (1995). Strictosidine synthase from *Catharanthus roseus*: Purification and characterization of multiple forms. *Biochemistry Journal, 306*, 571–580.
Dinesh-Kumar, S. P., Anandalakshmi, R., Marathe, R., Schiff, M., & Liu, Y. (2003). Virus-induced gene silencing. *Methods in Molecular Biology, 236*, 287–294.
Di Stilio, V. S., Kumar, R. A., Oddone, A. M., Tolkin, T. R., Salles, P., & McCarty, K. (2010). Virus-induced gene silencing as a tool for comparative functional studies in *Thalictrum*. *PLoS One, 5*, e12064.
Eisenreich, W., Rohdich, F., & Bacher, A. (2001). Deoxyxylulose phosphate pathway to terpenoids. *Trends in Plant Science, 6*, 78–84.
El-Sayed, M., Choi, Y. H., Frédérich, M., Roytrakul, S., & Verpoorte, S. (2004). Alkaloid accumulation in *Catharanthus roseus* cell suspension cultures fed with stemmadenine. *Biotechnology Letters, 26*, 793–798.
El-Sayed, M., & Verpoorte, R. (2007). *Catharanthus* terpenoid indole alkaloids: Biosynthesis and regulation. *Phytochemistry Reviews, 6*, 277–305.

Facchini, P. J. (2001). Alkaloid biosynthesis in plants: Biochemistry, cell biology, molecular regulation, and metabolic engineering. *Annual Review of Plant Physiology and Plant Molecular Biology*, *52*, 29–66.

Facchini, P. J., Bohlmann, J., Covello, P. S., De Luca, V., Mahadevan, R., Page, J. E., et al. (2012). Synthetic biosystems for the production of high-value plant metabolites. *Trends in Biotechnology*, *30*, 127–131.

Facchini, P. J., & De Luca, V. (2008). Opium poppy and Madagascar periwinkle: Model non-model systems to investigate alkaloid biosynthesis in plants. *The Plant Journal*, *54*, 763–784.

Fahn, W., Laussermair, E., Deus-Neumann, B., & Stöckigt, J. (1985). Late enzymes of vindoline biosynthesis *S*-adenosyl-L-methionine: 11-*O*-demethyl-17-*O*-deacetylvindoline 11-*O*-methylase and unspecific acetylesterase. *Plant Cell Reports*, *4*, 337–340.

Geerlings, A., Ibanez, M., Memelink, J., van der Heijden, R., & Verpoorte, R. (2000). Molecular cloning and analysis of strictosidine β-D-glucosidase, an enzyme in terpenoid indole alkaloid biosynthesis in *Catharanthus roseus*. *Journal of Biological Chemistry*, *275*, 3051–3056.

Gerasimenko, I., Sheludko, Y., Ma, X., & Stöckigt, J. (2002). Heterologous expression of a *Rauvolfia* cDNA encoding strictosidine glucosidase, a biosynthetic key to over 2000 monoterpenoid indole alkaloids. *European Journal of Biochemistry*, *269*, 2204–2213.

Geu-Flores, F., Sherden, N. H., Courdavault, V., Burlat, V., Glenn, W. S., Wu, C., et al. (2012). An alternative route to cylic terpenes by reductive cyclization in iridoid biosynthesis. *Nature*, *492*, 138–142.

Giddings, L. A., Liscombe, D. K., Hamilton, J. P., Childs, K. L., DellaPenna, D., Buell, C. R., et al. (2011). A stereoselective hydroxylation step of alkaloid biosynthesis by a unique cytochrome P450 in *Catharanthus roseus*. *Journal of Biological Chemistry*, *286*, 16751–16757.

Góngora-Castillo, E., Childs, K. L., Fedewa, G., Hamilton, J. P., Liscombe, D. K., Magallanes-Lundback, M., et al. (2012). Development of transcriptomic resources for interrogating the biosynthesis of monoterpene indole alkaloids in medicinal plant species. *PLoS One*, 7(12), e52506 10.1371.

Gould, B., & Kramer, E. M. (2007). Virus-induced gene silencing as a tool for functional analyses in the emerging model plant *Aquilegia* (columbine, Ranunculaceae). *Plant Methods*, *3*, 6.

Gueritte, F., Bac, N. V., Langlois, Y., & Potier, P. (1980). Biosynthesis of antitumour alkaloids from *Catharanthus roseus*. Conversion of 20'deoxyleurosidine into vinblastine. *Journal of the Chemical Society, Chemical Communications*, 452–453.

Guirimand, G., Burlat, V., Oudin, A., Lanoue, A., St-Pierre, B., & Courdavault, V. (2009). Optimization of the transient transformation of Catharanthus roseus cells by particle bombardment and its application to the subcellular localization of hydroxymethylbutenyl 4-diphosphate synthase and geraniol 10-hydroxylase. *Plant Cell Reports*, *28*, 1215–1234.

Guirimand, G., Courdavault, V., Lanoue, A., Mahroug, S., Guihur, A., & Blanc, N. (2010). Strictosidine activation in Apocynaceae: Towards a "nuclear time bomb"? *BMC Plant Biology*, *10*, 182.

Guirimand, G., Guihur, A., Poutrain, P., Hericourt, F., Mahroug, S., St-Pierre, B., et al. (2011). Spatial organization of the vindoline biosynthetic pathway in *Catharanthus roseus*. *Journal of Plant Physiology*, *168*, 549–557.

Hagel, J. M., & Facchini, P. J. (2010). Dioxygenases catalyze the O-demethylation steps of morphine biosynthesis in opium poppy. *Nature Chemical Biology*, *6*, 273–755.

Hemscheidt, T., & Zenk, M. H. (1985). Partial purification and characterization of a NADPH dependent tetrahydroalstonine synthase from *Catharanthus roseus* cell cultures. *Plant Cell Reports*, *4*, 216–219.

Hileman, L. C., Drea, S., Martino, G., Litt, A., & Irish, V. F. (2005). Virus-induced gene silencing is an effective tool for assaying gene function in the basal eudicot species *Papaver somniferum* (opium poppy). *The Plant Journal, 44*, 334–341.

Hong, S. B., Peebles, C. A., Shanks, J. V., San, K. Y., & Gibson, S. I. (2006). Expression of the *Arabidopsis* feedback-insensitive anthranilate synthase holoenzyme and tryptophan decarboxylase genes in *Catharanthus roseus* hairy roots. *Journal of Biotechnology, 122*, 28–38.

Hughes, E. H., Hong, S. B., Gibson, S. I., Shanks, J. V., & San, K. Y. (2004). Metabolic enginerring of the indole pathway in *Catharanthus roseus* hairy roots and increased accumulation of tryptamine and serpentine. *Metabolic Engineering, 6*, 268–276.

Ikeda, H., Esaki, N., Nakai, S., Hashimoto, K., Uesato, S., Soda, K., et al. (1991). Acyclic monoterpene primary alcohol: NADP+ oxidoreductase of *Rauwolfia serpentina* cells: The key enzyme in biosynthesis of monoterpene alcohols. *Journal of Biochemistry, 109*, 341–347.

Irmler, S., Schröder, G., St-Pierre, B., Crouch, N. P., Hotze, M., Schmidt, J., et al. (2000). Indole alkaloid biosynthesis in *Catharanthus roseus*: New enzyme activities and identification of cytochrome P450 *CYP72A1* as secologanin synthase. *The Plant Journal, 24*, 797–804.

Ishikawa, H., Colby, D. A., & Boger, D. L. (2008). Direct coupling of catharanthine and vindoline to provide vinblastine: Total synthesis of (+)- and ent-(−)-vinblastine. *Journal of the American Chemical Society, 130*, 420–421.

Kawai, H., Kuroyanagi, M., & Ueno, A. (1988). Iridoid glucosides from *Lonicera japonica* THUNB. *Chemical and Pharmaceutical Bulletin, 36*, 3664–3666.

Kuboyama, T., Yokoshima, S., Tokuyama, H., & Fukuyama, T. (2004). Stereocontrolled total synthesis of (+)-vincristine. *Proceedings of the National Academy of Sciences of the United States of America, 101*, 11966–11970.

Kutchan, T. M., Hampp, N., Lottspeich, F., Beyreuther, K., & Zenk, M. H. (1988). The cDNA clone for strictosidine synthase from *Rauwolfia serpentina*: DNA sequence determination and expression in *Escherichia coli*. *FEBS Letters, 237*, 40–44.

Laflamme, P., St-Pierre, B., & De Luca, V. (2001). Molecular and biochemical analysis of a Madagascar periwinkle root-specific minovincine-19-hydroxy-O-acetyltransferase. *Plant Physiology, 125*, 189–198.

Lange, B. M., & Croteau, R. (1999). Isopentenyl diphosphate biosynthesis via a mevalonate-independent pathway: Isopentenyl monophosphate kinase catalyzes the terminal enzymatic step. *Proceedings of the National Academy of Sciences of the United States of America, 96*, 13714–13719.

Lange, B. M., Wildung, M. R., Stauber, E. J., Sanchez, C., Pouchnik, D., & Croteau, R. (2000). Probing essential oil biosynthesis and secretion by functional evaluation of expressed sequence tags from mint glandular trichomes. *Proceedings of the National Academy of Sciences of the United States of America, 97*, 2934–2939.

Levac, D., Murata, J., Kim, W. S., & De Luca, V. (2008). Application of carborundum abrasion for investigating leaf epidermis: Molecular cloning of *Catharanthus roseus* 16-hydroxytabersonine-16-O-methyltransferase. *The Plant Journal, 53*, 225–236.

Li, R., Reed, D. W., Liu, E., Nowak, J., Pelcher, L. E., Page, J. E., et al. (2006). Functional genomic analysis of alkaloid biosynthesis in *Hyoscyamus niger* reveals a cytochrome P450 involved in littorine rearrangement. *Chemistry and Biology, 13*, 513–520.

Liscombe, D., & O'Connor, S. (2011). A virus-induced gene silencing approach to understanding alkaloid metabolism in *Catharanthus roseus*. *Phytochemistry, 72*, 1969–1977.

Liscombe, D. K., Usera, A. R., & O'Connor, S. E. (2010). Homolog of tocopherol C-methyltransferases catalyzes *N*-methylation in anticancer alkaloid biosynthesis. *Proceedings of the National Academy of Sciences of the United States of America, 107*, 18793–18798.

Liu, Y., Schiff, M., Marathe, R., & Dinesh-Kumar, S. P. (2002). Tobacco Rar1, EDS1 and *NPR1/NIM1* like genes are required for N-mediated resistance to tobacco mosaic virus. *The Plant Journal*, *30*, 415–429.

Lopez-Meyer, M., & Nessler, C. L. (1997). Tryptophan decarboxylase is encoded by two autonomously regulated genes in *Camptotheca acuminata* which are differentially expressed during development and stress. *The Plant Journal*, *11*, 1167–1175.

Luijendijk, T. J. C., Stevens, L. H., & Verpoorte, R. (1998). Purification and characterization of strictosidine beta-D-glucosidase from *Catharanthus roseus* cell suspension cultures. *Plant Physiology and Biochemistry*, *36*, 419–425.

Luijendijk, T. J. C., Vandermeijden, E., & Verpoorte, R. (1996). Involvement of strictosidine as a defensive chemical in *Catharanthus roseus*. *Journal of Chemical Ecology*, *22*, 1355–1366.

Ma, X., Panjikar, S., Koepke, J., Loris, E., & Stöckigt, J. (2006). The structure of *Rauvolfia serpentina* strictosidine synthase is a novel six-bladed beta-propeller fold in plant proteins. *The Plant Cell*, *18*, 907–920.

Madyastha, K. M., Guarnaccia, R., Baxter, C., & Coscia, C. J. (1973). *S-Adenosyl-L-methionine*: Loganic acid methyltransferase. *Journal of Biological Chemistry*, *248*, 2497–2501.

Madyastha, K. M., Ridgway, J. E., Dwyer, J. G., & Coscia, C. J. (1977). Subcellular localization of a cytochrome P-450-dependent monooxygenase in vesicles of the higher plant *Catharanthus roseus*. *The Journal of Cell Biology*, *72*, 303–313.

Magnotta, M., Murata, J., Chen, J., & De Luca, V. (2006). Identification of a low vindoline accumulating cultivar of *Catharanthus roseus* (L.) G. Don by alkaloid and enzymatic profiling. *Phytochemistry*, *67*, 1758–1764.

Magnotta, M., Murata, J., Chen, J., & De Luca, V. (2007). Expression of deacetylvindoline-4-O-acetyltransferase in *Catharanthus roseus* hairy roots. *Phytochemistry*, *68*, 1922–1931.

Maresh, J. J., Giddings, L. A., Friedrich, A., Loris, E. A., Panjikar, S., Trout, B. L., et al. (2008). Strictosidine synthase: Mechanism of a pictet-spengler catalyzing enzyme. *Journal of the American Chemical Society*, *130*, 710–723.

McCoy, E., & O'Connor, S. E. (2006). Directed biosynthesis of alkaloid analogs in the medicinal plant *Catharanthus roseus*. *Journal of the American Chemical Society*, *128*, 14276–14277.

McKnight, T. D., Bergey, D. R., Burnett, R. J., & Nessler, C. L. (1991). Expression of enzymatically active and correctly targeted strictosidine synthase in transgenic tobacco plants. *Planta*, *185*, 148–152.

Mcknight, T. D., Roessner, C. A., Devagupta, R., Scott, A. I., & Nessler, C. L. (1990). Nucleotide sequence of a cDNA encoding the vacuolar protein strictosidine synthase from *Catharanthus roseus*. *Nucleic Acids Research*, *18*, 4939.

Meehan, T. D., & Coscia, C. J. (1973). Hydroxylation of geraniol and nerol by a monooxygenase from *Vinca rosea*. *Biochemical and Biophysical Research Communications*, *53*, 1043–1048.

Memelink, J., & Gantet, P. (2007). Transcription factors involved in terpenoid indole alkaloid biosynthesis in *Catharanthus roseus*. *Phytochemistry Reviews*, *6*, 353–362.

Menke, F., Parchmann, S., Mueller, M., Kijne, J., & Memelink, J. (1999). Involvement of the octadecanoid pathway and protein phosphorylation in fungal elicitor-induced expression of terpenoid indole alkaloid biosynthetic genes in *Catharanthus roseus*. *Plant Physiology*, *119*, 1289–1296.

Mizutani, M., & Sato, F. (2011). Unusual P450 reactions in plant secondary metabolism. *Archives of Biochemistry and Biophysics*, *507*, 194–203.

Murata, J., Bienzle, D., Brandle, J. E., Sensen, C. W., & De Luca, V. (2006). Expressed sequence tags from Madagascar periwinkle (*Catharanthus roseus*). *FEBS Letters*, *580*, 4501–4507.

Murata, J., & De Luca, V. (2005). Localization of tabersonine 16-hydroxylase and 16-OH tabersonine-16-*O*-methyltransferase to leaf epidermal cells defines them as a major site of precursor biosynthesis in the vindoline pathway in *Catharanthus roseus*. *The Plant Journal*, *44*, 581–594.

Murata, J., Roepke, R., Gordon, H., & De Luca, V. (2008). The leaf epidermome of *Catharanthus roseus* reveals its biochemical specialization. *The Plant Cell*, *20*, 524–542.

Newman, J. D., & Chappell, J. (1999). Isoprenoid biosynthesis in plants: Carbon partitioning within the cytoplasmic pathway. *Critical Reviews in Biochemistry and Molecular Biology*, *34*, 95–106.

O'Connor, S. E. (2012). Strategies for engineering plant natural products: The iridoid-derived monoterpene indole alkaloids of *Catharanthus roseus*. *Methods in Enzymology*, *515*, 189–206.

O'Connor, S. E., & Maresh, J. J. (2006). Chemistry and biology of monoterpene indole alkaloid biosynthesis. *Natural Product Reports*, *23*, 532–547.

Oudin, A., Courtois, M., Rideau, M., & Clastre, M. (2007). The iridoid pathway in *Catharanthus roseus* alkaloid biosynthesis. *Phytochemistry Reviews*, *6*, 259–276.

Oudin, A., Mahroug, S., Courdavault, V., Hervouet, N., Zelwer, C., Rodríguez-Concepción, M., et al. (2007). Spatial distribution and hormonal regulation of gene products from methyl erythritol phosphate and monoterpene-secoiridoid pathways in *Catharanthus roseus*. *Plant Molecular Biology*, *65*, 13–30.

Papon, N., Vansiri, A., Gantet, P., Chenieux, J. C., Rideau, M., & Creche, J. (2004). Histidine-containing phosphotransfer domain extinction by RNA interference turns off a cytokinin signaling circuitry in *Catharanthus roseus* suspension cells. *FEBS Letters*, *558*, 85–88.

Pasquali, G., Gooddijn, O. J. M., de Waal, A., Verpoorte, A., Schilperoort, R. A., Hoge, J. H. C., et al. (1992). Coordinated regulation of two indole alkaloid biosynthetic genes from Catharnathus roseus by auxin and elicitors. *Plant Molecular Biology*, *18*, 1121–1131.

Pasquali, G., Porto, D. D., & Fett-Neto, A. G. (2006). Metabolic engineering of cell cultures versus whole plant complexity in production ofbioactive monoterpene indole alkaloids: Recent progress related to an old dilemma. *Journal of Bioscience and Bioengineering*, *101*, 287–296.

Pauw, B., Hilliou, F. A. O., Sandonis, V., Chatel, M. G., de Wolf, C. J. F., Champion, A., et al. (2004). Zinc finger proteins act as a transcriptional repressors of alkaloid biosynthesis genes in *Catharanthus roseus*. *Journal of Biological Chemistry*, *279*, 52940–52948.

Power, R., Kurz, W. G. W., & De Luca, V. (1990). Purification and characterization of acetylcoenzyme A: Deacetylvindoline 4-Oacetyltransferase from Catharanthus roseus. *Archives of Biochemistry and Biophysics*, *279*, 370–376.

Qureshi, A. A., & Scott, A. I. (1968). Interconversion of corynanthe, aspidosperma, and iboga alkaloids a model for indole alkaloid biosynthesis. *Chemical Communications*, 945–946.

Ramos-Valvidia, A. C., van der Heijden, R., & Verpoorte, R. (1997). Isopentenyl diphosphate isomerase: A core enzyme in isoprenoid biosynthesis. A review of its biochemistry and function. *Natural Product Reports*, *14*, 591–603.

Ratcliff, F., Martin-Hernandez, A. M., & Baulcombe, D. C. (2001). Tobacco rattle virus as a vector for analysis of gene function by silencing. *The Plant Journal*, *25*, 237–245.

Rischer, H., Oresic, M., Seppanen-Laakso, T., Katajamaa, M., Lammertyn, F., Ardiles-Diaz, W., et al. (2006). Gene-to-metabolite networks for terpenoid indole alkaloid biosynthesis in *Catharanthus roseus* cells. *Proceedings of the National Academy of Sciences of the United States of America*, *103*, 5614–5619.

Ro, D. K., Paradise, E. M., Ouellet, M., Fisher, K. J., & Newman, K. L. (2006). Production of the antimalarial drug precursor artermisinic acid in engineered yeast. *Nature*, *440*, 940–943.

Rodriguez, S., Compagnon, V., Crouch, N. P., St-Pierre, B., & De Luca, V. (2003). Jasmonate-induced epoxidation of tabersonine by cytochrome P-450 in hairy root cultures of *Catharanthus roseus*. *Phytochemistry*, *64*, 401–409.
Rodriguez-Conception, A., & Boronat, A. (2002). Elucidation of the methylerythritol phosphate pathway for isoprenoid biosynthesis in bacteria and plastids. A metabolic milestone achieved through genomics. *Plant Physiology*, *130*, 1079–1089.
Roepke, J., Salim, V., Wu, M., Thamm, A. M., Murata, J., Ploss, K., et al. (2010). Vinca drug components accumulate exclusively in leaf exudates of Madagascar periwinkle. *Proceedings of the National Academy of Sciences of the United States of America*, *107*, 15287–15292.
Roewer, I., Cloutier, N., Nessler, C., & De Luca, V. (1992). Transient induction of tryptophan decarboxylase (TDC) and strictosidine synthase (SS) genes in cell suspension cultures of *Catharanthus roseus*. *Plant Cell Reports*, *11*, 86–89.
Rohdich, F., Kis, K., Bacher, A., & Eisenreich, W. (2001). The non-mevalonate pathway of isoprenoids: Genes, enzymes and intermediates. *Current Opinion in Chemical Biology*, *5*, 535–540.
Rohmer, M. (1999). The discovery of a mevalonate-independent pathway for isoprenoid biosynthesis in bacteria, algae, and higher plants. *Natural Product Reports*, *16*, 565–574.
Runguphan, W., Maresh, J. J., & O'Connor, S. E. (2009). Silencing of tryptamine biosynthesis for production of nonnatural alkaloids in plant culture. *Proceedings of the National Academy of Sciences of the United States of America*, *106*, 13673–13678.
Runguphan, W., Qu, X., & O'Connor, S. E. (2010). Integrating carbon-halogen bond formation into medicinal plant metabolism. *Nature*, *468*, 461–464.
Sanchez-Iturbe, P., Galaz-Avalos, R., & Loyola-Vargas, V. (2005). Determination and partial purification of 10-oxogeranial:iridodial cyclase an enzyme catalyzing the synthesis of iridodial from Catharanthus roseus hairy roots. *Phyton*, *54*, 55–69.
Schilmiller, A. L., Last, R. L., & Pichersky, E. (2008). Harnessing plant trichome biochemistry for the production of useful compounds. *The Plant Journal*, *54*, 702–711.
Schröder, G., Unterbusch, E., Kaltenbach, M., Schmidt, J., Strack, D., De Luca, V., et al. (1999). Light-induced cytochrome P450-dependent enzyme in indole alkaloid biosynthesis: Tabersonine 16-hydroxylase. *FEBS Letters*, *458*, 97–102.
Senthil-Kumar, M., & Mysore, K. S. (2011). New dimension for VIGS in plant functional genomics. *Trends in Plant Science*, *16*, 656–665.
Shukla, A. K., Shasany, A. K., Gupta, M. M., & Khanuja, S. P. (2006). Transcriptome analysis in *Catharanthus roseus* leaves and roots for comparative terpenoid indole alkaloid profiles. *Journal of Experimental Botany*, *57*, 3921–3932.
Siberil, Y., Benhamron, S., Memelink, J., Giglioli-Guivarc'h, N., Thiersault, M., & Boisson, B. (2001). *Catharanthus roseus* G-box binding factors 1 and 2 act as repressors of strictosidine synthase gene expression in cell cultures. *Plant Molecular Biology*, *45*, 477–488.
Simkin, A. J., Miettinen, K., Claudel, P., Burlat, V., Guirimand, G., & Courdavault, V. (2013). Characterization of the plastidial geraniol synthase from Madagascar periwinkle which initiates the monoterpenoid branch of the alkaloid pathway in internal phloem associated parenchyma. *Phytochemistry*, *85*, 36–43.
Stevens, L. H., Blom, T. J. M., & Verpoorte, R. (1993). Subcellular localization of tryptophan decarboxylase, strictosidine synthase and strictosidine glucosidase in suspension-cultured cells of *Catharanthus roseus* and *Tabernaemontana divaricata*. *Plant Cell Reports*, *12*, 573–576.
Stöckigt, J., & Zenk, M. (1977). Strictosidine (isovincoside): The key intermediate in the biosynthesis on monoterpene indole alkaloids. *Journal of the Chemical Society, Chemical Communications*, 646–648.
St-Pierre, B., & De Luca, V. (1995). A cytochrome P-450 monooxygenase catalyzes the first step in the conversion of tabersonine to vindoline in *Catharanthus roseus*. *Plant Physiology*, *109*, 131–139.

St-Pierre, B., Laflamme, P., Alarco, A. M., & De Luca, V. (1998). The terminal *O*-acetyltransferase involved in vindoline biosynthesis defines a new class of proteins responsible for coenzyme A-dependent acyl transfer. *The Plant Journal, 14*, 703–713.

St-Pierre, B., Vázquez-Flota, F. A., & De Luca, V. (1999). Multicellular compartmentation of *Catharanthus roseus* alkaloid biosynthesis predicts intercellular translocation of a pathway intermediate. *The Plant Cell, 11*, 887–900.

Suttipanta, N., Pattanaik, S., Kulshrestha, M., Patra, B., Singh, S. K., & Yuan, L. (2011). The transcription factor CrWRKY1 positively regulates the terpenoid indole alkaloid biosynthesis in *Catharanthus roseus*. *Plant Physiology, 157*, 2081–2093.

Szabo, L. (2008). Molecular evolutionary lines in the formation of indole alkaloids derived from secologanin. *Arkivoc*, 167–181.

Treimer, J., & Zenk, M. (1979). Purification and properties of strictosidine synthase, the key enzyme in indole alkaloid formation. *European Journal of Biochemistry, 101*, 225–233.

Uesato, S., Ikeda, H., Fujita, T., Inouye, H., & Zenk, M. (1987). Elucidation of iridodial formation mechanism: Partial purification and characterization of the novel monoterpene cyclase from *Rauwolfia serpentina* cell suspension cultures. *Tetrahedron Letters, 28*, 4431–4434.

van der Fits, L., & Memelink, J. (2000). ORCA3, a jasmonate-responsive transcriptional regulator of plant primary and secondary metabolism. *Science, 289*, 295–297.

van der Heijden, R., Jacobs, D. I., Snoeijer, W., Hallard, D., & Verpoorte, R. (2004). The *Catharanthus* alkaloids: Pharmacognosy and biotechnology. *Current Medicinal Chemistry, 11*, 607–628.

Vazquez-Flota, F. A., De Carolis, E., Alarco, A. M., & De Luca, V. (1997). Molecular cloning and characterization of deacetoxyvindoline 4-hydroxylase, a 2-oxoglutarate dependent dioxygenase involved in the biosynthesis of vindoline in Catharanthus roseus (L.) G. Don. *Plant Molecular Biology, 34*, 935–948.

Vazquez-Flota, F., & De Luca, V. (1998a). Developmental and light regulation of desacetoxyvindoline 4-hydroxylase in *Catharanthus roseus* (L.) G. Don. *Plant Physiology, 117*, 1351–1361.

Vazquez-Flota, F., & De Luca, V. (1998b). Jasmonate modulates developmental and light regulated alkaloid biosynthesis in *Catharanthus roseus*. *Phytochemistry, 49*, 395–402.

Veau, B., Courtois, M., Oudin, A., Chenieux, J. C., Rideau, M., & Clastre, M. (2000). Cloning and expression of cDNAs encoding two enzymes of the MEP pathway in *Catharanthus roseus*. *Biochimica et Biophysica Acta, 1517*, 159–163.

Wege, S., Scholz, A., Gleissberg, S., & Becker, A. (2007). Highly efficient virus-induced gene silencing (VIGS) in California poppy (*Eschscholzia californica*): An evaluation of VIGS as a strategy to obtain functional data from nonmodel plants. *Annals of Botany, 100*, 641–649.

Wijekoon, C. P., & Facchini, P. J. (2012). Systematic knockdown of morphine pathway enzymes in opium poppy using virus-induced gene silencing. *The Plant Journal, 69*, 1052–1063.

Yamamoto, H., Katano, N., Ooi, A., & Inoue, K. (2000). Secologanin synthase which catalyzes the oxidative cleavage of loganin into secologanin is a cytochrome P450. *Phytochemistry, 53*, 7–12.

Yamazaki, Y., Sudo, H., Yamazaki, M., Aimi, N., & Saito, K. (2003). Camptothecin biosynthetic genes in hairy roots of *Ophiorrhiza pumila*: Cloning, characterization and differential expression in tissues and by stress compounds. *Plant & Cell Physiology, 44*, 395–403.

Zárate, R., & Verpoorte, R. (2007). Strategies for the genetic modification of the medicinal plant *Catharanthus roseus* (L.) G. Don. *Phytochemistry Reviews, 6*, 475–491.

Zhang, H., Hedhili, S., Montiel, G., Zhang, Y., Chatel, G., Pré, M., et al. (2011). The basic helix-loop-helix transcription factor CrMYC2 controls the jasmonate-responsive expression of the ORCA genes that regulate alkaloid biosynthesis in *Catharanthus roseus*. *The Plant Journal*, *67*, 61–71.

Zhao, J., & Verpoorte, R. (2007). Manipulating indole alkaloid production by *Catharanthus roseus* cell cultures in bioreactors: From biochemical processing to metabolic engineering. *Phytochemistry Reviews*, *6*, 435–457.

Zhu, J., Guggisberg, A., Kalt-Hadamowsky, M., & Hesse, M. (1990). Chemotaxonomic study of the genus *Tabernaemontana* (Apocynaceae) based on their indole alkaloid content. *Plant Systematics and Evolution*, *172*, 13–34.

Ziegler, J., & Facchini, P. J. (2008). Alkaloid biosynthesis: Metabolism and trafficking. *Annual Reviews of Plant Biology*, *59*, 735–769.

CHAPTER TWO

Increasing the Pace of New Discoveries in Tropane Alkaloid Biosynthesis

Jan Jirschitzka[*], **Franziska Dolke**[†], **John C. D'Auria**[*,1]
[*]Department of Biochemistry, Max Planck Institute for Chemical Ecology, Jena, Germany
[†]Department of Bioorganic Chemistry, Max Planck Institute for Chemical Ecology, Jena, Germany
[1]Corresponding author: e-mail address: dauria@ice.mpg.de

Contents

1. Introduction 40
2. Tropane Alkaloids in Plants: From Herbs to Trees 44
3. Biosynthesis of Tropane Alkaloids 49
4. Metabolic Engineering of Tropane Alkaloids 59
5. Conclusions 63
Acknowledgements 63
References 63

Abstract

Tropane alkaloids (TAs) are plant-derived natural products that have been exploited throughout history for their pharmaceutical properties. TAs are characterised by the presence of a tropane ring and are present in a variety of plant families including the Solanaceae, Convolvulaceae, Proteaceae, Rhizophoraceae, Brassicaceae and Erythroxylaceae. The structural genes and enzymes involved in TA biosynthesis have been characterised primarily in the Solanaceae, but several key steps remain unresolved. The scattered distribution of TAs in different angiosperm families suggests that there may be several alternative pathways for TA biosynthesis, and TA biosynthesis appears to have a polyphyletic origin. Molecular techniques are providing an ever-increasing amount of information on the nature and evolution of TA biosynthesis pathways and are aiding the development of techniques for increasing the production of useful tropanes through metabolic engineering.

ABBREVIATIONS

H6H hyoscyamine 6β-hydroxylase
MecgoR methylecgonone reductase
MeJA methyl jasmonate
MPO *N*-methylputrescine oxidase

Advances in Botanical Research, Volume 68
ISSN 0065-2296
http://dx.doi.org/10.1016/B978-0-12-408061-4.00002-X

ODC ornithine decarboxylase
PMT putrescine *N*-methyltransferase
SA salicylic acid
SAM *S*-adenosylmethionine
SDR short-chain dehydrogenase/reductase
TA(s) tropane alkaloid(s)
TR tropinone reductase

1. INTRODUCTION

Tropane alkaloids (TAs) are a large class of plant-derived secondary metabolites and are defined by their core structure, the 8-azabicyclo [3.2.1]octane nucleus (Fig. 2.1). There have been more than 200 structures reported in the literature (Lounasmaa & Tamminen, 1993), and many TAs are being identified in a range of plant families (de Oliveira et al., 2011; Eich, 2008; El Bazaoui, Bellimam, & Soulaymani, 2011; Griffin & Lin, 2000; Queiroz et al., 2009; Razzakov & Aripova, 2004; Sena-Filho et al., 2011). Like other alkaloids, TAs possess many potent pharmacological activities, including acting as cholinergic agents, deliriants and narcotic analgesics. Humans have exploited the pharmacological properties of TAs since ancient times.

One of the first domesticated plant species used exclusively for its medicinal properties is *Erythroxylum coca*, which produces the well-known TA cocaine (**42**). Recent archaeological evidence has found coca leaves in Peruvian house floors, dating the use of this species back at least 8000 years (Dillehay et al., 2010). Coca leaves are obtained from the two closely related species *E. coca* and *E. novogranatense* (Plowman, 1982). During the Incan empire (thirteenth to sixteenth century), coca leaves were used as a sacrificial offering and were additionally used by the aristocracy and religious elite for its medicinal properties (Naranjo, 1981). Interestingly, the practice of

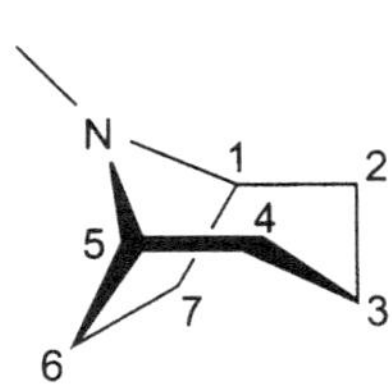

Figure 2.1 The tropane core structure. The basic tropane skeleton with the accepted carbon numbering system is depicted. Metabolites missing the methyl modification on the nitrogen are referred to as nortropanes.

chewing the leaves by the ruling classes was restricted to *E. novogranatense* because it contains high levels of methyl salicylate and thus is perceived to have a minty taste (Naranjo, 1981; Plowman & Rivier, 1983). In the Spanish colonial era, coca leaves were originally described as the "Devil's leaf" (Parkerson, 1983). However, conquistadores later recognised that chewing coca leaves increased stamina in addition to reducing thirst. Due to these observations, the coca leaf began to be used to increase the productivity of natives working as slaves in the mines (Plowman, 1984). In 1860, Albert Niemann pioneered the modern method for the isolation of cocaine (**42**) as a pure substance (Niemann, 1860). Pure cocaine (**42**) in its crystallised form then began to be consumed in large quantities for its euphoric and analgesic effects. The drug's rise to fame was partly due to its use and endorsement by famous individuals such as Sigmund Freud (Freud, 1884) and its inclusion as an ingredient in drinks such as Vin Mariani (wine) (Wink, 1998c) and Coca Cola (Plowman, 1981). The discovery of cocaine's (**42**) narcotic properties in the twentieth century quickly changed this "magic drug" into an illegal substance (Wink, 1998c). The modern abuse of the pure form of cocaine (**42**) has led to worldwide socioeconomic and health-related issues (Streatfeild, 2001).

Other well-known TAs that have been utilised throughout history include atropine (**1**) (Fig. 2.2) and scopolamine (**44**) (Fig. 2.7), which are

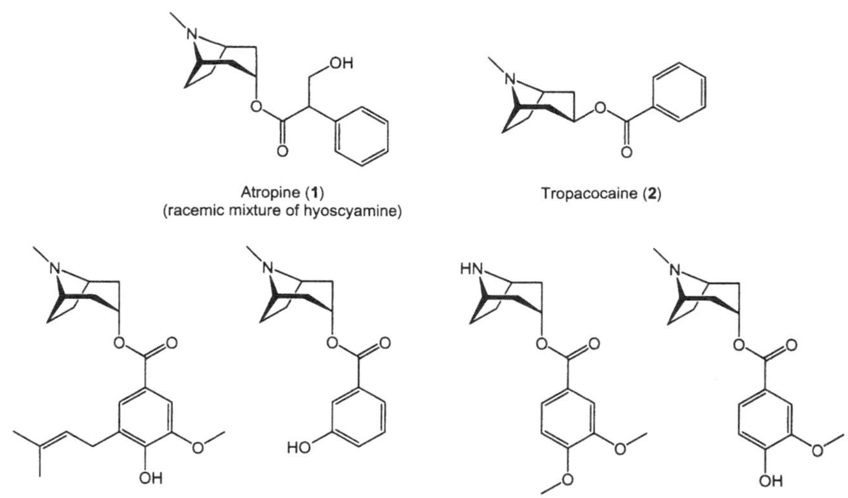

Figure 2.2 Examples of tropane alkaloids found in plants of the Solanaceae, Convolvulaceae, Brassicaceae and Phyllanthaceae.

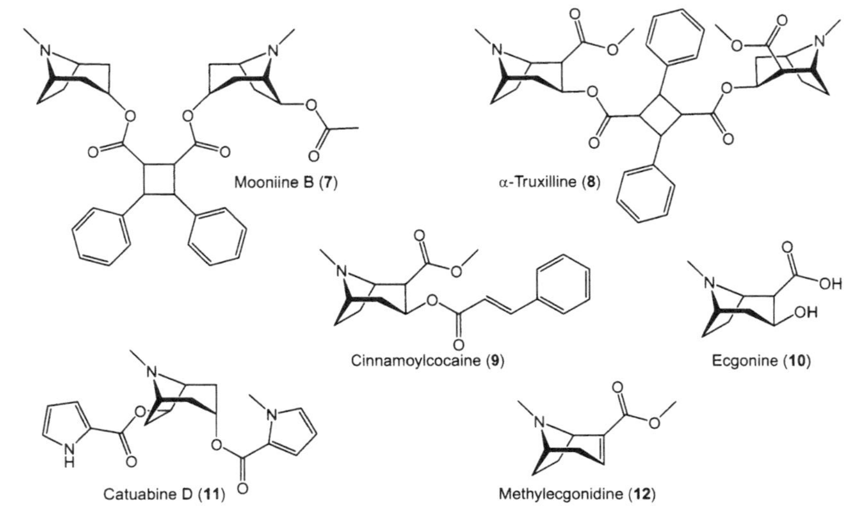

Figure 2.3 Examples of tropane alkaloids found in the Erythroxylaceae.

found in several genera of the Solanaceae. *Datura* species, for example, were described in the Ebers papyrus in about 1550 BC and were also used in Ayurvedic medicine in about 900 BC (Wink, 1998c). The plants were used to treat mental illness, fever, tumours, eczema and infections (Wink, 1998c). In medieval Europe, TAs from solanaceous plants were important in medicine, witchcraft and divination. The four solanaceous plants thorn apple (*Datura stramonium*), mandrake (*Mandragora officinarum*), henbane (*Hyoscyamus niger*) and belladonna (*Atropa belladonna*) were used as sedatives, sleep-inducing agents (henbane), aphrodisiacs and panaceas (mandrake). These plants were often ingredients added to witches' brews, and the association between witches and brooms may have its origin in the use of brooms to apply alkaloid-containing salves (Mann, 1992). The administration of plant salves was found to be particularly rapid when absorbed through the mucous membranes, including those of the female genitalia (Mann, 1992). Belladonna (Italian for beautiful woman) was used to enhance feminine beauty and sensuality by dropping plant extracts into the eyes of women and dilating their pupils (Schultes, 1976). In China, *Datura metel*, *D. stramonium* and *Anisodus tanguticus* (known as *Scopolia tanguticus*) are among the 50 essential Chinese medicinal plants (Schnorrenberger, 1978). Atropine (**1**) was first isolated in 1833 from the roots and leaves of *A. belladonna* while the first isolation of scopolamine (**44**) was performed using *Scopolia japonica* in 1888 (Geiger & Hesse, 1833; Mein, 1833; Schmidt & Henschke, 1888).

Solanaceous plants are often used for their hallucinogenic properties and their use in shamanistic and religious rituals. The majority of these plants contain high quantities of scopolamine (**44**), such as the red-flowered Peruvian plant known as Tonga, Huacacachu or grave plant (*Brugmansia sanguinea*). Tonga is reported to be used as an aid in contacting deceased ancestors (Schultes, 1976). Similarly, the main ingredient of Wysoccon, a drug used for the manhood initiation rituals carried out by Algonquin Indians that leads to the user "losing all memory of ever having been boys," is *D. stramonium* (Schultes, Hofmann, & Rätsch, 2001).

In contrast, plants of the Erythroxylaceae are more often valued for their medicinal properties. The leaves of Olokuto, a wild *Erythroxylum species* known as *Erythroxylum dekindtii*, are used to reduce fever in Angola. Olokuto contains ecgonine (**10**), methylecgonine (**38**), tropacocaine (**2**) and pseudotropine (**35**) (Campos Neves & Campos Neves, 1966). It is also very common for tropane alkaloid-producing plants to be used as a panacea. One example of this is *Erythroxylum emarginatum* from South and East Africa. This plant contains large quantities of methylecgonidine (**12**) as the main TA (Nishiyama et al., 2007) and its boiled leaf extracts are drunk to treat asthma, kidney problems, arthritis, child bearing problems and influenza (De Wet, 2011). In addition, the roots and leaves of *E. emarginatum* are used for pain relief, to increase alertness and also as an aphrodisiac (Nishiyama et al., 2007). Many other TA-containing plants are commonly used by different native tribes as aphrodisiacs. One of the most well-known plants in this category is *E. vacciniifolium*, also known as Catuaba. In Brazil, the bark and leaves of Catuaba are sold as a "natural Viagra." Catuabines (**11**) are the TA associated with many of these preparations (Graf & Lude, 1977; Kletter et al., 2004; Queiroz et al., 2009). Lastly, *E. moonii* from Sri Lanka does not fall into the traditional medicinal categories assigned to TA-containing plants. The leaves of this plant, which contain large amounts of mooniines (**7**), dimeric tropane alkaloids, are boiled and used as an anthelmintic for the treatment of parasitic roundworms (Atta-Ur-Rahman et al., 1998).

In contrast to the pharmacology of TAs, the natural role of these compounds in plants is poorly understood. While it is commonly assumed that TAs serve to defend plants against herbivores and pathogens, there is a paucity of studies available on this topic. Scopolamine (**44**) and hyoscyamine (**41**) appear to be utilised by solanaceous plants to mediate a wide array of ecological interactions with many types of insects. Scopolamine (**44**) has been implicated as a phagorepellent to adult lepidoptera including the silk moth (*Bombyx mori*) and the cabbage butterfly (*Pieris brassicae*) (Levinson,

1976). Further, scopolamine (**44**) has also been classified as a feeding deterrent for insects such as *Syntomismoga dorensis* and the honey bee (*Apis mellifera*) (Wink, 1998a). Besides serving as a deterrent, TAs can also function as toxic compounds. Cocaine (**42**) in concentrations similar to those found in coca leaves can serve as a natural insecticide for the tobacco hornworm (*Manduca sexta*) when sprayed on tomato leaves (Nathanson, Hunnicutt, Kantham, & Scavone, 1993). In addition, several studies have shown that scopolamine (**44**) increases mortality in the worm *Tubifex tubifex* (Wink, 1998b) and the insect *Spodoptera frugiperda* (Alves, Sartoratto, & Trigo, 2007). There is evidence that scopolamine (**44**) can also serve as an allelopathic compound, inhibiting the growth of the roots of the plant *Lepidium sativum* (Wink, 1998b).

Sequestration of plant-derived TAs in insects for use as defensive or toxic substances has also been observed in several instances. The winter cherry bug (*Acanthocoris sordidus*) is capable of absorbing scopolamine (**44**) produced by *Duboisia leichhardtii* and transforming it into atropine (**1**) via enzymatic de-epoxidisation (Kitamura, Tominaga, & Ikenaga, 2004). The aposematic butterfly *Placidula euryanassa* utilises scopolamine (**44**) obtained from *Brugmansia suaveolens* as a toxic compound to protect developing larvae from vertebrate predators (Freitas et al., 1996). The specialist lymantriid *Eloria noyesi* feeds exclusively on coca plants, sequesters cocaine (**42**) during the larval stage and retains cocaine (**42**) in the adult moths (Blum, Rivier, & Plowman, 1981). Polyhydroxylated TAs known as calystegines (**37**) are sequestered by several lepidopteran species and render the lepidopterans indigestible by inhibiting the glycosidases of potential predators (Nash & Watson, 1995).

2. TROPANE ALKALOIDS IN PLANTS: FROM HERBS TO TREES

In order to better understand the biological roles of TAs, it is critical to also understand the distribution of these metabolites across the plant kingdom. They are present in five major lineages of dicotyledons: the peripheral Eudicots (Proteaceae), the Malvid (Brassicaceae) and Fabid (Elaeocarpaceae, Erythroxylaceae, Moraceae, Phyllanthaceae, Rhizophoraceae) clusters of the Rosid lineage, the peripheral Asterids (Olacaceae) and the Lamid cluster of the Asterid lineage (Solanaceae, Convolvulaceae) (The Angiosperm Phylogeny Group, 2009). This includes seven different orders, which contain in total ten different families (Fig. 2.4). The scattered distribution pattern of TA-producing plant families raises the question of whether or not biosynthesis

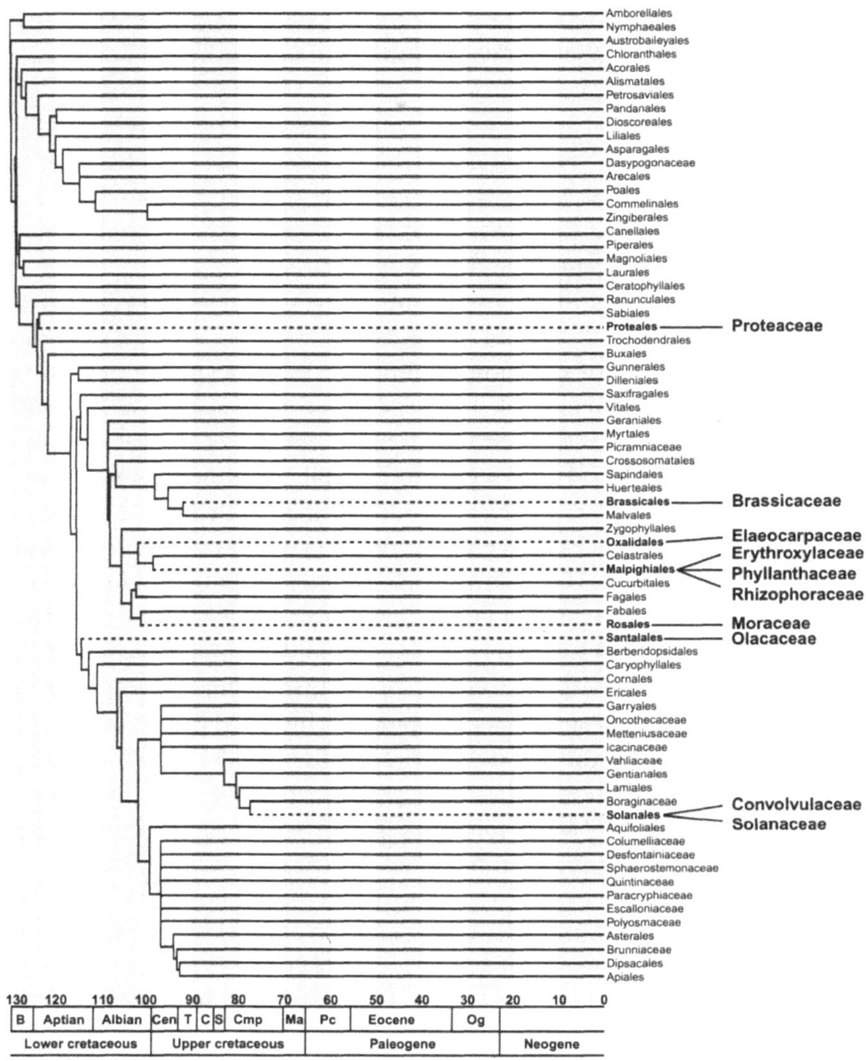

Figure 2.4 The scattered distribution of tropane producing among the angiosperms. Plant families that have been shown to produce tropane alkaloids are highlighted and the orders they belong to are displayed using a dashed line. The scale bar below represents millions of years. This phylogenetic tree has been modified and republished with permission of the Botanical Society of America, from Angiosperm diversification through time, Susana Magallon and Amanda Castillo, 96, 1, 2009; permission conveyed through Copyright Clearance Center, Inc.

of TAs is monophyletic or polyphyletic. The Solanaceae and Convolvulaceae, two closely related families of the Solanales, are separated from other major tropane alkaloid-containing families such as the Erythroxylaceae and Proteaceae by at least 120 million years (Fig. 2.4). Until recently, molecular data for genes and enzymes involved in TA biosynthesis were available only for members within the Solanaceae. However, new data on TA biosynthesis in *E. coca* suggest that TA biosynthesis originated at least twice during the course of angiosperm diversification (Jirschitzka et al., 2012).

The largest single family whose members are known to make TAs is the Solanaceae. There are 29 genera in this family with the ability to produce TAs. Some examples of the most common TAs found in the Solanaceae are shown in Fig. 2.2. In general, TAs present in the Solanaceae are esterified at the C-3 hydroxyl position, and the stereochemistry of this substituent is most often α. Atropine (**1**) and scopolamine (**44**), the most prominent representatives of TAs in this family, are both esterified with a tropic acid moiety. The major difference between these two metabolites is the epoxy group linking position C-6 and C-7 of the tropane skeleton. Both atropine (**1**) and scopolamine (**44**) are included in the World Health Organization's (WHO) essential drugs list (WHO, 2011). These alkaloids completely inhibit the action of the acetylcholine receptors of postganglionic parasympathetic nerves (Reas & Tsai, 1966). These receptors are involved in the constriction of the pupil, vasodilation and moderation of the heartbeat (Henderson & Roepke, 1937). As a result, these compounds are used in a wide variety of treatments including those for motion sickness, in ophthalmic surgery and as a treatment for bradycardia (Ebert, Siepmann, Oertel, Wesnes, & Kirch, 1998; Honkavaara & Pyykkö, 1999; Schwartz, de Roetth, & Papper, 1957). Recent studies have revealed that anisodamine (**43**), the biosynthetic intermediate between hyoscyamine (**41**) and scopolamine (**44**), is less toxic to the central nervous system than scopolamine. This led to the discovery that high doses of anisodamine (**43**) can ameliorate cognitive disorders and, therefore, it has been suggested as a novel treatment for Alzheimer's disease (Zhang et al., 2008).

The Convolvulaceae, also known as the morning glory family, contains 25 genera reported to make TAs. Like the TAs found in its sister family, the Solanaceae, the dominant TAs are tropine (3α-hydroxy) esters (Fig. 2.2). Convolvine (**5**), a dimethoxy benzoic acid ester of nortropine, was found in *Convolvulus pseudocantabricus* and was the first compound to be described from this family (Orechoff & Konowalowa, 1933). Convolvine (**5**) blocks the M-receptors of the heart and intestine while raising the sensitivity of the M-receptors of the salivary gland and the central nervous system. It has been suggested that this compound and its related substances maybe used as sedatives and nootropic agents (Mirzaev & Aripova, 1998). Some additional compounds found in this family are unique because they contain modifications that occur in the aromatic ester moiety. These compounds include merresectines (**3**) (Fig. 2.2) from the genus *Merrima,* which can be prenylated as well as glycosylated (Jenett-Siems et al., 2005).

A subclass of polyhydroxylated tropane alkaloids known as calystegines (**37**) were first discovered in the roots of *Calystegia sepium* (Convolvulaceae) (Goldmann et al., 1990; Tepfer et al., 1988). Unlike the other TAs described thus far, the calystegines (**37**) tend to be distributed across all plant tissue types. These compounds are proposed to function as glycosidase inhibitors like other monosaccharide-mimicking alkaloids (Asano, Nash, Molyneux, & Fleet, 2000). However, no concrete evidence for their use as a defensive compound in plants has been reported. Calystegines (**37**) have been found in nearly all families that make TAs, but the Moraceae is the only family that contains only calystegines (**37**) and no other type of TA (Asano, Oseki, Tomioka, Kizu, & Matsui, 1994; Asano, Tomioka, Kizu, & Matsui, 1994).

Cocaine (**42**) is one of the most well-known TAs and is exclusively found in members of the Erythroxylaceae. Since its isolation by Albert Niemann (Niemann, 1860), it has become infamous for its abuse as an illegal narcotic. This is due to the euphoria induced by its activity as a dopamine reuptake inhibitor (Galloway, 1988). Cocaine (**42**) has been found in 23 of the approximately 230 species in the genus *Erythroxylum* (Bieri, Brachet, Veuthey, & Christen, 2006). However, the difference in cocaine (**42**) concentration between wild and cultivated species can be more than 100-fold (Aynilian, Duke, Gentner, & Farnswor, 1974). It is believed that cocaine (**42**) is stored as a chlorogenic acid complex in the vacuoles found in the palisade parenchyma in the leaves (Ferreira, Duke, & Vaughn, 1998; Pardo Torre et al., 2013). One of the distinguishing characteristics of TAs found in the Erythroxylaceae is the common occurrence of a carbomethoxy group on the C-2 position of the tropane ring (Fig. 2.3). Like the TAs found in the Solanaceae, many of the TAs in the Erythoxylaceae are esterified at the C-3 position. However, the configuration at this position is predominantly β. Other common TAs found in the Erythroxylaceae include cinnamoylcocaine (**9**) and its dimeric derivatives, the truxillines (**8**) (Fig. 2.3). Truxillines (**8**) are thought to be the result of dimerisation due to UV radiation (Lydon et al., 2009). Despite the negative associations with cocaine (**42**), its derivatives have the potential to be legitimate medicines.

For example, modification of dopamine transporter function may help to alleviate some symptoms associated with Alzheimer's disease, Parkinson's disease, attention-deficit hyperactivity disorder, ageing and depression (Runyon & Carroll, 2006; Singh, 2000). The synthetic compound fluorotropacocaine can be used as a local anaesthetic and its ^{18}F-labelled derivative has been used as a muscarinic acetylcholine ligand for PET imaging (Kavanagh et al., 2012).

The model plant *Arabidopsis thaliana* has never been shown to contain TAs (Brock, Herzfeld, Paschke, Koch, & Drager, 2006), but TAs have been isolated from other members of the Brassicaceae. An alkaloid isolated from *Cochlearia arctica* (now *C. groenlandica*) and *Cochlearia officinalis*, cochlearine (**4**), has been shown to be a TA (Liebisch, Bernasch, & Schütte, 1973; Platonova & Kuzovkov, 1963). Since this discovery, a further 12 genera in the Brassicaceae have been reported to contain TAs (particularly calystegines (**37**); see above).

The Proteaceae represents the oldest tropane-producing family among the angiosperms (Fig. 2.4). There are five genera reported to contain tropanes and they are geographically limited to Australia and New Caledonia (Bick et al., 1981; Butler et al., 2000). Both 3α and 3β esters of tropine and pseudotropine, respectively, have been reported in the literature. The esters consist mainly of aromatic and aliphatic acids. Pyranotropanes contain a γ-pyrano group attached to the C-3 and C-4 position of the tropane ring (Fig. 2.5). Examples of these compounds include strobamine (**14**) and bellendine (**13**). In addition, the compounds ferruginine (**15**) and ferugine (**17**) found in the genus *Darlingia* have also been described and contain unique modifications at the C-4 position of the tropane ring. Both ferruginine (**15**) and ferugine (**17**) are nicotinic receptor antagonists and have been suggested to be used as a potential treatment for Alzheimer's disease (Lazny, Sienkiewicz, Olenski, Urbanczyk-Lipkowska, & Kalicki, 2012).

Very little data exist on the remaining TA-containing plant families. In fact, for some of these families, only one report has ever been published. *Bruguiera*, *Crossostylis* and *Pellacalyx*, three genera from the Rhizophoraceae, contain common tropanes such as tropinone (**30**), tropine (**34**) and various aromatic tropine esters (Arbain, Wiryani, & Sargent, 1991; Kato, 1975; Loder & Russell, 1966, 1969; Media, Pusset, Pusset, & Husson, 1983). Brugine (**16**) (Fig. 2.5), a unique dithiolane tropane alkaloid, has been identified from members belonging to the Rhizophoraceae. There is some controversy surrounding the description of TAs in *Peripentadenia mearsii*, a member of the family Elaeocarpaceae. The original study reported

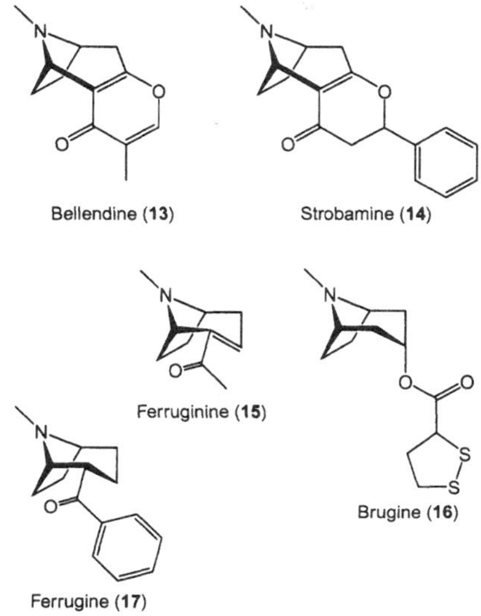

Figure 2.5 Examples of tropane alkaloids found in plants of the Proteaceae and Rhizophoraceae.

the discovery of 3α-acetoxy-6β-hydroxytropane, 2α-benzoyloxy-3β-hydroxynortropane, and 3β-benzoyloxytropane (Johns, Lamberto, & Sioumis, 1971). However, a year later, the same authors reported that they could not reproduce their original results (Johns, 1972). Therefore, a more detailed study of this species with a larger sample size is necessary to conclusively determine whether *P. mearsii* has the ability to synthesise TAs. Originally considered as a member of the Euphorbiaceae, the tropical plant *Phyllanthus discoides*, now a member of the family Phyllanthaceae (Nahar, Sarker, & Delazar, 2011), has been described to contain phyllalbine (**6**) (Parello, Longevialle, Vetter, & McCloskey, 1963). Scopolamine (**44**) was also found in *Heisteria olivae* (Olacaceae), which grows above 1000 m in the Andean highlands (Valera, De Budowski, Delle Monache, & Marini-Bettolo, 1977). This report should be taken with caution as there is no other description of tropanes in the Olacaceae, and scopolamine (**44**) has otherwise been reported only from plants in the Solanaceae.

3. BIOSYNTHESIS OF TROPANE ALKALOIDS

The biosynthesis of TAs has been a subject of study for nearly 200 years. The early studies were interested in the crystallisation of pharmaceutically important TAs such as atropine (**1**), scopolamine (**44**) and cocaine

(**42**) (Geiger & Hesse, 1833; Mein, 1833; Niemann, 1860; Schmidt & Henschke, 1888). Richard Willstätter's synthesis of ecgonine (**10**) in the beginning of the twentieth century established the first true foundations for applying chemical and other analytical tools to our understanding of how tropane alkaloids are made (Willstätter & Hollander, 1903). The predominant methods used to elucidate both structure and potential biosynthetic steps in the pathway have been radiolabelled feeding studies followed by chemical degradation analysis. Based on this type of investigation, a biosynthetic model has been established that can be attributed principally to Leete (1990). His predictions about the enzymatic steps in the biosynthesis in both cocaine (**42**) and other TAs have provided direction to biochemists and molecular biologists alike (Bjorklund & Leete, 1992). Because of the commercial interest in TAs of the Solanaceae and the genomic tools available for selected species, studies of the genes and enzymes involved in TA biosynthesis have focused on members of this family. As previously mentioned, the scattered distribution of TAs throughout the angiosperms suggests that their biosynthetic origins may be polyphyletic. Therefore, the current state of knowledge regarding the enzymes involved in TA production is heavily biased to a single family, and alternative pathways may be identified in other families.

In general, alkaloid biosynthesis begins with the recruitment of a nitrogen-containing metabolite of central metabolism. In many cases, amino acids serve as the initiating intermediate (Zulak, Liscombe, Ashihara, & Facchini, 2007). As early as 1954, the amino acids ornithine and arginine were predicted to be the starting substrates in the biosynthesis of TAs (Leete, Marion, & Spenser, 1954). van Soeren (1962), ^{14}C-proline fed to the roots of *A. belladonna*, showed that proline (**18**) could also be incorporated into the tropane ring. Several other studies using *D. metel* and *D. stramonium* also reported the incorporation of proline (**18**) into the compounds tropine (**34**) and scopolamine (**44**) (Liebisch & Schütte, 1967). Arginine (**20**), ornithine (**21**) and proline (**18**) are readily interconvertible via the shared intermediate pyrroline-5-carboxylate (**19**) (Fig. 2.6) (Delauney & Verma, 1993). Therefore, interpretation of the results following amino acid feeding has made it difficult in determining whether one or a combination of amino acids are truly responsible for entry into the TA pathway. Labelling studies on several different TA-producing plant species using [2-^{14}C]-ornithine have produced conflicting results: a symmetrical incorporation at positions C-1 and C-5 of the tropane ring has been reported for *Hyoscyamus albus* and *E. coca*, while an asymmetrical labelling (at C-5 only)

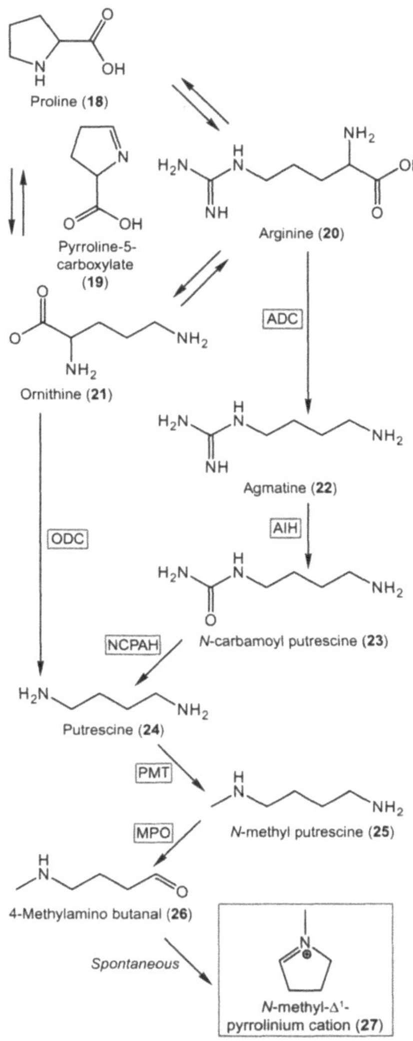

Figure 2.6 The initial steps of tropane biosynthesis leading to the formation of the *N*-methyl-Δ^1-pyrrolidinium cation. The amino acids proline, arginine and ornithine have all been implicated in the biosynthesis of tropane alkaloids. Their shared biosynthetic intermediate pyrroline-5-carboxylate has complicated the determination of which amino acid is the direct precursor. The production of putrescine can be directly formed via the decarboxylation of ornithine or indirectly from arginine. Following methylation, *N*-methylputrescine is oxidised to the intermediate 4-methylamino butanal, which spontaneously cyclizes to yield the *N*-methyl-Δ^1-pyrrolidinium cation. Tropane alkaloid biosynthetic enzymes that have been isolated and biochemically characterised appear in the shaded boxes. The abbreviations for these enzymes are as follows: ODC, ornithine decarboxylase; ADC, arginine decarboxylase; AIH, agmatine imino hydrolase; NCPAH, *N*-carbamoylputrescine amido hydrolase; PMT, putrescine *N*-methyltransferase; MPO, methyl putrescine oxidase.

was reported in *D. stramonium* and *D. metel* (Hashimoto, Yukimune, & Yamada, 1989; Leete, 1962, 1964, 1982; Liebisch, Ramin, Schoffin, & Schütte, 1965). It was suggested that selective methylation of ornithine (**21**) could explain the asymmetrical pattern observed in *Datura*. This hypothesis was tested by Ahmad and Leete by feeding DL-α-*N*-methyl-[2-^{14}C]-ornithine or DL-δ-*N*-methyl-[2-^{14}C]-ornithine to *D. stramonium* (Ahmad & Leete, 1970). Incorporation at a very low level was observed only in the case of feeding with DL-δ-*N*-methyl-[2-^{14}C]-ornithine providing evidence for the theory of selective methylation of ornithine.

It is possible to reach a symmetrical intermediate in one enzymatic step by conversion of ornithine (**21**) to putrescine (**24**), one of the simplest polyamines. This reaction is catalysed by ornithine decarboxylase (ODC), an enzyme isolated from several TA-producing plant species (Fig. 2.7) (Docimo et al., 2012; Imanishi et al., 1998; Michael, Furze, Rhodes, & Burtin, 1996). ODC is a pyridoxal phosphate-dependent decarboxylase. It is predicted to be a cytosolic enzyme (Sandmeier, Hale, & Christen, 1994) but appears to accumulate in the nucleus (Schipper, Cuijpers, de Groot, Thio, & Verhofstad, 2004).

An indirect route to putrescine (**24**) via arginine (**20**) has also been demonstrated. The first enzymatic step involves decarboxylation of arginine (**20**) by arginine decarboxylase (Fig. 2.6) (ADC). The decarboxylated product, agmatine (**22**), is then converted to *N*-carbamoyl putrescine (**23**) via the enzyme agmatine imino hydrolase. *N*-carbamoyl-putrescine amido hydrolase catalyses the final step resulting in the formation of putrescine (**24**). Biochemical studies performed by Malmberg revealed that putrescine (**24**) production via ODC is important for the supply of polyamines for primary metabolic processes such as cellular division, differentiation and development (Malmberg, Watson, Galloway, & Yu, 1998). In contrast, putrescine (**24**) supplied via the ADC-route is thought to be required for responses related to environmental stress (Malmberg et al., 1998).

The first committed step in TA biosynthesis is the formation of *N*-methyl putrescine (**25**). The methyl group attached to the nitrogen in the tropane skeleton is derived from methionine via the common methyl donor *S*-adenosylmethionine (SAM). The SAM-dependent methyl transferase responsible for this reaction is referred to as putrescine *N*-methyltransferase (Fig. 2.6) (PMT). The first PMT sequence isolated from plants was from tobacco (Hibi, Higashiguchi, Hashimoto, & Yamada, 1994). The biosynthesis of the pyrrolidine ring in both nicotine and TAs is thought to have the same origins within plants of the Solanaceae. Thus

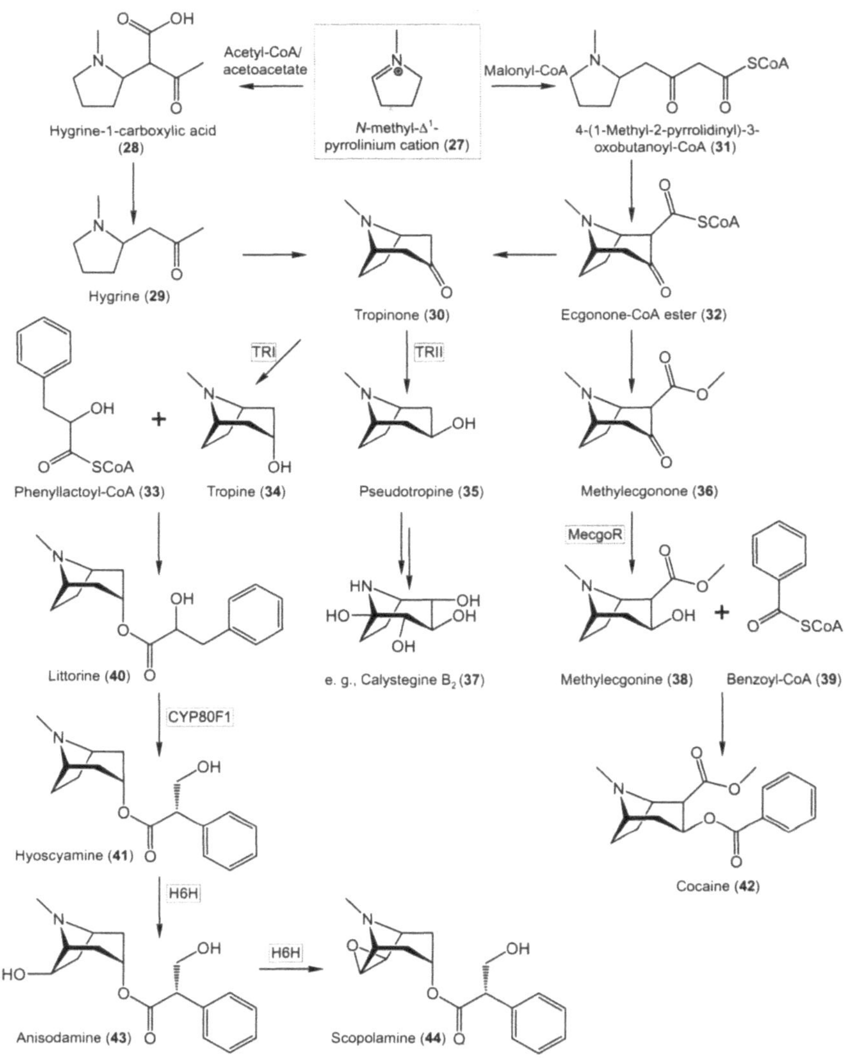

Figure 2.7 The mid- and late-biosynthetic reactions in tropane alkaloid production. Depicted here are two possible models for the condensation of the tropane ring. The utilisation of acetyl-CoA directly or indirectly via acetoacetate can yield hygrine-1-carboxylate. Alternatively, 4-(1-methyl-2-pyrrolidinyl)-3-oxobutanoyl-CoA can be formed via the decarboxylative condensation of two malonyl-CoA subunits. Once the ring is formed, the keto function at the C-3 position is reduced and the corresponding alcohol is esterified using an acyl-CoA substrate. The epoxidation of hyoscayamine is catalysed from one enzyme in a two-step process. Tropane alkaloid biosynthetic enzymes that have been isolated and biochemically characterised appear in the shaded boxes. The abbreviations for these enzymes are as follows: TR I, tropinone reductase I; TR II, tropinone reductase II, MecgoR, methylecgonone reductase; H6H, 6β-hydroxy hyoscyamine epoxidase.

far, PMTs from several TA-producing members of the Solanaceae and one member of the Convolvulaceae have been isolated (Kai, Zhang, et al., 2009; Liu, Zhu, Cheng, Meng, & Zhu, 2005; Teuber et al., 2007). PMT belongs to a large family of enzymes which are involved in the production of polyamines. Other members of this family include the spermidine and spermine synthases. There is evidence to suggest that PMT function evolved from an ancestral spermine or spermidine synthase as these enzymes are essential for supplying polyamines used in primary metabolism (Hashimoto, Tamaki, Suzuki, & Yamada, 1998). Immunolocalisation of both PMT and spermidine synthase demonstrated that both enzymes are present in the below-ground portions of the plant. This is consistent with the observation that TA biosynthesis in members of the Solanaceae occurs mainly in the roots (Nakajima & Hashimoto, 1999; Ziegler & Facchini, 2008). Although much attention has been given to *N*-methyl putrescine (**25**), at least one alternative intermediate has been identified. Leete (1985) discovered that *N*-methylspermidine can serve as an indirect precursor for the formation of the pyrrolidine ring of nicotine in *Nicotiana glutinosa*.

In order to convert *N*-methylputrescine (**25**) into the cyclic pyrrolidine ring, an oxidative deamination is required. The resulting compound, 4-methylaminobutanal (**26**), spontaneously cyclises to yield the *N*-methyl-Δ^1-pyrrolidinium cation (**27**) (Fig. 2.6) (Leete, 1990). Labelled 4-methylaminobutanal (**26**) was detected in *D. stramonium* plants following feeding with [2-^{14}C]-ornithine. Enzyme activities for this oxidation reaction have been described for several TA-producing members of the Solanaceae (Feth, Wray, & Wagner, 1985; Hashimoto, Mitani, & Yamada, 1990; Mizusaki, Tanabe, Noguchi, & Tamaki, 1973). The enzyme *N*-methylputrescine oxidase (MPO) was first characterised from *Nicotiana tabacum* and the corresponding gene has been isolated (Heim et al., 2007; Katoh, Shoji, & Hashimoto, 2007). MPO belongs to a class of copper-dependent diamine oxidases. The copper is required to oxidise a conserved tyrosine residue into a topaquinone, which is essential for enzyme catalysis (Matsuzaki, Fukui, Sato, Ozaki, & Tanizawa, 1994). Evidence suggests that MPO associates with other important enzymes involved in the biosynthesis of nicotine. This has led to the hypothesis that a metabolic channel exists in which a multi-enzyme complex is active. Although no MPO has yet been reported to exist in *E. coca*, a recent study using the remote isotope method has established that the oxidation of *N*-methylputrescine (**25**) in this species is stereoselective. The *pro*-S hydrogen atom is selectively removed in a ratio of 6–10:1 (Hoye, Bjorklund, Koltun, & Renner, 2000). A similar ratio has been reported for *N. tabacum* and *N. glutinosa,* which strongly suggests that an MPO homolog is present in *E. coca* (Wigle, Mestichelli, & Spenser, 1982).

The specific details of how the second ring in the tropane skeleton is formed are not yet known. It is clear, however, that some type of reaction has to occur with the *N*-methyl-Δ^1-pyrrolidinium cation (**27**). Evidence for the incorporation of *N*-methyl-Δ^1-pyrrolidinium (**27**) into the tropane core structure was observed by feeding [2-^{13}C,^{15}N]-*N*-methylpyrrolinium chloride to coca plants and analysing its incorporation into methylecgonine (**38**) (Leete, Bjorklund, Couladis, & Kim, 1991; Leete, Kim, & Rana, 1988). Several substrates for the condensation have been proposed (Fig. 2.7). Acetic acid, most likely in the form of acetyl-CoA, was observed to be incorporated into carbons C-2, C-3 and C-4 in hyoscyamine (**41**) and into carbons C-3 and C-9 of cocaine (**42**) (Kaczkowski, Schütte, & Mothes, 1961; Leete, 1983a; Liebisch, Peisker, Radwan, & Schütte, 1972). It was suggested at the time that two molecules of acetyl-CoA would first condense to yield acetoacetyl-CoA. The resulting condensation product, hygrine-1-carboxylic acid (**28**), a β-keto acid would have to spontaneously decarboxylate forming the alkaloid hygrine (**29**). An additional oxidation at the C-2 position of hygrine (**29**) would then be required for the final ring closure and subsequent formation of tropinone (**30**). This hypothesis lost favour because it was shown that hygrine (**29**) formation from the condensation of *N*-methyl-Δ^1-pyrrolidinium cation (**27**) with acetoacetyl-CoA can occur non-enzymatically (Endo, Hamaguchi, Hashimoto, & Yamada, 1988). In addition, feeding studies using stable isotopically labelled acetate in *D. stramonium* reported label incorporation only into position C-6 and C-7 of the pyrrolidine ring, instead of the previously reported positions C-2 and C-4 (Duran-Patron, O'Hagan, Hamilton, & Wong, 2000).

A more plausible hypothesis for the second ring closure involves the condensation of malonyl-CoA with the *N*-methyl-Δ^1-pyrrolidinium cation (**27**) (Fig. 2.7). Two successive rounds of decarboxylative condensation would yield the intermediate 4-(1-methyl-2-pyrrolidinyl)-3-oxobutanoyl-CoA (**31**). The evidence supporting this hypothesis comes from the feeding of racemic ethyl [2,3-$^{13}C_2$]-4-(*N*-methyl-2-pyrrolidinyl)-3-oxobutanoate to *Datura innoxia* (Abraham & Leete, 1995; Robins, Abraham, Parr, Eagles, & Walton, 1997). Further support for this hypothesis comes from the feeding of methyl (RS)-[1,2-$^{13}C_2$,1-^{14}C]-4-(1-methyl-2-pyrrolidinyl)-3-oxobutanoate to leaves of *E. coca* (Leete et al., 1991). This led to the labelling of incorporation in both cocaine (**42**) and methylecgonine (**38**). In plants, these types of condensation reactions are commonly catalysed by type III polyketide synthases (PKSs). However, stereotypical reactions catalysed by type III PKSs use a CoA-containing starter molecule. No reports in the literature exist for a PKS that would use a substrate such as the *N*-methyl-Δ^1-pyrrolidinium cation (**27**). The formation of hygrine (**29**) in this model can be explained by the spontaneous decarboxylation of 4-(1-methyl-2-pyrrolidinyl)-3-oxobutanoate. In addition, this model also explains the presence of the carbomethoxy group commonly found at the C-2 position in cocaine (**42**) and other TAs found in the Erythroxylaceae. Protection of the carboxylic acid found in the ecgonone-CoA ester (**32**) by the formation of the methyl ester would prevent spontaneous decarboxylation from occurring. The lack of this function in TAs found in the Solanaceae can be explained by the absence of the esterification reaction. Regardless of this functional group, the product formed following the second ring closure will contain a keto group at the C-3 position.

The reduction of the keto group found in tropinone (**30**) and methylecgonone (**36**) has been the subject of intense biochemical research. The first evidence for the stereospecific reduction of the keto function was observed by feeding [9-^{13}C,^{14}C,*O*-methyl-^{3}H]-2-carbomethoxy-3-tropinone to coca plants (Leete, 1983b). At the same time, an enzyme activity specific for the reduction of tropinone (**30**) into tropine (**34**) was purified from the roots of *D. stramonium* (Fig. 2.7) (Koelen & Gross, 1982). It was discovered that the enzyme requires NADPH as a cofactor. A second enzyme-producing pseudotropine (**35**) from the reduction of tropinone was purified from the roots of *H. niger* (Dräger, Hashimoto, & Yamada, 1988). At this time, it was discovered that pseudotropine (**35**) does not spontaneously isomerise to tropine (**34**) (Yamada et al., 1990).

Two genes encoding tropinone reductase I (TRI) and tropinone reductase II (TRII) specific for producing tropine (**34**) and pseudotropine (**35**), respectively, were isolated from the roots of *H. niger* (Hashimoto, Nakajima, Ongena, & Yamada, 1992). Both TRI and TRII belong to the short-chain dehydrogenase/reductase superfamily (SDR), members of which produce a wide variety of both primary and secondary metabolites. Enzymes of the SDR family share a common tertiary structure consisting of the "Rossmann" fold, a conserved motif consisting of six parallel β-sheets and two pairs of α-helices, have a dinucleotide cofactor-binding motif and an active site catalytical motif YxxxK (Moummou, Kallberg, Tonfack, Persson, & van der Rest, 2012). Since their initial isolation, many genes encoding TRI and TRII have been isolated from solanaceous species (Dräger & Schaal, 1994; Kai, Li, et al., 2009; Keiner, Kaiser, Nakajima,

Hashimoto, & Dräger, 2002; Portsteffen, Dräger, & Nahrstedt, 1994). Overall, TRI and TRII share amino acid sequence similarity of more than 50% and are assumed to have evolved from a common ancestor (Dräger, 2006). Although there is significant amino acid sequence differences between the two forms of tropinone reductases, a change of as little as five amino acids is required to change the stereospecificity of the reaction product (Nakajima, Kato, Oda, Yamada, & Hashimoto, 1999). Immunolocalisation experiments performed on potatoes revealed that TRs are localised to the tuber and roots (Kaiser et al., 2006). Recently, a tropinone reductase-like enzyme was discovered in *C. officinalis* (Brassicaceae), which reduces tropinone (**34**). Unlike the stereospecific tropinone reductases from the Solanaceae, this reductase is capable of producing both tropine (**34**) and pseudotropine (**35**) in equal ratios. Designated CoTR, the reductase from *C. officinalis* accepts a much broader range of substrates than other tropinone reductases (Brock, Brandt, & Drager, 2008). Further, phylogenetic analysis reveals that CoTR is more related to other members of the SDR family in the Brassicaeae than it is to either the TRIs or TRIIs found in the Solanaceae.

The first evidence for a polyphyletic origin for TA biosynthesis in plants came from the discovery of an alternate reductase enzyme for the reduction of methylecgonone (**36**) in *E. coca*. A homology-based approach using TR sequences from the Solanaceae identified several TR homologs within the coca transcriptome. However, all these enzymes failed to exhibit methylecgonone-reducing activity when heterologously expressed in *E. coli*. Using classical biochemical techniques, methylecgonone reductase (MecgoR) was purified from crude protein extracts from coca leaves (Jirschitzka et al., 2012). Unlike solanaceous TRs, MecgoR belongs to the aldo-keto reductase (AKR) superfamily of enzymes. MecgoR shares some similarity with chalcone reductase, responsible for the formation of deoxychalcones, as well as codeinone reductase, an enzyme involved in benzylisoquinoline biosynthesis in *Papaver somniferum* (opium poppy). In addition, MecgoR protein has been localised to the palisade parenchyma of young developing leaves. This is in contrast to the root localisation of TRs, catalysing the equivalent step in the roots of solanaceous plants. MecgoR is stereospecific for the production of the 3β-hydroxy-containing metabolite methylecgonine (**38**) (Fig. 2.7). MecgoR is also capable of using tropinone (**30**) as a substrate, however, it produces pseudotropine (**35**) exclusively. This is consistent with the presence of only 3β-hydroxy esters in *E. coca*.

The most common forms of TAs are esterified with either aromatic or aliphatic acids. The stereochemistry of the hydroxyl group is determined by the respective reductase. In *D. stramonium*, the accumulation of 3-acetyl-tropine occurs in cultured roots. Biochemical separation of this acyltransferase activity was performed, revealing that the enzymes responsible for the reaction utilise acetyl-CoA as the activated acid (Robins, Bachmann, Robinson, Rhodes, & Yamada, 1991). Further biochemical studies using *D. stramonium* successfully purified an acyltransferase, which utilised tigloyl-CoA to esterify the 3β-hydroxy group of pseudotropine (**35**) (Rabot, Peerless, & Robins, 1995). This 65-kDa enzyme was active in its monomeric form and could accept a wide variety of different acyl-CoA thioesters when using pseudotropine (**35**). However, attempts to use tropine (3α-hydroxytropane) (**34**) were unsuccessful.

Cocaine (**42**) is a benzoic acid ester of methylecgonine (**38**) (Fig. 2.7). Feeding studies using *trans*-[3-^{13}C,^{14}C]-cinnamic acid and the *N*-acetylcysteamine thioester of [3-^{13}C,^{14}C]-*trans*-cinnamic acid led to the prediction that the acyltransferase in *E. coca* utilises benzoyl-CoA (**39**) as the activated acid (Bjorklund & Leete, 1992). Although no gene sequences encoding acyltransferases involved in tropane ester formation have been reported, it is likely that these enzymes will be members of the BAHD family (D'Auria, 2006). This family of enzymes is responsible for the esterification of a wide range of plant-specific metabolites and uses CoA thioesters as co-substrates. In addition, BAHD-type enzymes are monomeric and similar in size to tigloyl-CoA:pseudotropine acyl transferase.

One step in TA side chain biosynthesis that has drawn particular interest is the rearrangement of the hydroxyl group of the phenyllactic acid moiety of littorine (**40**), which occurs in atropine (**1**) and scopolamine (**44**) formation (Fig. 2.7). In this process, a branched-chain residue, tropic acid, is formed from the straight chain phenyllactic acid. Many radiolabelled feeding studies have been performed to understand the mechanisms involved in this reaction (Ansarin & Woolley, 1994; Chesters, O'Hagan, & Robins, 1995; Leete, Kowanko, & Newmark, 1975; Robins, Bachmann, & Woolley, 1994; Robins et al., 1995). The pervading hypothesis for the conversion of the littorine (**40**) precursor into the hyoscyamine (**41**) product predicts that a cytochrome p450 coupled with an alcohol dehydrogenase is involved. These predictions are based on both feeding study results as well as quantum chemistry calculations (Sandala, Smith, & Radom, 2008). Using virus-induced gene silencing techniques, Li et al. (2006) were able to suppress the expression of the cytochrome p450 CYP80F1. This reduced the levels of hyoscyamine (**41**) and encouraged the accumulation of littorine (**40**). The involvement of a p450 was successfully probed by performing enzyme assays with synthetic deutero and arylfluoro analogues of littorine (Nasomjai et al., 2009).

The conversion of hyoscyamine (**41**) into the epoxide scopolamine (**44**) is catalysed by hyoscyamine 6β-hydroxylase (Fig. 2.7) (H6H). This enzyme was purified from *H. niger* and was shown to be a 2-oxoglutarate-dependent dioxygenase (Hashimoto, Matsuda, & Yamada, 1993; Hashimoto & Yamada, 1986). H6H catalyses a two-step reaction in which the hydroxylation at the C-6 position of hyoscyamine (**41**) is followed by epoxidation of the corresponding intermediate, anisodamine (6β-hydroxy hyoscyamine) (**43**). Further, the enzyme was determined to be localised exclusively to the pericycle of roots (Hashimoto et al., 1991). Information about the enzymes involved in the polyhydroxylation of the tropane ring skeleton resulting in the formation of calystegines (**37**) has not yet been determined. However, homology-based approaches using H6H as well as co-expression analysis with cytochrome p450s may be beneficial.

4. METABOLIC ENGINEERING OF TROPANE ALKALOIDS

In recent years, the majority of research on tropane alkaloid biosynthesis has focused on the engineering of increased levels of commercially important metabolites such as scopolamine (**44**) and atropine (**1**). These compounds are traditionally difficult to produce via chemical synthesis. Extensive reviews about this subject have been written by Zhang et al. (2005) and Palazon, Navarro-Ocana, Hernandez-Vazquez and Mirjalili (2008). All of these studies have used hairy root cultures rather than intact plants because TA biosynthesis in solanaceous species occurs in below-ground tissues (Zhang et al., 2005). This approach assumes that all the enzymatic machinery needed to supply important starting substrates is also present in these tissues. *Agrobacterium rhizogenes* is used to ensure that the roots can grow rapidly in hormone-free media and can also be used in genetic transformation (Chandra & Chandra, 2011). Because of our limited knowledge about the structural genes involved in TA biosynthesis, the choice of candidates has been restricted to ODC, PMT, TR and H6H. In an attempt to increase polyamine starter substrates to the pathway, an ODC gene isolated from mouse was overexpressed under the control of the CaMV 35S promoter in *D. innoxia*. The production of scopolamine (**44**) increased as much as six times over that observed in the controls.

The use of a non-plant-derived ODC may have avoided the common problem of transgene silencing that often occurs in genetically modified plants (Singh et al., 2011).

Transformation of a multigene construct for the overproduction of TAs was first reported by Zhang et al. (2004). Since the publication of the successful results of this attempt, several other groups have repeated this method in other solanaceous root cultures. Combining the overexpression of both PMT and H6H in *A. belladonna* has resulted in approximately a 2.5-fold increase of scopolamine above wild-type levels (Liu et al., 2010). A separate experiment, again using this same combination of genes, was able to increase hyoscyamine (**41**) levels to more than 24 times that found in wild-type *A. belladonna* (Yang et al., 2011). Using the plant *Scopolia parviflora*, a similar construct using PMT and H6H was used to make transgenic hairy roots. Increases in both hyoscyamine (**41**) and scopolamine (**44**) were observed in addition to an increase in the growth rate of the roots when compared with the no-insert transgenic controls. This growth increase could be attributed to the role of polyamines in the processes of growth and cell differentiation (Kang et al., 2011). A different combination of genes encoding PMT and TRI to make transgenic hairy root cultures of *Anisodus acutangulus* resulted in an overall increase in TAs that was 3–8 times higher than in the control lines (Kai et al., 2011). The same group, using a new construct consisting of the genes TRI and H6H, was able to observe a fivefold increase in total TA levels. The same construct also led to an improvement in the accumulation of anisodamine (**43**) when compared to both wild-type root cultures and transgenic root cultures using the PMT/TRI construct (Kai, Zhang, et al., 2012).

The induction of plant defense responses can also be beneficial to metabolic engineers wishing to increase the quantity of their targeted metabolites. For example, the treatment of hairy root cultures of *H. niger* with the phytohormone methyl jasmonate (MeJA) resulted in a fivefold increase in scopolamine (**44**) levels (Zhang et al., 2007). Table 2.1 summarises the known organic and inorganic elicitors used to increase TA production. Many other elicitors of plant defense metabolism have also been successful in increasing TA levels in root cultures. Besides MeJA, both yeast elicitor and abscisic acid treatment were capable of increasing TAs in *A. acutangulus* (Luo et al., 2012). Salicylic acid (SA), another common plant defense hormone, is often antagonistic to MeJA-induced responses (Niki, Mitsuhara, Seo, Ohtsubo, & Ohashi, 1998). While SA was not effective in increasing scopolamine levels in transgenic hairy root cultures of *Atropa*

Table 2.1 A summary of the biotic and abiotic elicitors used as treatments for increasing tropane alkaloid content in plants

Plant	Elicitor	Effect	Target	Reference
Atropa baetica (hairy root culture)	MeJa	Increase	Hyoscyamine	el Jaber-Vazdekis, Barres, Ravelo, and Zarate (2008)
	ASA	Increase	Anisodamine	
	SA	None	Scopolamine	
Datura metel (hairy root culture)	SA	Increase	Hyoscyamine	Ajungla, Patil, Barmukh, and Nikam (2009)
	Fungal extract	Increase	Scopolamine	
	Yeast extract	Increase		
	Aluminium chloride	Increase		
Hyoscyamus niger (hairy root culture)	Chitosan	Decrease	Hyoscyamine	Hong, Bhatt, Ping, and Keng (2012)
	Casein	None	Scopolamine	
	Yeast extract	Increase		
	D-sorbitol	None		
Anisodus acutangulus (hairy root culture)	MeJA	Increase	Hyoscyamine	Kai, Yang, et al. (2012)
	Silver nitrate	Increase	Anisodamine	
	Ethanol	Increase	Scopolamine	
	SA/ethanol	Decrease except anisodine	Anisodine	
Datura stramonium (hairy root culture)	Nitrate	Increase	Hyoscyamine	Amdoun et al. (2009)
	Phosphate	Increase		
	Calcium	Increase		
	Nitrate/calcium	Increase		
	Nitrate/phosphate	Decrease, none		
	Phosphate/calcium	Decrease, none		

Continued

Table 2.1 A summary of the biotic and abiotic elicitors used as treatments for increasing tropane alkaloid content in plants—cont'd

Plant	Elicitor	Effect	Target	Reference
Datura innoxia (cultured plantlets)	Aluminium chloride	Increase	Hyoscyamine	Karimi and Khataee (2012)
			Scopolamine	
Anisodus acutangulus (whole plant)	MeJA	Increase	Hyoscyamine	Luo et al. (2012)
	Yeast extract	Increase	Scopolamine	
	ABA	Increase except scopolamine	Anisodine	
Brugmansia suaveolens (whole plant)	MeJA	Increase	Scopolamine	Alves et al. (2007)

MeJA, methyl jasmonate; SA, salicylic acid; ASA, acetylsalicylic acid; ABA, abscisic acid.

baetica, acetyl salicylic acid (ASA) significantly increased gene transcript levels for several TA structural genes (el Jaber-Vazdekis et al., 2008). The use of heavy metal elicitors has also been shown to successfully increase the production of TAs. For example, both trivalent chromium [Cr(III)] and aluminium are capable of increasing the hyoscyamine (**41**) and scopolamine (**44**) content in *A. belladonna* and *D. innoxia* (respectively) (Karimi & Khataee, 2012; Vakili, Karimi, Sharifi, & Behmanesh, 2012). However, caution must be used when working with these types of elicitors because of their negative effects on plant growth and development.

Scaling-up of hairy root cultures will be required for any large-scale production of TAs. Bioreactors are therefore of particular interest, and both bubble column and stirred tank types have successfully been employed for the task of overproduction of TAs (Cardillo et al., 2010; Min et al., 2007). Optimisation of the media used in root culture must be performed in order to increase TA production while at the same time considering the growth rate of the culture. For example, increasing nitrate concentrations in *A. belladonna* hairy root cultures yielded higher amounts of TAs, at the same time reducing the overall growth of the culture (Chashmi, Sharifi, Karimi, & Rahnama, 2010). Other techniques that may prove helpful in the future for commercial production of TAs in culture include exploiting the process of exudation. This process has been documented to be successful for several other hairy root culture systems and includes the alteration of membrane permeability and development of constant extraction and removal of the metabolites of interest from the medium (Cai, Kastell, Knorr, & Smetanska, 2012). Other approaches that may be useful in increasing TA yield in culture include the use of specific transcription factors for the pathway and the employment of promoters appropriate for hairy roots (Jirschitzka, Mattern, Gershenzon, & D'Auria, 2013).

5. CONCLUSIONS

In contrast to other alkaloid classes such as the benzyl isoquinoline and terpene indole alkaloids, the knowledge base available for both the biosynthesis and molecular biology of tropane alkaloids is relatively small. Future prospects to increase this knowledge base will require a broadening of the model systems currently being used, especially those outside of the Solanaceae. The polyphyletic origin of TAs in plants, in conjunction with the multiple possibilities for starter substrates and the ring closure steps, strongly suggests that different enzyme classes have been recruited during the diversification of TA-producing plant lineages. The increasing ease of gene discovery through the constantly decreasing costs of high-throughput sequencing should facilitate detailed biochemical investigations into members of these other families.

ACKNOWLEDGEMENTS

This work was supported by the Max Planck Society. We would like to thank Dr. Jonathan Gershenzon and Dr. Sven Delaney for their proofreading of the manuscript. We would also like to thank Linda Maack for her help in obtaining and organising reference materials.

REFERENCES

Abraham, T. W., & Leete, E. (1995). New intermediate in the biosynthesis of the tropane alkaloids in *Datura innoxia*. *Journal of the American Chemical Society*, *117*, 8100–8105.

Ahmad, A., & Leete, E. (1970). Biosynthesis of tropine moiety of hyoscyamine from δ-*N*-methylornithine. *Phytochemistry*, *9*, 2345–2347.

Ajungla, L., Patil, P. P., Barmukh, R. B., & Nikam, T. D. (2009). Influence of biotic and abiotic elicitors on accumulation of hyoscyamine and scopolamine in root cultures of *Datura metel* L. *Indian Journal of Biotechnology*, *8*, 317–322.

Alves, M. N., Sartoratto, A., & Trigo, J. R. (2007). Scopolamine in *Brugmansia suaveolens* (Solanaceae): Defense, allocation, costs, and induced response. *Journal of Chemical Ecology*, *33*, 297–309.

Amdoun, R., Khelifi, L., Khelifi-Slaoui, M., Amroune, S., Benyoussef, E. H., Thi, D. V., et al. (2009). Influence of minerals and elicitation on *Datura stramonium* L. tropane alkaloid production: Modelization of the in vitro biochemical response. *Plant Science*, *177*, 81–87.

Ansarin, M., & Woolley, J. G. (1994). The biosynthesis of tropic acid. 5. The rearrangement of phenyllactate in the biosynthesis of tropic acid. *Phytochemistry, 35*, 935–939.
Arbain, D., Wiryani, R. D., & Sargent, M. V. (1991). A new tropane alkaloid from *Pellacalyx axillaris*. *Australian Journal of Chemistry, 44*, 1013–1015.
Asano, N., Nash, R. J., Molyneux, R. J., & Fleet, G. W. J. (2000). Sugar-mimic glycosidase inhibitors: Natural occurrence, biological activity and prospects for therapeutic application. *Tetrahedron-Asymmetry, 11*, 1645–1680.
Asano, N., Oseki, K., Tomioka, E., Kizu, H., & Matsui, K. (1994). *N*-containing sugars from *Morus alba* and their glycosidase inhibitory activities. *Carbohydrate Research, 259*, 243–255.
Asano, N., Tomioka, E., Kizu, H., & Matsui, K. (1994). Sugars with nitrogen in the ring isolated from the leaves of *Morus bombycis*. *Carbohydrate Research, 253*, 235–245.
Atta-Ur-Rahman, Khattak, K. F., Nighat, F., Shabbir, M., Hemalal, K. D., & Tillekeratne, L. M. (1998). Dimeric tropane alkaloids from *Erythroxylum moonii*. *Phytochemistry, 48*, 377–383.
Aynilian, G. H., Duke, J. A., Gentner, W. A., & Farnswor, N. R. (1974). Cocaine content of *Erythroxylum* species. *Journal of Pharmaceutical Sciences, 63*, 1938–1939.
Bick, I. R. C., Gillard, J. W., Leow, H. M., Lounasmaa, M., Pusset, J., & Sevenet, T. (1981). Biogenesis of proteaceous alkaloids. *Planta Medica, 41*, 379–385.
Bieri, S., Brachet, A., Veuthey, J. L., & Christen, P. (2006). Cocaine distribution in wild *Erythroxylum* species. *Journal of Ethnopharmacology, 103*, 439–447.
Bjorklund, J. A., & Leete, E. (1992). Biosynthesis of the tropane alkaloids and related-compounds. 49. Biosynthesis of the benzoyl moiety of cocaine from cinnamic acid via (*R*)-(+)-3-hydroxy-3-phenylpropanoic acid. *Phytochemistry, 31*, 3883–3887.
Blum, M. S., Rivier, L., & Plowman, T. (1981). Fate of cocaine in the Lymantriid *Eloria noyesi*, a predator of *Erythroxylum coca*. *Phytochemistry, 20*, 2499–2500.
Brock, A., Brandt, W., & Drager, B. (2008). The functional divergence of short-chain dehydrogenases involved in tropinone reduction. *The Plant Journal, 54*, 388–401.
Brock, A., Herzfeld, T., Paschke, R., Koch, M., & Drager, B. (2006). Brassicaceae contain nortropane alkaloids. *Phytochemistry, 67*, 2050–2057.
Butler, M. S., Katavic, P. L., Davis, R. A., Forster, P. I., Guymer, G. P., & Quinn, R. J. (2000). 10-Hydroxydarlingine, a new tropane alkaloid from the Australian proteaceous plant *Triunia erythrocarpa*. *Journal of Natural Products, 63*, 688–689.
Cai, Z. Z., Kastell, A., Knorr, D., & Smetanska, I. (2012). Exudation: An expanding technique for continuous production and release of secondary metabolites from plant cell suspension and hairy root cultures. *Plant Cell Reports, 31*, 461–477.
Campos Neves, M. T., & Campos Neves, A. (1966). Composition of leaves of *Erythroxylon dekindtii* of Angola. *Garcia Orta, 14*, 97–102.
Cardillo, A. B., Otalvaro, A. A. M., Busto, V. D., Talou, J. R., Velasquez, L. M. E., & Giulietti, A. M. (2010). Scopolamine, anisodamine and hyoscyamine production by *Brugmansia candida* hairy root cultures in bioreactors. *Process Biochemistry, 45*, 1577–1581.
Chandra, S., & Chandra, R. (2011). Engineering secondary metabolite production in hairy roots. *Phytochemistry Reviews, 10*, 371–395.
Chashmi, N. A., Sharifi, M., Karimi, F., & Rahnama, H. (2010). Differential production of tropane alkaloids in hairy roots and in vitro cultured two accessions of *Atropa belladonna* L. under nitrate treatments. *Zeitschrift für Naturforschung. Section C, 65*, 373–379.
Chesters, N. C. J. E., O'Hagan, D., & Robins, R. J. (1995). The biosynthesis of tropic acid—The (*R*)-D-phenyllactyl moiety is processed by the mutase involved in hyoscyamine biosynthesis in *Datura stramonium*. *Journal of the Chemical Society, Chemical Communications*, 127–128.
D'Auria, J. C. (2006). Acyltransferases in plants: A good time to be BAHD. *Current Opinion in Plant Biology, 9*, 331–340.

Delauney, A. J., & Verma, D. P. S. (1993). Proline biosynthesis and osmoregulation in plants. *The Plant Journal*, *4*, 215–223.
de Oliveira, S. L., Tavares, J. F., Branco, M. V. S. C., Lucena, H. F. S., Barbosa-Filho, J. M., de F. Agra, M., et al. (2011). Tropane alkaloids from *Erythroxylum caatingae* Plowman. *Chemistry & Biodiversity*, *8*, 155–165.
De Wet, H. (2011). Antibacterial activity of the five South African *Erythroxylaceae* species. *African Journal of Biotechnology*, *10*, 11511–11514.
Dillehay, T. D., Rossen, J., Ugent, D., Karathanasis, A., Vásquez, V., & Netherly, P. J. (2010). Early Holocene coca chewing in northern Peru. *Antiquity*, *84*, 939–953.
Docimo, T., Reichelt, M., Schneider, B., Kai, M., Kunert, G., Gershenzon, J., et al. (2012). The first step in the biosynthesis of cocaine in *Erythroxylum coca*: The characterization of arginine and ornithine decarboxylases. *Plant Molecular Biology*, *78*, 599–615.
Dräger, B. (2006). Tropinone reductases, enzymes at the branch point of tropane alkaloid metabolism. *Phytochemistry*, *67*, 327–337.
Dräger, B., Hashimoto, T., & Yamada, Y. (1988). Purification and characterization of pseudotropine forming tropinone reductase from *Hyoscyamus niger* root cultures. *Agricultural and Biological Chemistry*, *52*, 2663–2667.
Dräger, B., & Schaal, A. (1994). Tropinone reduction in *Atropa belladonna* root cultures. *Phytochemistry*, *35*, 1441–1447.
Duran-Patron, R., O'Hagan, D., Hamilton, J. T. G., & Wong, C. W. (2000). Biosynthetic studies on the tropane ring system of the tropane alkaloids from *Datura stramonium*. *Phytochemistry*, *53*, 777–784.
Ebert, U., Siepmann, M., Oertel, R., Wesnes, K. A., & Kirch, W. (1998). Pharmacokinetics and pharmacodynamics of scopolamine after subcutaneous administration. *Journal of Clinical Pharmacology*, *38*, 720–726.
Eich, E. (2008). *Solanaceae and Convolvulaceae: Secondary metabolites: Biosynthesis, chemotaxonomy, biological and economic significance (a handbook)*. Berlin, Germany: Springer-Verlag.
El Bazaoui, A., Bellimam, M. A., & Soulaymani, A. (2011). Nine new tropane alkaloids from *Datura stramonium* L. identified by GC/MS. *Fitoterapia*, *82*, 193–197.
el Jaber-Vazdekis, N., Barres, M. L., Ravelo, A. G., & Zarate, R. (2008). Effects of elicitors on tropane alkaloids and gene expression in *Atropa baetica* transgenic hairy roots. *Journal of Natural Products*, *71*, 2026–2031.
Endo, T., Hamaguchi, N., Hashimoto, T., & Yamada, Y. (1988). Non-enzymatic synthesis of hygrine from acetoacetic acid and from acetonedicarboxylic acid. *FEBS Letters*, *234*, 86–90.
Ferreira, J. F. S., Duke, S. O., & Vaughn, K. C. (1998). Histochemical and immunocytochemical localization of tropane alkaloids in *Erythroxylum coca* var. *coca* and *E. novogranatense* var. *novogranatense*. *International Journal of Plant Sciences*, *159*, 492–503.
Feth, F., Wray, V., & Wagner, K. G. (1985). Determination of methylputrescine oxidase by high-performance liquid-chromatography. *Phytochemistry*, *24*, 1653–1655.
Freitas, A. V. L., Trigo, J. R., Brown, K. S., Jr., Witte, L., Hartmann, T., & Barata, L. E. S. (1996). Tropane and pyrrolizidine alkaloids in the Ithomiines *Placidula euryanassa* and *Miraleria cymothoe* (Lepidoptera: Nymphalidae). *Chemoecology*, 7, 61–67.
Freud, S. (1884). Ueber Coca. *Centralblatt für die Gesammte Therapie*, *2*, 289–314.
Galloway, M. P. (1988). Neurochemical interactions of cocaine with dopaminergic systems. *Trends in Pharmacological Sciences*, *9*, 451–454.
Geiger, & Hesse (1833). Darstellung des Atropins. *Annalen der Pharmacie*, *5*, 43–81.
Goldmann, A., Milat, M. L., Ducrot, P. H., Lallemand, J. Y., Maille, M., Lepingle, A., et al. (1990). Tropane derivatives from *Calystegia sepium*. *Phytochemistry*, *29*, 2125–2127.
Graf, E., & Lude, W. (1977). Alkaloids from *Erythroxylum vacciniifolium* Martius, isolation of catuabine A, catuabine B, and catuabine C. *Archiv der Pharmazie*, *310*, 1005–1010.

Griffin, W. J., & Lin, G. D. (2000). Chemotaxonomy and geographical distribution of tropane alkaloids. *Phytochemistry*, *53*, 623–637.

Hashimoto, T., Hayashi, A., Amano, Y., Kohno, J., Iwanari, H., Usuda, S., et al. (1991). Hyoscyamine 6β-hydroxylase, an enzyme involved in tropane alkaloid biosynthesis, is localized at the pericycle of the root. *The Journal of Biological Chemistry*, *266*, 4648–4653.

Hashimoto, T., Matsuda, J., & Yamada, Y. (1993). Two-step epoxidation of hyoscyamine to scopolamine is catalyzed by bifunctional hyoscyamine 6β-hydroxylase. *FEBS Letters*, *329*, 35–39.

Hashimoto, T., Mitani, A., & Yamada, Y. (1990). Diamine oxidase from cultured roots of *Hyoscyamus niger*—Its function in tropane alkaloid biosynthesis. *Plant Physiology*, *93*, 216–221.

Hashimoto, T., Nakajima, K., Ongena, G., & Yamada, Y. (1992). Two tropinone reductases with distinct stereospecificities from cultured roots of *Hyoscyamus niger*. *Plant Physiology*, *100*, 836–845.

Hashimoto, T., Tamaki, K., Suzuki, K., & Yamada, Y. (1998). Molecular cloning of plant spermidine synthases. *Plant & Cell Physiology*, *39*, 73–79.

Hashimoto, T., & Yamada, Y. (1986). Hyoscyamine 6β-hydroxylase, a 2-oxoglutarate-dependent dioxygenase, in alkaloid-producing root cultures. *Plant Physiology*, *81*, 619–625.

Hashimoto, T., Yukimune, Y., & Yamada, Y. (1989). Putrescine and putrescine *N*-methyltransferase in the biosynthesis of tropane alkaloids in cultured roots of *Hyoscyamus albus*. 2. Incorporation of labeled precursors. *Planta*, *178*, 131–137.

Heim, W. G., Sykes, K. A., Hildreth, S. B., Sun, J., Lu, R. H., & Jelesko, J. G. (2007). Cloning and characterization of a *Nicotiana tabacum* methylputrescine oxidase transcript. *Phytochemistry*, *68*, 454–463.

Henderson, V. E., & Roepke, M. H. (1937). Drugs affecting parasympathetic nerves. *Physiological Reviews*, *17*, 373–407.

Hibi, N., Higashiguchi, S., Hashimoto, T., & Yamada, Y. (1994). Gene expression in tobacco low-nicotine mutants. *The Plant Cell*, *6*, 723–735.

Hong, M. L. K., Bhatt, A., Ping, N. S., & Keng, C. L. (2012). Detection of elicitation effect on *Hyoscyamus niger* L. root cultures for the root growth and production of tropane alkaloids. *Romanian Biotechnological Letters*, *17*, 7340–7351.

Honkavaara, P., & Pyykkö, I. (1999). Effects of atropine and scopolamine on bradycardia and emetic symptoms in otoplasty. *The Laryngoscope*, *109*, 108–112.

Hoye, T. R., Bjorklund, J. A., Koltun, D. O., & Renner, M. K. (2000). *N*-methylputrescine oxidation during cocaine biosynthesis: Study of prochiral methylene hydrogen discrimination using the remote isotope method. *Organic Letters*, *2*, 3–5.

Imanishi, S., Hashizume, K., Nakakita, M., Kojima, H., Matsubayashi, Y., Hashimoto, T., et al. (1998). Differential induction by methyl jasmonate of genes encoding ornithine decarboxylase and other enzymes involved in nicotine biosynthesis in tobacco cell cultures. *Plant Molecular Biology*, *38*, 1101–1111.

Jenett-Siems, K., Weigl, R., Bohm, A., Mann, P., Tofern-Reblin, B., Ott, S. C., et al. (2005). Chemotaxonomy of the pantropical genus *Merremia* (Convolvulaceae) based on the distribution of tropane alkaloids. *Phytochemistry*, *66*, 1448–1464.

Jirschitzka, J., Mattern, D. J., Gershenzon, J., & D'Auria, J. C. (2013). Learning from nature: New approaches to the metabolic engineering of plant defense pathways. *Current Opinion in Biotechnology*, *24*, 320–328.

Jirschitzka, J., Schmidt, G. W., Reichelt, M., Schneider, B., Gershenzon, J., & D'Auria, J. C. (2012). Plant tropane alkaloid biosynthesis evolved independently in the Solanaceae and Erythroxylaceae. *Proceedings of the National Academy of Sciences*, *109*, 10304–10309.

Johns, S. R. (1972). Corrigenda. *Australian Journal of Chemistry*, *25*, 1600.

Johns, S. R., Lamberto, J. A., & Sioumis, A. A. (1971). New tropane alkaloids, (+)-(3*R*,6*R*)-3α-acetoxy-6β – hydroxytropane and (+)-2α-benzoyloxy-3β-hydroxynortropane, from *Peripentadenia mearsii* (Euphorbiaceae). *Australian Journal of Chemistry*, *24*, 2399–2403.
Kaczkowski, J., Schütte, H. R., & Mothes, K. (1961). Die rolle des acetats in der biosynthese der tropanalkaloide. *Biochimica et Biophysica Acta*, *46*, 588–594.
Kai, G., Li, L., Jiang, Y., Yan, X., Zhang, Y., Lu, X., et al. (2009). Molecular cloning and characterization of two tropinone reductases in *Anisodus acutangulus* and enhancement of tropane alkaloid production in *AaTRI*-transformed hairy roots. *Biotechnology and Applied Biochemistry*, *54*, 177–186.
Kai, G., Yang, S., Luo, X., Zhou, W., Fu, X., Zhang, A., et al. (2011). Co-expression of *AaPMT* and *AaTRI* effectively enhances the yields of tropane alkaloids in *Anisodus acutangulus* hairy roots. *BMC Biotechnology*, *11*, 43–54.
Kai, G., Yang, S., Zhang, Y., Luo, X., Fu, X., Zhang, A., et al. (2012). Effects of different elicitors on yield of tropane alkaloids in hairy roots of *Anisodus acutangulus*. *Molecular Biology Reports*, *39*, 1721–1729.
Kai, G., Zhang, Y., Chen, J., Li, L., Yan, X., Zhang, R., et al. (2009). Molecular characterization and expression analysis of two distinct putrescine *N*-methyltransferases from roots of *Anisodus acutangulus*. *Physiologia Plantarum*, *135*, 121–129.
Kai, G., Zhang, A., Guo, Y., Li, L., Cui, L., Luo, X., et al. (2012). Enhancing the production of tropane alkaloids in transgenic *Anisodus acutangulus* hairy root cultures by over-expressing tropinone reductase I and hyoscyamine-6β-hydroxylase. *Molecular BioSystems*, *8*, 2883–2890.
Kaiser, H., Richter, U., Keiner, R., Brabant, A., Hause, B., & Dräger, B. (2006). Immunolocalisation of two tropinone reductases in potato (*Solanum tuberosum* L.) root, stolon and tuber sprouts. *Planta*, *225*, 127–137.
Kang, Y. M., Park, D. J., Min, J. Y., Song, H. J., Jeong, M. J., Kim, Y. D., et al. (2011). Enhanced production of tropane alkaloids in transgenic *Scopolia parviflora* hairy root cultures over-expressing putrescine *N*-methyl transferase (PMT) and hyoscyamine-6β-hydroxylase (H6H). *In Vitro Cellular & Developmental Biology—Plant*, 47, 516–524.
Karimi, F., & Khataee, E. (2012). Aluminum elicits tropane alkaloid production and antioxidant system activity in micropropagated *Datura innoxia* plantlets. *Acta Physiologiae Plantarum*, *34*, 1035–1041.
Kato, A. (1975). Brugine from *Bruguiera cylindrica*. *Phytochemistry*, *14*, 1458.
Katoh, A., Shoji, T., & Hashimoto, T. (2007). Molecular cloning of *N*-methylputrescine oxidase from tobacco. *Plant & Cell Physiology*, *48*, 550–554.
Kavanagh, P., Angelov, D., O'Brien, J., Fox, J., O'Donnell, C., Christie, R., et al. (2012). The syntheses and characterization 3β-(4-fluorobenzoyloxy)tropane (fluorotropacocaine) and its 3α isomer. *Drug Testing and Analysis*, *4*, 33–38.
Keiner, R., Kaiser, H., Nakajima, K., Hashimoto, T., & Dräger, B. (2002). Molecular cloning, expression and characterization of tropinone reductase II, an enzyme of the SDR family in *Solanum tuberosum* (L.). *Plant Molecular Biology*, *48*, 299–308.
Kitamura, Y., Tominaga, Y., & Ikenaga, T. (2004). Winter cherry bugs feed on plant tropane alkaloids and de-epoxidize scopolamine to atropine. *Journal of Chemical Ecology*, *30*, 2085–2090.
Kletter, C., Glasl, S., Presser, A., Werner, I., Reznicek, G., Narantuya, S., et al. (2004). Morphological, chemical and functional analysis of Catuaba preparations. *Planta Medica*, *70*, 993–1000.
Koelen, K. J., & Gross, G. G. (1982). Partial purification and properties of tropine dehydrogenase from root cultures of *Datura stramonium*. *Planta Medica*, *44*, 227–230.
Lazny, R., Sienkiewicz, M., Olenski, T., Urbanczyk-Lipkowska, Z., & Kalicki, P. (2012). Approaches to the enantioselective synthesis of ferrugine and its analogues. *Tetrahedron*, *68*, 8236–8244.

Leete, E. (1962). Stereospecific incorporation of ornithine into tropine moiety of hyoscyamine. *Journal of the American Chemical Society*, *84*, 55–57.
Leete, E. (1964). Biosynthesis of hyoscyamine: Proof that ornithine-2-C^{14} yields tropine labelled at C-1. *Tetrahedron Letters*, 1619–1622.
Leete, E. (1982). Biosynthesis of the pyrrolidine rings of cocaine and cuscohygrine from [5-^{14}C]-labeled ornithine via a symmetrical intermediate. *Journal of the American Chemical Society*, *104*, 1403–1408.
Leete, E. (1983a). Chemistry of the tropane alkaloids. 31. Biosynthesis of cocaine and cuscohygrine from [1-^{14}C]-labeled acetate and [4-^{3}H]-labeled phenylalanine in *Erythroxylon coca*. *Phytochemistry*, *22*, 699–704.
Leete, E. (1983b). Chemistry of the tropane alkaloids. 33. 2-Carbomethoxy-3-tropinone—An advanced intermediate in the biosynthesis of cocaine. *Journal of the American Chemical Society*, *105*, 6727–6728.
Leete, E. (1985). Spermidine: An indirect precursor of the pyrrolidine rings of nicotine and nornicotine in *Nicotiana glutinosa*. *Phytochemistry*, *24*, 957–960.
Leete, E. (1990). Recent developments in the biosynthesis of the tropane alkaloids. *Planta Medica*, *56*, 339–352.
Leete, E., Bjorklund, J. A., Couladis, M. M., & Kim, S. H. (1991). Late intermediates in the biosynthesis of cocaine: 4-(1-Methyl-2-pyrrolidinyl)-3-oxobutanoate and methyl ecgonine. *Journal of the American Chemical Society*, *113*, 9286–9292.
Leete, E., Kim, S. H., & Rana, J. (1988). Chemistry of the tropane alkaloids and related-compounds. 38. The incorporation of [2-^{13}C,^{14}C,^{15}N]-1-methyl-Δ^{1}-pyrrolinium chloride into cuscohygrine in *Erythroxylum coca*. *Phytochemistry*, *27*, 401–406.
Leete, E., Kowanko, N., & Newmark, R. A. (1975). Use of Carbon-13 nuclear magnetic-resonance to establish that biosynthesis of tropic acid involves an intramolecular rearrangement of phenylalanine. *Journal of the American Chemical Society*, *97*, 6826–6830.
Leete, E., Marion, L., & Spenser, I. D. (1954). The biogenesis of alkaloids: 12. The mode of formation of the tropine base of hyoscyamine. *Canadian Journal of Chemistry*, *32*, 1116–1123.
Levinson, H. Z. (1976). The defensive role of alkaloids in insects and plants. *Cellular and Molecular Life Sciences*, *32*, 408–411.
Li, R., Reed, D. W., Liu, E. W., Nowak, J., Pelcher, L. E., Page, J. E., et al. (2006). Functional genomic analysis of alkaloid biosynthesis in *Hyoscyamus niger* reveals a cytochrome P450 involved in littorine rearrangement. *Chemistry & Biology*, *13*, 513–520.
Liebisch, H. W., Bernasch, H., & Schütte, H. R. (1973). Zur Biosynthese der Tropanalkaloide. XII. Die Biosynthese des Cochlearins. *Zeitschrift für Chemie*, *13*, 372–373.
Liebisch, H. W., Peisker, K., Radwan, A. S., & Schütte, H. R. (1972). Zur biosynthese der tropanalkaloide. XI. Die bildung der C_3-brücke des tropins. *Zeitschrift für Pflanzenphysiologie*, *67*, 1–9.
Liebisch, H. W., Ramin, H., Schoffin, I., & Schütte, H. R. (1965). Zur Biosynthese der Tropanalkaloide. *Zeitschrift für Naturforschung. Part B: Chemie Biochemie Biophysik Biologie und Verwandten Gebiete*, *B 20*, 1183–1185.
Liebisch, H. W., & Schütte, H. R. (1967). Zur biosynthese der tropanalkaloide.VIII. Vorstufen des pyrrolidinringes. *Zeitschrift für Pflanzenphysiologie*, *57*, 434–439.
Liu, X. Q., Yang, C. X., Chen, M., Li, M. Y., Liao, Z. H., & Tang, K. X. (2010). Promoting scopolamine accumulation in transgenic plants of *Atropa belladonna* generated from hairy roots with over expression of *pmt* and *h6h* gene. *Journal of Medicinal Plants Research*, *4*, 1708–1713.
Liu, T., Zhu, P., Cheng, K. D., Meng, C., & Zhu, H. X. (2005). Molecular cloning and expression of putrescine *N*-methyltransferase from the hairy roots of *Anisodus tanguticus*. *Planta Medica*, *71*, 987–989.
Loder, J. W., & Russell, G. B. (1966). Tropine 1,2-dithiolane-3-carboxylate a new alkaloid from *Bruguiera sexangula*. *Tetrahedron Letters*, *51*, 6327–6329.

Loder, J. W., & Russell, G. B. (1969). Tumour inhibitory plants . Alkaloids of *Bruguiera sexangula* and *Bruguiera exaristata* (Rhizophoraceae). *Australian Journal of Chemistry, 22*, 1271–1275.

Lounasmaa, M., & Tamminen, T. (1993). The tropane alkaloids. In G. A. Cordell (Ed.), *The alkaloids*, Vol. 44, (pp. 1–114). New York: Academic Press.

Luo, X., Weng, S., Ni, X., Wang, X., Fu, X., Xiao, J., et al. (2012). The effects of elicitation on the expression of key enzyme genes and on production of tropane alkaloids in *Anisodus acutangulus* plant. *Biologia, 67*, 352–359.

Lydon, J., Casale, J. F., Kong, H., Sullivan, J. H., Daughtry, C. S. T., & Bailey, B. (2009). The effects of ambient solar UV radiation on alkaloid production by *Erythroxylum novogranatense* var. *novogranatense*. *Photochemistry and Photobiology, 85*, 1156–1161.

Malmberg, R. L., Watson, M. B., Galloway, G. L., & Yu, W. (1998). Molecular genetic analyses of plant polyamines. *Critical Reviews in Plant Sciences, 17*, 199–224.

Mann, J. (1992). *Murder, magic, and medicine*. New York: Oxford University Press.

Matsuzaki, R., Fukui, T., Sato, H., Ozaki, Y., & Tanizawa, K. (1994). Generation of the topa quinone cofactor in bacterial monoamine oxidase by cupric ion-dependent autooxidation of a specific tyrosyl residue. *FEBS Letters, 351*, 360–364.

Media, D. H. G., Pusset, M., Pusset, J., & Husson, H.-P. (1983). Alcaloides tropaniques de *Crossostylis* sp. *Journal of Natural Products, 46*, 398–400.

Mein (1833). Ueber die darstellung des atropins in weissen krystallen. *Annalen der Pharmacie, 6*, 67–72.

Michael, A. J., Furze, J. M., Rhodes, M. J. C., & Burtin, D. (1996). Molecular cloning and functional identification of a plant ornithine decarboxylase cDNA. *The Biochemical Journal, 314*, 241–248.

Min, J. Y., Jung, H. Y., Kang, S. M., Kim, Y. D., Kang, Y. M., Park, D. J., et al. (2007). Production of tropane alkaloids by small-scale bubble column bioreactor cultures of *Scopolia parviflora* adventitious roots. *Bioresource Technology, 98*, 1748–1753.

Mirzaev, Y. R., & Aripova, S. F. (1998). Neuro- and psychopharmacological investigation of the alkaloids convolvine and atropine. *Chemistry of Natural Compounds, 34*, 56–58.

Mizusaki, S., Tanabe, Y., Noguchi, M., & Tamaki, E. (1973). Phytochemical studies on tobacco alkaloids. 16. Changes in activities of ornithine decarboxylase, putrescine *N*-methyltransferase and *N*-methyl-putrescine oxidase in tobacco roots in relation to nicotine biosynthesis. *Plant & Cell Physiology, 14*, 103–110.

Moummou, H., Kallberg, Y., Tonfack, L. B., Persson, B., & van der Rest, B. (2012). The plant short-chain dehydrogenase (SDR) superfamily: Genome-wide inventory and diversification patterns. *BMC Plant Biology, 12*, 219–235.

Nahar, L., Sarker, S. D., & Delazar, A. (2011). Phytochemistry of the genus Phyllanthus. In R. Kuttan & K. B. Harikumar (Eds.), *Phyllanthus species* (pp. 119–138). Boca Raton: CRC Press.

Nakajima, K., & Hashimoto, T. (1999). Two tropinone reductases, that catalyze opposite stereospecific reductions in tropane alkaloid biosynthesis, are localized in plant root with different cell-specific patterns. *Plant & Cell Physiology, 40*, 1099–1107.

Nakajima, K., Kato, H., Oda, J., Yamada, Y., & Hashimoto, T. (1999). Site-directed mutagenesis of putative substrate-binding residues reveals a mechanism controlling the different stereospecificities of two tropinone reductases. *The Journal of Biological Chemistry, 274*, 16563–16568.

Naranjo, P. (1981). Social function of coca in pre-Columbian America. *Journal of Ethnopharmacology, 3*, 161–172.

Nash, R. J., & Watson, A. A. (1995). Inhibition of glycosidases by Lepidoptera; roles in the insects and leads to novel compounds? *Chemoecology, 5–6*, 167–171.

Nasomjai, P., Reed, D. W., Tozer, D. J., Peach, M. J. G., Slawin, A. M. Z., Covello, P. S., et al. (2009). Mechanistic insights into the cytochrome P450-mediated oxidation and rearrangement of littorine in tropane alkaloid biosynthesis. *ChemBioChem, 10,* 2382–2393.

Nathanson, J. A., Hunnicutt, E. J., Kantham, L., & Scavone, C. (1993). Cocaine as a naturally occurring insecticide. *Proceedings of the National Academy of Sciences of the United States of America, 90,* 9645–9648.

Niemann, A. (1860). Ueber eine neue organische Base in den Cocablättern. *Archiv der Pharmazie, 153,* 291–308.

Niki, T., Mitsuhara, I., Seo, S., Ohtsubo, N., & Ohashi, Y. (1998). Antagonistic effect of salicylic acid and jasmonic acid on the expression of pathogenesis-related (PR) protein genes in wounded mature tobacco leaves. *Plant & Cell Physiology, 39,* 500–507.

Nishiyama, Y., Moriyasu, M., Ichimaru, M., Sonoda, M., Iwasa, K., Kato, A., et al. (2007). Tropane alkaloids from *Erythroxylum emarginatum. Journal of Natural Medicines, 61,* 56–58.

Orechoff, A., & Konowalowa, R. (1933). Über die alkaloide von *Convolvulus pseudocantabricus* Schrenk. *Archiv der Pharmazie, 271,* 145–148.

Palazon, J., Navarro-Ocana, A., Hernandez-Vazquez, L., & Mirjalili, M. H. (2008). Application of metabolic engineering to the production of scopolamine. *Molecules, 13,* 1722–1742.

Pardo Torre, J. C., Schmidt, G. W., Paetz, C., Reichelt, M., Schneider, B., Gershenzon, J., et al. (2013). The biosynthesis of hydroxycinnamoyl quinate esters and their role in the storage of cocaine in *Erythroxylum coca. Phytochemistry, 91,* 177–186.

Parello, J., Longevialle, P., Vetter, W., & McCloskey, J. A. (1963). Structure de la phyllalbine. Application de la resonance magnetique nucleaire et de la spectrmetrie de masse a l'etude des derives du tropane. *Bulletin de la Société Chimique de France,* 2787–2793.

Parkerson, P. T. (1983). The Inca coca monopoly—Fact or legal fiction. *Proceedings of the American Philosophical Society, 127,* 107–123.

Platonova, T. F., & Kuzovkov, A. D. (1963). Alkaloids of *Cochlearia arctica. Medicinskaja promyslennost' SSSR, 17,* 19–20.

Plowman, T. (1981). Amazonian coca. *Journal of Ethnopharmacology, 3,* 195–225.

Plowman, T. (1982). The identification of coca (*Erythroxylum* species)—1860–1910. *Botanical Journal of the Linnean Society, 84,* 329–353.

Plowman, T. (1984). The ethnobotany of Coca (*Erythroxylum* spp., Erythroxylaceae). In G. T. Prance & J. A. Kallunki (Eds.), *Ethnobotany in the neotropics* (pp. 62–111). New York: New York Botanical Garden.

Plowman, T., & Rivier, L. (1983). Cocaine and cinnamoylcocaine content of *Erythroxylum species. Annals of Botany, 51,* 641–659.

Portsteffen, A., Dräger, B., & Nahrstedt, A. (1994). The reduction of tropinone in *Datura stramonium* root cultures by two specific reductases. *Phytochemistry, 37,* 391–400.

Queiroz, E. F., Zanolari, B., Marston, A., Guilet, D., Burgener, L., Paulo Mde, Q., et al. (2009). Two new tropane alkaloids from the bark of *Erythroxylum vacciniifolium* Mart. (Erythroxylaceae). *Natural Product Communications, 4,* 1337–1340.

Rabot, S., Peerless, A. C. J., & Robins, R. J. (1995). Tigloyl-CoA: Pseudotropine acyl transferase—An enzyme of tropane alkaloid biosynthesis. *Phytochemistry, 39,* 315–322.

Razzakov, N. A., & Aripova, S. F. (2004). Confolidine, a new alkaloid from the aerial part of *Convolvulus subhirsutus. Chemistry of Natural Compounds, 40,* 54–55.

Reas, H. W., & Tsai, T. H. (1966). The antagonism by atropine of the response of the nictitating membrane to sympathetic nerve stimulation. *The Journal of Pharmacology and Experimental Therapeutics, 152,* 186–196.

Robins, R. J., Abraham, T. W., Parr, A. J., Eagles, J., & Walton, N. J. (1997). The biosynthesis of tropane alkaloids in *Datura stramonium*: The identity of the intermediates between *N*-methylpyrrolinium salt and tropinone. *Journal of the American Chemical Society, 119,* 10929–10934.

Robins, R. J., Bachmann, P., Robinson, T., Rhodes, M. J., & Yamada, Y. (1991). The formation of 3α- and 3β-acetoxytropanes by *Datura stramonium* transformed root cultures involves two acetyl-CoA-dependent acyltransferases. *FEBS Letters, 292*, 293–297.

Robins, R. J., Bachmann, P., & Woolley, J. G. (1994). Biosynthesis of hyoscyamine involves an intramolecular rearrangement of littorine. *Journal of the Chemical Society, Perkin Transactions 1*, 615–619.

Robins, R. J., Chesters, N., Ohagan, D., Parr, A. J., Walton, N. J., & Woolley, J. G. (1995). The biosynthesis of hyoscyamine—The process by which littorine rearranges to hyoscyamine. *Journal of the Chemical Society, Perkin Transactions 1*, 481–485.

Runyon, S. P., & Carroll, F. I. (2006). Dopamine transporter ligands: Recent developments and therapeutic potential. *Current Topics in Medicinal Chemistry, 6*, 1825–1843.

Sandala, G. M., Smith, D. M., & Radom, L. (2008). The carbon-skeleton rearrangement in tropane alkaloid biosynthesis. *Journal of the American Chemical Society, 130*, 10684–10690.

Sandmeier, E., Hale, T. I., & Christen, P. (1994). Multiple evolutionary origin of pyridoxal-5′-phosphate-dependent amino-acid decarboxylases. *European Journal of Biochemistry, 221*, 997–1002.

Schipper, R. G., Cuijpers, V., de Groot, L., Thio, M., & Verhofstad, A. A. J. (2004). Intracellular localization of ornithine decarboxylase and its regulatory protein, antizyme-1. *The Journal of Histochemistry and Cytochemistry, 52*, 1259–1266.

Schmidt, E., & Henschke, H. (1888). Über die alkaloide der wurzel von *Scopolia japonica*. *Archiv der Pharmazie, 226*, 185–203.

Schnorrenberger, C. C. (1978). Die heilpflanzen phytotherapeutisch grundlegende applikationen. In *Handbuch der chinesischen Pflanzenheilkunde* (pp. 137–160). Freiburg im Breisgau: Hermann Bauer Verlag.

Schultes, R. E. (1976). *Hallucinogenic plants—A golden guide*. New York: Golden Press.

Schultes, R. E., Hofmann, A., & Rätsch, C. (2001). *Plants of the gods: Their sacred, healing, and hallucinogenic powers*. Lucerne: Healing Arts Press.

Schwartz, H., de Roetth, A., & Papper, E. M. (1957). Preanesthetic use of atropine and scopolamine in patients with glaucoma. *Journal of the American Medical Association, 165*, 144–146.

Sena-Filho, J. G., da Silva, M. S., Tavares, J. F., Oliveira, S. L., Romero, M. A. V., Xavier, H. S., et al. (2011). ChemInform abstract: Cytotoxic evaluation of pungencine (I): A new tropane alkaloid from the roots of *Erythroxylum pungens* O. E. Schulz. *ChemInform, 42*, 1742–1744.

Singh, S. (2000). Chemistry, design, and structure-activity relationship of cocaine antagonists. *Chemical Reviews, 100*, 925–1024.

Singh, A., Nirala, N. K., Das, S., Narula, A., Rajam, M. V., & Srivastava, P. S. (2011). Overexpression of *odc* (ornithine decarboxylase) in *Datura innoxia* enhances the yield of scopolamine. *Acta Physiologiae Plantarum, 33*, 2453–2459.

Streatfeild, D. (2001). *Cocaine: An unauthorized biography*. New York: St. Martin's Press.

Tepfer, D., Goldmann, A., Fleury, V., Maille, M., Message, B., Pamboukdjian, N., et al. (1988). Calystegins nutritional mediators in plant-microbe interactions. In R. Palacios & D. P. S. Verma (Eds.), *Molecular genetics of plant-microbe interactions, 1988* (pp. 139–144).

Teuber, M., Azemi, M. E., Namjoyan, F., Meier, A. C., Wodak, A., Brandt, W., et al. (2007). Putrescine *N*-methyltransferases—A structure-function analysis. *Plant Molecular Biology, 63*, 787–801.

The Angiosperm Phylogeny Group, (2009). An update of the angiosperm phylogeny group classification for the orders and families of flowering plants: APG III. *Botanical Journal of the Linnean Society, 161*, 105–121.

Vakili, B., Karimi, F., Sharifi, M., & Behmanesh, M. (2012). Chromium-induced tropane alkaloid production and H6H gene expression in *Atropa belladonna* L. (Solanaceae) *in vitro*-propagated plantlets. *Plant Physiology and Biochemistry, 52*, 98–103.

Valera, G. C., De Budowski, J., Delle Monache, F., & Marini-Bettolo, G. B. (1977). A new psychoactive drug: *Heisteria olivae* (Olacaceae). *Atti della Accademia Nazionale dei Lincei, Classe di Scienze Fisiche, Matematiche e Naturali, 62*, 363–364.

van Soeren, J. H. (1962). Bijdrage tot het onderzoek naar de biosynthese van de tropa-alkaloiden. *Pharmaceutisch Weekblad, 97*, 721–731.

WHO, (2011). *WHO model list of essential medicines, 17th list (March 2011)*. Geneva: World Health Organization.

Wigle, I. D., Mestichelli, L. J. J., & Spenser, I. D. (1982). ^{2}H NMR-spectroscopy as a probe of the stereochemistry of biosynthetic reactions—The biosynthesis of nicotine. *Journal of the Chemical Society, Chemical Communications*, 662–664.

Willstätter, R., & Hollander, C. (1903). Synthese der Ecgoninsäure. *Justus Liebigs Annalen der Chemie, 326*, 79–90.

Wink, M. (1998a). Chemical ecology of alkaloids. In M. F. Roberts & M. Wink (Eds.), *Alkaloids* (pp. 265–300). New York: Plenum Press.

Wink, M. (1998b). Modes of actions of alkaloids. In M. F. Roberts & M. Wink (Eds.), *Alkaloids* (pp. 301–326). New York: Plenum Press.

Wink, M. (1998c). A short history of alkaloids. In M. F. Roberts & M. Wink (Eds.), *Alkaloids* (pp. 11–44). New York: Plenum Press.

Yamada, Y., Hashimoto, T., Endo, T., Yukimune, Y., Kohno, J., Hamaguchni, N., et al. (1990). Biochemistry of alkaloid production in vitro. In B. V. Charlwood & M. J. C. Rhodes (Eds.), *Secondary products in plant tissue cultures* (pp. 227–341). Oxford: Oxford Science Publications.

Yang, C. X., Chen, M., Zeng, L. J., Zhang, L., Liu, X. Q., Lan, X. Z., et al. (2011). Improvement of tropane alkaloids production in hairy root cultures of *Atropa belladonna* by over-expressing *pmt* and *h6h* genes. *Plant Omics, 4*, 29–33.

Zhang, L., Ding, R., Chai, Y., Bonfill, M., Moyano, E., Oksman-Caldentey, K.-M. , et al. (2004). Engineering tropane biosynthetic pathway in *Hyoscyamus niger* hairy root cultures. *Proceedings of the National Academy of Sciences of the United States of America, 101*, 6786–6791.

Zhang, L., Kai, G., Lu, B.-B. , Zhang, H.-M. , Tang, K.-X. , Jiang, J.-H. , et al. (2005). Metabolic engineering of tropane alkaloid biosynthesis in plants. *Journal of Integrative Plant Biology, 47*, 136–143.

Zhang, W. W., Song, M. K., Cui, Y. Y., Wang, H., Zhu, L., Niu, Y. Y., et al. (2008). Differential neuropsychopharmacological influences of naturally occurring tropane alkaloids anisodamine versus scopolamine. *Neuroscience Letters, 443*, 241–245.

Zhang, L., Yang, B., Lu, B., Kai, G., Wang, Z., Xia, Y., et al. (2007). Tropane alkaloids production in transgenic *Hyoscyamus niger* hairy root cultures over-expressing putrescine *N*-methyltransferase is methyl jasmonate-dependent. *Planta, 225*, 887–896.

Ziegler, J., & Facchini, P. J. (2008). Alkaloid biosynthesis: Metabolism and trafficking. *Annual Review of Plant Biology, 59*, 735–769.

Zulak, K. G., Liscombe, D. K., Ashihara, H., & Facchini, P. J. (2007). Alkaloids. In A. Crozier, M. N. Clifford, & H. Ashihara (Eds.), *Plant secondary metabolites* (pp. 102–136). Oxford: Blackwell Publishing.

CHAPTER THREE

Deciphering the Evolution, Cell Biology and Regulation of Monoterpene Indole Alkaloids

Benoit St-Pierre[1], Sébastien Besseau, Marc Clastre, Vincent Courdavault, Martine Courtois, Joel Crèche, Eric Ducos, Thomas Dugé de Bernonville, Christelle Dutilleul, Gaëlle Glévarec, Nadine Imbault, Arnaud Lanoue, Audrey Oudin, Nicolas Papon, Olivier Pichon, Nathalie Giglioli-Guivarc'h

EA 2106 'Biomolécules et Biotechnologies Végétales', Université François-Rabelais de Tours, Tours, France

[1]Corresponding author: e-mail address: benoit.stpierre@univ-tours.fr

Contents

1. Introduction 74
   1.1 MIA structural diversity 75
   1.2 MIA distribution in angiosperm 76
   1.3 MIA biosynthetic origin 79
   1.4 MIA evolutionary origin 82
2. Regulation of MIA Biosynthesis Pathway 86
   2.1 Developmental control in plant 86
   2.2 Environmental factors 86
   2.3 Hormonal controls 87
   2.4 Intracellular signalling 90
3. Spatial Organization of MIA Biosynthesis 93
   3.1 Compartmentation of metabolites 93
   3.2 Compartmentation of MIA biosynthesis in specialised cells 94
   3.3 Subcellular organisation of the pathway 96
   3.4 Nuclear time bomb 99
   3.5 Biological function 101
4. Conclusions 101

Acknowledgements 102

References 102

Abstract

Monoterpene indole alkaloids (MIAs) constitute a large group of specialised metabolites with many potent pharmaceutical properties, including the antitumoral vinblastine and hypotensive ajmalicine. Hence a large body of phytochemical investigation delineates

Advances in Botanical Research, Volume 68
ISSN 0065-2296
http://dx.doi.org/10.1016/B978-0-12-408061-4.00003-1

the distribution and diversity of various MIA structural classes in Gentianales families. The biosynthetic pathway of these secondary metabolites involves several specific branches, including indole and monoterpenoid formations, secoiridoid assembly, central MIA biosynthesis and branch-specific reactions, as well as supply of primary metabolite precursors by the methylerythritol phosphate and shikimate pathways. Several genes and enzymatic activities involved in these pathways have been characterised, allowing detailed analysis of the molecular biology of this system in model plants such as *Catharanthus roseus* and *Rauvolfia serpentina*. With the prospects of improving production of MIAs in plant and cell culture, regulations of biosynthetic capacities have been thoroughly investigated. This pathway also presents a high degree of spatial organisation at the organ, cellular and subcellular levels. This chapter presents an overview of the structural diversity, the complexity of MIA biosynthesis, and regulation with an evolutionary perspective.

1. INTRODUCTION

Plant secondary metabolites, or plant natural products or specialised metabolites, now encompass about 100,000 chemically identified low molecular weight compounds. Alkaloids constitute a structurally diverse group of secondary metabolites, with an estimated 12,000 different molecules sharing as a unique common feature the presence of a nitrogen atom within a heterocyclic ring (Ziegler & Facchini, 2008). Many alkaloids disturb a specific cellular target in animals or microbes that feeds on plants. The toxicity of alkaloids contributes to the ability of plants to protect themselves. The level of protection is partially related to the diversity of alkaloids, the synergistic effect with other structural classes of natural product and the evolution of this molecular arsenal. The biological activity of alkaloids has long been used by humans, and some are well known for their health benefits. Because many of these alkaloids are present in low abundance or produced by rare or slowly growing plants, many research projects were developed towards understanding the architecture and the regulation of their long and complex biosynthetic pathways. These molecules are commonly synthesised in a plant-, organ- and even cell-specific manner. The purpose of this chapter is to give an overview of the biology of monoterpenoid indole alkaloids (MIAs), one of the major classes of alkaloids in plants, including the distribution in the plant kingdom, the evolutionary origin of this class of alkaloid, the regulatory networks of MIA biosynthesis and the architecture and spatial organisation of this biosynthetic pathway. One of the major research trends is the application of these discoveries to overcome the low yield of these natural products by biotechnological platform and metabolic engineering strategies. On the other hand, unusual biological mechanisms have been elucidated in this plant family, illustrating the fundamental interest of studying such secondary metabolic pathways beyond the primary interest for industrial application.

1.1. MIA structural diversity

More than 2500 MIAs were isolated, mainly from plants (Szabó, 2008). They belong to several structural subclasses and may be assembled in dimers. Several schemes contribute to the chemical diversity of the monomeric MIAs and related alkaloids.

a. The tryptamine subunit is extended into a quinoline subunit during the biosynthesis of the 'quinoindole' alkaloids group, which includes the campthotecan (prototype campthotecin) and the cinchonan (cinchonine) (Szabó, 2008; Fig. 3.1).

b. The tryptamine subunit may also be replaced by dopamine and provides the isoquinoline ring system of the tetrahydroisoquinoline

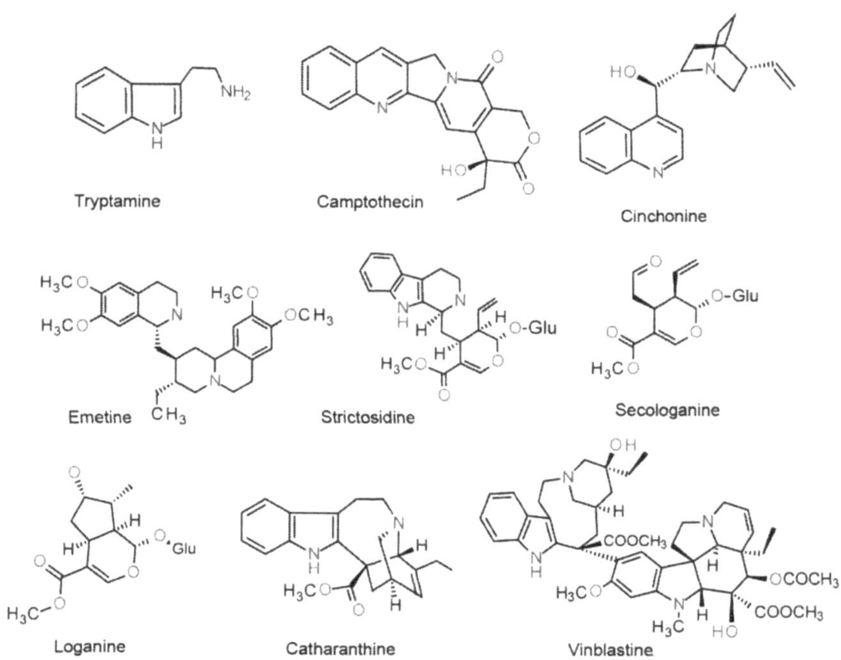

Figure 3.1 Selected monoterpenoid indole alkaloids and modified MIAs. (See colour plate.)

monoterpenoid alkaloids (Fig. 3.1), also called Ipecac alkaloids (e.g. emetine, cephaeline).

c. In most MIAs, the tryptamine skeleton is conserved. However, the secologanin subunit may be reorganised and cross-linked to the tryptamine moiety. This gives rise to a large number of the MIAs (Fig. 3.1). Several schemes of rearrangement lead to diverse structural subclasses, comprising three types of skeleton (Szabó, 2008).

The vincosans keep the type I skeleton of strictosidine, whereas the related and more diverse cyclovincosans are produced by ring closures of this skeleton. This primary cyclisation between the nucleophilic N-4 centre and one of the four electrophilic centres (C-17, C-19, C-21 and C-22) of the secologanin subunit gives various subclasses. The 'cyclovincosan' skeletons and some representatives are shown in Fig. 3.2. For instance, the antihypertensive drug ajmalicine belongs to the corynanthean subclass, the largest subclass of MIAs, with more than 1000 structures. Further cyclisation of the vincosan skeleton, both at C16–C2 and C21–N4, gives rise to the large subclass of the strychnan (>500 alkaloids), including strychnine.

More complex reorganisation of the secologanin subunit gives rise to the type II and type III skeletons, including the third largest subclass plumeran (type III β, 340 alkaloids), such as tabersonine, and the ibogan (type II, >80 alkaloids), such as catharanthine (Fig. 3.3). Furthermore, dimer formation is a common aspect of MIA biogenesis because about 300 dimers (or bisindole alkaloids) have been isolated, comprising two MIAs with either the same (homodimer) or different skeletons (heterodimers). The intermonomer link may concern either the tryptamine subunit, the secologanin subunit or both. Two of the most valuable alkaloids, vinblastine and vincristine, are isolated exclusively from *Catharanthus roseus* and widely used in anticancer chemotherapy (Fig. 3.1). These heterodimers are produced by coupling the plumeran vindoline to the ibogan catharanthine. This is not a rare process in *Catharanthus*: more than 20 homo- or heterodimers have been isolated from this medicinal plant (Szabó, 2008).

1.2. MIA distribution in angiosperm

More than 2500 chemical structures related to the MIAs described so far are widespread in a large number of plant species (Ziegler & Facchini, 2008). Some of these molecules present a high interest for human health such as the anticancer drugs vinblastine and vincristine and the antihypertensive drug ajmalicine specifically produced in *C. roseus*, the antiarrhythmic

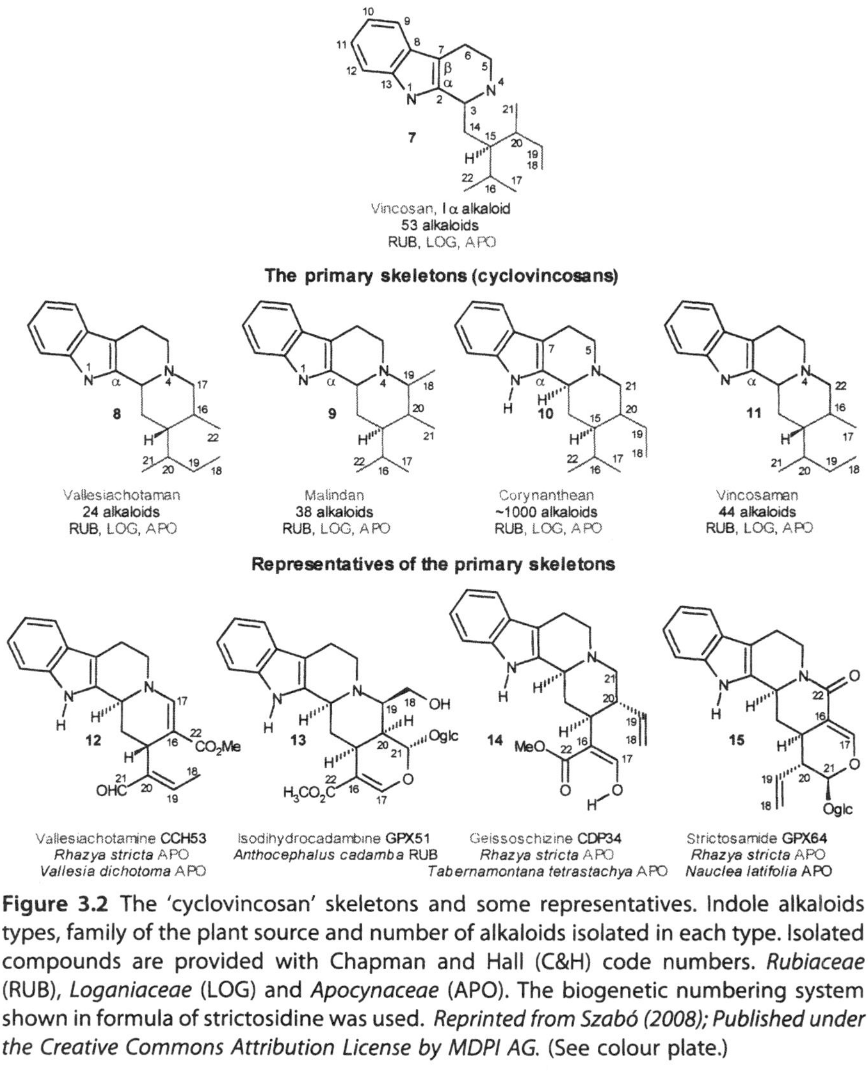

Figure 3.2 The 'cyclovincosan' skeletons and some representatives. Indole alkaloids types, family of the plant source and number of alkaloids isolated in each type. Isolated compounds are provided with Chapman and Hall (C&H) code numbers. *Rubiaceae* (RUB), *Loganiaceae* (LOG) and *Apocynaceae* (APO). The biogenetic numbering system shown in formula of strictosidine was used. *Reprinted from Szabó (2008); Published under the Creative Commons Attribution License by MDPI AG.* (See colour plate.)

ajmaline produced in *Rauvolfia serpentina* or the anticancer compound camptothecin mostly produced in *Camptotheca acuminata* (Guirimand, Courdavault, St-Pierre, & Burlat, 2010). These molecules are just a part of the large array of MIAs that a single plant species is able to produce. For example, there are more than 130 MIAs in *C. roseus* (van der Heijden, Jabos, Snoeijer, Hallard, & Verpoorte, 2004).

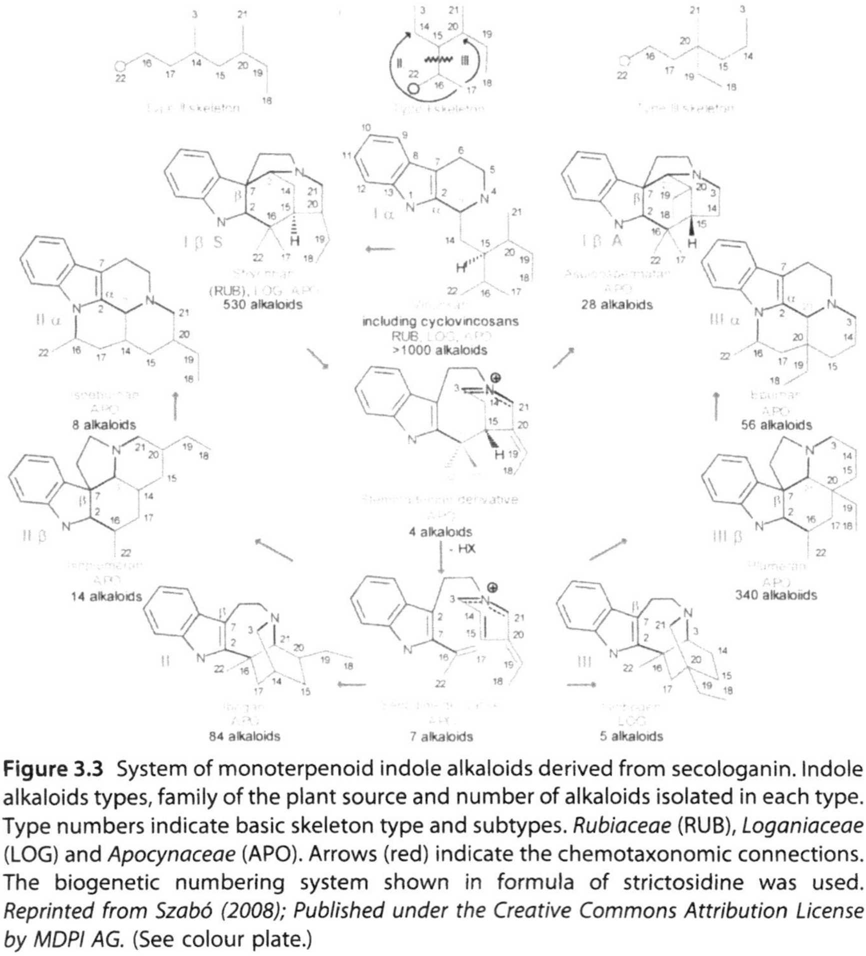

Figure 3.3 System of monoterpenoid indole alkaloids derived from secologanin. Indole alkaloids types, family of the plant source and number of alkaloids isolated in each type. Type numbers indicate basic skeleton type and subtypes. *Rubiaceae* (RUB), *Loganiaceae* (LOG) and *Apocynaceae* (APO). Arrows (red) indicate the chemotaxonomic connections. The biogenetic numbering system shown in formula of strictosidine was used. *Reprinted from Szabó (2008); Published under the Creative Commons Attribution License by MDPI AG.* (See colour plate.)

MIAs have been isolated from phylogenetically related taxons. They are mainly found in the Gentianales order but related Cornales and Garryales also bears indole alkaloid-producing families. The ‘quinoindole’ alkaloids group of MIAs represent about 100 natural products isolated from *Icacinaceae* (Garryales) and *Cornaceae* (Cornales, formerly *Nyssaceae*) families (APG, 2009; Szabó, 2008). Although *Icacinaceae* was associated with the distant Celastrales order in pre-APG taxonomy, it was later shown to be polyphyletic (Larsson, 2007). Camptothecin producing genera in *Icacinaceae* are now proposed to be part of *Icacinaceae sensu stricto* of the order Garryales, close to Gentianales (Fig. 3.4). The Ipecac alkaloids represent some 90 isolated natural products,

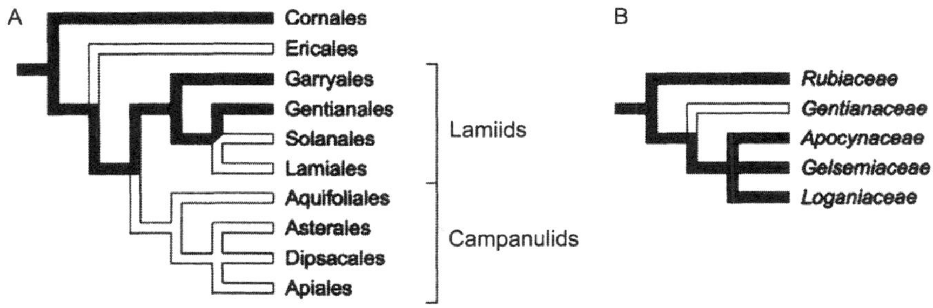

Figure 3.4 Mapping of camptothecin on the asterid orders (A) and the occurrence of monoterpene-indole alkaloids within the order Gentianales (B). MIAs are present in Gentianales families except in *Gentianaceae* and in the Apocynaceous genera previously treated as *Asclepiadacea. Reprinted from Larsson (2007). Copyright (2007); with permission from Elsevier.*

mainly in the *Rubiaceae* and *Cornaceae* (ex *Alangiaceae*) families. For instance, the roots of *Psychotria ipecacuanha* (*Rubiaceae*) are used as a source of emetic and an antiamoebic (Nomura, Quesada, & Kutchan, 2008).

The large group of MIAs with non-rearranged tryptamine moiety (more than 2000 natural products) were isolated mostly from the *Rubiaceae*, *Loganiaceae* and *Apocynaceae* families (Gentianales). Production of MIAs is a common character in Gentianales except in *Gentianaceae* and in the genera previously treated as the family *Asclepiadaceae* (Fig. 3.4; Larsson, 2007).

The primitive MIAs with a type I skeleton are found in *Rubiaceae*, *Loganiaceae* and *Apocynaceae*, while the more evolved MIAs with type II and III skeletons have a more restricted distribution. *Loganiaceae* produces Isoibogan type III MIAs, while all the other type II and type III MIAs were isolated from *Apocynaceae* (Fig. 3.3). Therefore, *Apocynaceae* is one of the major centres of MIA structural expansion.

1.3. MIA biosynthetic origin

Most alkaloids are derived through the decarboxylation of amino acid precursors to yield amines, or from anthranilic acid, nicotinic acid or primary amines. Complex alkaloids are elaborated by coupling these amines to different chemical partners, including organic acids and aldehydes, yielding central precursors of the diverse structural classes of complex alkaloids like isoquinoline, acridine, pyrrolizidine, tropane and MIAs. MIAs are designed around a common molecular skeleton, the central intermediate initiating their biosynthesis, strictosidine (Fig. 3.1). This first MIA results from the assembly of two precursors from different origins in the primary metabolism.

The indole skeleton is supplied by tryptamine following decarboxylation of tryptophan, a product of the shikimate pathway. The monoterpenoid moiety, secologanin, is provided by a complex reorganisation of the monoterpene geraniol produced from the primary metabolites DMAPP and IPP, which are supplied by the methyl erythritol phosphate (MEP) pathway. This reorganisation includes the cyclisation of a linear dialdehyde monoterpene, first into a bicyclic intermediate then into the monocyclic terpenoid-glucoside (or secoiridoid) secologanin. This conversion requires a set of unique cyclisation and ring opening reactions that produces the intermediate *cis–trans*-nepetalactol, and two glycosides, the iridoid loganin and secoiridoid secologanin. Iridoids constitute a diversified group of bicyclic monoterpene-glucoside widely distributed in the plant kingdom and well known for their herbivore deterrence activity.

The ongoing elucidation of some biosynthetic pathways illustrates the recruitment of enzymes belonging to recurrent multigene families such as cytochrome P450 monooxygenases, acyltransferases or methyltransferases. The elucidation of MIA biosynthetic pathways in *C. roseus* and *R. serpentina* has recently undergone major progress (Fig. 3.5). In these species, the pathways share a common origin with strictosidine synthase (STR) catalysing the stereospecific condensation of the indole precursor tryptamine with the terpenoid precursor secologanin to form the first MIA, (3α-H) strictosidine. The upstream biosynthesis of the indole precursor derivating from the shikimate pathway via tryptophan, and of the terpenoid precursor originating from the MEP pathway, is also shared within these plant species. Strictosidine β-glucosidase (SGD) catalysing the deglucosylation of strictosidine is the last common enzyme for the biosynthesis of 2000 MIAs, as the resulting aglycon is the starting point for many different, species-specific, lateral MIA pathways, with the observed possibility for a given species to harbour more than one of these pathways. It has to be noted that many enzymatic steps are yet to be discovered.

In contrast to the aerial plant parts of *C. roseus*, the roots do not produce vindoline but operate an alternate mechanism for tabersonine metabolism that results in the accumulation of lochnericine and horhammericine (Fig. 3.6; Laflamme, St-Pierre, & De Luca, 2001). The precise order of this root-specific tabersonine metabolism is not clear but includes two P450 monooxygenases for epoxidation by tabersonine 6, 7-epoxidase (Rodriguez, Compagnon, Crouch, St-Pierre, & De Luca, 2003) and side-chain hydroxylation by tabersonine 19-hydroxylase (Giddings et al., 2011), as well as a BAHD acyltransferase for the acetylation of the

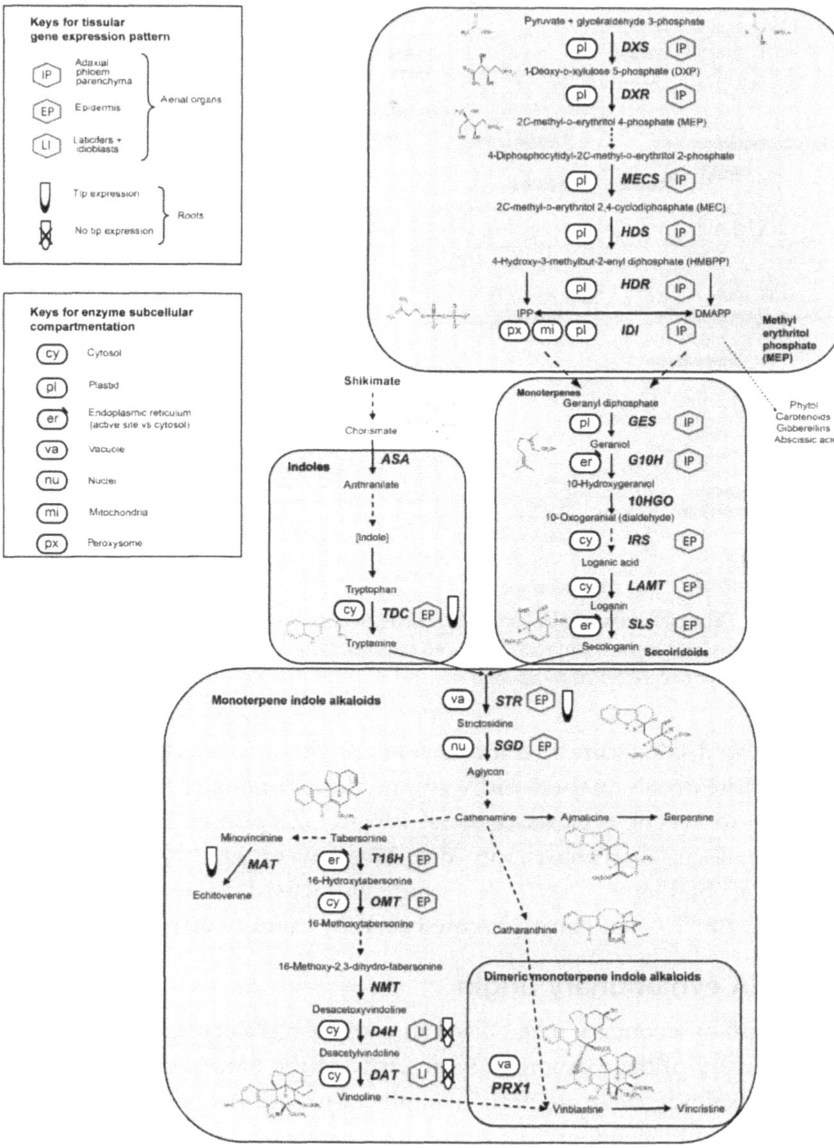

Figure 3.5 MIA biosynthetic pathway in *C. roseus*, subcellular compartmentation of enzymes and tissular localisation of transcripts. *(Data from Costa et al., 2008; Geu-Flores et al., 2012; Guirimand et al., 2009, 2012; Guirimand, Ginis, et al., 2011; Guirimand, Guihur, et al., 2011; Guirimand, Courdavault, Lanoue, et al., 2010; Levac et al., 2008; Mahroug et al., 2007; Murata et al., 2008; Oudin, Mahroug, et al., 2007; Simkin et al., 2013; Courdavault, unpublished results).* (See colour plate.)

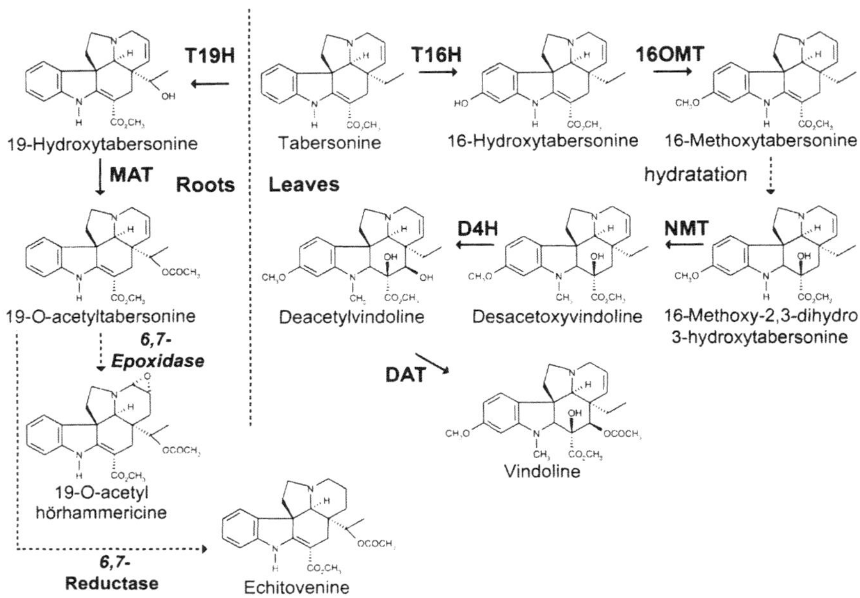

Figure 3.6 Tabersonine metabolism in *C. roseus*. In aerial parts, tabersonine is converted into vindoline biosynthesis. In roots and cell culture, tabersonine metabolism is oriented towards 19-hydroxylated derivatives.

hydroxylated side chain by minovincine acetyltransferase (Laflamme et al., 2001). More details on these pathways and the identified enzymatic steps are available in recent reviews (general review: Ziegler & Facchini, 2008; *C. roseus*-focused review: van der Heijden et al., 2004; Mahroug, Burlat, & St-Pierre, 2007; *R. serpentina*-focused review: Stöckigt & Panjikar, 2007; *C. accuminata*-focused review: Lorence & Nessler, 2004).

1.4. MIA evolutionary origin

The origin of secondary metabolism in evolution is a central question since the discovery of the extraordinary diversity of the natural products. In the cases of alkaloids, there are both examples of unique origin (monophyletic) and multiple independent origins (parallel/convergent) of their biosynthesis. An important step in the origin of a new alkaloid class is the acquisition of the metabolic activity for biosynthesis of its central intermediate. Monophyletic origin has been described as the acquisition of norcoclaurine synthase (NCS) at the origin of benzylisoquinoline alkaloids (Liscombe, MacLeod, Loukanina, Nandi, & Facchini, 2005). In contrast, the pyrrolizidine alkaloids' biosynthetic potential is an example of non-monophyletic origin. The common precursor homospermidine can be produced by deoxyhypusine synthase (DHS) present in all eukaryotes, an enzyme involved in activation of eukaryotic initiation factor 5A (Ober & Hartmann, 1999). Several distant families evolved independently a specific homospermidine synthase that is derived from the corresponding DHS (Reimann, Nurhayati, Backenköhler, & Ober, 2004).

Molecular evolution of key enzymatic steps in MIAs has not been thoroughly investigated but its restricted distribution in plant taxa argues for a monophyletic origin. The dual precursor required for MIA assembly requires a combination of molecular events. Production of tryptamine from tryptophan by tryptophan decarboxylase (TDC) is not a rare event in plants because it is part of the auxin biosynthesis pathway (Mano & Nemoto, 2012; Zhao, 2012). TDC is also involved in the formation of the monoamine serotonin (Grosse & Klapheck, 1979), and of simple alkaloids such as the β-carbolines, which are found in several plant families (Berlin, Rügenhagen, Kuzovkina, Fecker, & Sasse, 1994). Biosynthetic capability for the second precursor secologanin has a more restricted distribution in the plant kingdom although it is at the origin of a large number of nitrogen-free natural compounds. Secoiridoids are characteristic of the whole order Gentianales, including *Apocynaceae*, *Gentianaceae*, *Loganiceae*, *Rubiaceae* (Jensen, 1992; Jensen & Schripsema, 2002). They are also present in the order Garryales (*Icacinaceae sensu stricto*), Scrophulariales (*Oleaceae*), Dipsacales (*Caprifoliaceae*), Lamiales (*Verbenaceae*) and Cornales (*Hydrangaceae*, *Cornaceae* including *Nyssaceae*) (taxa according to APGIII taxonomy; Dinda, Debnath, & Harigaya, 2007b). These taxa represent a source of huge chemodiversity and have phylogenetic congruence. Simple secoiridoids, terpene-conjugated secoiridoids, aromatic-conjugated secoiridoids and bis-, tris- and tetrakis-secoiridoids have been isolated (Dinda et al., 2007b). Many are potent herbivore deterrents.

A single cytochrome P450 enzyme is required for production of secologanin in an iridoid-producing species. In *C. roseus*, this enzyme is a member of the CYP72 family (Irmler et al., 2000). The CYP72 family belongs to a large multifamily clan of P450s in plants and is associated with the metabolism of a diversity of compounds including terpenoids (Nelson & Werck-Reichhart, 2011). One member of this clan has acquired a new function, likely following duplication of an iridoids-specific P450, to catalyse this unique ring opening reaction.

Because secologanin biosynthesis requires the iridoid loganin, the metabolic activities to produce an iridoid were essential in the evolution of MIAs. Iridoids are more widespread than secoiridoids in plants. Iridoids are present in several families of Asterids like *Verbenaceae*, *Scrophulariaceae*, *Rubiaceae*, *Caprifoliaceae*, *Bignoniaceae*, *Lamiaceae*, *Oleaceae*, *Acanthaceae*, *Plantaginaceae*, *Loganiaceae*, *Gentianaceae*, *Valerianaceae* (Dinda, Debnath, & Harigaya, 2007a). Iridoid cyclase is the key step for production of an iridoid from a widespread monoterpene precursor. This unique enzyme was recently shown in *C. roseus* to have a distinct origin from the regular terpene cyclase (Geu-Flores et al., 2012). Terpene cyclic structures are normally formed by reactions using prenyl diphosphates as the substrate. In contrast, iridoid cyclase use 10-oxogeranial and probably couples an initial NAD(P)H-dependent reduction step with a subsequent cyclisation step to form the ring structure. A short-chain reductase was likely recruited early in the evolution of the iridoid biosynthesis. Following step in the molecular evolution of MIA biosynthesis is expected to be the acquisition of an enzymatic activity to associate tryptamine and secologanin in a stereospecific enzymatic 'Pictet–Spengler-type' condensation that forms strictosidine. STR exhibits high substrate specificity, accepting only a few substrates (Stöckigt, Barleben, Panjikar, & Loris, 2008). A conservative view would associate this acquisition at the evolution of Gentianales. Gentianales may be some 89–83 million year old (Wikström, Savolainen, & Chase, 2001), or older (109–93 million year old) for the stem group (Janssens, Knox, Huysmans, Smets, & Merckx, 2009). STR sequences from *R. serpentina*, *C. roseus* and *Ophiorhiza pumila* are highly conserved (50–79% identity). Crystal structure of STR from *R. serpentina* represents an example of a six-bladed, four-stranded β-propeller fold (Ma, Panjikar, Koepke, Loris, & Stöckigt, 2006). STR has also distant sequence similarity to a large family of conserved sequences in plant and animal genomes, which form various groups. One of them carries diverse catalytic activity including paraoxonase; lactonase and related hydrolytic reactions, while the so-called strictosidine synthase-like (SSL) subgroup has many representatives in all plant genomes but has no known function (Hicks et al., 2011). Functional divergence at the origin of the SSL subgroup includes the loss of metal-coordinating residues which are conserved in the hydrolases group. A large number of SSL sequences have been annotated as STR or SSL. However, most members of the SSL group likely catalyse hydrolytic reactions rather than the condensation typical in STR (Hicks et al., 2011). While a SSL from *Arabidopsis* might be active on secologanin, yielding an unknown product proposed to be a dimer (Kibble et al., 2009), a grape SSL has probably hydrolytic activity (Hicks et al., 2011). In the course of STR evolutionary origin, the ancestral sequence appears to have acquired distinct properties from the ubiquitous *SSL* genes.

Pictet–Spengler-type condensations are rare in metabolism; only two other examples of enzymes catalysing this reaction are known (Stöckigt et al., 2008). The first involves the condensation of dopamine to secologanin, the central step of tetrahydroisoquinoline monoterpenoid (Ipecac) alkaloids. It is catalysed by deacetylipecoside synthase (DIS) and deacetylisoipecoside synthase, leading, respectively, to 1-(*R*)-deacetylipecoside and 1-(*S*)-deacetylisoipecoside. Although their amino acid sequence is not known, DIS shares several biochemical characteristics with STR, except for the amine substrate specificity (De-Eknamkul, Suttipanta, & Kutchan, 2000). DIS is therefore expected to share sequence similarity to STR. The second example is NCS, which is the first committed step in the biosynthesis of benzylisoquinoline alkaloids. NCS catalyses the condensation of dopamine and 4-hydroxyphenylacetaldehyde, leading to (S)-norcoclaurine (Samanani, Liscombe, & Facchini, 2004). Although the reaction mechanism seems to be analogous to STR, amino acid sequence alignment shows no sequence identity between NCS and STR (Stöckigt et al., 2008). NCS gene appears to have been recruited from the PR10a family of plant defence proteins.

The last common step in monoterpenoid indole alkaloid biosynthesis involves hydrolysis of strictosidine yielding a reactive aglycone with two aldehyde functional groups. SGD stereoselectively catalyses the deglucosylation of strictosidine, but not its 3β(R)-epimer vincoside, which is consistent with the occurrence of only strictosidine in *Catharanthus* and *Rauvolfia*. Genes encoding SGD have been cloned (Geerlings, Ibanñez, Memelink, van der Heijden, & Verpoorte, 2000; Gerasimenko, Sheludko, Ma, & Stöckigt, 2002). The stereospecific glucosidase was found most similar to the Ipecac alkaloid β-D-glucosidase from *P. ipecacuanha* (Nomura et al., 2008). Phylogenetic analysis with other plant family 1 glycosyl hydrolases showed closer relationship within the group of Gentianales alkaloid β-glucosidases compared to more distant enzymes involved in cyanogenic heterosides hydrolysis.

The molecular bases of later enzymatic steps in MIA biosynthesis are less well known, with a few exceptions. The molecular rearrangement and complex cyclic organisation of strictosidine aglycone into ajmaline (ajmalan, type I skeleton) is known in details in *R. serpentina*. In addition, the terminal stages of vindoline biosynthesis, from tabersonine, a plumeran (type III β skeleton) is almost completely described in *C. roseus*.

Although a large body of phytochemical analysis documents the widespread distribution of vincosan-type skeleton within Gentianales, and the occurrence of type II and type III skeletons, essentially in the *Apocynaceae* family, the enzymatic reactions involved in the conversion of type I precursors into type II and III skeletons are poorly characterised. The extraordinary chimiodiversification of MIAs in Gentianales and Cornales shows a gradient of specialisation from the widespread molecular skeletons to the more restricted rearranged skeletons. The adaptative forces behind the chemical diversification of MIAs might be the reflect of the extraordinary evolutionary success of these plant families, *Rubiaceae* (611 genera, 13,548 species) and *Apocynaceae* (415 genera, 5031 species) rank both in the top 10 largest families in the plant kingdom (www.theplantlist.org).

2. REGULATION OF MIA BIOSYNTHESIS PATHWAY

2.1. Developmental control in plant

MIA biosynthesis genes are transcribed early in the *C. roseus* organ development of both shoot and roots. The pathway is activated in young seedlings, in primordia of floral organs, in the very young stem and in young leaves, as well as in the proliferation zone of the root apex (St-Pierre, Vázquez-Flota, & De Luca, 1999). In addition, transcripts are less abundant in mature tissues, which leads to a gradient of expression, for instance, between the tip and base of young leaves, as well as between developing and mature organs. However, enzyme and enzymatic activity persist in tissues for some time after transcript down-regulation (St-Pierre et al., 1999). In addition to a developmental control at organ level, the MIA biosynthetic pathway is subjected to cell- and tissue-specific regulation (see Section 3).

2.2. Environmental factors

2.2.1 Light

Specific branches of MIA biosynthesis have been shown to be regulated by light under a phytochrome-dependent process. The tabersonine to vindoline conversion is regulated by light at different developmental stages of the plants (Campos-Tamayo et al., 2008; De Luca et al., 1986; Vazquez-Flota, St-Pierre & De Luca, 2000). Etiolated seedlings accumulate tabersonine and 16-methoxytabersonine, while shifting to light induces enzymatic activities in cotyledons required for the conversion into vindoline. (St-Pierre, Laflamme, Alarco, & De Luca, 1998; Vazquez-Flota & De Luca, 1998; Vazquez-Flota, De Carolis, Alarco, & De Luca, 1997). UV light is also known to induce MIA accumulation through activation of biosynthetic genes in *Catharanthus* cell culture (Ouwerkerk,

Hallard, Verpoorte, & Memelink, 1999). Accumulation of MIAs in the leaf epidermis might act as a UV filter to protect internal tissues (Mahroug, Courdavault, Thiersault, St-Pierre, & Burlat, 2006).

2.2.2 Biotic stress

Fungal elicitors, like yeast extract and culture filtrate of phytopathogenic fungus *Pythium aphanidermatum*, induces MIA biosynthesis and biosynthetic enzymes in cell culture, linking MIAs to plant defence reactions (DiCosmo et al., 1987; Eilert, Constabel, & Kurz, 1986; Menke, Parchmann, Mueller, Kijne, & Memelink, 1999). Wounding also induces MIA biosynthesis in seedlings (Vázquez-Flota, Carrillo-Pech, Minero-García, & Miranda-Ham, 2004).

2.3. Hormonal controls

Due to the complexity of the MIA biosynthetic pathway and considering the large number of enzymatic steps and their low biosynthetic rates *in planta* (Mahroug et al., 2007; Oudin, Courtois, Rideau, & Clastre, 2007), simplified models of *C. roseus,* such as cell suspension, have emerged as powerful tools to study the regulation of MIA biosynthesis, including the role of hormonal signalling (Giglioli-Guivarch et al., 2006; Hedhili, Courdavault, Giglioli-Guivarc'h, & Gantet, 2007; Fig. 3.7). For example, in the *C. roseus* C20D cell line, auxin provided as 2,4-dichlorophenoxyacetic acid (2,4-D) inhibits MIA production (Arvy, Imbault, Naudascher, Thiersault, & Doireau, 1994) as a consequence of the down-regulation of the first two MEP pathway genes (Chahed et al., 2000; Veau et al., 2000), coding for 1-deoxy-D-xylulose 5-phosphate synthase (DXS) and 1-deoxy-D-xylulose 5-phosphate reductoisomerase (DXR), and the down-regulation of the secoiridoid pathway gene-encoding G10H (Papon et al., 2005). In contrast, in auxin-starved medium, cytokinins increase MIA biosynthesis (Décendit et al., 1992) and enhance at least *DXS, DXR* and *G10H* expression (Oudin, Mahroug, et al., 2007). Furthermore, synergistic interaction between cytokinin and ethylene transduction pathways has been reported, because the addition of these two hormones further enhances G10H expression (Papon et al., 2005).

Jasmonate (JA) is well known to induce a variety of secondary metabolites in various plant species, including alkaloids, terpenoids and phenylpropanoids (Memelink, Verpoorte, & Kijne, 2001). JA or its methyl ester, methyl jasmonate (MeJA) also appears to function as a central regulator of MIA biosynthesis in *C. roseus* (Gantet, Imbault, Thiersault, & Doireau,

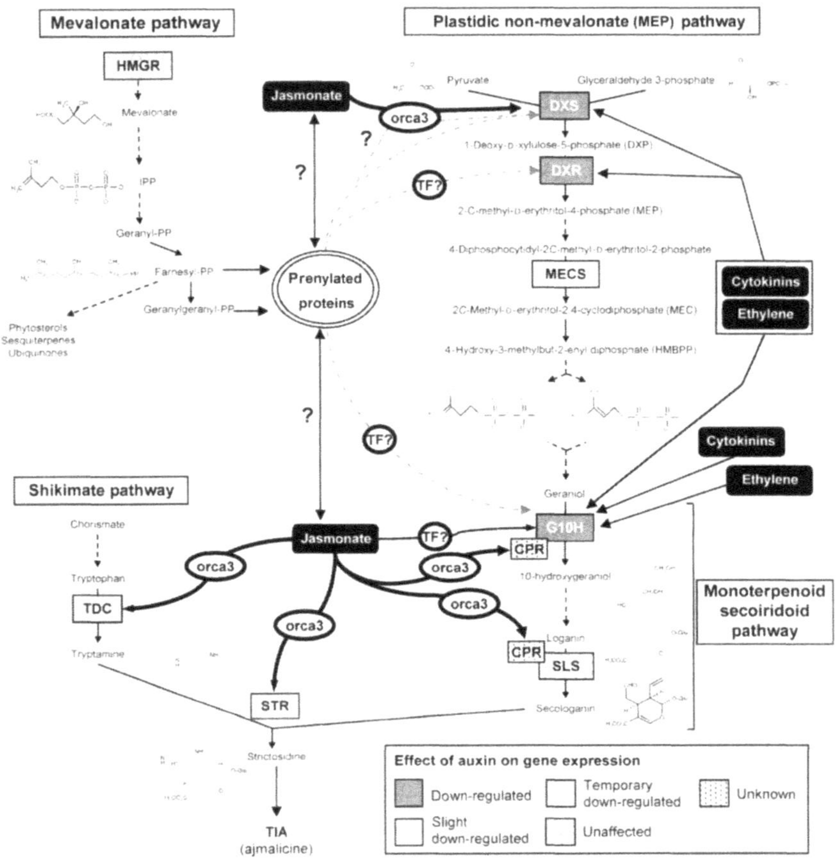

Figure 3.7 Proposed regulatory network between mevalonate biosynthesis pathway, protein prenyltransferases, monoterpenoid indole alkaloids biosynthesis pathway, hormonal signals and gene regulation in *Catharanthus roseus*. In the regulatory network, arrows represent a component of the network activating another component of the network. Auxin effects are indicated by shading of boxes around the enzyme names. Orca3, Orca3 transcription factor; TF, hypothetical transcription factor. Hypothetical links between components of the network are notified with '?'. *Reprinted from Hedhili et al. (2007); with kind permission from Springer Science and Business Media.*

1998). Additionally, both plants and seedlings of *C. roseus* respond to JA with an increased production of MIA (Aerts, Gisi, De Carolis, De Luca, & Baumann, 1994; El-Sayed & Verpoorte, 2005), whereas inhibition of JA biosynthesis in suspension cell lines blocks MIA production (Gantet et al., 1998). In *C. roseus* cell suspension, application of exogenous MeJA leads to a coordinate up-regulation of all characterised genes associated with the biosynthesis of the terpenoid precursor of MIA (Oudin, Mahroug, et al., 2007; van der Fits & Memelink, 2000) and to the subsequent increase in MIA biosynthesis. The way by which JA regulates MIA gene expression has been partially elucidated by the isolation of two specific AP2/ERF-domain transcription factors, ORCA2 and ORCA3 (Memelink et al., 2001; Menke, Champion, Kijne, & Memelink, 1999; van der Fits & Memelink, 2000). ORCA3 exhibits the most pleiotropic effects by enhancing the expression of genes involved in many steps of MIA biosynthesis, such as terpenoid and indole precursor biosynthesis, strictosidine synthesis and modification (van der Fits & Memelink, 2000). In the model plant *Arabidopsis*, JA signalling involves SCF^{COI1}-mediated ubiquitination of regulatory proteins that control the transcription of JA-responsive genes (Devoto & Turner, 2005; Howe & Jander, 2008). The JA intracellular active form, (+)-7-iso-JA-L-Ile, promotes ubiquitination and subsequent degradation of JAZ repressor proteins via the 26S proteasome (26S), resulting in the derepression of transcription factors and the expression of early response genes (Thines et al., 2007). Degradation of JAZ repressor releases AtMYC2 from an inhibitory complex with JAZ. In this context, JAZ protein was shown to interact with AtMYC2, a bHLH transcription factor that regulates JA-early response genes. In the particular case of *C. roseus*, the bHLH protein CrMYC2 is able to regulate ORCA gene expression (Zhang et al., 2011), which has consequence on MIA genes regulation.

In the *C. roseus* cell line, cytokinins were shown to stimulate MIA biosynthesis. The addition of zeatin to an auxin-depleted medium of *C. roseus* cell suspension enhanced the MIA production while zeatin was unable to overcome the inhibition of MIA biosynthesis induced by auxin (Décendit et al., 1992). This enhancement of MIA biosynthesis by zeatin was correlated with a great increase of G10H expression (Papon et al., 2005). In contrast, in these experimental conditions, the expression of DXS, DXR and MECS remained constant. MIA biosynthesis is also enhanced in suspension cells treated with ethylene (Yahia et al., 1998). Moreover, a strong synergic effect of these two hormones on the activation of *G10H* gene expression level was observed. When zeatin and ethylene were added together to the culture medium, the mRNA level of MEP pathway genes coordinately increased in suspension cells (Papon et al., 2005). These data suggested that these two hormones stimulate the MIA biosynthesis by two distinct mechanisms (Yahia et al., 1998). In addition, zeatin stimulates the bioconversion of exogenous secologanin in the MIA ajmalicine, suggesting that cytokinins may also act on other downstream enzymatic steps of the MIA biosynthesis pathway (Décendit et al., 1992).

In plants, cytokinin signal transduction operates through a multistep phosphorelay involving cytokinin receptors (HKs), phospho-transfer proteins (HPTs) and response regulators (RRs) among which a subfamily (type-β RRs) corresponds to transcription factors regulating the transcription of several genes (Hwang, Chen, & Sheen, 2002; To & Kieber, 2008). Recent findings suggest a possible link between the cytokinin signalling cascade and the regulation of genes implicated in the MIA biosynthesis. In periwinkle cell suspensions, the down regulation of CrHPt1, a histidine phosphotransferase protein, prevents the up-regulation by cytokinins of *CrDXR* and *CrG10H*, two genes encoding key enzymes of the MEP and the secoiridoid pathways, respectively (Amini, Andreu, Glévarec, Rideau, & Crèche, 2012). Additionally, CrRR5, a type-β RR drives the expression of *CrHDS*, encoding the hydroxymethylbutenyl diphosphate synthase, another major enzyme of the MEP pathway (Ginis et al., 2012). CrRR5 is the first identified transcription factor mediating the CK signalling that targets a gene from the MEP pathway involved in isoprenoid metabolism.

Gibberelic acid was found to antagonise, in a dose-dependent manner, the stimulation of ajmalicine biosynthesis by cytokinins in cell suspension (Amini, Glévarec, Andreu, Rideau, & Crèche, 2009). Moreover, low concentrations of the gibberellin biosynthesis inhibitor paclobutrazol could reverse the inhibitory effects of low auxin levels on ajmalicine accumulation in the cells. Gibberellic acid also inhibited the accumulation of vinblastine, vindoline and catharanthine in the *C. roseus* plant (Pan et al., 2010). Abscisic acid (ABA) could stimulate the accumulation of catharanthine and vindoline in *C. roseus* (Smith, Smart, Kurz, & Misawa, 1987).

2.4. Intracellular signalling

2.4.1 Calcium signalling

Because auxin inhibits MIA biosynthesis in cell culture, predominantly at secologanin biosynthesis level, downstream signalling has been investigated (Poutrain, Mazars, Thiersault, Rideau, & Pichon, 2009). Changes in free cytoplasmic calcium concentration ($[Ca^{2+}]_{cyt}$) occur during many physiological processes and particularly in auxin signalling (Singla, Chugh, Khurana, & Khurana, 2006). A pharmacological approach deciphered the role of calcium as a second messenger in the transduction pathway leading to the inhibitory effect of 2,4-D, in regulating MIA biosynthesis in *C. roseus* cells (Poutrain et al., 2009). Auxin-dependent MIA biosynthesis was shown to be differentially regulated by two distinct calcium-release components from internal stores showing pharmacological profiles similar to those displayed by animal RyR and IP3 channels. MIA biosynthesis is stimulated by caffeine (Ca^{2+}-release activator through RyR channels) and by heparin and TMB8 (Ca^{2+}-release inhibitors of IP3 channels), whereas MIA biosynthesis is inhibited by mastoparan (Ca^{2+}-release activator of IP3 channels) and by ruthenium red and DHBP (Ca^{2+}-release inhibitors of RyR channels). Furthermore, calcium acts on MIA biosynthesis by regulating the monoterpene moiety of the MIA biosynthesis pathway because calcium channel modulators preferentially modulate *G10H* expression, a gene involved in the secoiridoid monoterpene pathway. This finding suggests an opposite and coordinated action of multiple Ca^{2+}-release pathways in 2,4-D signal transduction, adding a new level of complexity to calcium signalling in plants.

2.4.2 Involvement of protein prenylation events in MIA biosynthesis regulation

Several evidences indicate that isoprenylated proteins are involved in the regulation of MIA biosynthesis. Previous work, designed to study the role of the MVA pathway in MIA biosynthesis showed, unexpectedly, that MIA biosynthesis in the *C. roseus* cell line was completely inhibited when the MVA pathway was blocked with an inhibitor of HMG-CoA reductase (Imbault, Thiersault, Dupéron, Benabdelmouna, & Doireau, 1996). Because mevalonate is not a precursor for the biosynthesis of the isoprenoid moiety dedicated to MIA (Contin, van der Heijden, Lefebre, & Verpoorte, 1998), the HMG-CoA reductase inhibitor was expected to have an indirect effect on MIA biosynthesis. However, this class of inhibitors is also known to interfere with protein isoprenylation by depleting the endogenous pool of prenyl precursors.

Protein isoprenylation in animals has received considerable attention because several oncogenic forms of isoprenylated proteins depend on prenyl modifications for their cellular multiplication effects and for tumour development (Gelb et al., 2006). Protein isoprenylation consists of a post-translational modification by the formation of cysteine thioether bonds with a farnesyl (C_{15}) or geranylgeranyl (C_{20}) moiety at the carboxy terminus. The corresponding terpenoid substrates of protein prenyltransferase are farnesyl diphosphate (FPP) and/or geranylgeranyl diphosphate (GGPP). Although FPP and GGPP precursors for prenylation originate from the MVA pathway in animals, in the plant system the situation may be different for GGPP. Recently, it was demonstrated that geranylgeranylation in tobacco BY-2 cells clearly depends on the MEP pathway (Gerber et al., 2009). Two different protein geranylgeranyl transferases (PGGT-I and Rab PGGT-II) and a single protein farnesyl transferase (PFT) catalyse the isoprenylation reactions in animal, yeast and plant cells (Rodríguez-Concepción et al., 1999; Schafer & Rine, 1992). PFT and PGGT-I are classified in a common CaaX-prenyltransferase (CaaX-PTase) family. In plants, PFT has been implicated in several regulation mechanisms such as cell division, meristem cellular differentiation, flower development and in the auxin signalling pathway (Courdavault, Clastre, Simkin, & Giglioli-Guivarc'h, 2013). In contrast, the physiological functions of PGGT-I remains less understood. PGGT-I negatively regulates ABA signalling in guard cells and auxin signalling leading to lateral root initiation without affecting other ABA or auxin response pathways (Courdavault et al., 2013).

Involvement of protein isoprenylation in the regulation of MIA biosynthesis has been investigated by applying a CaaX-PTase downexpression strategy. The *C. roseus* genes encoding the β subunits of each CaaX-PTase have been cloned. Subsequently, *C. roseus* cell lines were generated for RNA interference targeting PFT or PGGT-I: RNAi PFT and RNAi PGGT-I cell lines (Courdavault, Burlat, St-Pierre, & Giglioli-Guivarc'h, 2005). Both CaaX-PTase activities were required for the expression of *DXS*, *DXR* and *G10H*, involved in the early steps of MIA biosynthesis, while no decrease in the central steps of MIA biosynthesis, secologanin synthase (*SLS*), *STR* and *TDC* transcript levels was observed (Courdavault, Thiersault, Courtois, Gantet et al., 2005). These results are consistent with the role of PFT- and PGGT-I-dependent protein isoprenylation in the cascade of events leading to an efficient transcriptional activation of early steps in monoterpenoid biosynthesis. It can be also noted that in the aerial organs of young *C. roseus* plants, the prenylated protein-dependent genes *DXS*, *DXR* and *G10H*, were found to be expressed specifically in internal phloem parenchyma cells (Burlat et al., 2004), whereas *SLS*, *TDC* and *STR* expression was restricted to epidermal cells (Irmler et al., 2000; St-Pierre et al., 1999). Such tissue-specific expression in phloem parenchyma potentially involves a coordinated regulatory process, which may require CaaX-PTase activity.

In an attempt to better understand the involvement of CaaX-PTases in MIA biosynthesis, the potential role of protein isoprenylation in the JA signalling cascade was investigated. Isoprenylated proteins have been suggested to be involved in part of the JA signalling pathway (Trusov et al., 2006) through the characterisation of *Arabidopsis* αβγ heterotrimeric G protein mutants. The two G protein γ subunits (Gγ) of *Arabidopsis* are small proteins that bear a PGGT-I-specific CaaX motif. The activity of heterotrimeric G proteins depends on the isoprenylation status of Gγ as *Arabidopsis* mutants, bearing a CaaX-truncated γ subunit, present altered responses to the physiological process involving heterotrimeric G proteins (Chakravorty & Botella, 2007). In *C. roseus*, the inhibition of endogenous protein isoprenylation by applying the CaaX-PTase inhibitor S-perillyl alcohol, blocked the up-regulation of *ORCA3* induced by MeJA (Courdavault et al., 2005). These data point to a specific role of protein isoprenylation in JA signalling as ORCA3 was specifically involved in the early steps of JA signalling. To determine which CaaX-PTase activity is involved in JA signalling, MeJa signal transduction was tested in the specific RNAi PFT or RNAi PGGT-I inhibited *C. roseus* cell lines. Silencing of PFT in *C. roseus* cell lines activity does not alter *ORCA3* expression, indicating that farnesylated proteins are not likely to play a part in JA signalling. In contrast, the depletion of PGGT-I activity leads to a down-regulation of *ORCA3* expression to a similar extent as previously observed in *S*-perillyl alcohol-treated cells (Courdavault, Burlat, St-Pierre, & Giglioli-Guivarc'h, 2009). This points to a positive and specific action in JA signalling of proteins that are isoprenylated by PGGT-I.

3. SPATIAL ORGANIZATION OF MIA BIOSYNTHESIS

3.1. Compartmentation of metabolites

Early works suggested that MIAs accumulate in specialised cells in *C. roseus*. Yoder and Mahlberg (1976) used chemical indicators to identify laticifers and 'specialised parenchyma cells' as the sites of alkaloid accumulation in *C. roseus*. Latex could be collected from *C. roseus* fruits and was shown to contain various MIAs (Eilert et al., 1985). Direct observation of *C. roseus* leaves by epifluorescence microscopy showed the random distribution of cells throughout the mesophyll that displayed distinctive autofluorescent properties, including laticifer and idioblast (Mersey & Cutler, 1986). Leaf sections and protoplast preparations revealed the presence of larger yellow autofluorescent cells with few chloroplasts, compared to the surrounding red autofluorescent mesophyll cells. These 'idioblast' cells, which occur in several plant families, may be associated with the biosynthesis and accumulation of secondary products (Mersey & Cutler, 1986). In *C. roseus*, the protoplasts isolated from idioblast and laticifer cells were shown to be enriched in MIAs compared to the other mesophyll cells (Mersey & Cutler, 1986). More recently, epidermis abrasion with carborundum and surface extraction were used to reveal alkaloid localisation in *C. roseus* leaves. Analysis of abraded leaves showed enrichment of tabersonine and 16-methoxytabersonine in epidermis extract while levels of vindoline and catharanthine were very low in comparison to whole leaf extracts, suggesting their likely accumulation in the central part of the leaf (Murata & De Luca, 2008; Murata, Roepke, Gordon, & De Luca, 2005). This is in agreement with mesophyll localisation of the last two steps of vindoline biosynthesis within laticifer and idioblast cells. However, a recent study of the *C. roseus* leaf surface composition by selective extraction of metabolites following dipping in organic solvent have shown that catharanthine is mostly present as exudates in the wax layer of the leaf surface (Roepke et al., 2010), together with triterpene ursolic acid (Murata & De Luca, 2008). The chloroform soluble fraction of leaf surface contained nearly 100% of catharanthine content along with 3–5% of vindoline in comparison to the whole leaf extract.

3.2. Compartmentation of MIA biosynthesis in specialised cells

Recently, the spatial organisation of alkaloid biosynthesis has been extensively investigated using *in situ* hybridisation and immunocytochemistry methods (De Luca & St-Pierre, 2000; Kutchan, 2005; Mahroug et al., 2007; Ziegler & Facchini, 2008). An astonishing complexity has been uncovered showing multicellular organisations as a recurrent common feature. These types of organisation implicate the necessity of intercellular translocation processes. In *C. roseus,* a series of publications showed the sequential involvement of three types of tissues during MIA biosynthesis: (1) internal phloem-associated parenchyma (IPAP), (2) epidermis and (3) laticifers–idioblasts (Fig. 3.5; Burlat et al., 2004; Irmler et al., 2000; Oudin, Mahroug, et al., 2007; St-Pierre et al., 1999). The IPAP cells harbour the expression of genes involved in early steps of monoterpenoid biosynthesis, i.e., four MEP pathway genes and *geraniol 10-hydroxylase* (G10H, CYP76B6) encoding the first committed enzyme in monoterpenoid biosynthesis (Burlat et al., 2004; Oudin, Mahroug, et al., 2007). The intermediate steps leading to the synthesis of the two MIA precursors, tryptamine and secologanin, and to their subsequent condensation to form the first MIA strictosidine occur within the epidermis (Irmler et al., 2000; St-Pierre et al., 1999). Finally, the last two steps in the biosynthesis of vindoline, one of the monomeric MIA precursors of the dimeric MIA vinblastine, are localised to specialised laticifer–idioblast cells (St-Pierre et al., 1999). Recently, these results were elegantly enhanced by an RT-PCR analysis of laser-captured microdissected *C. roseus* leaf cells (Murata & De Luca, 2008) and by an EST analysis study of epidermis-enriched fractions obtained using an original carborundum abrasion technique (Levac, Murata, Kim, & De Luca, 2008; Murata, Roepke, Gordon, & De Luca, 2005). Furthermore, the accumulation and secretion of catharanthine in the leaf wax layer suggest that catharanthine biosynthesis operates in leaf epidermis (Roepke et al., 2010). Taken together, these results suggest that to insure a continuity in the metabolic flux along the MIA pathway, it is necessary to consider the translocation of an unknown monoterpenoid intermediate from IPAP to epidermis and the translocation of an unknown MIA intermediate from epidermis to laticifer–idioblast (Fig. 3.5). The identification of these shuttling intermediates necessitates the localisation of two subsequent enzymatic steps within two different cell types.

In contrast to the aerial parts, few studies have investigated the compartmentation of MIA biosynthesis in underground tissues. In the root, TDC, STR and MAT transcripts are localised in protoderm and cortical cells around root apical meristem (Laflamme et al., 2001; St-Pierre et al., 1999). Coexpression of TDC and STR along with MAT within cortical tissue of the transformed hairy root lines of *C. roseus* suggests that the complete MIA pathway for tabersonine synthesis might be operational in a single cell type. This root unicellular compartmentation model would contrast with the multicellular cooperation for MIA biosynthesis in aerial organs. A unicellular compartmentation model would also better explain how dedifferentiated cell cultures have MIA biosynthetic potential. However, this model needs to be confirmed by localisation of MEP pathway and other enzymatic steps in *C. roseus* roots.

The spatial organisation of alkaloid biosynthesis was recently investigated in *C. acuminata*. This medicinal plant from the *Nyssaceae* family produced alkaloids with structural similarity to MIAs, however, the spatial organisation of camptothecin biosynthesis was found to be unique. Camptothecin accumulates in idioblast cells of leaf parenchymatic and epidermal tissues and also in glandular trichomes but not in laticifer cells (Pasqua, Monacelli, & Valletta, 2004). However, gene expression of TDC, involved in camptothecin synthesis, was detected in leaf idioblast cells but not in glandular trichomes, which suggests translocation of camptothecin between cells (Valletta, Trainotti, Santamaria, & Pasqua, 2010). Although, the spatial organisation of MIA biosynthesis in *C. roseus* does not appear to be conserved in species producing isoquinoline terpenoid alkaloids, the synthesis of alkaloids at the plant surface is preserved at the epidermis and accompanied by above-surface structures like the trichome.

3.3. Subcellular organisation of the pathway

Alkaloid biosynthesis also involves complex subcellular compartmentation of biosynthetic enzymes and transmembrane movement of metabolites. Beside cytosol, organelles such as ER (either lumen or membranes), plastid (either stroma or thylakoids), mitochondrion and vacuole have been implicated in various alkaloid biosynthetic pathways (for reviews, see Facchini & St-Pierre, 2005; Mahroug et al., 2007; Ziegler & Facchini, 2008). Some of these results are based on *in silico* analysis of biosynthetic enzyme sequences but also on more formal experimental evidences. Initially, localisation was based on organelle fractionation in density gradient and direct localisation of enzymes by immunogold (De Luca & Cutler, 1987; McKnight, Bergey, Burnett, & Nessler, 1991), more recently GFP-fusion imaging analysis allowed a systematic (re)evaluation of the subcellular localisation of all the enzymes available in a given alkaloid pathway. A more complete spatial compartmentation model now integrates both cellular and subcellular levels (Costa et al., 2008; Guirimand et al., 2009, 2012; Guirimand, Ginis, et al., 2011; Guirimand, Guihur, et al., 2011; Guirimand, Courdavault, Lanoue, et al., 2010; Guirimand, Courdavault, St-Pierre, & Burlat, 2010).

Besides cytosol, five subcellular compartments have been implicated in MIA biosynthesis, namely: plastids, vacuole, endoplasmic reticulum, mitochondria and nucleus (Fig. 3.5). TDC, IRS, LAMT, 16OMT, desacetoxyvindoline-4-hydroxylase (D4H) and deacetylvindoline-4-*O*-acetyltransferase (DAT) enzymes essentially operate in the cytosol (De Luca & Cutler, 1987; Guirimand, Ginis, et al., 2011; Guirimand, Guihur, et al., 2011; Vazquez-Flota et al., 1997).

Plastids appear to be pivotal as the source of the two primary precursors required for MIA biosynthesis. Tryptophan synthesis in plants is known to be localised exclusively in the plastids, and IPP formed by the MEP pathway for biosynthesis for monoterpene residue has also been localised in this compartment following immunolocalisation of HDS, and targeting of GFP-fusion protein (Guirimand et al., 2009, 2012; Oudin, Mahroug, et al., 2007). Formation of the monoterpene precursor by geraniol synthase has also been assigned to the plastid by functional analysis of targeting signals

(Simkin et al., 2013). Tryptophan and geraniol need, then, to be transported out of the plast by an unknown process. Formation of long tubular extensions budding from plastids (stromules), loaded with MEP pathway enzymes, might increase the surface/volume ratio for transmembrane exchanges with the cytosol (Guirimand et al., 2009; Schattat, Barton, Baudisch, Klösgen, & Mathur, 2011). Out of the plastid, geraniol is metabolised into secologanin by a series of reactions taking place in the cytosol, either by soluble enzymes or endoplasmic reticulum-anchored enzymes with a cytosolic active site. Geraniol is first hydroxylated at the external surface of ER by G10H (Guirimand et al., 2009). This ER localisation is also consistent with subcellular localisation of cytochrome P450 reductase in ER for electron transfer to cytochrome P450 enzymes. After reduction by an unknown oxidoreductase into 10-oxogeranial, formation of the iridoid ring scaffold is catalysed by the cytosolic iridoid cyclase (Geu-Flores et al., 2012). The next known step, loganic acid methyltransferase, is also a soluble protein localised in the cytosol by GFP-fusion analysis (Guirimand, Ginis, et al., 2011). The final cytosolic step, resulting in secologanin synthesis by a cytochrome P450, targets GFP-fusion to the ER (Guirimand, Ginis, et al., 2011). Following the exit of tryptophan from the plastid, the localisation of tryptophan decarboxylase into cytosol by GFP-fusion assigned synthesis of tryptamine in this compartment (Guirimand, Ginis, et al., 2011). Therefore, after initial elaboration in the plastid, the synthesis of both indolic and terpenoid precursors for strictosidine elaboration are achieved and completed by cytosolic reactions. Interestingly, using bimolecular fluorescence complementation assays and yeast two-hybrid analysis, all of the soluble cytosolic enzymes which have been characterised in the secologanin pathway, as well as TDC, forms dimers or higher-order structure that excluded them from passive diffusion into the nucleus (Geu-Flores et al., 2012; Guirimand, Ginis, et al., 2011). For strictosidine synthesis, secologanin and tryptamine transport to the vacuole is required as the synthase is targeted, following entry in the ER to this endomembrane system (Guirimand, Courdavault, Lanoue, et al., 2010; McKnight et al., 1991). However, for the next steps initiating synthesis of various structural classes of MIAs, strictosidine is compartmented away from the vacuolar pool of strictosidine, which may build up to the millimolar range (Guirimand, Courdavault, Lanoue, et al., 2010). In fact, the stereospecific glucosidase that hydrolyses strictosidine has an unusual targeting. SGD was first proposed to be localised in ER on the basis of *in vivo* apparition of strictosidine-induced yellow fluorescence around the nucleus, presumed to be the *trans*-nuclear golgi network, and presence of a putative ER-anchoring sequence (Geerlings et al., 2000). However, GFP imaging refuted this hypothesis by localising GFP-SGD into the nucleus (Guirimand, Courdavault, Lanoue, et al., 2010). A C-terminal bipartite nuclear localisation signal was also identified. Both *C. roseus* and *R. serpentina* SGD were shown to accumulate as highly stable supramolecular aggregates within the nucleus. Interestingly, in the reverse fusion experiment, the construct without free C-terminal end SGD-GFP was also targeted to the nucleus, but with a non-aggregated diffuse fluorescence pattern. This unusual localisation of SGD in the nucleus may be essential for its physical separation from the accumulated strictosidine pool in the vacuole under normal physiological conditions. In the condition leading to MIA biosynthesis, the sequestration of strictosidine biosynthesis within the vacuole and its subsequent deglucosylation within the nucleus by a stable supramolecular SGD complex implies that an unknown transportation system of strictosidine across the tonoplast plays an important role in the control of the MIA biosynthetic flux. This unknown transportation step appears as rate limiting during the MIA biosynthetic pathway as MeJa and Ethephon treatments lead to an important increase of the strictosidine pool, whereas the level of vindoline and catharanthine slightly decrease (Guirimand, Courdavault, Lanoue, et al., 2010). At the moment, apart from an efficient physical separation of SGD from the vacuole-accumulated strictosidine which circumvents the potential deleterious effect of a massive activation of the strictosidine pool, the physiological reason for the intriguing SGD nuclear sequestration appears unclear. The role of SGD multimerisation may be to prevent both a potential leakage of SGD into the cytoplasm by passive diffusion and a means to stabilise its enzymatic activity upon potential proteolysis attack (see Section 3.4).

Following reactions, from strictosidine aglycone formation down to catharanthine or tabersonine biosynthesis, are not known at the gene level. However, the conversion of tabersonine into vindoline synthesis is now fairly well understood. At the subcellular level, by combining GFP imaging, bimolecular fluorescence complementation assays and yeast two-hybrid analysis, the first biosynthetic enzyme, tabersonine 16-hydroxylase (T16H), was established to be anchored to the ER as a monomer via a putative N-terminal helix (St-Pierre & De Luca, 1995; Guirimand, Guihur, et al., 2011). 16OMT was also shown to homodimerize in the cytoplasm, allowing its exclusion from the nucleus and thus facilitating the uptake of T16H conversion product, although no T16H/16OMT interactions occur (Guirimand, Guihur, et al., 2011). Moreover, the last two biosynthetic enzymes, D4H and DAT, were shown to operate as monomers that reside in the nucleocytoplasmic compartment following passive diffusion to the nucleus due to small protein size (Guirimand, Guihur, et al., 2011). Finally, coupling of catharanthine and vindoline to form bisindole alkaloids can be catalysed *in vitro* by non-specific peroxidases. Because the most abundant leaf peroxidases are present in the vacuolar compartment (AVLBS/Prx1), this compartment was suggested to be the site of bisindole alkaloid biosynthesis (Costa et al., 2008).

Taken together the multicellular compartmentation of MIA biosynthesis highlights the importance of the inter- and intracellular translocations of intermediates and their potential regulatory role. Members of the ATP-binding cassette transporter and MATE superfamilies have been demonstrated to translocate various alkaloids such as nicotine or berberine (Nour-Eldin & Halkier, 2013; Yazaki, 2005; Yazaki, Sugiyama, Morita, & Shitan, 2008). The possibility that alkaloid intermediates and/or enzymes flow through the symplasm for intercellular exchanges should also be considered.

3.4. Nuclear time bomb

Glycoside hydrolysis by specific sequestrated glycosidases activates many glycosylated secondary metabolites leading to plant defence strategies against herbivores (Morant et al., 2008) such as those observed in the *Brassicaceae* glucosinolate–myrosinase systems (Kissen, Rossiter, & Bones, 2009; Kliebenstein, Kroymann, & Mitchell-Olds, 2005). Although the differential compartmentation has not been elucidated in every model, the accumulating glucosylated metabolites must be physically separated (either at the cellular level or at the subcellular level) from the activating β-glucosidases (Kissen et al., 2009). The activation of toxic or repulsive metabolites occurs following enzyme–substrate reunion during herbivore feeding (Morant et al., 2008). Interestingly, studies on *Ligustrum obtusifolium* leaves showed that an unidentified sequestrated β-glucosidase was able to activate oleuropein, a phenolic secoiridoid glucoside chemically related to strictosidine, leading to the production of a highly reactive dialdehyde that acts as a strong protein cross-linker with a potent chemical defence role (Konno, Hirayama, Yasui, & Nakamura, 1999). Protein cross-linking renders the plant proteins less digestible and reduces the nutritional value of plant tissue for the herbivore. Such an activation mechanism has been proposed for strictosidine with the formulated hypothesis that upon cell damage, SGD would rapidly convert strictosidine into an aglycon, which has been shown to have anti-microbial activity (Geerlings et al., 2000; Gerasimenko et al., 2002; Luijendijk, van der Meijden, & Verpoorte, 1996).

Further evidence of the presence of 'detonator' activity in the nucleus of *Catharanthus* cells have recently been published, the so-called 'nuclear time bomb' model (Guirimand, Courdavault, Lanoue, et al., 2010). Electrophoretic-mobility shift assays clearly show that the strictosidine deglucosylation product has *in vitro* protein cross-linking and precipitating properties that strictosidine does not have. A significant pool of strictosidine, in the mM range, was observed in *C. roseus* young leaves and is rapidly 10-fold increased following hormonal treatment mimicking herbivore attack (MeJa+Ethephon). Using *in situ* hybridisation and GFP-imaging approaches, the cellular and subcellular localisation of STR and SGD indicated physical separation of both enzymes. In the leaf, both proteins are expressed exclusively in the epidermis; however, vacuolar localisation of STR advocate for a pool of strictosidine in the vacuole separated from SGD, which is sequestered in aggregates within the nucleus.

We suggested that massive activation of the strictosidine vacuolar pool by the nuclear SGD complex may occur following cellular disruption, for instance, during herbivore feeding or pathogen attack. The strictosidine aglycone possesses two reactive aldehyde functions and induces protein reticulation and precipitation. Activation of this nuclear time bomb could help the plant to deter some herbivores from their feeding habit in a similar manner to the *Ligustrum*/oleuropein system (Konno et al., 1999). In the latter case, the specialist herbivore *Brahmaea wallichii* avoids this plant defence strategy by an adaptive evolution. High concentration of free glycine is found in the larval digestive juice which quenches the protein cross-linking effect of the activated oleuropein aglycon (Konno, Okada, & Hirayama, 2001). This illustrates the selective pressure that such a defence system puts on non-specialist herbivores.

Such an activation of strictosidine is probably only a part of the plant defence strategies developed in these species given their metabolic complexity including numerous directly toxic compounds (Chockalingam, Sundari, & Thenmozhi, 1989; Luijendijk et al., 1996; Meisner, Weissenberg, Palevitch, & Aharonson, 1981).

Therefore, under normal physiological conditions, the control of strictosidine vacuolar efflux could insure that the level of activated dialdehyde, produced following strictosidine deglucosylation in the nucleus, is not too high to be fully metabolised by the following steps of the MIA biosynthetic pathway.

3.5. Biological function

Many secondary metabolites function in defence responses against pathogens and herbivores. For MIA, strictosidine, and especially its aglycone product, were shown to possess antifungal activity *in vitro* (Luijendijk et al., 1996). Roepke et al. (2010) further showed the antifungal activity of catharanthine against *Phytopthora nicotianiae*. Catharanthine was found to inhibit the fungal growth at concentrations found on the leaf surface. The antiherbivory effect of catharanthine was also prospected by feeding the *Catharanthus* leaves or supplementing catharanthine in diets of various lepideptoran larvae (Roepke et al., 2010). The presence of catharanthine on the leaf surface was suggested to be an important deterrent to insect herbivory by causing decreased feeding or through its toxicity that appears to lead to insect death.

4. CONCLUSIONS

The recent studies of MIAs have progressed from the phytochemical characterisation of the amazing structural diversity of this class of complex alkaloids, the discovery of chemical synthesis schemes, towards the establishment of a biological platform to produce MIA in bioreactors, the elucidation of MIA metabolic pathways, the deciphering of unique multicellular and subcellular levels of organisation in their assembly line and the discovery of the signals and signalling mechanisms governing the expression of biosynthetic genes. Although considerable progress has been accomplished in these areas, a number of great problems remain to be solved. The origin of MIA biosynthesis in angiosperms, like any plant-specialised metabolisms, is a fascinating puzzle that has only been partially investigated. The restricted distribution of MIAs in the plant kingdom, within a larger group of iridoids producing taxa, is an interesting context in which to study the progressive acquisition of unique biosynthetic capacities. In link with their evolution, the elucidation of the MIAs' biological functions, most likely protective, and their interactions with plant predators, has lagged behind the interest for the powerful pharmacological properties of many MIAs. Considerable progress in identification of genes within the central stages of MIA biosynthesis partially hides the poor characterisation of most enzymatic steps leading to MIA skeleton reorganisation following aglycone formation. Unlike substitution reactions, which are catalysed by multifamily proteins with conserved domains, tracking the molecular determinant of these unusual reactions will probably require combined metabolomic, transcriptomic and proteomic approaches as well as functional tools like virus-induced gene silencing. Mapping of gene expression in *C. roseus* organs and protein targeting in cell compartments revealed a distinct paradigm for secondary product biosynthesis which requires biochemical capacity of three cell types to elaborate the most complex alkaloids and several subcellular compartments. The molecular determinants of intermediates shuffling are potential control points for MIA complexity and important tools to achieve synthetic pathways in a microbial platform.

ACKNOWLEDGEMENTS

Financial support was provided by the 'Ministère de l'Enseignement Supérieur et de la Recherche' (MESR, France), the University of Tours, the 'Région Centre', and the 'Ligue Contre le Cancer, comité d'Indre et Loire'.

REFERENCES

Aerts, R., Gisi, D., De Carolis, E., De Luca, V., & Baumann, T. W. (1994). Methyl jasmonate vapor increases the developmentally controlled synthesis of alkaloids in *Catharanthus* and *Cinchona* seedlings. *The Plant Journal, 5*, 635–643.

Amini, A., Andreu, F., Glévarec, G., Rideau, M., & Crèche, J. (2012). Down-regulation of the CrHPT1 histidine phosphotransfer protein prevents cytokinin-mediated up-regulation of *CrDXR*, and *CrG10H* transcript levels in periwinkle cell cultures. *Molecular Biology Reports, 39*, 8491–8496.

Amini, A., Glévarec, G., Andreu, F., Rideau, M., & Crèche, J. (2009). Low levels of gibberellic acid control the biosynthesis of Ajmalicine in *Catharanthus roseus* cell suspension cultures. *Planta Medica, 75*, 187–191.

APG III, THE ANGIOSPERM PHYLOGENY GROUP (2009). An update of the Angiosperm Phylogeny Group classification for the orders and families of flowering plants: APG III. *Botanical Journal of the Linnean Society, 161*, 105–121.

Arvy, M.-P., Imbault, N., Naudascher, F., Thiersault, M., & Doireau, P. (1994). 2,4-D and alkaloid accumulation in periwinkle cell suspensions. *Biochimie, 76*, 410–416.

Berlin, J., Rügenhagen, C., Kuzovkina, I. N., Fecker, L. F., & Sasse, F. (1994). Are tissue cultures of *Peganum harmala* a useful model system for studying how to manipulate the formation of secondary metabolites? *Plant Cell, Tissue and Organ Culture, 38*, 289–297.

Burlat, V., Oudin, A., Courtois, M., Rideau, M., & St-Pierre, B. (2004). Co-expression of three MEP pathway genes and geraniol 10-hydroxylase in internal phloem parenchyma of *Catharanthus roseus* implicates multicellular translocation of intermediates during the

biosynthesis of monoterpene indole alkaloids and isoprenoid-derived primary metabolites. *The Plant Journal*, *38*, 131–141.
Campos-Tamayo, F., Hernandez-Dominguez, E., & Vazquez-Flota, F. A. (2008). Vindoline formation in shoot cultures of *Catharanthus roseus* is synchronously activated with morphogenesis through the last biosynthetic step. *Annals of Botany*, *102*, 409–415.
Chahed, K., Oudin, A., Guivarc'h, N., Hamdi, S., Chénieux, J. C., Rideau, M., et al. (2000). 1-Deoxy-D-xylulose 5-phosphate synthase from periwinkle: cDNA identification and induced gene expression in terpenoid indole alkaloid-producing cells. *Plant Physiology and Biochemistry*, *38*, 559–566.
Chakravorty, D., & Botella, J. R. (2007). Over-expression of a truncated Arabidopsis thaliana heterotrimeric G protein γ subunit results in a phenotype similar to α and β subunit knockouts. *Gene*, *393*, 163–170.
Chockalingam, S., Sundari, M. S. N., & Thenmozhi, S. (1989). Impact of the extract of *Catharanthus roseus* on feeding and enzymatic digestive activities of *Spodoptera litura*. *Journal of Environmental Biology*, *10*, 303–307.
Contin, A., van der Heijden, R., Lefebre, A. W., & Verpoorte, R. (1998). The iridoid glucoside secologanin is derived from the novel triose phosphate/pyruvate pathway in a *Catharanthus roseus* cell culture. *FEBS Letters*, *4*, 413–416.
Costa, M. M., Hilliou, F., Duarte, P., Pereira, L. G., Almeida, I., Leech, M., et al. (2008). Molecular cloning and characterization of a vacuolar class III peroxidase involved in the metabolism of anticancer alkaloids in *Catharanthus roseus*. *Plant Physiology*, *146*, 403–417.
Courdavault, V., Burlat, V., St-Pierre, B., & Giglioli-Guivarc'h, N. (2005). Characterisation of CaaX-prenyltransferases in *Catharanthus roseus*: Relationships with the expression of genes involved in the early stages of monoterpenoid biosynthetic pathway. *Plant Science*, *168*, 1097–1107.
Courdavault, V., Burlat, V., St-Pierre, B., & Giglioli-Guivarc'h, N. (2009). Proteins prenylated by type I protein geranylgeranyltransferase act positively on the jasmonate signalling pathway triggering the biosynthesis of monoterpene indole alkaloids in *Catharanthus roseus*. *Plant Cell Reports*, *28*, 83–93.
Courdavault, V., Clastre, M., Simkin, A. D., & Giglioli-Guivarc'h, N. (2013). Prenylated proteins are required for methyl-jasmonate-induced monoterpenoid indole alkaloids biosynthesis in *Catharanthus roseus*. In T. J. Bach & M. Rhomer (Eds.), *Isoprenoid synthesis in plants and microorganisms* (pp. 285–296). New York: Springer.
Courdavault, V., Thiersault, M., Courtois, M., Gantet, P., Oudin, A., Doireau, P., et al. (2005). CaaX-prenyltransferases are essential for expression of genes involved in the early stages of monoterpenoid biosynthetic pathway in *Catharanthus roseus* cells. *Plant Molecular Biology*, *57*, 855–870.
De Luca, V., & Cutler, A. J. (1987). Subcellular localization of enzymes involved in indole alkaloid biosynthesis in *Catharanthus roseus*. *Plant Physiology*, *85*, 1099–1102.
De Luca, V., Balsevich, J., Tyler, R. T., Eilert, U., Panchuk, B. D., & Kurz, W. G. W. (1986). Biosynthesis of indole alkaloids: Developmental regulation of the biosynthetic pathway from tabersonine to vindoline in *Catharanthus roseus*. *Journal of Plant Physiology*, *125*, 147–156.
De Luca, V., & St-Pierre, B. (2000). The cell and developmental biology of alkaloid biosynthesis. *Trends in Plant Sciences*, *5*, 168–173.
Décendit, A., Liu, D., Ouelhazi, L., Doireau, P., Mérillon, J. M., & Rideau, M. (1992). Cytokinin-enhanced accumulation of indole alkaloids in *Catharanthus roseus* cell cultures. The factors affecting the cytokinin response. *Plant Cell Reports*, *11*, 400–403.
De-Eknamkul, W., Suttipanta, N., & Kutchan, T. M. (2000). Purification and characterization of deacetylipecoside synthase from *Alangium lamarckii* Thw. *Phytochemistry*, *55*, 177–181.

Devoto, A., & Turner, J. G. (2005). Jasmonate-regulated Arabidopsis stress signalling network. *Physiologia Plantarum*, *123*, 161–172.

DiCosmo, F., Quesnel, A., Misawa, M., & Tallevi, S. G. (1987). Increased synthesis of ajmalicine and catharanthine by cell suspension cultures of *Catharanthus roseus* in response to fungal culture-filtrates. *Applied Biochemistry and Biotechnology*, *14*, 101–106.

Dinda, B., Debnath, S., & Harigaya, Y. (2007a). Naturally occurring iridoids. A review, part 1. *Chemical and Pharmaceutical Bulletin*, *55*, 159–222.

Dinda, B., Debnath, S., & Harigaya, Y. (2007b). Naturally occurring secoiridoids and bioactivity of naturally occurring iridoids and secoiridoids—A review, part 2. *Chemical and Pharmaceutical Bulletin*, *55*, 689–728.

Eilert, U., Constabel, F., & Kurz, W. G. W. (1986). Elicitor-stimulation of monoterpene indole alkaloids formation in suspension cultures of *Catharanthus roseus*. *Journal of Plant Physiology*, *126*, 11–22.

Eilert, U., Nesbitt, L. R., & Constabel, F. (1985). Laticifers and latex in fruits of periwinkle, Catharanthus roseus. *Canadian Journal of Botany*, *63*, 1540–1546.

El-Sayed, M., & Verpoorte, R. (2005). Methyljasmonate accelerates catabolism of monoterpenoid indole alkaloids in *Catharanthus roseus* during leaf processing. *Fitoterapia*, *76*, 83–90.

Facchini, P. J., & St-Pierre, B. (2005). Synthesis and trafficking of alkaloid biosynthetic enzymes. *Current Opinion in Plant Biology*, *8*, 657–666.

Gantet, P., Imbault, N., Thiersault, M., & Doireau, P. (1998). Necessity of a functional octadecanoic pathway for indole alkaloid synthesis by *Catharanthus roseus* cell suspensions cultured in an auxin-starved medium. *Plant & Cell Physiology*, *39*, 220–225.

Geerlings, A., Ibanñez, M. M.-L., Memelink, J., van der Heijden, R., & Verpoorte, R. (2000). Molecular cloning and analysis of strictosidine β-*D*-glucosidase, an enzyme in terpenoid indole alkaloid biosynthesis in *Catharanthus roseus*. *Journal of Biological Chemistry*, *275*, 3051–3056.

Gelb, M. H., Brunsveld, L., Hrycyna, C. A., Michaelis, S., Tamanoi, F., Van Voorhis, W. C., et al. (2006). Therapeutic intervention based on protein prenylation and associated modifications. *Nature Chemical Biology*, *2*, 518–528.

Gerasimenko, I., Sheludko, Y., Ma, X., & Stöckigt, J. (2002). Heterologous expression of a Rauvolfia cDNA encoding strictosidine glucosidase, a biosynthetic key to over 2000 monoterpenoid indole alkaloids. *European Journal of Biochemistry*, *269*, 2204–2213.

Gerber, E., Hemmerlin, A., Hartmann, M., Heintz, D., Hartmann, M. A., Mutterer, J., et al. (2009). The plastidial 2-C-methyl-d-erythritol 4-phosphate pathway provides the isoprenyl moiety for protein geranylgeranylation in tobacco BY-2 cells. *The Plant Cell*, *21*, 285–300.

Geu-Flores, F., Sherden, N. H., Courdavault, V., Burlat, V., Glenn, W. S., Wu, C., et al. (2012). An alternative route to cyclic terpenes by reductive cyclization in iridoid biosynthesis. *Nature*, *492*, 138–142.

Giddings, L. A., Liscombe, D. K., Hamilton, J. P., Childs, K. L., DellaPenna, D., Buell, C. R., et al. (2011). A stereoselective hydroxylation step of alkaloid biosynthesis by a unique cytochrome P450 in *Catharanthus roseus*. *Journal of Biological Chemistry*, *286*, 16751–16757.

Giglioli-Guivarch, N., Courdavault, V., Oudin, A., Crèche, J., Crèche, J., & St-Pierre, B. (2006). Madagascar periwinkle, an attractive model for studying the control of the biosynthesis of terpenoid derivative compounds. In J. A. Texeira da Silva (Ed.), (1st ed.). *Floriculture, ornamental and plant biotechnology: Advances and topical issues*, Vol. 2. Ikenobe, Japan: Global Science Books.

Ginis, O., Oudin, A., Guirimand, G., Chebbi, M., Courdavault, V., Glévarec, G., et al. (2012). A type-B response regulator drives the expression of the hydroxymethylbutenyl diphosphate synthase gene in periwinkle. *Journal of Plant Physiology*, *169*, 1571–1574.

Grosse, W., & Klapheck, S. (1979). Enzymic studies on the biosynthesis of serotonin and the regulation in walnuts (*Juglans regia* L.). *Zeitschrift für Pflanzenphysiologie, 93*, 359–363.
Guirimand, G., Burlat, V., Oudin, A., Lanoue, A., St-Pierre, B., & Courdavault, V. (2009). Optimization of the transient transformation of *Catharanthus roseus* cells by particle bombardment and itsapplication to the subcellular localization of hydroxymethylbutenyl 4-diphosphate synthase and geraniol 10-hydroxylase. *Plant Cell Reports, 28*, 1215–1234.
Guirimand, G., Courdavault, V., Lanoue, A., Mahroug, S., Guihur, A., Blanc, N., et al. (2010). Strictosidine activation in Apocynaceae: Towards a "nuclear time bomb?" *BMC Plant Biology, 10*, 182.
Guirimand, G., Courdavault, V., St-Pierre, B., & Burlat, V. (2010). Biosynthesis and regulation of alkaloids. In E.-C. Pua & M. R. Davey (Eds.), *Plant developmental biology—Biotechnological perspectives*: Vol. 2, (pp. 139–160). Berlin, Heidelberg: Springer-Verlag.
Guirimand, G., Ginis, O., Guihur, A., Poutrain, P., Hericourt, F., Oudin, A., et al. (2011). The subcellular organization of strictosidine biosynthesis in *Catharanthus roseus* epidermis highlights several transtonoplast translocations of intermediate metabolites. *FEBS Journal, 278*, 749–763.
Guirimand, G., Guihur, A., Phillips, M. A., Oudin, A., Glévarec, G., Melin, C., et al. (2012). A single gene encodes isopentenyl diphosphate isomerase isoforms targeted to plastids, mitochondria and peroxisomes in *Catharanthus roseus*. *Plant Molecular Biology, 79*, 443–459.
Guirimand, G., Guihur, A., Poutrain, P., Hericourt, F., Mahroug, S., St-Pierre, B., et al. (2011). Spatial organization of the vindoline biosynthetic pathway in *Catharanthus roseus*. *Journal of Plant Physiology, 168*, 519–628.
Hedhili, S., Courdavault, V., Giglioli-Guivarc'h, N., & Gantet, P. (2007). Regulation of the terpene moiety biosynthesis of *Catharanthus roseus* terpene indole alkaloids. *Phytochemistry Reviews, 6*, 341–351.
Hicks, M. A., Barber, A. E., Giddings, L. A., Caldwell, J., O'Connor, S. E., & Babbittet, P. C. (2011). The evolution of function in strictosidine synthase-like proteins. *Proteins: Structure, Function and Bioinformatics, 79*, 3082–3098.
Howe, G. A., & Jander, G. (2008). Plant immunity to insect herbivores. *Annual Review of Plant Biology, 59*, 41–66.
Hwang, I., Chen, H. C., & Sheen, J. (2002). Two-component signal transduction pathways in *Arabidopsis*. *Plant Physiology, 129*, 500–515.
Imbault, N., Thiersault, M., Dupéron, P., Benabdelmouna, A., & Doireau, P. (1996). Pravastatin: A tool for investigating the availability of mevalonate metabolites for primary and secondary metabolism in *Catharanthus roseus* cell suspension. *Physiologia Plantarum, 98*, 803–809.
Irmler, S., Schroder, G., St-Pierre, B., Crouch, N. P., Hotze, M., Schmidt, J., et al. (2000). Indole alkaloid biosynthesis in *Catharanthus roseus*: New enzyme activities and identification of cytochrome P450 CYP72A1 as secologanin synthase. *The Plant Journal, 24*, 797–804.
Janssens, S. B., Knox, E. B., Huysmans, S., Smets, E. F., & Merckx, V. S. F. T. (2009). Rapid radiation of Impatiens (Balsaminaceae) during Pliocene and Pleistocene: Results of a global climate change. *Molecular Phylogenetics and Evolution, 52*, 806–824.
Jensen, S. R. (1992). Systematic implications of the distribution of iridoids and other chemical compounds in the Loganiaceae and other families of the Asteridae. *Annals of the Missouri Botanical Garden, 79*, 284–302.
Jensen, S. R., & Schripsema, J. (2002). Chemotaxonomy and pharmacology of Gentianaceae. In L. Struwe & V. A. Albert (Eds.), *Gentianaceae: Systematics and natural history*. (pp. 573–632). Cambridge: Cambridge University Press.
Kibble, N. A. J., Sohani, M. M., Shirley, N., Byrt, C., Roessner, U., Bacic, A., et al. (2009). Phylogenetic analysis and functional characterisation of strictosidine synthase-like genes in Arabidopsis thaliana. *Functional Plant Biology, 36*, 1098–1109.

Kissen, R., Rossiter, J. T., & Bones, A. M. (2009). The 'mustard oil bomb': Not so easy to assemble?! Localization, expression and distribution of the components of the myrosinase enzyme system. *Phytochemistry Reviews*, *8*, 69–86.
Kliebenstein, D. J., Kroymann, J., & Mitchell-Olds, T. (2005). The glucosinolate-myrosinase system in an ecological and evolutionary context. *Current Opinion in Plant Biology*, *8*, 264–271.
Konno, K., Hirayama, C., Yasui, H., & Nakamura, M. (1999). Enzymatic activation of oleuropein: A protein crosslinker used as a chemical defense in the privet tree. *Proceedings of the National Academy of Sciences of the United States of America*, *96*, 9159–9164.
Konno, K., Okada, S., & Hirayama, C. (2001). Selective secretion of free glycine, a neutralizer against a plant defense chemical, in the digestive juice of the privet moth larvae. *Journal of Insect Physiology*, *47*, 1451–1457.
Kutchan, T. M. (2005). A role for intra- and intercellular translocation in natural product biosynthesis. *Current Opinion in Plant Biology*, *8*, 292–300.
Laflamme, P., St-Pierre, B., & De Luca, V. (2001). Molecular and biochemical analysis of a Madagascar periwinkle root-specific minovincinine-19-hydroxy-O-acetyltransferase. *Plant Physiology*, *125*, 189–198.
Larsson, S. (2007). The "new" chemosystematics: Phylogeny and phytochemistry. *Phytochemistry*, *68*, 2903–2907.
Levac, D., Murata, J., Kim, W. S., & De Luca, V. (2008). Application of carborundum abrasion for investigating the leaf epidermis: Molecular cloning of *Catharanthus roseus* 16-hydroxytabersonine-16-O-methyltransferase. *The Plant Journal*, *53*, 225–236.
Liscombe, D. K., MacLeod, B. P., Loukanina, N., Nandi, O. I., & Facchini, P. J. (2005). Evidence for the monophyletic evolution of benzylisoquinoline alkaloid biosynthesis in angiosperms. *Phytochemistry*, *66*, 2501–2520.
Lorence, A., & Nessler, C. L. (2004). Camptothecin, over four decades of surprising findings. *Phytochemistry*, *65*, 2735–2749.
Luijendijk, T. J. C., van der Meijden, E., & Verpoorte, R. (1996). Involvement of strictosidine as a defensive chemical in *Catharanthus roseus*. *Journal of Chemical Ecology*, *22*, 1355–1366.
Ma, X. Y., Panjikar, S., Koepke, J., Loris, E., & Stöckigt, J. (2006). The structure of *Rauvolfia serpentina* strictosidine synthase is a novel six-bladed β-propeller fold in plant proteins. *The Plant Cell*, *18*, 907–920.
Mahroug, S., Burlat, V., & St-Pierre, B. (2007). Cellular and sub-cellular organization of the monoterpenoid indole alkaloid pathway in *Catharanthus roseus*. *Phytochemistry Reviews*, *6*, 363–381.
Mahroug, S., Courdavault, V., Thiersault, M., St-Pierre, B., & Burlat, V. (2006). Epidermis is a pivotal site of at least four secondary metabolic pathways in *Catharanthus roseus* aerial organs. *Planta*, *223*, 1191–1200.
Mano, Y., & Nemoto, K. (2012). The pathway of auxin biosynthesis in plants. *Journal of Experimental Botany*, *63*, 2853–2872.
McKnight, T. D., Bergey, D. R., Burnett, R. J., & Nessler, C. L. (1991). Expression of enzymatically active and correctly targeted strictosidine synthase in transgenic tobacco plants. *Planta*, *185*, 148–152.
Meisner, J., Weissenberg, M., Palevitch, D., & Aharonson, N. (1981). Phagodeterrency induced by leaves and leaf extracts of *Catharanthus roseus* in the larva of *Spodoptera littoralis* (Lepidoptera, Noctuidae). *Journal of Economic Entomology*, *74*, 131–135.
Memelink, J., Verpoorte, R., & Kijne, J. W. (2001). ORCAnization of jasmonate-responsive gene expression in alkaloid metabolism. *Trends in Plant Science*, *6*, 212–219.
Menke, F. L., Champion, A., Kijne, J. W., & Memelink, J. (1999). A novel jasmonate- and elicitor-responsive element in the periwinkle secondary metabolite biosynthetic gene Str

interacts with a jasmonate- and elicitor-inducible AP2-domain transcription factor ORCA2. *EMBO Journal, 18*, 4455–4463.
Menke, F. L., Parchmann, S., Mueller, M. J., Kijne, J. W., & Memelink, J. (1999). Involvement of the octadecanoid pathway and protein phosphorylation in fungal elicitor-induced expression of terpenoid indole alkaloid biosynthclic genes in *Catharanthus roseus*. *Plant Physiology, 119*, 1289–1296.
Mersey, B. G., & Cutler, A. J. (1986). Differential distribution of specific indole alkaloids in leaves of *Catharanthus roseus*. *Canadian Journal of Botany, 64*, 1039–1045.
Morant, A. V., Jorgensen, K., Jorgensen, C., Paquette, S. M., Sanchez-Perez, R., Moller, B. L., et al. (2008). β-glucosidases as detonators of plant chemical defense. *Phytochemistry, 69*, 1795–1813.
Murata, J., & De Luca, V. (2008). The leaf epidermome of *Catharanthus roseus* reveals its biochemical specialization. *The Plant Cell, 20*, 524–542.
Murata, J., Roepke, J., Gordon, H., & De Luca, V. (2005). Localization of tabersonine 16-hydroxylase and 16-OH-tabersonine-16-O-methyltransferase to leaf epidermal cells defines them as a major site of precursor biosynthesis in the vindoline pathway in *Catharanthus roseus*. *The Plant Journal, 44*, 581–594.
Nelson, D., & Werck-Reichhart, D. (2011). A P450-centric view of plant evolution. *The Plant Journal, 66*, 194–211.
Nomura, T., Quesada, A. L., & Kutchan, T. M. (2008). The new β-D-glucosidase in terpenoid-isoquinoline alkaloid biosynthesis in *Psychotria ipecacuanha*. *Journal of Biological Chemistry, 283*, 34650–34659.
Nour-Eldin, H. H., & Halkier, B. A. (2013). The emerging field of transport engineering of plant specialized metabolites. *Current Opinion in Biotechnology, 24*, 263–270.
Ober, D., & Hartmann, T. (1999). Homospermidine synthase, the first pathway-specific enzyme of pyrrolizidine alkaloid biosynthesis, evolved from deoxyhypusine synthase. *Proceedings of the National Academy of Sciences of the United States of America, 96*, 14777–14782.
Oudin, A., Courtois, M., Rideau, M., & Clastre, M. (2007). The iridoid pathway in *Catharanthus roseus* alkaloid biosynthesis. *Phytochemistry Reviews, 6*, 259–276.
Oudin, A., Mahroug, S., Courdavault, V., Hervouet, N., Zelwer, C., Rodríguez-Concepción,- M., et al. (2007). Spatial distribution and hormonal regulation of gene products from methyl erythritol phosphate and monoterpene-secoiridoid pathways in *Catharanthus roseus*. *Plant Molecular Biology, 65*, 13–30.
Ouwerkerk, P. B. F., Hallard, D., Verpoorte, R., & Memelink, J. (1999). Identification of UV-B light-responsive regions in the promoter of the tryptophan decarboxylase gene from *Catharanthus roseus*. *Plant Molecular Biology, 41*, 491–503.
Pan, Q., Chen, Y., Wang, Q., Yuan, F., Xing, S., Tian, Y., et al. (2010). Effect of plant growth regulators on the biosynthesis of vinblastine, vindoline and catharanthine in *Catharanthus roseus*. *Plant Growth Regulation, 60*, 133–141.
Papon, N., Bremer, J., Vansiri, A., Andreu, F., Rideau, M., & Crèche, J. (2005). Cytokinin and ethylene control indole alkaloid production at the level of the MEP/terpenoid pathway in *Catharanthus roseus* suspension cells. *Planta Medica, 71*, 572–574.
Pasqua, G., Monacelli, B., & Valletta, A. (2004). Cellular localisation of the anti-cancer drug camptothecin in *Camptotheca acuminata* Decne (Nyssaceae). *European Journal of Histochemistry, 48*, 321–328.
Poutrain, P., Mazars, C., Thiersault, M., Rideau, M., & Pichon, O. (2009). Two distinct intracellular Ca^{2+}-release components act in opposite ways in the regulation of the auxin- dependent MIA biosynthesis in *Catharanthus roseus* cells. *Journal of Experimental Botany, 60*, 1387–1398.
Reimann, A., Nurhayati, N., Backenköhler, A., & Ober, D. (2004). Repeated evolution of the pyrrolizidine alkaloid-mediated defense system in separate angiosperm lineages. *The Plant Cell, 16*, 2772–2784.

Rodriguez, S., Compagnon, V., Crouch, N. P., St-Pierre, B., & De Luca, V. (2003). Jasmonate-induced epoxidation of tabersonine by a cytochrome P-450 in hairy root cultures of *Catharanthus roseus*. *Phytochemistry*, *64*, 401–409.

Rodríguez-Concepción, M., Yalovsky, S., & Gruissem, W. (1999). Protein prenylation in plants: Old friends and new targets. *Plant Molecular Biology*, *39*, 865–870.

Roepke, J., Salim, V., Wu, M., Thamm, A. M. K., Murata, J., Ploss, K., et al. (2010). Vinca drug components accumulate exclusively in leaf exudates of Madagascar periwinkle. *Proceedings of the National Academy of Sciences of the United States of America*, *107*, 15287–15292.

Samanani, N., Liscombe, D. K., & Facchini, P. J. (2004). Molecular cloning and characterization of norcoclaurine synthase, an enzyme catalyzing the first committed step in benzylisoquinoline biosynthesis. *The Plant Journal*, *40*, 302–313.

Schafer, W. R., & Rine, J. (1992). Protein prenylation: Genes, enzymes, targets and functions. *Annual Review of Genetics*, *26*, 209–237.

Schattat, M., Barton, K., Baudisch, B., Klösgen, R. B., & Mathur, J. (2011). Plastid stromule branching coincides with contiguous endoplasmic reticulum dynamics. *Plant Physiology*, *155*, 1667–1677.

Simkin, A. J., Miettinen, K., Claudel, P., Burlat, V., Guirimand, G., Courdavault, V., et al. (2013). Characterization of the plastidial geraniol synthase from Madagascar periwinkle which initiates the monoterpenoid branch of the alkaloid pathway in internal phloem associated parenchyma. *Phytochemistry*, *85*, 36–43.

Singla, B., Chugh, A., Khurana, J. P., & Khurana, P. (2006). An early auxinresponsive Aux/IAA gene from wheat (*Triticum aestivum*) is induced by epibrassinolide and differentially regulated by light and calcium. *Journal of Experimental Botany*, *57*, 4059–4070.

Smith, J. I., Smart, N. J., Kurz, W. G. W., & Misawa, M. (1987). Stimulation of indole alkaloid production in cell suspension cultures of *Catharanthus roseus* by abscisic acid. *Planta Medica*, *53*, 470–474.

Stöckigt, J., Barleben, L., Panjikar, S., & Loris, E. A. (2008). 3D-Structure and function of strictosidine synthase-the key enzyme of monoterpenoid indole alkaloid biosynthesis. *Plant Physiology and Biochemistry*, *46*, 340–355.

Stöckigt, J., & Panjikar, S. (2007). Structural biology in plant natural product biosynthesis—Architecture of enzymes from monoterpenoid indole and tropane alkaloid biosynthesis. *Natural Product Reports*, *24*, 1382–1400.

St-Pierre, B., & De Luca, V. (1995). A cytochrome P-450 monooxygenase catalyzes the first step in the conversion of tabersonine to vindoline in *Catharanthus roseus*. *Plant Physiology*, *109*, 131–139.

St-Pierre, B., Laflamme, P., Alarco, A. M., & De Luca, V. (1998). The terminal *O*-acetyltransferase involved in vindoline biosynthesis defines a new class of proteins responsible for coenzyme A-dependent acyl transfer. *The Plant Journal*, *14*, 703–713.

St-Pierre, B., Vázquez-Flota, F. A., & De Luca, V. (1999). Multicellular compartmentation of *Catharanthus roseus* alkaloid biosynthesis predicts intercellular translocation of a pathway intermediate. *The Plant Cell*, *11*, 887–900.

Szabó, L. F. (2008). Rigorous biogenetic network for a group of indole alkaloids derived from strictosidine. *Molecules*, *13*, 1875–1896.

Thines, B., Katsir, L., Melotto, M., Niu, Y., Mandaokar, A., Liu, G., et al. (2007). JAZ repressor proteins are targets of the SCF^{COI1} complex during jasmonate signalling. *Nature*, *448*, 661–665.

To, J. P., & Kieber, J. J. (2008). Cytokinin signaling: Two-components and more. *Trends in Plant Science*, *13*, 85–92.

Trusov, Y., Rookes, J. E., Chakravorty, D., Armour, D., Schenk, P. M., & Botella, J. R. (2006). Heterotrimeric G proteins facilitate Arabidopsis resistance to necrotrophic pathogens and are involved in jasmonate signaling. *Plant Physiology*, *140*, 210–220.

Valletta, A., Trainotti, L., Santamaria, A. R., & Pasqua, G. (2010). Cell-specific expression of tryptophan decarboxylase and 10-hydroxygeraniol oxidoreductase, key genes involved in camptothecin biosynthesis in *Camptotheca acuminata* Decne (Nyssaceae). *BMC Plant Biology, 10*, 69.

van der Fits, L., & Memelink, J. (2000). ORCA3, a jasmonate-responsive transcriptional regulator of plant primary and secondary metabolism. *Science, 289*, 295–297.

van der Heijden, R., Jabos, D. J., Snoeijer, W., Hallard, D., & Verpoorte, R. (2004). The Catharanthus alkaloids: Pharmacognosy and biotechnology. *Current Medicinal Chemistry, 11*, 607–628.

Vázquez-Flota, F., Carrillo-Pech, M., Minero-García, Y., & Miranda-Ham, M. L. (2004). Alkaloid metabolism in wounded *Catharanthus roseus* seedlings. *Plant Physiology and Biochemistry, 42*, 623–628.

Vazquez-Flota, F. A., De Carolis, E. D., Alarco, A. M., & De Luca, V. (1997). Molecular cloning and characterization of desacetoxyvindoline-4-hydroxylase, a 2-oxoglutarate dependent-dioxygenase involved in the biosynthesis of vindoline in *Catharanthus roseus* (L.) G.Don. *Plant Molecular Biology, 34*, 935–948.

Vazquez-Flota, F., & De Luca, V. (1998). Jasmonate modulates development and light regulated alkaloid biosynthesis in *Catharanthus roseus*. *Phytochemistry, 49*, 395–402.

Vazquez-Flota, F. A., St-Pierre, B., & De Luca, V. (2000). Light activation of vindoline biosynthesis does not require cytomorphogenesis in *Catharanthus roseus* seedlings. *Phytochemistry, 55*, 531–536.

Veau, B., Courtois, M., Oudin, A., Chenieux, J. C., Rideau, M., & Clastre, M. (2000). Cloning and expression of cDNAs encoding two enzymes of the MEP pathway in *Catharanthus roseus*. *Biochimica et Biophysica Acta, 1517*, 159–163.

Wikström, N., Savolainen, V., & Chase, M. W. (2001). Evolution of the angiosperms: Calibrating the family tree. *Proceedings of the Royal Society B Biological Sciences, 268*, 2211–2220.

Yahia, A., Kevers, C., Gaspar, T., Chénieux, J. C., Rideau, M., & Crèche, J. (1998). Cytokinins and ethylene stimulate indole alkaloid accumulation in cell suspension cultures of *Catharanthus roseus* by two distinct mechanisms. *Plant Sciences, 133*, 9–15.

Yazaki, K. (2005). Transporters of secondary metabolites. *Current Opinion in Plant Biology, 8*, 301–307.

Yazaki, K., Sugiyama, A., Morita, M., & Shitan, N. (2008). Secondary transport as an efficient membrane transport mechanism for plant secondary metabolites. *Phytochemistry Reviews*, 7, 513–524.

Yoder, L. R., & Mahlberg, P. G. (1976). Reactions of alkaloid and histochemical indicators in laticifers and specialized parenchyma cells of *Catharanthus roseus* (Apocynaceae). *American Journal of Botany, 63*, 1167–1173.

Zhang, H., Hedhili, S., Montiel, G., Zhang, Y., Chatel, G., Pré, M., et al. (2011). The basic helix-loop-helix transcription factor CrMYC2 controls the jasmonate-responsive expression of the ORCA genes that regulate alkaloid biosynthesis in *Catharanthus roseus*. *The Plant Journal, 67*, 61–71.

Zhao, Y. (2012). Auxin biosynthesis: A simple two-step pathway converts tryptophan to indole-3-acetic acid in plants. *Molecular Plant, 5*, 334–338.

Ziegler, J., & Facchini, P. (2008). Alkaloid biosynthesis: Metabolism and trafficking. *Annual Review of Plant Biology, 59*, 735–769.

CHAPTER FOUR

Biosynthesis and Catabolism of Purine Alkaloids

Hiroshi Ashihara[*,1], **Takao Yokota**[†], **Alan Crozier**[‡]
[*]Department of Biological Sciences, Ochanomizu University, Otsuka, Bunkyo-ku, Tokyo, Japan
[†]Department of Biosciences, Teikyo University, Utsunomiya, Japan
[‡]School of Medicine, College of Medical, Veterinary and Life Sciences, University of Glasgow, Glasgow, United Kingdom
[1]Corresponding author: e-mail address: ashihara.hiroshi@ocha.ac.jp

Contents

1. Introduction 112
2. Occurrence of Purine Alkaloids in Plant Kingdom 113
3. Biosynthesis of Purine Alkaloids 115
   3.1 Biosynthesis of purine alkaloids from xanthosine 115
   3.2 The biosynthesis of caffeine from *de novo* purine biosynthesis 120
   3.3 Purine alkaloid synthesis from cellular purine nucleotide pools 123
   3.4 The SAM route of caffeine biosynthesis 123
   3.5 Estimation of the activity of purine alkaloid synthesis using radio-labelled precursors 123
   3.6 Genes and molecular structure of *N*-methyltransferases 124
   3.7 Regulation of caffeine biosynthesis 125
4. Catabolism of Caffeine 126
5. Distribution of Purine Alkaloids in Tissues and Cells 128
   5.1 Distribution in tissues 128
   5.2 Subcellular distribution 128
6. Biotechnology of Purine Alkaloids 129
7. *In Planta* Function of Purine Alkaloids 131
8. Conclusions and Perspectives 133
References 133

Abstract

A limited number of plant species accumulate purine alkaloids, such as caffeine and theobromine, which are synthesized from xanthosine, a catabolite of purine nucleotides. The main biosynthetic pathway is a sequence consisting of xanthosine → 7-methylxanthosine → 7-methylxanthine → theobromine → caffeine. This review summarizes the occurrence of purine alkaloids in the plant kingdom, the caffeine biosynthesis routes from purine precursors, the enzymes and genes of *N*-methyltransferases, key enzymes of caffeine biosynthesis, caffeine catabolism and the possible ecological role of caffeine. Finally, we introduce transgenic plants in which caffeine production is either

Advances in Botanical Research, Volume 68
ISSN 0065-2296
http://dx.doi.org/10.1016/B978-0-12-408061-4.00004-3

suppressed or induced by the introduction of caffeine encoded genes. Such plants have the potential to be used for the production of decaffeinated coffee and tea or as natural pesticides in agriculturally important crops.

1. INTRODUCTION

Purine alkaloids, methylxanthines and methyluric acids, are secondary plant metabolites derived from purine nucleotides. These structures are based on xanthine or uric acid skeletons (Fig. 4.1). Methyl groups originating from *S*-adenosyl-L-methionine (SAM) are attached to nitrogen atoms at positions 1, 3, 7, 9 or an oxygen atom at position 2. The most widely distributed methylxanthine in the plant kingdom is caffeine (1,3,7-trimethylxanthine) which accumulates in leaves and seeds of tea (*Camellia sinensis*), coffee (*Coffea arabica*) and a limited number of other species. Sizable amounts of theobromine (3,7-dimethylxanthine) are stored in the seeds of cacao (*Theobroma cacao*; Zheng, Koyama, Nagai, & Ashihara, 2004), while theacrine (1,3,7,9-tetramethyl uric acid) accumulates in the leaves of a novel wild Chinese tea, kucha (*Camellia assamica* var. *kucha*; Lu et al., 2009; Zheng, Ye, Kato, Crozier, & Ashihara, 2002).

Although caffeine was isolated from tea and coffee in the early 1820s, the biosynthetic pathway of caffeine from purine nucleotides was not fully established until a highly purified *N*-methyltransferase (caffeine synthase)

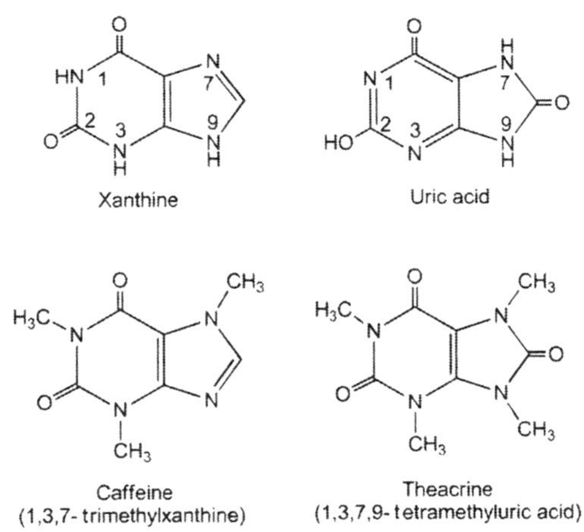

Figure 4.1 Purine alkaloid structures based on xanthine and uric acid. Structures of a major methylxanthine (caffeine) and methyluric acid (theacrine) are shown.

was obtained from tea leaves and a gene encoding this enzyme was cloned ~180 years later (Kato, Mizuno, Crozier, Fujimura, & Ashihara, 2000; Kato et al., 1999). The biosynthetic pathway of caffeine from xanthosine is now well understood in coffee and tea plants. Caffeine catabolism usually starts with the formation of theophylline, although little is known about the demethylases involved in the conversion. The recent advances in caffeine research involving the use of caffeine synthase genes to produce transgenic plants have opened up the possibilities of making decaffeinated coffee and tea plants as well as the use of caffeine as a natural pesticide in species of agricultural importance.

In this chapter, we summarize current information on the occurrence, biosynthesis and catabolism of purine alkaloids. The ecological role of caffeine, as well as the potential value of genetically modified caffeine-reduced and caffeine-induced plants, is also discussed. Purine alkaloid biosynthesis in plants has been reviewed recently (Ashihara, Kato, & Crozier, 2011; Ashihara, Ogita, & Crozier, 2011; Ashihara, Sano, & Crozier, 2008). Consumption of purine alkaloids can have a diversity of impacts on human health (Lean, Ashihara, Clifford, & Crozier, 2012). The metabolism and pharmacological function of purine alkaloids in animals, humans and plants and their impact on human nutrition as dietary constituents of tea, cacao, coffee and many soft drinks are discussed in detail in books edited by Fredholm (2011) and Crozier, Ashihara, and Tomas-Barberan (2012).

2. OCCURRENCE OF PURINE ALKALOIDS IN PLANT KINGDOM

Accumulation of purine alkaloids occurs in several plant species used for beverages and foods. In an earlier review, we noted that purine alkaloids, including caffeine, had been detected in at least 80 species in 13 orders of plant kingdom (Ashihara & Crozier, 1999). This was based on a review by Kihlman (1977) that quoted Willaman and Schubert (1961) and O'Connell (1969), who reported that caffeine occurs in more than 63 species which were distributed among 17 families and 28 genera. More recently, Stewart (1985) and Kretschmar and Baumann (1999) reported the occurrence of caffeine in flowers and leaves of citrus cultivars. Table 4.1 shows the classification of purine alkaloid-containing plants compiled by Mazzafera, Baumann, Shimizu, and Silvarolla (2009). These authors included only reports from the literature where identifications were based on up-to-date analytical techniques such as HPLC and HPLC–MS, and

Table 4.1 Occurrence of caffeine and related purine alkaloids in the plant kingdom

Core eudicots clade	Order	Family	Genus	Typical species	Purine alkaloids
Eurosids II	Malvales	Malvaceae *s.l.*	*Cola*	*C. acuminata* (kola nut)	Cf > Tb
			Herrania	*H. purpurea*	Mu
			Theobroma	*T. cacao* (cacao)	Tb > Cf
	Sapindales	Rutaceae	*Citrus*	*C. paradisi* (grapefruit)	Cf > Tp > Tb > Px
			Poncirus	*P. trifoliata* (trifoliate orange)	Cf > Tp > Tb, Px
		Sapindaceae	*Paullinia*	*P. cupana* (guarana)	Cf > Tb > Tp
Asterids	Ericales	Theaceae	*Camellia*	*C. sinensis* (tea)	Cf
Euasterids I	Gentianales	Rubiaceae	*Coffea*	*C. arabica* (coffee)	Cf
Euasterids II	Aquifoliales	Aquifoliaceae	*Ilex*	*I. paraguariensis* (maté)	Cf

Cf, caffeine; Mu, methyluric acid; Px, paraxanthine; Tb, theobromine; Tp, theophylline.
Adapted from Mazzafera et al. (2009) with slight modification.

as a consequence the distribution of purine alkaloids is restricted to species in six families of higher plants. The occurrence of caffeine is confined to Malvales and Sapindales of Eurosids II, to Ericales of Asterids, and to Gentianales and Aquifoliales of Euasterids I and II, respectively. More than 10 species of *Coffea* and *Theobroma* contain purine alkaloids (Hammerstone, Romanczyk, & Aitken, 1994; Mazzafera & Carvalho, 1992). Other species, especially those that are not of economic significance, have received little attention, and may well contain at least small quantities of caffeine. However, many of the claimed identifications in the literature need to be confirmed with the use of modern analytical methods as opposed to the spectrometric and paper chromatography-based identifications that prevailed in older publications. In this context, the occurrence of theobromine synthase genes in apparently non-purine alkaloid accumulating species of *Camellia*, such as *C. japonica* and *C. kissii* (Table 4.2; Ishida, Kitao, Mizuno, Tanikawa, & Kato, 2009), is of interest. Clearly, further studies are required to obtain a realistic picture of the distribution of purine alkaloids in the plant kingdom.

3. BIOSYNTHESIS OF PURINE ALKALOIDS

The xanthine and uric acid skeletons of purine alkaloids are derived from purine nucleotides. Results from studies on *in situ* metabolism of radioactive precursors and from the identification of enzymes and genes have established that the main caffeine biosynthetic pathway is a four-step sequence consisting of three methylation reactions and a nucleosidase reaction starting with xanthosine acting as the initial purine substrate (Fig. 4.2).

3.1. Biosynthesis of purine alkaloids from xanthosine

In a narrow sense, purine alkaloid biosynthesis means caffeine formation from xanthosine. Less than four enzymes are involved in the reactions (Table 4.2). Typically, the most abundant purine alkaloid is caffeine. Other purine alkaloids found in some plant species are intermediates in pathways associated with the biosynthesis and catabolism of caffeine (Figs. 4.2 and 4.4). Methyluric acids that occur in Kucha, and some *Coffea* species, are formed by the oxidation of caffeine, while tetramethyluric acids are produced by an additional methylation at the 9N position (Fig. 4.1; Baumann, Oechslin, & Wanner, 1976; Petermann, Baumann, & Wanner, 1977; Zheng et al., 2002).

Table 4.2 Enzymes and encoding genes involved in caffeine biosynthesis

EC number	Enzyme name	Substrates	Gene name (#)	Source
2.1.1.158	7-Methylxanthosine synthase (*S*-adenosyl-L-methionine: xanthosine *N*7-methyltransferase)	Xanthosine	*CmXRS1* (AB034699)	*Coffea arabica*
			CaXMT (AB048793)	*C. arabica*
2.1.1.159	Theobromine synthase (*S*-adenosyl-L-methionine:7-methylxanthine *N*3-methyltransferase)	7-Methylxanthine (paraxanthine)	*CTS1* (AB034700)	*C. arabica*
			CTS2 (AB054841)	*C. arabica*
			CaMXMT1 (AB048794)	*C. arabica*
			CaMXMT2 (AB084126)	*C. arabica*
			BTS1 (AB096699)	*Theobroma cacao*
			ICS1 (AB056108)	*Camellia irrawadiensis*
			PCS1 (AB207817)	*Camellia ptilophylla*
			CkCS1 (AB362884)	*Camellia kissii*[a]
			CjCS1 (AB297451)	*Camellia japonica*[a]
2.1.1.160	Caffeine synthase (*S*-adenosyl-L-methionine:3,7-dimethylxanthine *N*1-methyltransferase)	Paraxanthine	*CCS1* (AB086414)	*C. arabica*
		Theobromine	*CtCS7* (AB086415)	*C. arabica*
			CaDXMT1 (AB084125)	*C. arabica*
			TCS1 (AB031280)	*C. sinensis*
3.2.2.25	*N*-methyl nucleosidase (7-methylxanthosine ribohydrolase)	7-Methylxanthosine (3- or 7-methyl-purine nucleosides)	Not registered	*C. sinensis*

[a]Apparently non-purine alkaloid accumulating species

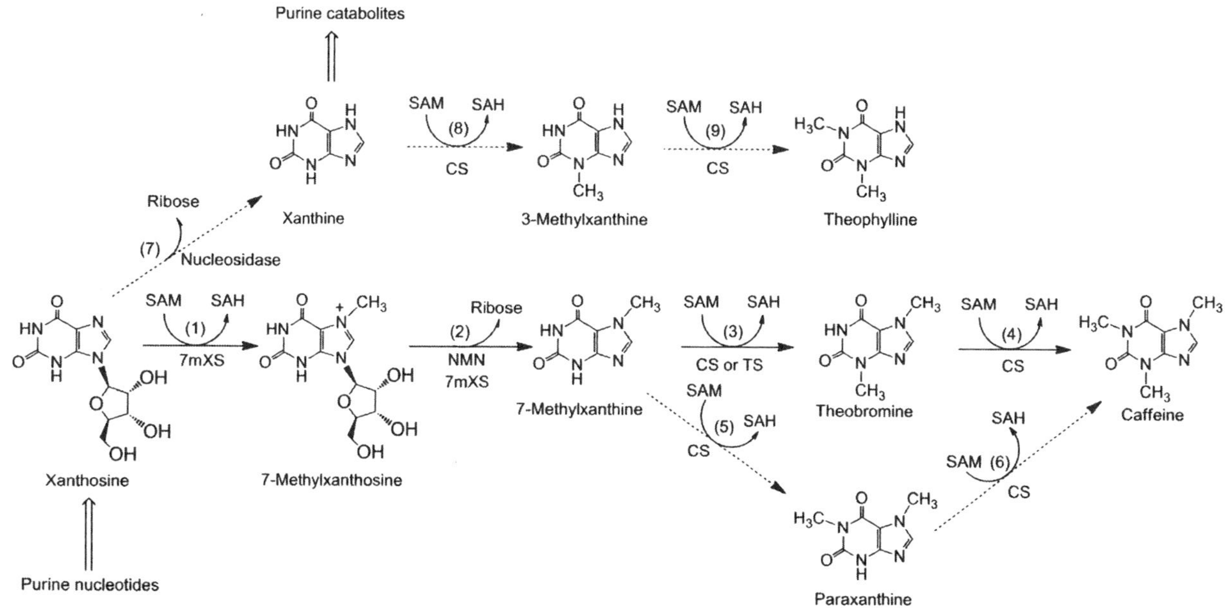

Figure 4.2 The biosynthetic pathways of caffeine from xanthosine. The major pathway consisting of four steps are shown in solid arrows (steps 1–4). The enzymes involved are as follows: 7-methylxanthosine synthase (7mXS, EC 2.1.1.158) (steps 1 and 2); *N*-methylnucleosidase (NMN, EC 3.2.2.25) (step 2); theobromine synthase (TS, EC 2.1.1.159) (step 3) and caffeine synthase (CS, EC 2.1.1.160) (steps 3 and 4). Minor pathways, shown with dotted arrows, may occur because of the broad substrate specificities of the caffeine synthase. SAM, *S*-adenosyl-L-methionine; SAH *S*-adenosyl-L-homocysteine.

3.1.1 Conversion of xanthosine to 7-methylxanthine

Caffeine biosynthesis is initiated by the conversion of xanthosine to 7-methylxanthosine, a reaction catalysed by 7-methylxanthosine synthase (EC 2.1.1.158) (step 1 in Fig. 4.2). The second step involves a nucleosidase which catalyzes the hydrolysis of 7-methylxanthosine resulting in the formation of 7-methylxanthine (step 2 in Fig. 4.2). *N*-Methylnucleosidase (EC 3.2.2.25), which was discovered in tea leaves (Negishi, Ozawa, & Imagawa, 1988), may participate in this reaction. Although the substrate specificity of this enzyme has not been rigorously established, it can hydrolyse not only 7-methylxanthosine but also 3- and 7-methylpurine nucleosides, such as 3-methylxanthosine, 3-methyladenosine and 7-methylguanosine.

Recent structural studies of a recombinant *Coffea canephora* 7-methylxanthosine synthase indicate that both the methyl transfer and nucleoside cleavage are coupled and catalysed by a single enzyme (McCarthy & McCarthy, 2007). Although highly purified native enzyme preparations of 7-methylxanthosine synthase have not yet been isolated, this activity has been detected in crude extracts and partially purified enzyme preparations (Fujimori, Suzuki, & Ashihara, 1991; Negishi, Ozawa, & Imagawa, 1985). Recombinant enzyme proteins of 7-methylxanthosine synthase prepared using the coffee gene sequences have been produced and some biochemical properties are characterized (Table 4.2; McCarthy & McCarthy, 2007; Mizuno et al., 2003; Uefuji, Ogita, Yamaguchi, Koizumi, & Sano, 2003). The 7-methylxanthine synthase uses xanthosine as a substrate, but not xanthosine monophosphate (XMP) (Mizuno et al., 2003). The results do not support the hypothesis proposed by Schulthess, Morath, and Baumann (1996) that caffeine biosynthesis starts with the metabolically channelled formation of 7-methyl-XMP. The gene-encoding 7-methylxanthosine synthase in *C. sinensis* has as yet not been cloned.

3.1.2 Conversion of 7-methylxanthine to theobromine

The second methylation step in the caffeine biosynthesis pathway (step 3 in Fig. 4.2) is also catalysed by SAM-dependent *N*-methyltransferases but the participating enzymes are distinct from the 7-methylxanthosine synthase involved in the first step of caffeine biosynthesis (Mösli Waldhauser, Gillies, Crozier, & Baumann, 1997). Attempts to obtain highly purified native *N*-methyltransferase(s) from tea and coffee plants were initially unsuccessful (Kato et al., 1996; Mazzafera, Wingsle, Olsson, & Sandberg, 1994; Mösli Waldhauser, Kretschmar, & Baumann, 1997) as only partial purification was possible because of the instability of the enzyme activity. However, in 1999 highly purified caffeine synthase (EC 2.1.1.160) was obtained from 100 g fresh weight of young tea leaves by Kato et al. (1999). The enzyme was purified 520-fold to apparent homogeneity by ammonium sulphate fractionation and hydroxyapatite, anion-exchange, adenosine agarose and gel-filtration chromatography. The enzyme displayed a sharp pH optimum at 8.5. The final preparation exhibited 3N- and 1N-methyltransferase activity with rather broad substrate specificity, showing high activity towards paraxanthine (1,7-dimethylxanthine), 7-methylxanthine, and theobromine and low activity with 3-methylxanthine and 1-methylxanthine. However, the enzyme did not 7-methylate either xanthosine or XMP. The 20-amino acid *N*-terminal sequence was determined and using the sequence, a gene-encoding caffeine synthase was cloned (Kato et al., 2000).

Plural genes encoding methylxanthine *N*-methyltransferases possessing the activity of theobromine synthase and/or caffeine synthase have been cloned (Table 4.2). The recombinant enzymes obtained from these genes show different substrate specificities. The recombinant coffee caffeine synthase (EC 2.1.1.160) can utilize paraxanthine, theobromine and 7-methylxanthine as substrates. In *C. arabica* and *C. sinensis*, this dual-functional caffeine synthase participates in the last two steps of caffeine biosynthesis, that is, 7-methylxanthine → theobromine → caffeine (steps 3 and 4 in Fig. 4.2).

In addition to the dual-functional caffeine synthase, genes encoding theobromine synthase (EC 2.1.1.159) have been cloned; the recombinant enzyme is specific for the conversion of 7-methylxanthine to theobromine (step 3 in Fig. 4.2). The enzyme obtained from coffee genes catalyzes 3N- but not 1N-methylation. Furthermore, the activity of 3N-methylation of 7-methylxanthine (step 3 in Fig. 4.2) was much greater than the 3N-methylation activity of paraxanthine (step 5 in Fig. 4.2; Mizuno, Tanaka, Kato, Ashihara, & Fujimura, 2001; Ogawa, Herai, Koizumi, Kusano, & Sano, 2001). The theobromine synthase appears to participate principally in theobromine synthesis in theobromine-accumulating plants, such as *T. cacao*, *Camellia ptilophylla* and *Camellia irrawadiensis* (Yoneyama et al., 2006). In contrast, the principal role of theobromine synthase in *C. arabica* and *C. sinensis* is not well understood.

3.1.3 Conversion of theobromine to caffeine

Conversion of theobromine to caffeine (step 4 in Fig. 4.2) is performed by the dual-functional caffeine synthase (EC 2.1.1.160). The methylation of the

1N position of theobromine (caffeine formation) by the caffeine synthase (step 4) is slower than that of the 3N position of 7-methylxanthine (theobromine formation) (step 3). This may cause transient accumulation of theobromine in caffeine-synthesizing young tissues of *C. sinensis* (Ashihara & Kubota, 1986). In *in situ* ^{14}C-tracer experiments using leaf disks, an initial temporary accumulation of ^{14}C-theobromine was observed even in caffeine-synthesizing tissues (Ashihara, Gillies, & Crozier, 1997; Ashihara, Monteiro, Gillies, & Crozier, 1996; Ashihara, Takasawa, & Suzuki, 1997). This indicates that conversion of theobromine to caffeine is slow. No single functional 1N-methyltransferase which catalyzes only the conversion of theobromine to caffeine (step 4) has been reported.

In addition to the main caffeine biosynthesis pathway, some minor routes of purine alkaloid biosynthesis may also operate (Fig. 4.2). The presence of these alternative pathways appears to be caused by the broad substrate specificities of the *N*-methyltransferases. Caffeine synthase catalyzes 3N- and 1N-methylations, thus, 3N-methyl reactions (steps 3 and 8 in Fig. 4.2) and 1N-methyl reactions (steps 4, 5 and 9) may take place. Paraxanthine is the most suitable substrate of caffeine synthase (Kato et al., 1999), but only limited amounts of paraxanthine accumulate in plant tissues, because 1N-methylation of 7-methylxanthine (step 5) is much slower than the 3N-methylation of paraxanthine (step 6). Paraxanthine formed in this manner is almost certainly rapidly converted to caffeine. The production of caffeine via paraxanthine probably represents only a minor route in the overall biosynthesis of caffeine. Small amounts of theophylline may be produced by the side reactions (steps 7–9). If this does occur, most of the resultant theophylline will be degraded by the catabolic pathway (see Section 4).

3.2. The biosynthesis of caffeine from *de novo* purine biosynthesis

Plants synthesize purine nucleotides by the *de novo* and salvage (reutilization) pathways. In the *de novo* route, purine nucleotides are synthesized from small molecules, such as glycine, glutamine and aspartate, 5-phosphoribosyl-1-pyrophosphate, 10-formyl tetrahydrofolate and carbon dioxide. The *de novo* biosynthetic pathway of purine nucleotides, AMP and GMP is shown in Fig. 4.3. The complete *Arabidopsis* genome (Arabidopsis-Genome-Initiative, 2000; van der Graaff et al., 2004) and the available sequence information from the rice genome (International-Rice-Genome-Sequencing-Project, 2005) have revealed that plants synthesize inosine-5′-monophosphate (IMP), AMP and GMP using similar

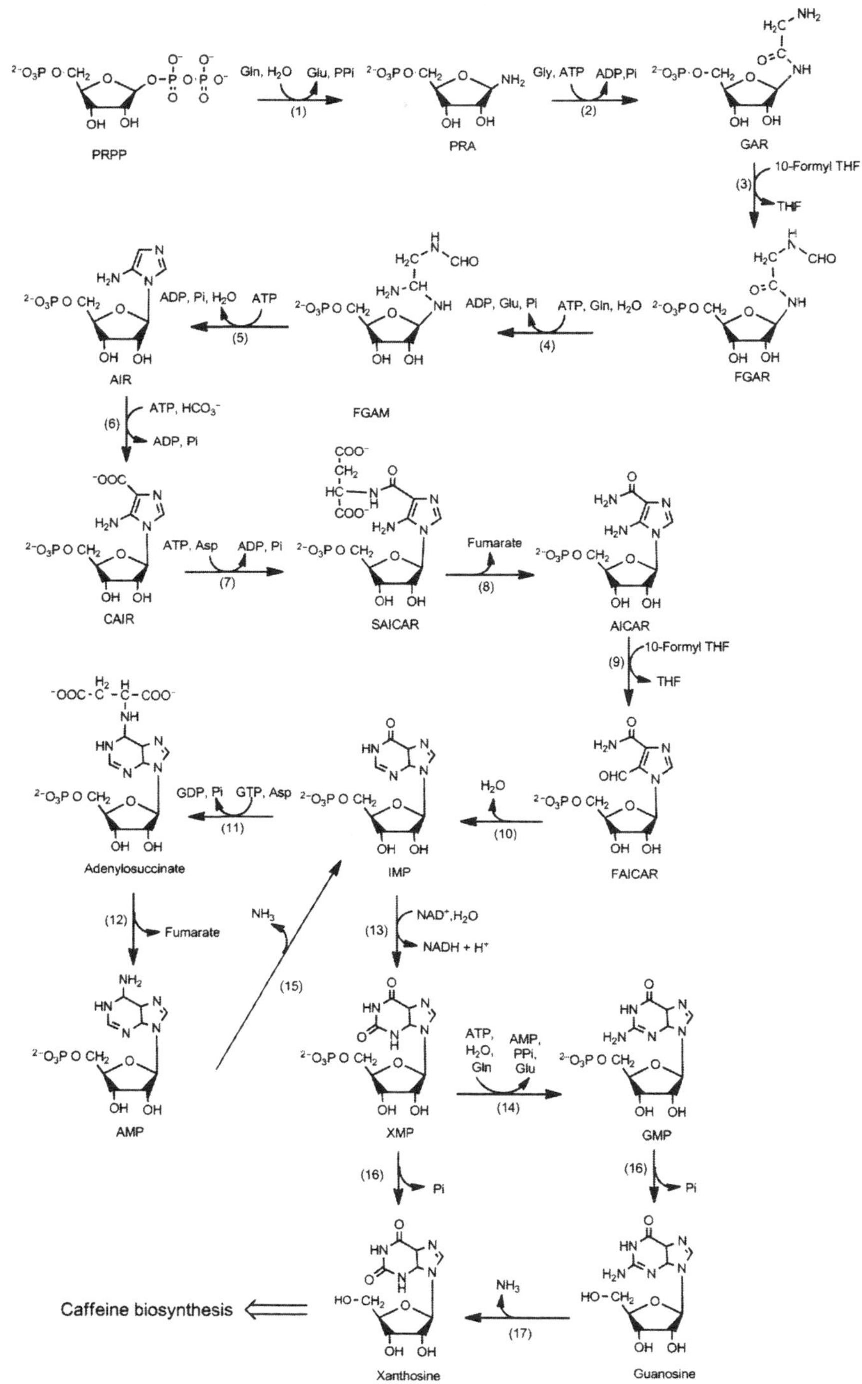
PRPP
PRA
GAR
FGAR
FGAM
AIR
CAIR
SAICAR
AICAR
FAICAR
IMP
Adenylosuccinate
AMP
XMP
GMP
Xanthosine
Guanosine
Gln, H2O
Glu, PPi
Gly, ATP
ADP,Pi
10-Formyl THF
THF
ADP, Glu, Pi
ATP, Gln, H2O
ADP, Pi, H2O
ATP
ATP, HCO3-
ADP, Pi
ATP, Asp
ADP, Pi
Fumarate
H2O
GDP, Pi
GTP, Asp
Fumarate
NH3
NAD+,H2O
NADH + H+
ATP, H2O, Gln
AMP, PPi, Glu
Pi
NH3
Caffeine biosynthesis

reactions to those found in microorganisms and animals. It is very well known that the final products, ATP and GTP, are utilized as the building blocks of nucleic acids as well as the high-energy compounds which participate in cell metabolism.

IMP, an intermediate of the *de novo* purine biosynthetic pathway, may be directly incorporated into the caffeine biosynthesis pathway rather than being converted to adenine and guanine nucleotides. It was reported in the 1960s that precursors of purine *de novo* synthesis, such as serine, glycine, formaldehyde and formate, are converted to caffeine (Anderson & Gibbs, 1962; Proiser & Serenkov, 1963). The contribution of *de novo* purine biosynthesis to caffeine biosynthesis was demonstrated in the young tea leaf disks using ^{15}N-glycine, ^{14}C-labelled precursors and inhibitors of *de novo* purine biosynthesis (Ito & Ashihara, 1999). Xanthosine, the initial precursor of purine alkaloid synthesis, is produced from IMP by the reactions catalysed by IMP dehydrogenase (EC 1.1.1.205, step 13) and nucleotidase (EC 3.1.3.5, step 16). Ribavirin, an inhibitor of IMP dehydrogenase, reduces the rate of caffeine biosynthesis in tea and coffee plants (Keya, Crozier, & Ashihara, 2003). These findings confirm that *de novo* purine synthesis contributes to the caffeine biosynthesis *in planta*. However, it has not yet been established whether *de novo* purine biosynthesis specific for purine alkaloid formation is present or if the common *de novo* pathway of purine biosynthesis is functional for both nucleotide and purine alkaloid synthesis in plants.

Figure 4.3 *De novo* biosynthetic pathway of purine nucleotides in plants. Enzymes (EC numbers) shown are: (1) PRPP amidotransferase (EC 2.4.2.14); (2) GAR synthetase (EC 6.3.4.13); (3) GAR formyl transferase (EC 2.1.2.2); (4) FGAM synthetase (EC 6.3.5.3); (5) AIR synthetase (EC 6.3.3.1); (6) AIR carboxylase (EC 4.1.1.21); (7) SAICAR synthetase (EC 6.3.2.6); (8) adenylosuccinate lyase (EC 4.3.2.2); (9) AICAR formyl transferase (EC 2.1.2.3); (10) IMP cyclohydrolase (EC 3.5.4.10); (11) SAMP synthetase (EC 6.3.4.4); (12) adenylosuccinate lyase (EC 4.3.2.2); (13) IMP dehydrogenase (EC 1.1.1.205); (14) GMP synthetase (EC 6.3.5.2); (15) AMP deaminase (EC 3.5.4.6); (16) 5′-nucleotidase (EC 3.1.3.5) and (17) guanosine deaminase (EC 3.5.4.15). Metabolites: PRA, 5-phosphoribosyl amine; GAR, glycineamide ribonucleotide; FGAR, formylglycineamide ribonucleotide; FGRAM, formylglycine amidine ribonucleotide; AIR, 5-aminoimidazole ribonucleotide; CAIR, 5-aminoimidazole 4-carboxylate ribonucleotide; SCAIR, 5-aminoimidazole-4-N-succinocarboxyamide ribonucleotide; AICAR, 5-aminoimidazole-4-carboxyamide ribonucleotide; FAICAR, 5-formamidoimidazole-4-carboxyamide ribonucleotide; XMP, xanthosine-5′-monophosphate. *Adapted from Zrenner and Ashihara (2011) with modifications.*

3.3. Purine alkaloid synthesis from cellular purine nucleotide pools

A portion of the xanthosine used for caffeine biosynthesis is derived from the adenine and guanine nucleotide pools. The AMP → IMP → XMP → xanthosine route (steps 15, 13 and 16 in Fig. 4.3) is likely to predominate in the conversion of adenine nucleotides to xanthosine. The activity of all enzymes involved in the conversions has been detected in tea leaves (Koshiishi, Kato, Yama, Crozier, & Ashihara, 2001). Xanthosine utilized for caffeine biosynthesis is also supplied from guanine nucleotides by a GMP → guanosine → xanthosine pathway (steps 16 and 17). Nucleotidase (EC 3.1.3.5) and guanosine deaminase (EC 3.5.4.15) participate in this conversion (Negishi, Ozawa, & Imagawa, 1994).

3.4. The SAM route of caffeine biosynthesis

The SAM pathway is a variation in the route from adenine nucleotides discussed above. SAM is the methyl donor for methylation reactions in the caffeine biosynthetic pathway. In the process, SAM is converted to *S*-adenosyl-L-homocysteine, which is then hydrolysed to homocysteine and adenosine. Homocysteine is recycled via the SAM cycle to replenish SAM levels, and adenosine released from the cycle is converted to AMP and utilized for caffeine biosynthesis by the AMP route. Since 3 mol of *S*-adenosyl-L-homocysteine are produced via the SAM cycle for each mole of caffeine that is synthesized, in theory this pathway has the capacity to be the sole source of both the purine skeleton and the methyl groups required for caffeine biosynthesis in young tea leaves (Ashihara & Crozier, 2001; Koshiishi et al., 2001).

3.5. Estimation of the activity of purine alkaloid synthesis using radio-labelled precursors

^{14}C-Labelled purine bases and nucleosides have been used to estimate purine alkaloid biosynthesis activity in leaves and fruit tissues. Among exogenously administered purine bases and nucleobases, adenine and adenosine are the most effective precursors for the biosynthesis of caffeine (Deng & Ashihara, 2010; Suzuki, Ashihara, & Waller, 1992). Exogenously applied [^{14}C]adenine and [^{14}C]adenosine are readily converted to AMP by the so-called purine salvage pathway catalysed by adenine phosphoribosyltransferase (EC 2.4.2.7) and adenosine kinase (EC 2.7.1.20) (Stasolla, Katahira, Thorpe, & Ashihara, 2003; Zrenner & Ashihara, 2011;

Zrenner, Stitt, Sonnewald, & Boldt, 2006). The resultant [^{14}C]AMP is utilized for caffeine biosynthesis. No direct catabolic pathways of adenine and adenosine are present in plants. In contrast, other purine nucleosides and bases, such as inosine, hypoxanthine and xanthine, are intermediates of the purine catabolic pathway as well as being substrates of the salvage pathways (Ashihara & Crozier, 1999; Deng & Ashihara, 2010). Therefore, significant amounts of these purine compounds are catabolised when they are exogenously applied to plant tissues. As a consequence, these compounds are not ideal precursors for the estimation of the rate of purine alkaloid biosynthesis. Nor are purine nucleotides, such as IMP, AMP and GMP, suitable compounds for assessing purine alkaloid biosynthesis activity because their rate of uptake by plant cells and tissues is very low. Consequently, only limited amounts of purine bases and nucleosides are likely to be produced and enter the cell, arguably as a result of hydrolysis of nucleotides in the cell wall by nucleotidases, phosphatases and nucleosidases. Examples of estimates based on this type of feeding study are found in reports on the changes in purine alkaloid biosynthesis during growth and development of leaves, flowers and fruits (Fujimori & Ashihara, 1990; Fujimori et al., 1991; Koshiro, Zheng, Wang, Nagai, & Ashihara, 2006; Koyama, Tomoda, Kato, & Ashihara, 2003; Terrasaki, Suzuki, & Ashihara, 1994).

3.6. Genes and molecular structure of *N*-methyltransferases

The *N*-methyltransferase gene associated with caffeine synthase was first obtained from leaves of *C. sinensis* (Kato et al., 2000). To obtain the caffeine synthase gene, the rapid amplification of complementary DNA ends (RACE) technique with degenerate gene-specific primers based on the *N*-terminal sequence of caffeine synthase was used. Subsequently, genes of *C. arabica* and other plants were isolated by polymerase chain reaction and library screening methods. This involved designing degenerated primers based on conserved regions in tea caffeine synthase (AB031280) and unknown proteins of *Arabidopsis thaliana*. Details of the cloning and identification of genes are reviewed elsewhere (Ashihara et al., 2008; Kato & Mizuno, 2004).

Comparisons of the amino acid sequences of the caffeine synthase family in *Coffea* and *Camellia* plants have been reported. Four highly conserved regions, motif A, motif B′, motif C and the YFFF, occur in the caffeine synthase amino acid sequence. Three conserved motifs of the binding site of the methyl donor of SAM (motifs A, B and C) have been reported in the majority of plant SAM-dependent *O*-methyltransferases

(Joshi & Chiang, 1998). The motif B′ and YFFF regions of the motif B′ methyltransferase family (also called the SABATH family) contains many specific hydrophobic amino acids. Most members of this newly characterized motif B′ methyltransferase family catalyse the formation of small, volatile methyl esters by using SAM as a methyl donor and substrates with a carboxyl group as the methyl acceptor. Members of this family include salicylic acid carboxyl methyltransferase, benzoic acid carboxyl methyltransferase, jasmonic acid carboxyl methyltransferase, farnesoic acid carboxyl methyltransferase, indole-3-acetic acid methyltransferase, gibberellic acid methyltransferase and loganic acid carboxyl methyltransferase (Ashihara, Kato, et al., 2011; Kato & Mizuno, 2004). The evolutionary relationship of caffeine synthase family enzymes has been summarized elsewhere (Ashihara, Kato, et al., 2011; Ishida et al., 2009).

Crystallographic data on salicylic acid carboxyl methyltransferase from *Clarkia breweri* suggest that members of this family exist as dimers in solution (Zubieta et al., 2003). Analysis of 7-methylxanthosine synthase and caffeine synthase from *C. canephora* has also revealed a dimeric structure (McCarthy & McCarthy, 2007).

Despite of the marked similarity in amino acid sequences of *N*-methyltransferases, each enzyme catalyzes the methylation of specific substrate(s). Some reports suggest that a single amino acid residue of the *N*-methyltransferases decides the substrate specificity (Ogawa et al., 2001; Yoneyama et al., 2006). Crystallography analyses of the structures of recombinant 7-methylxanthine synthase and caffeine synthase of *C. canephora* have revealed several elements that appear critical for substrate selectivity. Serine-316 in caffeine synthase appears central to the recognition of xanthosine. Likewise, glutamine-161 in 7-methylxanthosine synthase and histidine-160 in caffeine synthase are also important for substrate binding (McCarthy & McCarthy, 2007; McCarthy et al., 2007).

3.7. Regulation of caffeine biosynthesis

As with the synthesis of other plant secondary metabolites, caffeine biosynthesis is regulated primarily at steps in gene expression and the subsequent synthesis of enzymes. Changes in the expression rates of genes (transcript level) and quantity (activity) of key enzymes of the caffeine biosynthesis were often associated with various physiological changes. For example, the so-called 'coarse control' has been found during growth and development of leaves of *C. sinensis* (Fujimori et al., 1991; Li, Ogita,

Keya, & Ashihara, 2008; Mohanpuria et al., 2009) and fruits of *C. arabica* and *C. canephora* (Koshiro et al., 2006), and in stressed leaves of *T. cacao* (Bailey, Bae, et al., 2005; Bailey, Strem, Bae, de Mayolo, & Guiltinan, 2005).

Caffeine biosynthesis may also be regulated by the 'fine control' comprising the supply of substrates and/or methyl donor and feedback control of enzyme activity. The results obtained with tea tissue culture indicate that precursors of purine nucleotides, adenosine, guanosine or hypoxanthine added to the culture media influenced the rate of growth but not caffeine accumulation. Therefore, caffeine biosynthesis is not controlled by the availability of purine precursors (Deng, Li, Ogita, & Ashihara, 2008). In contrast, addition of paraxanthine (a preferred substrate of caffeine synthase) doubled the caffeine level compared to controls. This indicates that availability of SAM is not a principal factor in the control of caffeine biosynthesis (Deng et al., 2008).

4. CATABOLISM OF CAFFEINE

Caffeine appears to be the end product in most purine alkaloid-forming plants. However, limited amounts of caffeine are very slowly degraded with the removal of the three methyl groups, resulting in the formation of xanthine (Fig. 4.4). Catabolism of caffeine has been studied using ^{14}C-labelled caffeine (Ashihara, Gillies, et al., 1997; Ashihara, Takasawa, et al., 1997; Mazzafera, 2004; Suzuki & Waller, 1984a, 1984b). Caffeine catabolism begins with its conversion to theophylline in a reaction catalysed by a 7N-demethylase (step 1 in Fig. 4.4). This conversion is the rate-limiting step in purine alkaloid catabolism, because in contrast to caffeine, theophylline is readily catabolised. Interestingly, rapid catabolism of theophylline occurs only in purine alkaloid-forming plant species. Ito, Crozier, and Ashihara (1997) compared the metabolism of [8-^{14}C]theophylline in the purine alkaloid-forming plants *C. sinensis*, *C. irrawadiensis* and *Ilex paraguariensis* and in the non-producing plant species *Avena sativa*, *Vigna mungo* and *Catharanthus roseus*. Extensive uptake and metabolism of [8-^{14}C]theophylline were found only in the purine alkaloid-forming plants. The major route of caffeine catabolism is a caffeine → theophylline → 3-methylxanthine → xanthine → uric acid → allantoin → allantoic acid pathway (steps 1–6 in Fig. 4.4). Allantoic acid is degraded to CO_2 and NH_3. The xanthine catabolic pathway is present in all plants, but plural catabolic pathways to allantoic acid have been reported (Zrenner & Ashihara, 2011).

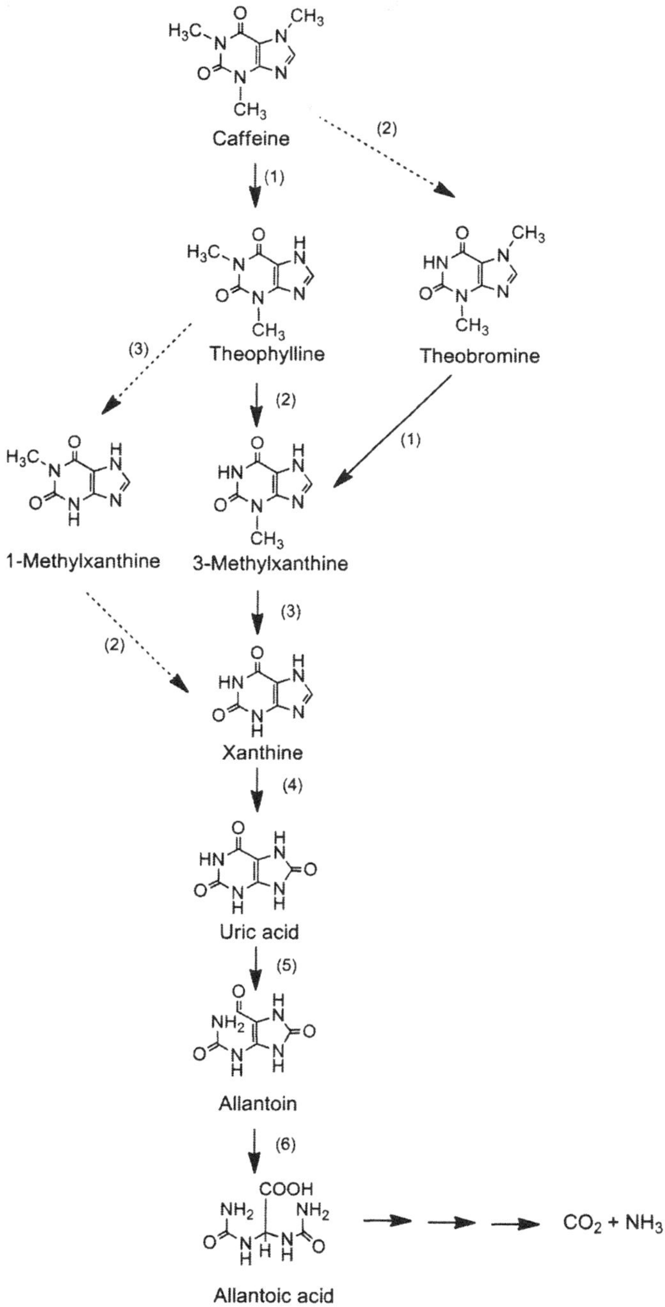

Figure 4.4 Possible routes for the catabolism of caffeine. After demethylation, xanthine enters the conventional oxidative purine catabolism pathway and is degraded to CO_2

In contrast to theophylline, theobromine is a precursor of caffeine biosynthesis in caffeine-producing plants. Degradation of theobromine has been observed in a theobromine-accumulating plant, *T. cacao* (Koyama et al., 2003; Zheng et al., 2004) where it was converted to 3-methylxanthine by a 7N-demethylase and then catabolised by the pathways discussed above.

5. DISTRIBUTION OF PURINE ALKALOIDS IN TISSUES AND CELLS

5.1. Distribution in tissues

In young tea leaves, immune-histochemical localization with primary anti-caffeine antibodies and conjugated secondary antibodies on leaf sections proved at the tissue level caffeine is localized within vascular bundles, mainly the precursor phloem (van Breda, Merwe, Robbertse, & Apostolides, 2013). Photosynthetic cells, that is, palisade and spongy parenchyma, also contain caffeine but in much lower concentrations, that is, amounts undetectable by immune-labelling and confocal scanning microscopy analysis. This could be misleading as the area of palisade and spongy parenchyma was much larger than that of the vascular tissues. Thus, more caffeine might be distributed over this area opposed to the smaller vascular area where it is concentrated and as a consequence gives a higher fluorescence signal.

Using RNA *in situ* hybridisation technique, Li, Gu, and Ye (2007) reported that the caffeine synthase gene was expressed mainly in the palisade parenchyma and the epicuticle of tea leaves and less so in the spongy parenchyma and hypoderm.

5.2. Subcellular distribution

In general, water-soluble secondary metabolites including many alkaloids are stored in vacuoles (Wink, 2010). Although no conclusive results on the vacuolar localization of caffeine has been demonstrated, it has been suggested that in coffee, caffeine is present in vacuoles as a complex with

and NH_3. The conversion of caffeine to theophylline is the rate-limiting step in caffeine-accumulating species such as *Coffea arabica* and *Camellia sinensis*. Solid arrows indicate major routes and dotted arrow minor conversions. The pathways are based on data obtained from the feeding experiments. As yet, no *N*-demethylase enzymes involved in caffeine degradation in plants have been characterized. Enzymes: (1) 7N-demethylase, (2) 1N-demethylase, (3) 3N-demethylase, (4) xanthine dehydrogenase (EC 1.1.1.204), (5) uricase (EC 3.5.1.5) and (6) allantoinase (EC 3.5.2.5).

chlorogenic acids (Mösli Waldhauser & Baumann, 1996). In tea and coffee seedlings, caffeine is synthesized exclusively in the green chlorophyll-containing tissues, but biosynthetic activity seemingly absent in roots (Ashihara & Kubota, 1986; Zheng & Ashihara, 2004). In addition, theobromine and caffeine biosynthesis occurs in cotyledons of developing cacao fruits (Zheng et al., 2004) and in young coffee seeds (Koshiro et al., 2006; Mazzafera, Crozier, & Sandberg, 1994). In tea leaves, using a biochemical fractionation with a Percoll density gradient of intracellular organelles, the activity of caffeine synthase, SAH hydrolase, adenosine nucleosidase, adenine phosphoribosyltransferase and adenosine kinase was associated with a purified chloroplast preparation from young tea leaves (Kato, Crozier, & Ashihara, 1998; Koshiishi et al., 2001). This suggests that, at least in tea leaves, caffeine biosynthesis is localized in chloroplasts where *de novo* and salvage purine nucleotide synthesis occur, and most members of the SAM cycle enzymes are present.

A molecular approach to investigate the localisation of caffeine synthase in coffee plants has been carried out by two groups. Ogawa et al. (2001) investigated the localisation using a cDNA fragments covering the entire coding region of the caffeine synthase gene fused to pGFP2. When the resulting plasmid was introduced into the epidermal layer of onions by particle bombardment, green fluorescence was detected in the cytoplasm.

Kumar et al. (2007) investigate the localisation of caffeine synthase in coffee endosperm cells using the promoter for one of the *N*-methyltransferase gene families involved in caffeine biosynthesis. These constructs and pCAMBIA 1301 bearing the intron *uidA* gene driven by the cauliflower mosaic virus (CaMV) *35S* promoter were electroporated into coffee endosperm, and the activity of β-glucuronidase (GUS) localized. In tissues transformed with the construct-containing promoter and first exon, enzymatic activity was localized on the outer surface of the vacuole. Antibodies to the coffee caffeine synthase were also specifically localized in the same region. In tissues bearing either the caffeine synthase-GUS construct without the first exon or pCAMBIA 1301 with intron GUS, GUS activity was spread throughout the cytoplasm. The results suggest that *N*-methyltransferase is targeted to the external surface of the vacuole.

6. BIOTECHNOLOGY OF PURINE ALKALOIDS

Using the gene sequences of *N*-methyltransferases involved in caffeine biosynthesis, two types of transgenic plants have been established. One is the

Table 4.3 Transgenic plants using genes involved in caffeine biosynthesis

Plant species	Transferred gene	Method	Purine alkaloid content	Purpose	References
Coffea canephora	*CaMXMT1*	Antisence, RNAi	3–10 mg/g FW	Decaffeinated coffee	Ogita et al. (2003)
Coffea arabica					
Camellia sinensis	*Caffeine synthase* (*FJ554589*)	RNAi	15–40 mg/g DW	Decaffeinated tea	Mohanpuria et al. (2011)
Nicotiana tabacum	*CaMXMT1*, *CaMXMTMT1/2*, *CaDXM1*	Sence	3–5 μg/g FW	Endogenous pesticides	Uefuji et al. (2005)
Chrysanthemum (*Dendranthema* × *grandiflorum* cv. Shinba)	*CaMXMT1*, *CaMXMTMT1/2*, *CaDXM1*	Sence	3–4 μg/g FW	Fungal resistance	Kim, Lim, Kang, et al. (2011) and Kim, Lim, Yoda, et al. (2011)

construction of genetically modified decaffeinated coffee and tea plants, in which caffeine production is suppressed. The other is the introduction of caffeine biosynthesis into non-caffeine-producing plants (Table 4.3). The first approach involves introducing antisense or the double-stranded RNA interference (RNAi) constructs for the caffeine synthase gene into coffee or tea (Mohanpuria, Kumar, Ahuja, & Yadav, 2011; Ogita, Uefuji, Yamaguchi, Nozomu, & Sano, 2003). The second was achieved using a multi-gene transfer system, involving cDNAs for all three *N*-methyltransferases of coffee plants, and the resulting transgenic plants successfully synthesized caffeine in leaves (Uefuji et al., 2005). It has been shown that transgenic caffeine-producing tobacco plants are resistant against tobacco cutworms (*Spodoptera litura*) and pathogenic microbes including *Pseudomonas syringe* and tobacco mosaic virus (Kim & Sano, 2008; Kim, Uefuji, Ogita, & Sano, 2006). Caffeine-producing chrysanthemum plants are resistant against herbivores, *Lepidoptera caterpillars* and aphids, which are among the most serious pests in agriculture (Kim, Lim, Kang, et al., 2011; Kim, Lim, Yoda, et al., 2011). While these finding are of interest, as yet no transgenic plants with a modified caffeine content has been marketed commercially. Technical details of transgenic plants with an altered caffeine content are reviewed elsewhere (Ashihara, Ogita, et al., 2011; Ogita, Uefuji, Morimoto, & Sano, 2005).

7. *IN PLANTA* FUNCTION OF PURINE ALKALOIDS

It has long been thought that purine alkaloids are the waste end products of purine nucleotides. Degradation of caffeine in most species is relatively slow even in aged leaves, and it appears not to act as a nitrogen reserve since considerable amounts remain in detached leaves following abscission (Suzuki et al., 1992). It has, however, also been proposed that purine alkaloids have an ecological role providing a chemical defence *in planta* against insect and vertebrate herbivores as well as fungi, bacteria and viruses. In such circumstances, the accumulation of caffeine in young leaves, fruits and flower buds of tea and coffee plants may act as a defence to protect young soft tissues from pathogens and herbivores. The results currently obtained from the transgenic caffeine-forming plants supports this hypothesis. It has been reported that transgenic caffeine-producing tobacco and chrysanthemum plants, respectively, have repellent effects against tobacco cutworm (*Spodoptera litura*) and fungal resistance (Kim, Lim, Kang, et al., 2011; Kim, Lim, Yoda, et al., 2011; Kim et al., 2006; Uefuji et al., 2005).

Kim and Sano (2008) found that caffeine-producing transgenic plants constitutively expressed defence-related genes encoding pathogenesis-related (PR)-1a and proteinase inhibitor II under non-stressed conditions. Transgenic tobacco plants were highly resistant against the pathogens, tobacco mosaic virus and *Pseudomonas syringae*. Expression of PR-1a and PR-2 was higher in transgenic plants than in wild-type plants following infection. Exogenous caffeine applied to wild-type tobacco leaves conferred similar resistance properties. These findings indicate that caffeine acts as a signal molecule activating the defence system of host plants by directly or indirectly initiating gene expression. Recently, it was shown that caffeine induces the production of a mildly toxic secondary metabolite *in planta* which stimulates endogenous self-defence systems, thereby conferring tolerance or resistance against biotic stresses (Kim, Choi, & Sano, 2010). These results, of what are essentially greenhouse-based studies, strongly suggest that caffeine can have a key role in chemical defence of plants. However, its ecological function in caffeine-forming plants in a natural ecosystem remains to be determined.

The chemical defence against neighbouring plants of the same or different species is described as allelopathic or autotoxic function. Caffeine in seed coats and falling leaves is released into the soil and inhibits germination of seeds around the parent plants. For example, Chou and Waller (1980) reported that aqueous extracts of leaves, stems and roots of *C. arabica*, in which caffeine was a major constituent, inhibited the germination and radicle growth of rye, lettuce and fescue seed. Caffeine may, therefore, be act as an allelopathic substance. It is, however, unclear to what extent caffeine is involved in allelopathy in natural ecosystems, especially as soil bacteria such as *Pseudomonas putida* can degrade purine alkaloids (Gluck & Lingens, 1988; Hohnloser, Osswald, & Lingens, 1980).

Recently, an interesting hypothesis has been proposed which associates caffeine in the nectar of flowers of *Coffea* and *Citrus* species with the memory of pollinating insects (Wright et al., 2013). When honeybees and other pollinators learn to associate a floral scent with food (i.e. nectar) while foraging, they are more likely to visit flowers bearing the same scent signals. It was shown that bees were three times more likely to remember a floral scent when it was associated with <0.3 mM caffeine, concentrations which prevail in the nectar of flowers of *Coffea* and *Citrus* species. Caffeine appears to potentiate responses of mushroom body neurons involved in olfactory learning and memory by acting as an adenosine receptor antagonist. Such behaviour increases the foraging efficiency of the bees while concomitantly leading to more effective pollination.

8. CONCLUSIONS AND PERSPECTIVES

It is over a decade since our first review article on purine alkaloids was published in the *Advances in Botanical Research* (Ashihara & Crozier, 1999). Subsequently, there has been significant progress in the cloning of genes encoding *N*-methyltransferases involved in caffeine biosynthesis from tea and coffee plants which has provided to support the presence of the biosynthetic pathway initiated from xanthosine; xanthosine → 7-methylxanthosine → 7-methylxanthine → theobromine → caffeine. Detailed properties of the *N*-methyltransferases involved in caffeine biosynthesis were revealed using recombinant enzyme proteins. In contrast, caffeine demethylases, key enzymes of caffeine catabolism, have not yet been isolated. There have been few advances in basic research topics such as the distribution of purine alkaloids in plant kingdom, the control mechanism of caffeine biosynthesis and catabolism, the cellular metabolic organization of caffeine biosynthesis links to *de novo* and salvage purine nucleotide biosynthesis, intercellular translocation and accumulation mechanisms. Much of the widespread interest of the general public in caffeine research is focused on applied area of research. Sano and his co-workers produced several transgenic plants using caffeine synthase genes including low caffeine-containing *Coffea* plants which have the potential to be used to produce genetically modified, decaffeinated coffee beans. Transgenic caffeine-producing tobacco and chrysanthemum plants were shown to offer resistance to animal, fungal and bacterial pests with purine alkaloid acting as an internal pesticide. These findings are fascinating but to date they have not been developed commercially.

Over the last decade, studies on purine alkaloids have made a number of key advances but at this juncture more research is needed both for fundamental research as well as topics that have the potential to offer substantial benefits to agriculture.

REFERENCES

Anderson, L., & Gibbs, M. (1962). The biosynthesis of caffeine in the coffee plant. *The Journal of Biological Chemistry*, *237*, 1941–1944.

Arabidopsis-Genome-Initiative, (2000). Analysis of the genome sequence of the flowering plant *Arabidopsis thaliana*. *Nature*, *408*(6814), 796–815.

Ashihara, H., & Crozier, A. (1999). Biosynthesis and metabolism of caffeine and related purine alkaloids in plants. *Advances in Botanical Research*, *30*, 117–205.

Ashihara, H., & Crozier, A. (2001). Caffeine: A well known but little mentioned compound in plant science. *Trends in Plant Science*, *6*, 407–413.

Ashihara, H., Gillies, F. M., & Crozier, A. (1997). Metabolism of caffeine and related purine alkaloids in leaves of tea (*Camellia sinensis* L.). *Plant & Cell Physiology*, *38*, 413–419.

Ashihara, H., Kato, M., & Crozier, A. (2011). Distribution, biosynthesis and catabolism of methylxanthines in plants. In B. B. Fredholm (Ed.), *Methylxanthines (handbook of experimental pharmacology)*, Vol. 200, (pp. 11–31). Berlin, Germany: Springer.

Ashihara, H., & Kubota, H. (1986). Patterns of adenine metabolism and caffeine biosynthesis in different parts of tea seedlings. *Physiologia Plantarum*, *68*, 275–281.

Ashihara, H., Monteiro, A. M., Gillies, F. M., & Crozier, A. (1996). Biosynthesis of caffeine in leaves of coffee. *Plant Physiology*, *111*, 747–753.

Ashihara, H., Ogita, S., & Crozier, A. (2011). Purine alkaloid metabolism. In H. Ashihara, A. Crozier, & A. Komamine (Eds.), *Plant metabolism and biotechnology* (pp. 163–189). Chichester, UK: John Wiley & Sons, Ltd.

Ashihara, H., Sano, H., & Crozier, A. (2008). Caffeine and related purine alkaloids: Biosynthesis, catabolism, function and genetic engineering. *Phytochemistry*, *69*, 841–856.

Ashihara, H., Takasawa, Y., & Suzuki, T. (1997). Metabolic fate of guanosine in higher plants. *Physiologia Plantarum*, *100*, 909–916.

Bailey, B. A., Bae, H., Strem, M. D., Bowers, J. H., Antúnez De Mayolo, G., Guiltinan, M. J., et al. (2005). Developmental expression of stress response genes in *Theobroma cacao* leaves and their response to Nep1 treatment and a compatible infection by *Phytophthora megakarya*. *Plant Physiology and Biochemistry*, *43*, 611–622.

Bailey, B. A., Strem, M. D., Bae, H., de Mayolo, G. A., & Guiltinan, M. J. (2005). Gene expression in leaves of *Theobroma cacao* in response to mechanical wounding, ethylene, and/or methyl jasmonate. *Plant Science*, *168*, 1247–1258.

Baumann, T. W., Oechslin, M., & Wanner, H. (1976). Caffeine and methylated uric acids: Chemical patterns during vegatative development of *Coffea liberica*. *Biochemie und Physiologie der Pflanzen*, *170*, 217–225.

Chou, C.-H., & Waller, G. (1980). Possible allelopathic constituents of *Coffea arabica*. *Journal of Chemical Ecology*, *6*, 643–654.

Crozier, A., Ashihara, H., & Tomas-Barberan, F. (2012). *Teas, cocoa and coffee: Plant secondary metabolites and health*. Oxford, UK: Wiley-Blackwell.

Deng, W.-W., & Ashihara, H. (2010). Profiles of purine metabolism in leaves and roots of *Camellia sinensis* seedlings. *Plant & Cell Physiology*, *51*, 2105–2118.

Deng, W.-W., Li, Y., Ogita, S., & Ashihara, H. (2008). Fine control of caffeine biosynthesis in tissue cultures of *Camellia sinensis*. *Phytochemistry Letters*, *1*, 195–198.

Fredholm, B. B. (2011). *Methylxanthines (handbook of experimental pharmacology)* (Vol. 200). Berlin, Germany: Springer.

Fujimori, N., & Ashihara, H. (1990). Adenine metabolism and the synthesis of purine alkaloids in flowers of *Camellia* plants. *Phytochemistry*, *29*, 3513–3516.

Fujimori, N., Suzuki, T., & Ashihara, H. (1991). Seasonal variations in biosynthetic capacity for the synthesis of caffeine in tea leaves. *Phytochemistry*, *30*, 2245–2248.

Gluck, M., & Lingens, F. (1988). Heteroxanthinedemethylase, a new enzyme in the degradation of caffeine by *Pseudomonas putida*. *Applied Microbiology and Biotechnology*, *28*, 59–62.

Hammerstone, J. F., Romanczyk, L. J., & Aitken, W. M. (1994). Purine alkaloid distribution within *Herrania* and *Theobroma*. *Phytochemistry*, *35*, 1237–1240.

Hohnloser, W., Osswald, B., & Lingens, F. (1980). Enzymological aspects of caffeine demethylation and formaldehyde oxidation by *Pseudomonas putida* C1. *Hoppe-Seyler's Zeitschrift für Physiologishe Chemie*, *361*, 1763–1766.

International-Rice-Genome-Sequencing-Project (2005). The map-based sequence of the rice genome. *Nature*, *436*, 793–800.

Ishida, M., Kitao, N., Mizuno, K., Tanikawa, N., & Kato, M. (2009). Occurrence of theobromine synthase genes in purine alkaloid-free species of *Camellia* plants. *Planta*, *229*, 559–568.

Ito, E., & Ashihara, H. (1999). Contribution purine nucleotide biosynthesis *de novo* to the formation of caffeine in young tea (*Camellia sinensis*) leaves. *Journal of Plant Physiology, 154*, 145–151.

Ito, E., Crozier, A., & Ashihara, H. (1997). Theophylline metabolism in higher plants. *Biochimica et Biophysica Acta, General Subjects, 1336*, 323–330.

Joshi, C. P., & Chiang, V. L. (1998). Conserved sequence motifs in plant S-adenosyl-L-methionine-dependent methyltransferases. *Plant Molecular Biology, 37*, 663–674.

Kato, A., Crozier, A., & Ashihara, H. (1998). Subcellular localization of the *N*-3 methyltransferase involved in caffeine biosynthes in tea. *Phytochemistry, 48*, 777–779.

Kato, M., Kanehara, T., Shimizu, H., Suzuki, T., Gillies, F. M., Crozier, A., et al. (1996). Caffeine biosynthesis in young leaves of *Camellia sinensis*: *In vitro* studies on *N*-methyltransferase activity involved in the conversion of xanthosine to caffeine. *Physiologia Plantarum, 98*, 629–636.

Kato, M., & Mizuno, K. (2004). Caffeine synthase and related methyltransferases in plants. *Frontiers in Bioscience, 9*, 1833–1842.

Kato, M., Mizuno, K., Crozier, A., Fujimura, T., & Ashihara, H. (2000). A gene encoding caffeine synthase from tea leaves. *Nature, 406*, 956–957.

Kato, M., Mizuno, K., Fujimura, T., Iwama, M., Irie, M., Crozier, A., et al. (1999). Purification and characterization of caffeine synthase from tea leaves. *Plant Physiology, 120*, 579–586.

Keya, C. A., Crozier, A., & Ashihara, H. (2003). Inhibition of caffeine biosynthesis in tea (*Camellia sinensis*) and coffee (*Coffea arabica*) plants by ribavirin. *FEBS Letters, 554*, 473–477.

Kihlman, B. A. (1977). *Occurence and biosynthesis of methylated oxypurines in plants Caffeine and Chromosome*. (pp. 11–18), Amsterdam, The Netherlands: Elsevier.

Kim, Y. S., Choi, Y. E., & Sano, H. (2010). Plant vaccination: Stimulation of defense system by caffeine production in planta. *Plant Signaling & Behavior, 5*, 489–493.

Kim, Y. S., Lim, S., Kang, K. K., Jung, Y. J., Lee, Y. H., Choi, Y. E., et al. (2011). Resistance against beet armyworms and cotton aphids in caffeine-producing transgenic chrysanthemum. *Plant Biotechnology, 28*, 393–395.

Kim, Y. S., Lim, S., Yoda, H., Choi, C. S., Choi, Y. E., & Sano, H. (2011). Simultaneous activation of salicylate production and fungal resistance in transgenic chrysanthemum producing caffeine. *Plant Signaling & Behavior, 6*, 409–412.

Kim, Y. S., & Sano, H. (2008). Pathogen resistance of transgenic tobacco plants producing caffeine. *Phytochemistry, 69*, 882–888.

Kim, Y. S., Uefuji, H., Ogita, S., & Sano, H. (2006). Transgenic tobacco plants producing caffeine: A potential new strategy for insect pest control. *Transgenic Research, 15*, 667–672.

Koshiishi, C., Kato, A., Yama, S., Crozier, A., & Ashihara, H. (2001). A new caffeine biosynthetic pathway in tea leaves: Utilisation of adenosine released from the S-adenosyl-L-methionine cycle. *FEBS Letters, 499*, 50–54.

Koshiro, Y., Zheng, X. Q., Wang, M., Nagai, C., & Ashihara, H. (2006). Changes in content and biosynthetic activity of caffeine and trigonelline during growth and ripening of *Coffea arabica* and *Coffea canephora* fruits. *Plant Science, 171*, 242–250.

Koyama, Y., Tomoda, Y., Kato, M., & Ashihara, H. (2003). Metabolism of purine bases, nucleosides and alkaloids in theobromine-forming *Theobroma cacao* leaves. *Plant Physiology and Biochemistry, 41*, 977–984.

Kretschmar, J. A., & Baumann, T. W. (1999). Caffeine in citrus flowers. *Phytochemistry, 52*, 19–23.

Kumar, V., Satyanarayana, K. V., Ramakrishna, A., Chandrashekar, A., & Ravishankar, G. A. (2007). Evidence for localization of *N*-methyltransferase (MMT) of caffeine biosynthetic

pathway in vacuolar surface of *Coffea canephora* endosperm elucidated through localization of GUS reporter gene driven by NMT promoter. *Current Science, 93*, 383–389.
Lean, M. E. J., Ashihara, H., Clifford, M. N., & Crozier, A. (2012). Purine alkaloids: A focus on caffeine and related compounds in beverages. In A. Crozier, H. Ashihara, & F. Tomas-Barberan (Eds.), *Teas, cocoa and coffee: Plant secondary metabolites and health* (pp. 24–44). Oxford, UK: Wiley-Blackwell.
Li, Y. H., Gu, W., & Ye, S. (2007). Expression and location of caffeine synthase in tea plants. *Russian Journal of Plant Physiology, 54*, 698–701.
Li, Y., Ogita, S., Keya, C. A., & Ashihara, H. (2008). Expression of caffeine biosynthesis gene in tea (*Camellia sinensis*). *Zeitschrift für Naturforschung. C, A journal of Biosciences, 63c*, 267–270.
Lu, J.-L., Wang, D.-M., Shi, X.-G., Yang, D.-P., Zheng, X.-Q., & Ye, C.-X. (2009). Determination of purine alkaloids and catechins in different parts of *Camellia assamica* var. kucha by HPLC-DAD/ESI-MS/MS. *Journal of the Science of Food and Agriculture, 89*, 2024–2029.
Mazzafera, P. (2004). Catabolism of caffeine in plants and microorganisms. *Frontiers in Bioscience, 9*, 1348–1359.
Mazzafera, P., Baumann, T., Shimizu, M., & Silvarolla, M. (2009). Decaf and the steeplechase towards decaffito—The coffee from caffeine-free Arabica plants. *Tropical Plant Biology, 2*, 63–76.
Mazzafera, P., & Carvalho, A. (1992). Breeding for low seed caffeine content of coffee (*Coffea* L.) by interspecific hybridization. *Euphytica, 59*, 55–60.
Mazzafera, P., Crozier, A., & Sandberg, G. (1994). Studies on the metabolic control of caffeine turnover in developing endosperms and leaves of *Coffea arabica* and *Coffea dewevrei*. *Journal of Agricultural and Food Chemistry, 42*, 1423–1427.
Mazzafera, P., Wingsle, G., Olsson, O., & Sandberg, G. (1994). *S*- Adenosyl-L-methionine: theobromine 1-*N*-methyltransferase, An enzyme catalysing the synthesis of caffeine in coffee. *Phytochemistry, 17*, 1577–1584.
McCarthy, A. A., Biget, L., Lin, C., Petiard, V., Tanksley, S. D., & McCarthy, J. G. (2007). Cloning, expression, crystallization and preliminary X-ray analysis of the XMT and DXMT *N*-methyltransferases from *Coffea canephora* (robusta). *Acta Crystallographica. Section F, Structural Biology and Crystallization Communications, 63*, 304–307.
McCarthy, A. A., & McCarthy, J. G. (2007). The structure of two *N*-methyltransferases from the caffeine biosynthetic pathway. *Plant Physiology, 144*, 879–889.
Mizuno, K., Kato, M., Irino, F., Yoneyama, N., Fujimura, T., & Ashihara, H. (2003). The first committed step reaction of caffeine biosynthesis: 7-methylxanthosine synthase is closely homologous to caffeine synthases in coffee (*Coffea arabica* L.). *FEBS Letters, 547*, 56–60.
Mizuno, K., Tanaka, H., Kato, M., Ashihara, H., & Fujimura, T. (2001). cDNA cloning of caffeine (theobromine) synthase from coffee (*Coffea arabica* L.). *Colloque Scientifique International sur le Café, 19*, 815–818.
Mohanpuria, P., Kumar, V., Ahuja, P. S., & Yadav, S. K. (2011). Producing low-caffeine tea through post-transcriptional silencing of caffeine synthase mRNA. *Plant Molecular Biology, 76*, 523–534.
Mohanpuria, P., Kumar, V., Joshi, R., Gulati, A., Ahuja, P., & Yadav, S. (2009). Caffeine biosynthesis and degradation in tea [*Camellia sinensis* (L.) O. Kuntze] is under developmental and seasonal regulation. *Molecular Biotechnology, 43*, 104–111.
Mösli Waldhauser, S. S., & Baumann, T. W. (1996). Compartmentation of caffeine and related purine alkaloids depends exclusively on the physical chemistry of their vacuolar complex formation with chlorogenic acids. *Phytochemistry, 42*, 985–996.
Mösli Waldhauser, S. S., Gillies, F. M., Crozier, A., & Baumann, T. W. (1997). Separation of the *N*-7 methyltransferase, the key enzyme in caffeine biosynthesis. *Phytochemistry, 45*, 1407–1414.

Mösli Waldhauser, S. S., Kretschmar, J. A., & Baumann, T. W. (1997). *N*-Methyltransferase activities in caffeine biosynthesis: biochemical characterization and time course during leaf development of *Coffea arabica*. *Phytochemistry, 44*, 853–859.

Negishi, O., Ozawa, T., & Imagawa, H. (1985). Methylation of xanthosine by tea-leaf extracts and caffeine biosynthesis. *Agricultural and Biological Chemistry, 49*, 887–890.

Negishi, O., Ozawa, T., & Imagawa, H. (1988). *N*-methylnucleosidase from tea leaves. *Agricultural and Biological Chemistry, 52*, 169–175.

Negishi, O., Ozawa, T., & Imagawa, H. (1994). Guanosine deaminase and guanine deaminase from tea leaves. *Bioscience, Biotechnology, and Biochemistry, 58*, 1277–1281.

O'Connell, F. D. (1969). Isolation of caffeine from *Banisteriopsis inebrians* (Malpighiaceae). *Naturwissenschaften, 56*, 139.

Ogawa, M., Herai, Y., Koizumi, N., Kusano, T., & Sano, H. (2001). 7-Methylxanthine methyltransferase of coffee plants. Gene isolation and enzymatic properties. *The Journal of Biological Chemistry, 276*, 8213–8218.

Ogita, S., Uefuji, H., Morimoto, M., & Sano, H. (2005). Metabolic engineering of caffeine production. *Plant Biotechnology, 22*, 461–468.

Ogita, S., Uefuji, H., Yamaguchi, Y., Nozomu, K., & Sano, H. (2003). Producing decaffeinated coffee plants. *Nature, 423*, 823.

Petermann, J., Baumann, T. W., & Wanner, H. (1977). A new tetramethyluric acid from *Coffea liberica* and *C. Dewevre*. *Phytochemistry, 16*, 620–621.

Proiser, E., & Serenkov, G. P. (1963). Caffeine biosynthesis in tea leaves. *Biokhimiya (Moscow), 28*, 857–861.

Schulthess, B. H., Morath, P., & Baumann, T. W. (1996). Caffeine biosynthesis starts with the metabolically channelled formation of 7-methyl-XMP—A new hypothesis. *Phytochemistry, 41*, 169–175.

Stasolla, C., Katahira, R., Thorpe, T. A., & Ashihara, H. (2003). Purine and pyrimidine nucleotide metabolism in higher plants. *Journal of Plant Physiology, 160*, 1271–1295.

Stewart, I. (1985). Identification of caffeine in citrus flowers and leaves. *Journal of Agricultural and Food Chemistry, 33*, 1163–1165.

Suzuki, T., Ashihara, H., & Waller, G. R. (1992). Purine and purine alkaloid metabolism in *Camellia* and *Coffea* plants. *Phytochemistry, 31*, 2575–2584.

Suzuki, T., & Waller, G. R. (1984a). Biodegradation of caffeine: Formation of theophylline and caffeine in mature *Coffea arabica* fruits. *Journal of the Science of Food and Agriculture, 35*, 66–70.

Suzuki, T., & Waller, G. R. (1984b). Biosynthesis and biodegradation of caffeine, theobromine, and theophylline in *Coffea arabica* L. fruits. *Journal of Agricultural and Food Chemistry, 32*, 845–848.

Terrasaki, Y., Suzuki, T., & Ashihara, H. (1994). Purine metabolism and the biosynthesis of purine alkaloids in tea fruits during development. *Plant Physiology (Life Science Advances), 13*, 135–142.

Uefuji, H., Ogita, S., Yamaguchi, Y., Koizumi, N., & Sano, H. (2003). Molecular cloning and functional characterization of three distinct *N*-methyltransferases involved in the caffeine biosynthetic pathway in coffee plants. *Plant Physiology, 132*, 372–380.

Uefuji, H., Tatsumi, Y., Morimoto, M., Kaothien-Nakayama, P., Ogita, S., & Sano, H. (2005). Caffeine production in tobacco plants by simultaneous expression of three coffee *N*-methyltrasferases and its potential as a pest repellant. *Plant Molecular Biology, 59*, 221–227.

van Breda, S. V., Merwe, C. F., Robbertse, H., & Apostolides, Z. (2013). Immunohistochemical localization of caffeine in young *Camellia sinensis* (L.) O. Kuntze (tea) leaves. *Planta, 237*, 849–858. http://dx.doi.org/10.1007/s00425-012-1804-x.

van der Graaff, E., Hooykaas, P., Lein, W., Lerchl, J., Kunze, G., Sonnewald, U., et al. (2004). Molecular analysis of "*de novo*" purine biosynthesis in solanaceous species and in *Arabidopsis thaliana*. *Frontiers in Bioscience, 9*, 1803–1816.

Willaman, J. J., & Schubert, B. G. (1961). Alkaloid-bearing plants and their contained alkaloid. *Technical Bulletin, Agricultural Reserch Service, US Department of Agriculture* No. 1234, 1.

Wink, M. (2010). Introduction: Biochemistry, physiology and ecological functions of secondary metabolites. In M. Wink (Ed.), *Biochemistry of plant secondary metabolism. Annual plant reviews*: vol. 40, (pp. 1–19). Chichester, UK: Wiley-Blackwell.

Wright, G. A., Baker, D. D., Palmer, M. J., Stabler, D., Mustard, J. A., Power, E. F., et al. (2013). Caffeine in floral nectar enhances a pollinator's memory of reward. *Science, 339*, 1202–1204.

Yoneyama, N., Morimoto, H., Ye, C. X., Ashihara, H., Mizuno, K., & Kato, M. (2006). Substrate specificity of *N*-methyltransferase involved in purine alkaloids synthesis is dependent upon one amino acid residue of the enzyme. *Molecular Genetics and Genomics, 275*, 125–135.

Zheng, X. Q., & Ashihara, H. (2004). Distribution, biosynthesis and function of purine and pyridine alkaloids in *Coffea arabica* seedlings. *Plant Science, 166*, 807–813.

Zheng, X. Q., Koyama, Y., Nagai, C., & Ashihara, H. (2004). Biosynthesis, accumulation and degradation of theobromine in developing *Theobroma cacao* fruits. *Journal of Plant Physiology, 161*, 363–369.

Zheng, X. Q., Ye, C. X., Kato, M., Crozier, A., & Ashihara, H. (2002). Theacrine (1,3,7,9-tetramethyluric acid) synthesis in leaves of a Chinese tea, kucha (*Camellia assamica* var. kucha). *Phytochemistry, 60*, 129–134.

Zrenner, R., & Ashihara, H. (2011). Nucleotide metabolism. In H. Ashihara, A. Crozier, & A. Komamine (Eds.), *Plant metabolism and biotechnology* (pp. 135–162). Chichester, UK: John Wiley & Sons, Ltd.

Zrenner, R., Stitt, M., Sonnewald, U., & Boldt, R. (2006). Pyrimidine and purine biosynthesis and degradation in plants. *Annual Review of Plant Biology, 57*, 805–836.

Zubieta, C., Ross, J. R., Koscheski, P., Yang, Y., Pichersky, E., & Noel, J. P. (2003). Structural basis for substrate recognition in the salicylic acid carboxyl methyltransferase family. *The Plant Cell, 15*, 1704–1716.

CHAPTER FIVE

Camptothecin: Biosynthesis, Biotechnological Production and Resistance Mechanism(s)

Supaart Sirikantaramas*, Mami Yamazaki†, Kazuki Saito†,‡,1
*Department of Biochemistry, Faculty of Science, Chulalongkorn University, Bangkok, Thailand
†Graduate School of Pharmaceutical Sciences, Chiba University, Chuo-ku, Chiba, Japan
‡RIKEN Center for Sustainable Resource Science, Tsurumi-ku, Yokohama, Japan
[1]Corresponding author: e-mail address: ksaito@faculty.chiba-u.jp

Contents

1. Introduction 140
2. Camptothecin Biosynthetic Pathway 143
   2.1 Tryptamine and secologanin pathways 143
   2.2 Post-strictosidine pathway 147
3. Biotechnological Production of Camptothecin and Its Derivatives 150
   3.1 New plant sources 150
   3.2 Callus and cell suspension culture 151
   3.3 Hairy roots 152
   3.4 Endophytic fungi 153
4. Resistance Mechanism(s) in Camptothecin-Producing Plants 154
5. Conclusions 155
Acknowledgements 156
References 156

Abstract

Camptothecin (CPT) is a water insoluble and cytotoxic monoterpene indole alkaloid, which is used as the substrate to form water-soluble derivatives (such as topotecan and irinotecan) for use as anti-cancer drugs. CPT has been found in at least 16 different plant species belonging to 3, 5 and 13 unrelated plant orders, families and genera, respectively, across the plant kingdom and also in endophytic fungi associated with these CPT-producing plants. Increasing demand for CPT to satisfy chemotherapy requirements and a shortage of *Camptotheca acuminata* and *Nothapodytes foetida* used as the commercial sources of CPT are driving the need to find alternative sources for its production. Although the biosynthetic pathway of CPT remains poorly understood, limiting the powerful approach via metabolic engineering, several different biotechnological production technologies for CPT have been reported using plant tissue/organ

Advances in Botanical Research, Volume 68
ISSN 0065-2296
http://dx.doi.org/10.1016/B978-0-12-408061-4.00005-5

cultures. In this chapter, the current understanding of the CPT biosynthetic pathway, including the biosynthetic genes and intermediate metabolites, is outlined. Then the different natural and biotechnological sources for CPT production are discussed. Finally, how CPT-producing organisms resist their own toxic metabolite and how this knowledge may potentially be of use in CPT-resistance management is covered.

ABBREVIATIONS

ASA anthranilate synthase
CPR NADPH:cytochrome P450 reductases
CPT camptothecin
DMAPP dimethylallyl diphosphate
DNA deoxyribonucleic acid
DXP 1-deoxy-D-xylulose-5-phosphate
DXR 1-deoxy-D-xylulose-5-phosphate reductoisomerase
DXS 1-deoxy-D-xylulose-5-phosphate synthase
G10H geraniol 10-hydroxylase
GPP geranyl diphosphate
GS geraniol synthase
HDR 1-hydroxy-2-methyl-2(*E*)-butenyl-4-diphosphate reductase
HIF-1 hypoxia-inducible factor 1
HMBPP 1-hydroxy-2-methyl-2(*E*)-butenyl-4-diphosphate
IPP isopentenyl diphosphate
MEP 2C-methyl-D-erythritol-4-phosphate
MVA mevalonate
SLS secologanin synthase
STR strictosidine synthase
TDC tryptophan decarboxylase
TIA monoterpenoid indole alkaloid
TSA the alpha-subunit of tryptophan synthase
TSB the beta-subunit of tryptophan synthase

1. INTRODUCTION

The water-insoluble camptothecin (CPT) is a tryptophan-derived quinoline alkaloid with a unique pentacyclic structure (Fig. 5.1) that is synthesized via the monoterpenoid indole alkaloid (TIA) pathway from tryptophan and terpenoid precursors. It was first identified from the Chinese Happy tree ("Xi Shu"), *Camptotheca acuminata* (Wall et al., 1966) and found to exhibit anti-cancer activity by inhibiting the religation function of DNA topoisomerase I (Hsiang, Hertzberg, Hecht, & Liu, 1985), an enzyme with

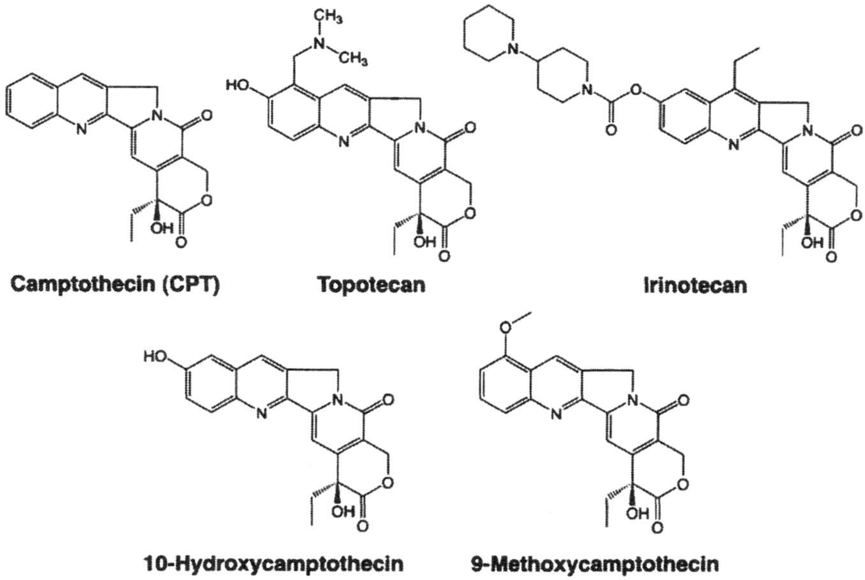

Figure 5.1 Structures of camptothecin and its derivatives.

deoxyribonucleic acid (DNA) topology modification via single strand breaks to, for example, remove supercoils. CPT is thought to stabilize the topoisomerase I–DNA covalent complex and so selectively inhibit the topoisomerase I activity (reviewed in Pommier, 2006, 2009). These CPT-stabilized covalent cleavage complexes act as physical barriers to DNA synthesis (replication and repair) and chromatin structure and gene transcription control and kill cells as a result of replication fork collision. Accordingly, CPT transforms the enzyme into an intracellular cytotoxic poison and hence CPTs are named topoisomerase "poisons" rather than conventional enzyme inhibitors (Svejstrup, Christiansen, Gromova, Andersen, & Westergaard, 1991). Note that CPTs also act as potent inhibitors of replication, transcription and packing of viruses that require the host topoisomerase I such as double-stranded DNA-containing adenoviruses, papova viruses and herpes viruses or the single-stranded DNA-containing autonomous parvoviruses (reviewed in Pantazis, Han, Chatterjee, & Wyche, 1999).

In addition, CPT functions as an anti-cancer agent through its inhibition of the hypoxia-inducible factor 1 (HIF-1), a master regulator of the ability of many cancer cells to survive under conditions of oxygen deprivation.

Indeed, in a screen for such inhibitors, of the four compounds that were found to exhibit HIF-1 inhibitory activity, three of these were CPTs (Rapisarda et al., 2002).

Due to its low aqueous solubility and inactivation at physiological pH, several water-soluble CPT derivatives have been developed and subsequently approved for clinical use (reviewed in Pommier, Leo, Zhang, & Marchand, 2010). These include the Food and Drug Administration-approved and clinically administered topotecan (Hycamtin) and irinotecan (Camptosar or CPT-11; Fig. 5.1), which have a significant activity against adult and paediatric solid tumours and have been used for the treatment of colon, ovarian and small-cell lung cancers (Douillard et al., 2000; Rodriguez-Galindo et al., 2000; Stewart, 2004). Therefore, the topoisomerase I–drug interactions have been extensively studied in order to gain more information to improve the efficiency of inhibitory drugs for cancer treatment (Koster, Palle, Bot, Bjornsti, & Dekker, 2007) or as fungicides.

Because of the increasing demand for CPT, the search for other CPT-producing sources has been of great interest. In fact, at least 16 CPT-producing plants have currently been reported, including the first reported plants *C. acuminata* and *Nothapodytes foetida*, the two species it is commercially harvested from, and other diverse tree/plant species including *Camptotheca lowreyana*, *C. yunnanensis*, *Ervatamia heyneana*, *Merrilliodendron megacarpum*, *Mostuea brunonis*, *Ophiorrhiza filistipula*, *O. fucosa*, *O. harrisiana*, *O. kuroiwai*, *O. leukiuensis*, *O. mungos*, *O. plumbea*, *O. prostrata*, *O. pumila*, *O. ridleyana*, *O.* sp. 35 and *Pyrenacantha klaineana* (Aimi et al., 1989; Aiyama, Nagai, Nokata, Shinohara, & Sawada, 1988; Arisawa, Gunasekera, Cordell, & Farnsworth, 1981; Asano et al., 2004; Beegum et al., 2007; Dai, Cardellina, & Boyd, 1999; Gunasekera, Badawi, Cordell, Farnsworth, & Chitnis, 1979; Li, Yi, Wang, Zhang, & Beasley, 2002; Saito et al., 2001; Tafur, Nelson, DeLong, & Svoboda, 1976; Viraporn et al., 2011; Wall et al., 1966; Zhou et al., 2000).

This review begins in Section 2 by outlining the current understanding of the CPT biosynthetic pathway, which has not yet been fully elucidated, and the advanced technologies that are necessary to identify the remaining genes and metabolites involved in the CPT production. Then, in Section 3, the biotechnological production systems that are being developed to enhance the production of CPT and its derivatives from different sources are reviewed. Finally, in Section 4, an outline of the current understanding of the self-resistance mechanism(s) of CPT-producing plants and how this could potentially be used to either develop more efficient drugs or overcome drug resistance is outlined.

2. CAMPTOTHECIN BIOSYNTHETIC PATHWAY

Although CPT was first isolated almost 50 years ago, its biosynthetic pathway still remains poorly understood (Fig. 5.2). Broadly speaking, CPT biosynthesis follows the shikimate pathway of (i) the common primary metabolism of tryptophan from chorismate, and then branching into the common secondary TIA pathway starting from the decarboxylation of tryptophan to tryptamine, in parallel with (ii) the non-mevalonate 2C-methyl-D-erythritol-4-phosphate (MEP) pathway leading to the formation of secologanin, from pyruvate and glyceraldehyde-3-phosphate, and then the merging of these in (iii) the final specific pathway to CPT starting from the formation of strictosidine from tryptamine and secologanin. However, in the second (common TIA anabolism) and especially the third (CPT-specific anabolism) pathways, many of the biosynthetic enzymes and intermediates leading to CPT are evidently missing. The pathway leading to strictosidine, the universal precursor for TIAs, is mostly understood and the best defined part, but in contrast beyond this, the pathway is poorly resolved. Due to the complexity of the overall pathway, here for convenience we have divided it into the two main parts of the (i) pre-strictosidine event (tryptamine and secologanin pathway) and the (ii) post-strictosidine pathway.

2.1. Tryptamine and secologanin pathways

2.1.1 Chorismate to tryptamine (shikimate pathway)

Hutchinson, Heckendorf, Straughn, Daddona, and Cane (1979) first demonstrated the involvement of tryptamine and secologanin in CPT

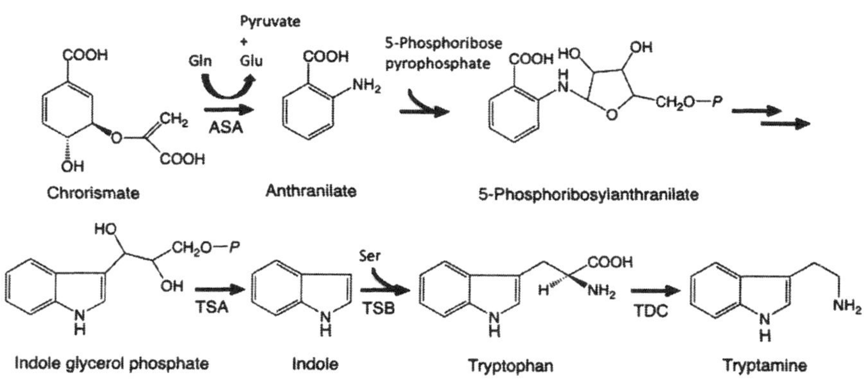

Figure 5.2 Tryptamine biosynthetic pathway. ASA, anthranilate synthase; TSA, the alpha-subunit of tryptophan synthase; TSB, the beta-subunit of tryptophan synthase; TDC, tryptophan decarboxylase; Gln, glutamine; Glu, glutamic acid; Ser, serine; *P*, phosphate group.

biosynthesis in *C. acuminata* using radioactive precursor feeding experiments. Three biosynthetic enzymes involved in the pathway leading to tryptamine have been characterized in CPT-producing plants (Fig. 5.2). Tryptophan biosynthesis begins with the conversion of chorismate to anthranilate by the alpha-subunit of anthranilate synthase (ASA). Although there are two non-identical copies of *ASA* genes in *C. acuminata* that are differently regulated, gene expression analysis supports that one of these is likely to be involved in the early indole pathway and CPT biosynthesis (Lu, Gorman, & McKnight, 2005).

After the production of the intermediates 5-phosphoribosylanthranilate and indole glycerol phosphate, the alpha-subunit of tryptophan synthase (TSA) produces indole, which is then condensed with serine to yield tryptophan. This reaction is catalyzed by the beta-subunit of tryptophan synthase (TSB). TSB from *C. acuminata* has been cloned and characterized (Lu & McKnight, 1999), revealing that TSB mRNA and protein levels are strongly correlated with the CPT contents in all the different plant organs examined, supporting its involvement in CPT production.

Tryptophan conversion to tryptamine by decarboxylation is then catalyzed by tryptophan decarboxylase (TDC; EC 4.1.1.28; Noé, Mollenschott, & Berlin, 1984), which is the branching point from primary metabolism into the TIA-specific secondary pathway. The cDNA clone encoding TDC was isolated from the Madagascar periwinkle plant, *Catharanthus roseus* (De Luca, Marineau, & Brisson, 1989) and subsequently from *C. acuminata* and *O. pumila* (Lopez-Meyer & Nessler, 1997; Yamazaki, Sudo, Yamazaki, Aimi, & Saito, 2003). Two different isoforms of TDC were reported in *C. acuminata* (the 502 and 498 amino acid gene products of *TDC*1 and *TDC2*, respectively) that have an 84.1% amino acid sequence identity to each other. The TDC1 isoform is developmentally regulated, while the TDC2 isoform might be involved in the plant's defence system (Lopez-Meyer & Nessler, 1997). However, only one copy of *TDC* has been isolated from *O. pumila* with a gene product showing 72% and 74% amino acid identity to the *C. acuminata* TDC1 and TDC2 isoforms and 67% identity to the TDC from *C. roseus* (Yamazaki, Sudo, et al., 2003). The *O. pumila* TDC shows a different induction pattern after exposure of the plant to the stress compounds compared with that observed in *C. acuminata*, suggesting that the coordinated regulation of expression of the *TDC* genes is different between these two CPT-producing plants. Moreover, the cell-specific expression of the *C. acuminata TDC1* and *TDC2* isoforms do not match the accumulation sites of CPT, indicating the likely intercellular transport of CPT *in planta* (Valletta, Trainotti, Santamaria, & Pasqua, 2010).

2.1.2 MEP pathway

2.1.2.1 Isopentenyl diphosphate formation

The formation of isopentenyl diphosphate (IPP), the precursor of terpenoid biosynthesis, can occur via both the mevalonate (MVA) pathway and the methylerythritol phosphate (MEP) pathway. Both isoprenoid pathways are operative simultaneously in higher plants. The MVA pathway enzymes are located in the cytoplasm where they supply precursors of triterpenes, sesquiterpenes and sterols, whereas the enzymes of the MEP pathway are believed to be located in plastids where they produce precursors for monoterpenes, some sesquiterpenes, diterpenes and carotenoids (Eisenreich, Rohdich, & Bacher, 2001; Rohdich, Kiss, Bacher, & Eisenreich, 2001). However, the MVA and MEP pathways are not completely independent, with likely crosstalk by communal sharing of intermediates and substrates (Hemmerlin et al., 2003).

The biosynthetic pathway leading to secologanin formation was shown to principally originate through the MEP pathway (Fig. 5.3) using radioactive labelling and pathway-specific inhibitors (Yamazaki et al., 2004). Accordingly the MEP pathway to secolaganin is outlined hereafter.

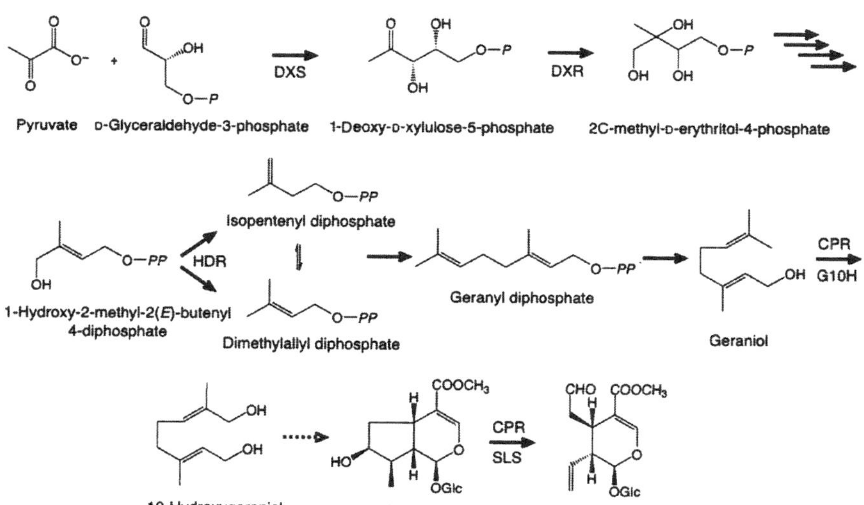

Figure 5.3 Secologanin biosynthetic pathway. DXS, 1-deoxy-D-xylulose-5-phosphate synthase; DXR, 1-deoxy-D-xylulose-5-phosphate reductoisomerase; HDR, 1-hydroxy-2-methyl-2(*E*)-butenyl-4-diphosphate reductase; G10H, geraniol 10-hydroxylase; SLS, secologanin synthase; CPR, NADPH:cytochrome P450 reductase; *P*, phosphate group; *PP*, diphosphate group. The arrow with a dotted shaft represents the involvement of multiple enzymatic steps.

The initial step to IPP biosynthesis is formed from the condensation of glyceraldehyde-3-phosphate and pyruvate by 1-deoxy-D-xylulose-5-phosphate synthase (DXS) resulting in 1-deoxy-D-xylulose-5-phosphate. The 1-deoxy-D-xylulose-5-phosphate (DXP) is then converted to 2-C-methyl-D-erythritol-4-phosphate *via* DXP reductoisomerase (DXR) before further enzymatic catalyzed stages leading to 1-hydroxy-2-methyl-2 (*E*)-butenyl-4-diphosphate (HMBPP), from which a further process catalyzed by HMBPP reductase (HDR) leads to IPP and its isomer dimethylallyl diphosphate (DMAPP).

The genes encoding DXR and HDR in the MEP pathway have been cloned from *C. acuminata* and characterized (Wang et al., 2008; Yao et al., 2008). These two genes constitutively express in the stem, leaf and root of the plant. The similar induction pattern, including when plants are treated with stress compounds, such as methyl jasmonate, suggests a common expression control system (Wang et al., 2008; Yao et al., 2008).

2.1.2.2 IPP to secologanin

Secologanin is formed through several steps starting from the condensation of IPP and DMAPP, the intermediate products from the MEP pathway, to yield geranyl diphosphate (GPP). GPP is then converted to geraniol by geraniol synthase (GS). Although GS has not been studied in CPT-producing plants, characterization of the GS isoform involved in the production of TIA in *C. roseus* has been reported recently (Simkin et al., 2013). At least several cytochrome P450 monooxygenases are involved in the formation of secologanin from geraniol, including geraniol 10-hydroxylase (G10H), deoxyloganin 7-hydroxylase and secologanin synthase (SLS). Although the gene *CYP72A1* encoding SLS from *C. roseus* was identified many years ago (Irmler et al., 2000), that from *C. acuminata* was reported only recently (Sun et al., 2011).

Several NADPH:cytochrome P450 reductases (CPRs; EC 1.6.2.4) that are essential for the activities of those cytochrome P450 proteins have been isolated from *O. pumila* and *N. foetida* (Huang, Sung, Do, & Huang, 2012; Yamazaki, Urano, et al., 2003) and show the potential to be indirectly involved in CPT biosynthesis. G10H, which hydroxylates the monoterpenoid geraniol at the C-10 position in the first committed step in the formation of secologanin, is one such cytochrome P450 monooxygenase (Collu et al., 2001). The final reaction for the biosynthesis of secologanin is also catalyzed by a P450 protein, SLS (EC 1.3.3.9; Irmler et al., 2000, Yamamoto, Katano, Ooi, & Inoue, 2000). The enzyme CPR is essential for the activity of G10H and other cytochrome P450 monooxygenases, as it functions in electron transfer from NADPH to cytochrome P450 (Meijer et al., 1993).

Very recently, iridoid synthase was characterized from *C. roseus* (Geu-Flores et al., 2012). This is a new enzyme that generates the iridoid ring scaffold from the linear monoterpene 10-oxogeranial as the substrate and probably couples an initial NAD(P)H-dependent reduction step with a subsequent cyclization step via a Diels–Alder cycloaddition or a Michael addition.

2.2. Post-strictosidine pathway

2.2.1 Strictosidine synthesis

The initial step of TIA biosynthesis involves the strictosidine synthase (STR; EC 4.3.3.2)-mediated condensation of tryptamine with the iridoid glucoside secologanin to yield strictosidine (Stöckigt & Ruppert, 1999; Stöckigt & Zenk, 1977; Fig. 5.4), a universal precursor to a wide range of TIAs. STR was first isolated from the snakeroot, *Rauvolfia serpentina* (Kutchan, Hampp, Lottspeich, Beyreuther, & Zenk, 1988), and subsequently from *C. roseus* (McKnight, Roessner, Devagupta, Scott, & Nessler, 1990). The gene encoding the vacuolar-localized 351 amino acid STR involved in CPT biosynthesis in *O. pumila* was cloned from the hairy

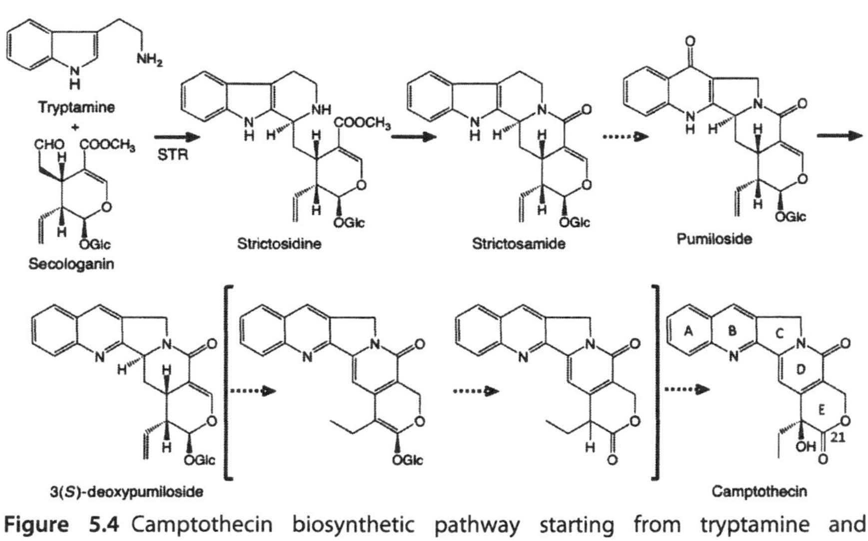

Figure 5.4 Camptothecin biosynthetic pathway starting from tryptamine and secologanin. STR, strictosidine synthase; Glc, glucose moiety. Arrows with a dotted shaft represent the involvement of multiple enzymatic steps. The intermediates in the bracket are presumed to be involved in the pathway.

roots using homology-based approaches (Yamazaki, Sudo, et al., 2003), and when heterologously expressed in *Escherichia coli*, the recombinant product exhibited STR activity. A high expression level of *STR* transcripts and protein was observed in the stems and roots of *O. pumila* and was found to be correlated with the enzyme activities in both organs (Yamazaki, Urano, et al., 2003). The STR from the non-CPT-producing *Ophiorrhiza japonica* has been cloned and characterized; it has a 96% amino acid identity to the *O. pumila* STR and a very high amino acid identity to that from other TIA-producing plants, including *R. serpentina* and *C. roseus* at 55% and 51% identity, respectively (Lu et al., 2009). Elicitor treatments, such as methyl jasmonate and salicylic acid, induced the expression of *O. japonica STR* transcripts, but in contrast methyl jasmonate had no effect upon the *STR* expression level in *O. pumila*, whilst yeast extract, 1-napthaleneacetic acid and salicylic acid all repressed *STR* expression (Lu et al., 2009; Yamazaki, Sudo, et al., 2003). However, it should be noted that the plant tissues used were different between *O. japonica* and *O. pumila*, and so, in addition to a difference in the *STR* regulation of alkaloid production between the CPT-producing and CPT-non-producing *Ophiorrhiza* plants, this could also simply represent differential tissue-specific regulation patterns. Moreover, in *C. roseus* cell suspensions, the *STR*, *TDC* and *CPR* genes are co-ordinately induced by fungal elicitors and methyl jasmonate (Collu et al., 2001; Meijer et al., 1993) and are repressed by auxin (Pasquali, Goddijn, de Waal, & Verpoorte, 1992). In contrast, in *O. pumila* hairy root cultures, neither methyl jasmonate nor fungal elicitor enhanced CPT production (Saito et al., 2001). It is apparent that the cell suspensions and hairy roots are different type of tissues but this might suggest a possible different regulatory mechanism in TIA biosynthesis between *O. pumila* and *C. roseus*.

STR has been somewhat extensively studied in other alkaloid-producing plants. The structural analysis of the reported crystal structure of STR from *R. serpentina* revealed the likely essential roles of certain amino acid residues involved in catalysis (Ma, Panjikar, Koepke, Loris, & Stöckigt, 2006). This study opens the way for the production of novel (unnatural) alkaloid compounds using metabolic engineering, as shown by the work of Runguphan and O'Connor (2009). The STR mutant (Val214Met) over-expressed in the hairy root cell cultures of *C. roseus* produced a novel (unnatural) strictosidine intermediate and alkaloid products in the presence of tryptamine analogs. Although the complete CPT biosynthetic pathway is still elusive, these studies suggest the possibility to reengineer the identified biosynthetic enzymes for the production of novel unnatural CPT derivatives that might be of benefit in terms of an enhanced anti-cancer activity.

2.2.2 Strictosidine to CPT biosynthesis

Intramolecular cyclization of strictosidine yields strictosamide, a penultimate precursor of CPT (Hutchinson et al., 1979). However, the remaining details and precise intermediates between strictosamide and CPT are not completely defined. It has been postulated that CPT could be formed from strictosamide by the sequential (i) oxidation–recyclization of the B- and C-rings, (ii) oxidation of the D-ring and removal of the C-21 glucose moiety and (iii) oxidation of the ring E (Fig. 5.4; Hutchinson et al., 1979).

3(*S*)-Pumiloside and 3(*S*)-deoxypumiloside were isolated from *Ophiorrhiza* species as possible intermediates leading to the formation of CPT (Aimi et al., 1989; Kitajima et al., 2005), while 3(*S*)-pumiloside was also detected in *C. acuminata* (Carte et al., 1990). Recently, Asano et al. (2013) employed a metabolomics approach to identify two possible post-strictosidine intermediates based on their molecular masses and their positive correlation with *TDC* gene expression following RNAi-mediated *TDC* gene suppression in the hairy roots of *O. pumila*. This study demonstrated the power of combined metabolomic and transgenic approaches to discover intermediates in the biosynthetic pathway. However, the unidentified intermediates might be difficult to discover if they are unstable and need a protein complex or metabolon to catalyze the reaction (Asano et al., 2013).

A transcriptomic-based approach was used to catalog the transcriptome in the young leaves of *C. acuminata* (Sun et al., 2011), from which 520 ESTs representing 20 biosynthetic genes that are involved in the biosyntheses of tryptamine and secologanin were identified. From these, *G10H*, *SLS*, *STR*, six cytochrome P450 transcripts and one glucosidase gene that might be involved in the post-strictosidine pathway were identified and characterized (Sun et al., 2011). Utilizing next generation sequencing technology, the large-scale transcriptome sequence and expression profiles for three related TIA-producing plants (*C. acuminata*, *C. roseus* and *R. serpentina*) and the comparative transcriptome in the different developmental tissues of the callus and root cultures of *C. acuminata* were profiled (Gongora-Castillo et al., 2012). By a similar strategy, Yamazaki et al. (2013) used the combined analyses of deep transcriptome and metabolome analysis of hairy roots and cell suspension culture cells of *O. pumila* to identify likely candidates that are involved in the synthesis of secondary metabolites, such as CPT-related alkaloids and anthraquinones. The data from these different plant species could become a great resource for the discovery of unidentified genes in the CPT biosynthetic pathway. From such studies, the complete biosynthetic pathway may be revealed in the near future and this could create an inventive platform for the production of CPT.

3. BIOTECHNOLOGICAL PRODUCTION OF CAMPTOTHECIN AND ITS DERIVATIVES

Nowadays, the water-soluble CPT analogs with an intact lactone ring, topotecan and irinotecan, are chemically synthesized from CPT that is still extracted from the bark and seeds of *C. acuminata* and *N. foetida*. Although all parts of *C. acuminata* contain some CPT, the highest level is reported to be found in young leaves (4–5 mg/g dry weight), at some 1.5- and 2.5-fold higher than that in the seeds and bark, respectively (reviewed in Lorence & Nessler, 2004). Thus, the continual harvesting of young leaves without destruction of the trees may improve the obtained CPT yield per plant per year.

However, these plants are listed as endangered species and so, with the increasing demand for these drugs, it is therefore critical to establish an alternative and sustainable biosource for CPT production. Searching for novel CPT-producing plants would also provide additional sources that could be used for CPT extraction and relieve the pressure on *C. acuminata* and *N. foetida* (Section 3.1). In addition, there has been much interest in trying to develop alternative production systems, such as callus and cell suspension cultures (Section 3.2), as well as from cultures of CPT-producing-plant-associated endophytic fungi (Section 3.3). These systems could help to resolve the limited CPT supply for the future increasing demand for it. Certainly, the development of hairy root cultures of *O. pumila* and *C. acuminata* (Section 3.4) and the cloning and characterization of genes encoding key enzymes of the pathway leading to CPT formation in plants will open new possibilities for alternative and sustainable production systems for CPT.

3.1. New plant sources

With respect to new natural plant sources, the genus *Ophiorrhiza* (Family Rubiaceae) contains the largest number of members (~429 nominal species) compared with those in the genera *Camptotheca* and *Nothapodytes* at ~5 and ~15 nominal species, respectively, (http://www.theplantlist.org), of which several are already known to be prominent CPT-producing sources. Although not all species currently tested within the genus *Ophiorrhiza* produce CPT (or at least to the same prominent level), extrapolating from this data (e.g. 5/8 or ~60% of species in Thailand; Viraporn et al., 2011) suggests that many plants in this genus could potentially be an alternative source for CPT. *Ophiorrhiza mungos* was the first plant in this genus reported to contain

CPT and its natural derivatives (Tafur et al., 1976), with currently over 10 other species in this genus from different countries reported to contain CPT. It is noteworthy that members within this group of plants produce different amounts of CPT and its derivatives.

Although an increasing number of plant species are known to produce CPT, the cultivation of these plants is not usually suitable (viable) for the commercial production of CPT, for example, because of poor seed germination or low yields. One potential way to circumvent this is through the clonal propagation of selected high yield cultivars via shoot bud cultures for the rapid propagation of highly productive trees (Jain & Nessler, 1996).

In addition, there is an increasing interest in developing the biotechnological production of CPT. Micropropagation using suitable plant hormone combinations has been used to generate a large number of *in vitro* plants. Namdeo, Priya, and Bhosale (2012) reported that *in vitro*-propagated *O. mungos* contain a higher level of CPT compared with naturally grown plants. Shoot multiplication can be successfully obtained from leaf and node explants of *O. alata* using varying amounts of kinetin and α-naphthalene acetic acid (Ya-ut, Chareonsap, & Sukrong, 2011). Besides CPT produced *in planta*, several biotechnological methods for CPT production have been reported, and these are outlined below.

3.2. Callus and cell suspension culture

The induction of callus and cell suspension cultures from parts of *C. acuminata* stems cultured in either MS or Gamborg B5 media with a combination of plant hormones led to the intracellular production of CPT, with the CPT level being dependent on the medium and growth regulators used (van Hengel, Harkes, Wichers, Hesselink, & Buitelaar, 1992). In addition, the CPT derivative 10-hydroxycamptothecin (Fig. 5.1) could be produced in the callus of *C. acuminata* (Wiedenfeld, Furmanowa, Roeder, Guzewska, & Gustowski, 1997). In general, the level of alkaloid production obtained in culture was 100- to 1000-fold lower than that from soil-grown plants, but cell suspension cultures from stem parts of *N. foetida* can also produce almost the same amount of both CPT and 9-methoxycamptothecin (Fig. 5.1) in the medium (Fulzele, Satdive, & Pol, 2001), with only a trace amount of these alkaloids being found inside the cells, which makes for an easier and higher yield enrichment. Thengane et al. (2003) also reported a high level of CPT production (1.3% by dry weight) in callus cultures derived from the cotyledons of *N. foetida*, but with only a trace amount of

9-methoxycamptothecin. It has been clearly shown that using different plant growth regulators can significantly affect the production of CPT and its derivatives (Roja & Heble, 1994; Thengane et al., 2003). Therefore, different growth regulator concentrations and combinations must be tested for the optimization of production yields, which may then approach viable yield levels. Interestingly, the suspension culture of *O. pumila* was found to not produce CPT (Kitajima et al., 1998).

The addition of elicitors, such as yeast extract or methyl jasmonate, to cell suspension cultures can significantly increase the CPT production level (Song & Byun, 1998). In addition, the sugar concentration was also shown to affect the CPT production, where a high (6% (w/v)) sucrose concentration in the culture medium resulted in an 11-fold higher CPT production from cell suspension cultures of *C. acuminata* compared with that with 2% (w/v) sucrose (Kim, Hur, & Byun, 1999). It has also been reported that certain inorganic microelements, such as I^-, Cu^{2+}, Co^{2+} and $MoO_4{}^{2-}$, as well as the nitrogen source supply in the medium can influence the CPT production level with up to a two- to threefold increased CPT production at optimal supplement levels (Pan, Shi, Liu, Gao, & Lu, 2004; Pan, Xu, Liu, Gao, & Lu, 2004). These reports suggest that CPT production in the callus or cell suspension cultures can be manipulated via the addition of different compounds, although this will probably require optimization for each tissue type and culture developmental stage using multivariate analysis for each plant species/cultivar.

3.3. Hairy roots

The induction of hairy root cultures on several CPT-producing plants has been reported, and hairy root cultures induced by transformation with *Agrobacterium rhizogenes* are an attractive option as they provide long-term aseptic root clones, that are often genetically stable, with growth rates comparable with those of the fastest growing cell suspension cultures.

Although the callus culture of *O. pumila* did not produce CPT, in contrast its hairy root culture produced substantial amounts of CPT in both the root cells and in the culture medium (Saito et al., 2001). However, in contrast to that found in the callus culture, treatment with yeast extract and methyl jasmonate did not significantly increase the level of CPT production (Saito et al., 2001). The addition of polystyrene resin (Diaion HP-20) to the media, which reversibly absorbs the CPT, increased the amount of excreted CPT making the isolation step easier. This may then imply an external concentration-dependent regulation of secretion (or concentration-gradient-dependent passive transport) of CPT. Regardless, this method can potentially be commercially applicable for the large-scale CPT production, as shown by the successful upscale production of CPT from *O. pumila* hairy roots in a 3-L bioreactor (Sudo, Yamakawa, Yamazaki, Aimi, & Saito, 2002). The establishment of a hairy root culture of *O. alata* that produces CPT has been reported recently (Ya-ut et al., 2011), while *C. acuminata* hairy root cultures with the capability to produce both CPT and 10-hydroxycamptothecin have also been reported (Lorence, Medina-Bolivar, & Nessler, 2004).

3.4. Endophytic fungi

Endophytic fungi with the ability of producing plant bioactive compounds have received a fair amount of attention since the discovery that a fungal endophyte isolated from the phloem of the Pacific yew-producing taxol was reported (Stierle, Strobel, & Stierle, 1993). Several endophytic fungi isolated from different CPT-producing plants have subsequently been identified including the CPT-producing fungal endophyte, *Entrophosphora infrequens*, isolated from the inner bark of the CPT-producing evergreen tree *N. foetida* in liquid medium (Puri, Verma, Amna, Qazi, & Spiteller, 2005). Up-scaled CPT production from this or another endophytic fungi isolated from the twigs of *N. foetida* in a bioreactor yielded a relatively high CPT production level of 4.96 ± 0.73 mg/100 g of dry cell mass when cultured for 48 h (Amna et al., 2006). The endophytic *Neurospora* sp., isolated from the seeds of *N. foetida* (Rehman et al., 2008), and two endophytic fungal strains of *Fusarium solani*, isolated from the CPT-producing White Pear tree, *Apodytes dimmidiata*, were reported to be able to produce CPT, 9-methoxycamptothecin and 10-hydroxycamptothecin (Shweta et al., 2010). These two *F. solani* strains were able to produce different CPT analogs, but the obtained yield was much lower compared with those produced from the endophytic fungi isolated from the twigs of *N. foetida* cultured in a bioreactor described earlier.

Although the endophytic fungus isolated from the inner bark of *C. acuminata* was shown to produce CPT, 9-methoxycamptothecin and 10-hydroxycamptothecin (Kusari, Zuhlke, & Spiteller, 2009; Liu, Ding, Deng, & Chen, 2010), their production levels substantially decreased on repeated subculturing. The reason for the drastic decrease in CPT production in the endophytic fungal culture is because the endophytes do not have the complete CPT biosynthetic pathway, but require at least STR from the host plant (Kusari, Zuhlke, & Spiteller, 2011). During successive generations in isolated subculture, the fungus depletes the host-plant-derived STR enzyme. Moreover, the expression of several fungal genes involved in the CPT biosynthesis, including *G10H*, *SLS* and *TDC*, appear to be genetically unstable in tissue culture (Kusari, Zuhlke, et al., 2011). This questions the validity of endophytic fungi as an alternative source for plant bioactive compound production, especially if a new fungal starter culture is required frequently to maintain viable production yields.

4. RESISTANCE MECHANISM(S) IN CAMPTOTHECIN-PRODUCING PLANTS

How CPT-producing plants resist or detoxify CPT is quite different from the reported self-resistance mechanisms in many plants producing other toxic compounds (Sirikantaramas, Yamazaki, & Saito, 2008a). In *O. pumila* hairy root cultures, CPT excretion into the culture medium operates by a passive transporter-independent mechanism where the intracellular concentration of CPT determines the excretion rate (Sirikantaramas, Sudo, Asano, Yamazaki, & Saito, 2007). These cells seem to be resistant to CPT due to amino acid substitutions in their topoisomerase I enzyme, the cellular target of CPT, that change the ability of CPT to covalently bind to the topoisomerase I, and so result in complete self-resistance (Sirikantaramas, Yamazaki, & Saito, 2008b).

In accordance, it is known that mutations in human topoisomerase I is one of mechanisms for CPT resistance in cancer cells, where the CPT-resistant human leukaemia cell line was found to possess two amino acid substitutions (Met370Thr and Asn722Ser) in its topoisomerase I (Fujimori, Harker, Kohlhagen, Hoki, & Pommier, 1995). It was clarified that Met370Thr mutation is not involved in CPT resistance since threonine at position 370 is found in the CPT-sensitive *Saccharomyces cerevisiae* topoisomerase I (Bjornsti, Benedetti, Viglianti, & Wang, 1989). Another point mutation, Asn722Ser, conferring CPT resistance in the leukaemia cell line is the same as that found in the topoisomerase I from several CPT-producing plants, including *C. acuminata* and several CPT-producing *Ophiorrhiza* species, suggesting the convergent evolution of a common CPT-resistance mechanism (Sirikantaramas et al., 2008b; Viraporn et al., 2011). The Asn722Ser mutation leads to the elimination of a water-mediated contact between topoisomerase I and CPT, preventing the interaction with CPT and its poisoning of topoisomerase I (Chrencik et al., 2004).

With both closely related CPT-producing and non-producing species being found in the genus *Ophiorrhiza*, this provides a group of plants for investigating how CPT-resistant topoisomerase I may have evolved. It is proposed that CPT-producing plants were partially primed for CPT resistance before the complete evolution of the CPT biosynthetic pathway (Sirikantaramas et al., 2008b, Sirikantaramas, Yamazaki, & Saito, 2009). Evidence from an *in vivo* CPT sensitivity assay in *Saccharomyces cerevisiae* showed that the topoisomerase I from the CPT-lacking *O. japonica* is already partially resistant to CPT. These suggest the possibility of co-evolution between the CPT biosynthetic pathway and the target topoisomerase I in the producing plants.

The finding of potential convergent evolution in CPT resistance between human and CPT-producing plants suggests the possibility for predicting drug resistance in other organisms. This includes the prediction for future point mutations in human topoisomerase I that confer CPT resistance since there are other point mutations in the topoisomerase I of CPT-producing plants that have not been identified in human topoisomerase I yet (Sirikantaramas et al., 2008b). In addition, Kusari, Kosuth, Cellarova, and Spiteller (2011) proposed the same target-based mutation as a self-resistance mechanism in the CPT-producing fungal endophyte, *F. solani*. Therefore, the same mutations of topoisomerase I appear to be a universal mechanism for organisms to develop resistance, and an understanding of this could potentially be applied for the further development of high-efficiency CPT analogs to overcome the CPT resistance caused by mutations of those identified amino acid residues in different organisms.

5. CONCLUSIONS

With an ever-increasing demand for CPT supply, it is essential to establish a feasible system for its sustainable and economically viable large-scale production. Several anti-cancer drugs such as paclitaxel and podophyllotoxin have been produced commercially in plant cell cultures (Wilson & Roberts, 2012). Although possible techniques to achieve CPT production using either plant tissue/organ cultures or endophytic fungi have been demonstrated, the yields currently obtained from these cultures are not high enough to be viably developed for commercialization. Advances in metabolic engineering and synthetic biology are, therefore, promising alternative approaches. With the combination of advanced genomics, proteomics and metabolomics, we believe that the CPT biosynthetic pathway will be fully unravelled and enable the engineering of high-level production systems as well as product modification for drug improvement. The latter will be aided by the information learned from CPT-resistance mechanisms in the CPT-producing plants.

ACKNOWLEDGEMENTS

The parts of authors' study have been supported by Grant-in-Aids from The Ministry of Education, Culture, Sports, Science and Technology (MEXT) and the Japan Society for the Promotion of Science (JSPS) (to M.Y., and K.S.), and Special Task Force for Activating Research (STAR), Ratchadaphiseksomphot Endowment Fund of Chulalongkorn University 56-007-23-004 (to S.S). The authors thank Dr. Robert Butcher, Faculty of Science, Chulalongkorn University, for editing the manuscript.

REFERENCES

Aimi, N., Nishimura, M., Miwa, A., Hoshino, H., Sakai, H., & Haginiwa, J. (1989). Pumiloside and deoxypumiloside; plausible intermediates of camptothecin biosynthesis. *Tetrahedron Letters*, *30*, 4991–4994.

Aiyama, R., Nagai, H., Nokata, K., Shinohara, C., & Sawada, S. (1988). A camptothecin derivative from *Nothapodytes foetida*. *Phytochemistry*, *27*, 3663–3664.

Amna, T., Puri, S. C., Verma, V., Sharma, J. P., Khajuria, R. K., Musarrat, J., et al. (2006). Bioreactor studies on the endophytic fungus *Entrophospora infrequens* for the production of an anticancer alkaloid camptothecin. *Canadian Journal of Microbiology*, *52*, 189–196.

Arisawa, M., Gunasekera, S. P., Cordell, G. A., & Farnsworth, N. R. (1981). Plant anticancer agents XXI. Constituents of *Merrilliodendron megacarpum*. *Planta Medica*, *43*, 404–407.

Asano, T., Kobayashi, K., Kashihara, E., Sudo, H., Sasaki, R., Iijima, Y., et al. (2013). *Suppression of camptothecin biosynthetic genes results in metabolic modification of secondary products in hairy roots of Ophiorrhiza pumila*. *Phytochemistry*, *91*, 128–139.

Asano, T., Watase, K., Sudo, H., Kitajima, M., Takayama, H., Aimi, N., et al. (2004). Camptothecin production by in vitro cultures of *Ophiorrhiza liukiuensis* and *O. kuroiwai*. *Plant Biotechnology*, *21*, 275–281.

Beegum, A. S., Martin, K. P., Zhang, C. L., Nishitha, I. K., Slater, L. A., & Madhusoodanan, P. V. (2007). Organogenesis from leaf and internode explants of *Ophiorrhiza prostrata*, an anticancer drug (camptothecin) producing plant. *Electronic Journal of Biotechnology*, *10*, 114–123.

Bjornsti, M.-A., Benedetti, P., Viglianti, G. A., & Wang, J. C. (1989). Expression of human DNA topoisomerase I in yeast cells lacking yeast DNA topoisomerase I: Restoration of sensitivity of the cells to the antitumor drug camptothecin. *Cancer Research*, *49*, 6319–6323.

Carte, B. K., DeBrosse, C., Eggleston, D., Hemling, M., Mentzer, M., Poehland, B., et al. (1990). Isolation and characterization of a presume biosynthetic precursor of camptothecin from extracts of *Camptotheca acuminata*. *Tetrahedron*, *46*, 2747–2760.

Chrencik, J. E., Staker, B. L., Burgin, A. B., Pourquier, P., Pommier, Y., Stewart, L., et al. (2004). Mechanisms of camptothecin resistance by human topoisomerase I mutations. *Journal of Molecular Biology*, *339*, 773–784.

Collu, G., Unver, N., Peltenburg-Looman, A. M. G., van der Heijden, R., Verpoorte, R., & Memelink, J. (2001). Geraniol 10-hydroxylase, a cytochrome P450 enzyme involved in terpenoid indole alkaloid biosynthesis. *FEBS Letters*, *508*, 215–220.

Dai, J. R., Cardellina, J. H., & Boyd, M. R. (1999). 20-O-β-Glucopyranosyl camptothecin from Mostuea brunonis: A potential camptothecin pro-drug with improved solubility. *Journal of Natural Products*, *62*, 1427–1429.

De Luca, V., Marineau, C., & Brisson, N. (1989). Molecular cloning and analysis of cDNA encoding a plant tryptophan decarboxylase: Comparison with animal dopa decarboxylases. *Proceedings of the National Academy of Sciences of the United States of America*, *86*, 2582–2586.

Douillard, H. Y., Cunningham, D., Roth, A. D., Navarro, M., James, R. D., Karasek, P., et al. (2000). Irinotecan combined with fluorouracil compared with fluorouracil alone as first-line treatment for metastatic colorectal cancer: A multicentre randomised trial. *Lancet*, *355*, 1041–1047.

Eisenreich, W., Rohdich, F., & Bacher, A. (2001). Deoxyxylulose phosphate pathway to terpenoids. *Trends in Plant Science*, *6*, 78–84.

Fujimori, A., Harker, W. G., Kohlhagen, G., Hoki, U., & Pommier, Y. (1995). Mutation at the catalytic site of topoisomerase I in CEM/C2, a human leukemia cell line resistant to camptothecin. *Cancer Research*, *55*, 1339–1346.

Fulzele, D. P., Satdive, R. K., & Pol, B. B. (2001). Growth and production of camptothecin by cell suspension cultures of *Nothapodytes foetida*. *Planta Medica*, *67*, 150–152.

Geu-Flores, F., Sherden, N. H., Courdavault, V., Burlat, V., Glenn, W. S., Wu, C., et al. (2012). An alternative route to cyclic terpenes by reductive cyclization in iridoid biosynthesis. *Nature*, *492*, 138–142.

Gongora-Castillo, E., Childs, K. L., Fedewa, G., Hamilton, J. P., Liscombe, D. K., Magallanes-Lundback, M., et al. (2012). Development of transcriptomic resources for interrogating the biosynthesis of monoterpene indole alkaloids in medicinal plant species. *PLoS One*, 7, e52506.

Gunasekera, S. P., Badawi, M. M., Cordell, G. A., Farnsworth, N. R., & Chitnis, M. (1979). Plant anticancer agents X. Isolation of camptothecin and 9-methoxycamptothecin from *Ervatamia heyneana*. *Journal of Natural Products*, *42*, 475–477.

Hemmerlin, A., Hoeffler, J.-F., Meyer, O., Tritsch, D., Kagan, I. A., Grosdemange-Billiard, C., et al. (2003). Cross-talk between the cytosolic mevalonate and the plastidial methylerythritol phosphate pathways in tobacco bright yellow-2 cells. *The Journal of Biological Chemistry*, *278*, 26666–26676.

Hsiang, Y. H., Hertzberg, R., Hecht, S., & Liu, L. F. (1985). Camptothecin induces protein-linked DNA breaks via mammalian DNA topoisomerase I. *The Journal of Biological Chemistry*, *260*, 14873–14878.

Huang, F.-C., Sung, P.-H., Do, Y.-Y., & Huang, P.-L. (2012). Differential expression and functional characterization of the NADPH cytochrome P450 reductase genes from *Nothapodytes foetida*. *Plant Science*, *190*, 16–23.

Hutchinson, C. R., Heckendorf, A. H., Straughn, J. L., Daddona, P. E., & Cane, D. E. (1979). Biosynthesis of camptothecin. 3. Definition of strictosamide as the penultimate biosynthetic precursor assisted by carbon-13 and deuterium NMR spectroscopy. *Journal of the American Chemical Society*, *101*, 3358–3369.

Irmler, S., Schröder, G., St-Pierre, B., Crouch, N. P., Hotze, M., Schmidt, J., et al. (2000). Indole alkaloid biosynthesis in *Catharanthus roseus*: New enzyme activities and identification of cytochrome P450 CYP72A1 as secologanin synthase. *The Plant Journal*, *24*, 797–804.

Jain, A. K., & Nessler, C. L. (1996). Clonal propagation of *Camptotheca acuminata*. *Plant Cell, Tissue and Organ Culture*, *44*, 229–233.

Kim, S. H., Hur, B. K., & Byun, S. Y. (1999). Effect of sugar concentration on camptothecin production in cell suspension cultures of *Camptotheca acuminata*. *Biotechnology and Bioprocess Engineering*, *4*, 277–280.

Kitajima, M., Fischer, U., Nakamura, M., Ohsawa, M., Ueno, M., Takayama, H., et al. (1998). Anthraquinone from *Ophiorrhiza pumila* tissue and cell cultures. *Phytochemistry*, *48*, 107–111.

Kitajima, M., Fujii, N., Yoshino, F., Sudo, H., Saito, K., Aimi, N., et al. (2005). Camptothecins and two new monoterpene glucosides from *Ophiorrhiza liukiuensis*. *Chemical and Pharmaceutical Bulletin*, *53*, 1355–1358.

Koster, D. A., Palle, K., Bot, E. S. M., Bjornsti, M.-A., & Dekker, N. H. (2007). Antitumour drugs impede DNA uncoiling by topoisomerase I. *Nature*, *448*, 213–217.

Kusari, S., Kosuth, J., Cellarova, E., & Spiteller, M. (2011). Survival-strategies of endophytic *Fusarium solani* against indigenous camptothecin biosynthesis. *Fungal Ecology*, *4*, 219–223.

Kusari, S., Zuhlke, S., & Spiteller, M. (2009). An endophytic fungus from *Camptotheca acuminata* that produces camptothecin and analogues. *Journal of Natural Products*, *72*, 2–7.

Kusari, S., Zuhlke, S., & Spiteller, M. (2011). Effect of artificial reconstitution of the interaction between the plant *Camptotheca acuminata* and the fungal endophyte *Fusarium solani* on camptothecin biosynthesis. *Journal of Natural Products*, *74*, 764–775.

Kutchan, T. M., Hampp, N., Lottspeich, F., Beyreuther, K., & Zenk, M. H. (1988). The cDNA clone for strictosidine synthase from *Rauvolfia serpentina* DNA sequence determination and expression in *Escherichia coli*. *FEBS Letters*, *237*, 40–44.

Li, S., Yi, Y., Wang, Y., Zhang, Z., & Beasley, R. S. (2002). Camptothecin accumulation and variations in *Camptotheca*. *Planta Medica*, *68*, 1010–1016.

Liu, K., Ding, X., Deng, B., & Chen, W. (2010). 10-Hydroxycamptothecin produced by a new endophytic *Xylaria* sp., M20, from *Camptotheca acuminata*. *Biotechnology Letters*, *32*, 689–693.

Lopez-Meyer, M., & Nessler, C. L. (1997). Tryptophan decarboxylase is encoded by two autonomously regulated genes in *Camptotheca acuminata* which are differentially expressed during development and stress. *The Plant Journal*, *11*, 1167–1175.

Lorence, A., Medina-Bolivar, F., & Nessler, C. L. (2004). Camptothecin and 10-hydroxycamptothecin from *Camptotheca acuminata* hairy roots. *Plant Cell Reports*, *22*, 437–441.

Lorence, A., & Nessler, C. L. (2004). Camptothecin, over four decades of surprising findings. *Phytochemistry*, *65*, 2735–2749.

Lu, H., Gorman, E., & McKnight, T. D. (2005). Molecular characterization of two anthranilate synthase alpha subunit genes in *Camptotheca acuminata*. *Planta*, *221*, 352–360.

Lu, H., & McKnight, T. D. (1999). Tissue-specific expression of the beta-subunit of tryptophan synthase in *Camptotheca acuminata*, an indole alkaloid-producing plant. *Plant Physiology*, *120*, 43–51.

Lu, Y., Wang, H., Wang, W., Qian, Z., Li, L., Wang, J., et al. (2009). Molecular characterization and expression analysis of a new cDNA encoding strictosidine synthase from Ophiorrhiza japonica. *Molecular Biology Reports*, *36*, 1845–1852.

Ma, X., Panjikar, S., Koepke, J., Loris, E., & Stöckigt, J. (2006). The structure of *Rauvolfia serpentina* strictosidine synthase is a novel six-bladed beta-propeller fold in plant proteins. *The Plant Cell*, *18*, 907–920.

McKnight, T. D., Roessner, C. A., Devagupta, R., Scott, A. I., & Nessler, C. L. (1990). Nucleotide sequence of a cDNA encoding the vacuolar protein strictosidine synthase from *Catharanthus roseus*. *Nucleic Acids Research*, *18*, 4939.

Meijer, A. H., Lopes Cardoso, M. I. L., Voskuilen, J. T., de Waal, A., Verpoorte, R., & Hoge, J. H. C. (1993). Isolation and characterization of a cDNA clone from *Catharanthus roseus* encoding NADPH:cytochrome P-450 reductase, an enzyme essential for reactions catalysed by cytochrome P-450 mono-oxygenases in plants. *The Plant Journal*, *4*, 47–60.

Namdeo, A. G., Priya, T., & Bhosale, B. B. (2012). Micropropagation and production of camptothecin from in vitro plants of *Ophiorrhiza mungos*. *Asian Pacific Journal of Tropical Biomedicine, 2*, S662–S666.
Noé, W., Mollenschott, C., & Berlin, J. (1984). Tryptophan decarboxylase from *Catharanthus roseus* cell suspension cultures: Purification, molecular and kinetic data of the homogenous protein. *Plant Molecular Biology, 3*, 281–288.
Pan, X.-W., Shi, Y.-Y., Liu, X., Gao, X., & Lu, Y.-T. (2004). Influence of inorganic microelements on the production of camptothecin with suspension cultures of *Camptotheca acuminata*. *Plant Growth Regulation, 44*, 59–63.
Pan, X.-W., Xu, H.-H., Liu, X., Gao, X., & Lu, Y.-T. (2004). Improvement of growth and camptothecin yield by altering nitrogen source supply in cell suspension cultures of *Camptotheca acuminata*. *Biotechnology Letters, 26*, 1745–1748.
Pantazis, P., Han, Z., Chatterjee, D., & Wyche, J. (1999). Water-insoluble camptothecin analogues as potential antiviral drugs. *Journal of Biomedical Science, 6*, 1–7.
Pasquali, G., Goddijn, O. J. M., de Waal, A., & Verpoorte, R. (1992). Coordinated regulation of two indole alkaloid biosynthetic genes from *Catharanthus roseus* by auxin and elicitors. *Plant Molecular Biology, 18*, 1121–1131.
Pommier, Y. (2006). Topoisomerase I inhibitors: Camptothecins and beyond. *Nature Reviews. Cancer, 6*, 789–802.
Pommier, Y. (2009). DNA topoisomerase I inhibitors: Chemistry, biology, and interfacial inhibition. *Chemical Reviews, 109*, 2894–2902.
Pommier, Y., Leo, E., Zhang, H., & Marchand, C. (2010). DNA topoisomerases and their poisoning by anticancer and antibacterial drugs. *Chemistry & Biology, 17*, 421–433.
Puri, S. C., Verma, V., Amna, T., Qazi, G. N., & Spiteller, M. (2005). An endophytic fungus from Nothapodytes foetida that produces camptothecin. *Journal of Natural Products, 68*, 1717–1719.
Rapisarda, A., Uranchimeg, B., Scudiero, D. A., Selby, M., Sausville, E. A., et al. (2002). Identification of small molecule inhibitors of hypoxia-inducible factor 1 transcriptional activation pathway. *Cancer Research, 62*, 4316–4324.
Rehman, S., Shawl, A. S., Kour, A., Andrabi, R., Sudan, P., Sultan, P., et al. (2008). An endophytic *Neurospora* sp. from *Nothapodytes foetida* producing camptothecin. *Applied Biochemistry and Microbiology, 44*, 203–209.
Rodriguez-Galindo, C., Radomski, K., Stewart, C. F., Furman, W., Santana, V. M., & Houghton, P. J. (2000). Clinical use of topoisomerase I inhibitors in anticancer treatment. *Medical and Pediatric Oncology, 35*, 385–402.
Rohdich, F., Kiss, K., Bacher, A., & Eisenreich, W. (2001). The non-mevalonate pathway of isoprenoids: Genes, enzymes and intermediates. *Current Opinions in Chemistry and Biology, 5*, 535–540.
Roja, G., & Heble, M. R. (1994). Alkaloid synthesis by root cultures of Rauwolfia serpentina transformed by Agrobacterium rhizogenes. *Phytochemistry, 36*, 381–383.
Runguphan, W., & O'Connor, S. E. (2009). Metabolic reprogramming of periwinkle plant culture. *Nature Chemical Biology, 5*, 151–153.
Saito, K., Sudo, H., Yamazaki, M., Koseki-Nakamura, M., Kitajima, M., Takayama, H., et al. (2001). Feasible production of camptothecin by hairy root culture of *Ophiorrhiza pumila*. *Plant Cell Reports, 20*, 267–271.
Shweta, S., Zuehlke, S., Ramesha, B. T., Priti, V., Kumar, P. M., Ravikanth, G., et al. (2010). Endophytic fungal strains of *Fusarium solani*, from *Apodytes dimidiata* E. Mey. ex Arn (Icacinaceae) produce camptothecin, 10-hydroxycamptothecin and 9-methoxycamptothecin. *Phytochemistry, 71*, 117–122.
Simkin, A. J., Miettinen, K., Claudel, P., Burlat, V., Guirimand, G., Courdavault, V., et al. (2013). Characterization of the plastidial geraniol synthase from Madagascar periwinkle

which initiates the monoterpenoid branch of the alkaloid pathway in internal phloem associated parenchyma. *Phytochemistry*, *85*, 36–43.
Sirikantaramas, S., Sudo, H., Asano, T., Yamazaki, M., & Saito, K. (2007). Transport of camptothecin in hairy roots of *Ophiorrhiza pumila*. *Phytochemistry*, *68*, 2881–2886.
Sirikantaramas, S., Yamazaki, M., & Saito, K. (2008a). Mechanisms of resistance to self-produced toxic secondary metabolites in plants. *Phytochemistry Reviews*, 7, 467–477.
Sirikantaramas, S., Yamazaki, M., & Saito, K. (2008b). Mutations in topoisomerase I as a self-resistance mechanism coevolved with the production of the anticancer alkaloid camptothecin in plants. *Proceedings of the National Academy of Sciences of the United States of America*, *105*, 6782–6786.
Sirikantaramas, S., Yamazaki, M., & Saito, K. (2009). A survival strategy: The coevolution of the camptothecin biosynthetic pathway and self-resistance mechanism. *Phytochemistry*, *70*, 1894–1898.
Song, S. H., & Byun, S. Y. (1998). Elicitation of camptothecin production in cell cultures of *Camptotheca acuminata*. *Journal of Microbiology and Biotechnology*, *8*, 631–638.
Stewart, D. J. (2004). Topotecan in the first-line treatment of small cell lung cancer. *The Oncologist*, *9*, 33–42.
Stierle, A., Strobel, G., & Stierle, D. (1993). Taxol and taxane production by *Taxomyces andreanae*, an endophytic fungus of Pacific yew. *Science*, *260*, 214–216.
Stöckigt, J., & Ruppert, M. (1999). Strictosidine—The biosynthetic key to monoterpenoid indole alkaloids. In O. Meth-Cohn, D. H. R. Barton, & K. Nakanishi (Series Eds.) & J. W. Kelly (Vol. Ed.), *Comprehensive natural products chemistry: Vol. 4. Amino acids, peptides, porphyrins, and alkaloids* (pp. 109–138). Amsterdam: Elsevier.
Stöckigt, J., & Zenk, M. H. (1977). Strictosidine (isovincoside): The key intermediate in the biosynthesis of monoterpenoid indole alkaloids. *Journal of the Chemical Society, Chemical Communications*, 646–648.
Sudo, H., Yamakawa, T., Yamazaki, M., Aimi, N., & Saito, K. (2002). Bioreactor production of camptothecin by hairy root cultures of *Ophiorrhiza pumila*. *Biotechnology Letters*, *24*, 359–363.
Sun, Y., Luo, H., Li, Y., Sun, C., Song, J., Niu, Y., et al. (2011). Pyrosequencing of the *Camptotheca acuminata* transcriptome reveals putative genes involved in camptothecin biosynthesis and transport. *BMC Genomics*, *12*, 533.
Svejstrup, J. Q., Christiansen, K., Gromova, I. I., Andersen, A. H., & Westergaard, O. (1991). New technique for uncoupling the cleavage and religation reactions of eukaryotic topoisomerase I: The mode of action of camptothecin at a specific recognition site. *Journal of Molecular Biology*, *222*, 669–678.
Tafur, S., Nelson, J. D., DeLong, D. C., & Svoboda, G. H. (1976). Antiviral components of *Ophiorrhiza mungos*. Isolation of camptothecin and 10-methoxycamptothecin. *Lloydia*, *39*, 261–262.
Thengane, S. R., Kulkarni, D. K., Shrikhande, V. A., Joshi, S. P., Sonawane, K. B., & Krishnamurthy, K. V. (2003). Influence of medium composition on callus induction and camptothecin(s) accumulation in *Nothapodytes foetida*. *Plant Cell, Tissue and Organ Culture*, *72*, 247–251.
Valletta, A., Trainotti, L., Santamaria, A. R., & Pasqua, G. (2010). Cell-specific expression of tryptophan decarboxylase and 10-hydroxygeraniol oxidoreductase, key genes involved in camptothecin biosynthesis in *Camptotheca acuminata* Decne (Nyssaceae). *BMC Plant Biology*, *10*, 69.
van Hengel, A. J., Harkes, M. P., Wichers, H. J., Hesselink, P. G. M., & Buitelaar, R. M. (1992). Characterization of callus formation and camptothecin production by cell lines of *Camptotheca acuminata*. *Plant Cell, Tissue and Organ Culture*, *28*, 11–18.
Viraporn, V., Yamazaki, M., Saito, K., Denduangboripant, J., Chayamarit, K., Chuanasa, T., et al. (2011). Correlation of camptothecin-producing ability and phylogenetic relationship in the genus *Ophiorrhiza*. *Planta Medica*, *77*, 759–764.

Wall, M. E., Wani, M. C., Cook, C. E., Palmer, K. H., McPhail, A. T., & Sim, G. A. (1966). Plant antitumor agents. I. The isolation and structure of camptothecin, a novel alkaloidal leukemia and tumor inhibitor from *Camptotheca acuminata. Journal of the American Chemical Society, 88*, 3888–3890.

Wang, Q., Pi, Y., Hou, R., Jiang, K., Huang, Z., Hsieh, M., et al. (2008). Molecular cloning and characterization of 1-hydroxy-2-methyl-2-(E)-butenyl 4-diphosphate reductase (CaHDR) from *Camptotheca acuminata* and its functional identification in *Escherichia coli. BMB Reports, 41*, 112–118.

Wiedenfeld, H., Furmanowa, M., Roeder, E., Guzewska, J., & Gustowski, W. (1997). Camptothecin and 10-hydroxycamptothecin in callus and plantlets of *Camptotheca acuminata. Plant Cell, Tissue and Organ Culture, 49*, 213–218.

Wilson, S. A., & Roberts, S. C. (2012). Recent advances towards development and commercialization of plant cell culture processes for the synthesis of biomolecules. *Plant Biotechnology Journal, 10*, 249–268.

Yamamoto, H., Katano, N., Ooi, A., & Inoue, K. (2000). Secologanin synthase which catalyzes the oxidative cleavage of loganin into secologanin is a cytochrome P450. *Phytochemistry, 53*, 7–12.

Yamazaki, Y., Kitajima, M., Arita, M., Takayama, H., Sudo, H., Yamazaki, M., et al. (2004). Biosynthesis of camptothecin. In silico and in vivo tracer study from [1-13C]glucose. *Plant Physiology, 134*, 161–170.

Yamazaki, M., Mochida, K., Asano, T., Nakabayashi, R., Chiba, M., Udomson, N., et al. (2013). Coupling deep transcriptome analysis with untargeted metabolic profiling in *Ophiorrhiza pumila* to further the understanding of the biosynthesis of the anti-cancer alkaloid camptothecin and anthraquinones. *Plant & Cell Physiology, 54*, 686–696.

Yamazaki, Y., Sudo, H., Yamazaki, M., Aimi, N., & Saito, K. (2003). Camptothecin biosynthetic genes in hairy roots of *Ophiorrhiza pumila*: Cloning, characterization and differential expression in tissues and by stress compounds. *Plant & Cell Physiology, 44*, 395–403.

Yamazaki, Y., Urano, A., Sudo, H., Kitajima, M., Takayama, H., Yamazaki, M., et al. (2003). Metabolite profiling of alkaloids and strictosidine synthase activity in camptothecin producing plants. *Phytochemistry, 62*, 461–470.

Yao, H., Gong, Y., Zuo, K., Ling, H., Qiu, C., Zhang, F., et al. (2008). Molecular cloning, expression profiling and functional analysis of a DXR gene encoding 1-deoxy-D-xylulose 5-phosphate reductoisomerase from *Camptotheca acuminata. Journal of Plant Physiology, 165*, 203–213.

Ya-ut, P., Chareonsap, P., & Sukrong, S. (2011). Micropropagation and hairy root culture of *Ophiorrhiza alata* Craib for camptothecin production. *Biotechnology Letters, 33*, 2519–2526.

Zhou, B. N., Hoch, J. M., Johnson, R. K., Mattern, M. R., Eng, W. K., Ma, J., et al. (2000). Use of COMPARE analysis to discover new natural product drugs: Isolation of camptothecin and 9-methoxycamptothecin from a new source. *Journal of Natural Products, 63*, 1273–1276.

CHAPTER SIX

Improved Production of Plant Isoquinoline Alkaloids by Metabolic Engineering

Fumihiko Sato[1]
Division of Integrated Life Science, Graduate School of Biostudies, Kyoto University, Oiwake-cho, Kitashirakawa, Sakyo-ku, Kyoto, Japan
[1]Corresponding author: e-mail address: fsato@lif.kyoto-u.ac.jp

Contents

1. Introduction 164
2. Isoquinoline Alkaloid Biosynthesis and Pathway Characterization 165
3. Metabolic Pathway Engineering 171
   3.1 Pathway engineering with overexpression of the rate-limiting enzyme and transcription factors in alkaloid biosynthesis 171
   3.2 Pathway engineering with trimming of pathways and the introduction of new branch pathways 173
4. Perspectives 176
5. Conclusion 177
Acknowledgement 177
References 177

Abstract

Higher plants produce diverse low-molecular-weight chemicals such as alkaloids, terpenoids and phenylpropanoid compounds. Among these chemicals, alkaloids are particularly important in medicine due to their high biological activities. However, the low yield of metabolites, especially alkaloids, in plants limits large-scale development of the plant natural product industry. In this chapter, I describe the metabolic engineering of plants to improve the yield and quality of useful secondary metabolites and their potentials. Among alkaloids, I focus on the production of isoquinoline alkaloids (IQAs), as their biosynthesis has been most intensively studied at the molecular level and their application in metabolic engineering has also been intensively examined. Based on our understanding of IQA engineering, I discuss its application to the production of other alkaloids through similar approaches.

Advances in Botanical Research, Volume 68
ISSN 0065-2296
http://dx.doi.org/10.1016/B978-0-12-408061-4.00006-7

ABBREVIATIONS

4′OMT 3′-hydroxy-*N*-methyl-coclaurine 4′OMT
6OMT norcoclaurine 6OMT
BBE berberine bridge enzyme
Cj *Coptis japonica*
CNMT coclaurine *N*-methyltransferase
CoOMT columbamine *O*-methyltransferase
COR codeinone reductase
CYP719A1 canadine synthase
CYP719A2/A3 stylopine synthase
CYP719A5 chelanthifoline synthase
CYP719B1 salutaridine synthase
CYP80A1 berbamunine synthase
CYP80B1 *N*-methyl-coclaurine 3′-hydroxylase
CYP82N2v2 (P6H) protopine 6-hydroxylase
CYP82G2 corytuberine synthase
IQAs isoquinoline alkaloids
MeJA methyl jasmonate
NCS norcoclaurine synthase
OMT *O*-methyltransferase
RNAi RNA interference
SalAT salutaridinol-7-*O*-acetyltransferase
SalR salutaridine reductase
SMT (*S*)-scoulerine 9-*O*-methyltransferase
THBO tetrahydroprotoberberine oxidase

1. INTRODUCTION

Higher plants produce diverse low-molecular-weight chemicals such as alkaloids, terpenoids and phenylpropanoid compounds (Croteau, Kutchan, & Lewis, 2000). Although these compounds are widely used for health and nutrition in humans, alkaloids are particularly important in medicine due to their high biological activities. However, the low yield of metabolites, especially alkaloids, in plants limits large-scale development of the plant natural product industry. Thus, many research groups have tried to establish production systems using the micropropagation of elite plant clones, cell culture and organ culture based on incomplete information regarding the biosynthesis of these metabolites. Over the past 20 years, our understanding of the biochemistry and molecular biology of natural products, including alkaloids, has rapidly expanded. Based on the knowledge regarding biosynthetic enzymes and their corresponding genes, the modification of productivity and metabolic profiles has become possible (Sato, Inai, & Hashimoto, 2007; Sato & Yamada, 2008). Furthermore, the microbial production of plant metabolites through the reconstruction of plant biosynthetic pathways has become feasible (Chow & Sato, 2013; Hawkins & Smolke, 2008; Minami et al., 2008; Nakagawa et al., 2011; see also Chapter 7). Metabolic engineering in plant cells has advantages and disadvantages. In this chapter, I discuss its potentials and pitfalls based on our experience.

2. ISOQUINOLINE ALKALOID BIOSYNTHESIS AND PATHWAY CHARACTERIZATION

Due to technological difficulties in the cultivation and relatively slow growth of medicinal plants, plant cell and tissue cultures have been intensively examined for the production of secondary metabolites (Sato & Yamada, 2008; Verpoorte, van Der Heijden, van Gulik, & Ten Hoopen, 1991). Whereas the selection of high-yield lines and treatment with an elicitor such as methyl jasmonate has enabled the industrial production of some metabolites, for example, shikonin production in selected *Lithospermum erythrorhizon* cells, or paclitaxel production in Pacific yew cell cultures with methyl jasmonate, it is quite difficult to achieve both high productivity and stability for industrial application (Sato & Yamada, 2008). While the development of hairy roots by the transformation of plant cells with *Agrobacterium rhizogenes* provides a solution with unlimited growth, morphological differentiation of roots for physical strength, high potential for the mass production of valuable secondary metabolites and greater genetic stability, for fermentation, we need a much faster growth of cells/tissues, higher productivity and improved quality of metabolites (for details of cell culture techniques, please see Sato & Yamada, 2008).

Recent advances in molecular biology have provided tools for metabolic engineering. The identification of many genes of biosynthetic enzymes and characterization of the spatial and developmental regulation of their expression have clarified their roles in the biosynthesis of secondary metabolites and provided strategies for metabolic engineering such as the trimming of undesired pathways, the enhancement of rate-limiting steps and the introduction of new pathways to produce novel compounds (Sato, Inai, & Hashimoto, 2007; Sato, Inui, & Takemura, 2007).

Isoquinoline alkaloid (IQA) biosynthesis provides a good model in metabolic engineering, as it has been intensively studied at the molecular level and experimentally challenged. The biosynthetic pathway for IQAs starts with L-tyrosine, as shown in Fig. 6.1. Among IQAs, the biosynthesis of benzylisoquinoline alkaloid from norcoclaurine has been the most investigated to date. All of the enzyme genes of the total nine enzymatic reaction steps from norcoclaurine to berberine have been characterized at the DNA level (Fig. 6.1): a norcoclaurine synthase (NCS; an entry enzyme in IQA biosynthesis, PR10 family such as *Tf*NCS from *Thalictrum flavum* (Lee & Facchini, 2010; Samanani, Liscombe, & Facchini, 2004) and a novel dioxygenase-like protein, *Cj*NCS1, from *Coptis japonica* (Minami, Dubouzet, Iwasa, & Sato, 2007)), an *N*-methyltransferase (coclaurine *N*-methyltransferase (CNMT, Choi, Morishige, Shitan, Yazaki, & Sato, 2002)), three *O*-methyltransferases (OMTs; norcoclaurine 6OMT, 3′-hydroxy-*N*-methyl-coclaurine 4′OMT (6OMT, 4′OMT, Morishige, Tsujita, Yamada, & Sato, 2000), (*S*)-scoulerine 9-*O*-methyltransferase (SMT, Takeshita et al., 1995)), a hydroxylase (CYP80B1, Pauli & Kutchan, 1998), a berberine bridge enzyme (BBE, Dittrich & Kutchan, 1991), a methylenedioxy ring-forming enzyme (canadine synthase; CYP719A1, Ikezawa et al., 2003) and a tetrahydroprotoberberine oxidase (THBO, Gesell et al., 2011; Matsushima, Minami, Hori, & Sato, 2012).

The number of available enzyme genes is growing rapidly (Figs. 6.1 and 6.2): the genes of corytuberine synthase in aporphine biosynthesis (CYP82G2, Ikezawa, Iwasa, & Sato, 2008); those of salutaridine synthase (CYP719B1, Gesell et al., 2009), salutaridine reductase (SalR, Ziegler et al., 2006), salutaridinol-7-*O*-acetyltransferase (SalAT, Grothe, Lenz, & Kutchan, 2001), thebaine 6-*O*-demethylase and codeine *O*-demethylase (T6ODM, CODM, Hagel & Facchini, 2010) and NADPH-dependent codeinone reductase (COR, Unterlinner, Lenz, & Kutchan, 1999) in morphinan alkaloid biosynthesis; those of chelanthifoline synthase (CYP719A5, Ikezawa, Iwasa, & Sato, 2009), stylopine synthase (CYP719A2/A3, Ikezawa, Iwasa, & Sato, 2007), (*S*)-cis-*N*-methylstylopine 14-hydroxylase (Beaudoin & Facchini, 2013), protopine 6-hydroxylase (CYP82N2v2, Takemura, Ikezawa, Iwasa, & Sato, 2013), sanguinarine reductase (Vogel, Lawson, Sippl, Conarad, & Roos, 2010) and dihydrobenzophenanthridine oxidase (DBOX, Hagel et al., 2012) in benzophenanthridine biosynthesis; those of berbamunine synthase (CYP80A1, Kraus & Kutchan, 1995) in the bis-benzylisoquinoline alkaloid pathway; those of *O*-methyltransferases (Dang & Facchini, 2012) and a 10-gene cluster (three *O*-methyltransferases (PSMT1-3), four cytochrome P450s (CYP82X1/2, CYP82Y1 & CYP719A21), an acetyltransferase (PSAT1), a carboxylesterase (PSCXE1)

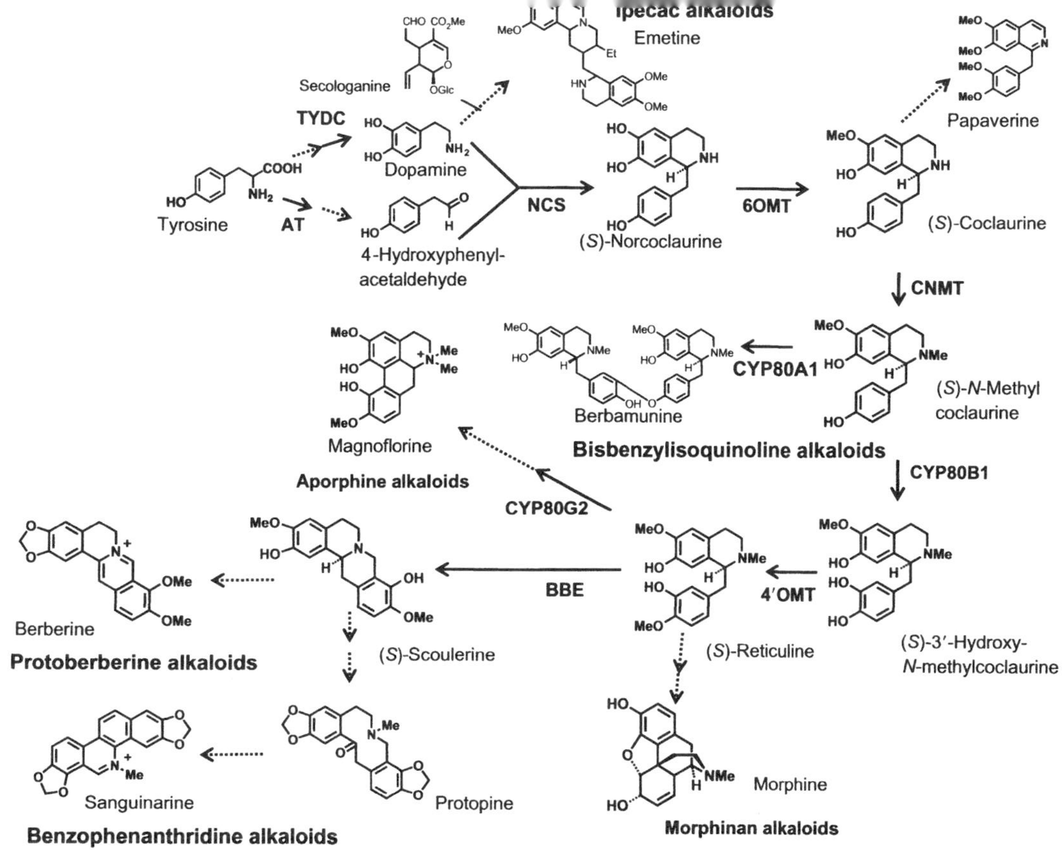

Figure 6.1 Main pathway in isoquinoline alkaloid biosynthesis. Major isoquinoline alkaloids, such as aporphine-, benzophenanthridine- and protoberberine-type isoquinoline alkaloids, are synthesized through a common pathway derived from norcoclaurine to reticuline. Aporphine alkaloids are synthesized from (*S*)-reticuline by the reaction of corytuberine synthase (CYP80G2). Protoberberine and benzophenanthridine alkaloids are synthesized from (*S*)-scoulerine produced from (*S*)-reticuline by the reaction of berberine bridge enzyme (BBE). Ipecac and some other IQAs are synthesized from dopamine without norcoclaurine biosynthesis. The names of the biosynthetic enzymes for which cDNA has been isolated and characterized are shown. Abbreviations for biosynthetic enzymes are defined in the text, except for TYDC (tyrosine/DOPA decarboxylase; Facchini, Huber-Allanach, & Tari, 2000) and TyrAT (tyrosine aminotransferase; Lee & Facchini, 2011).

A

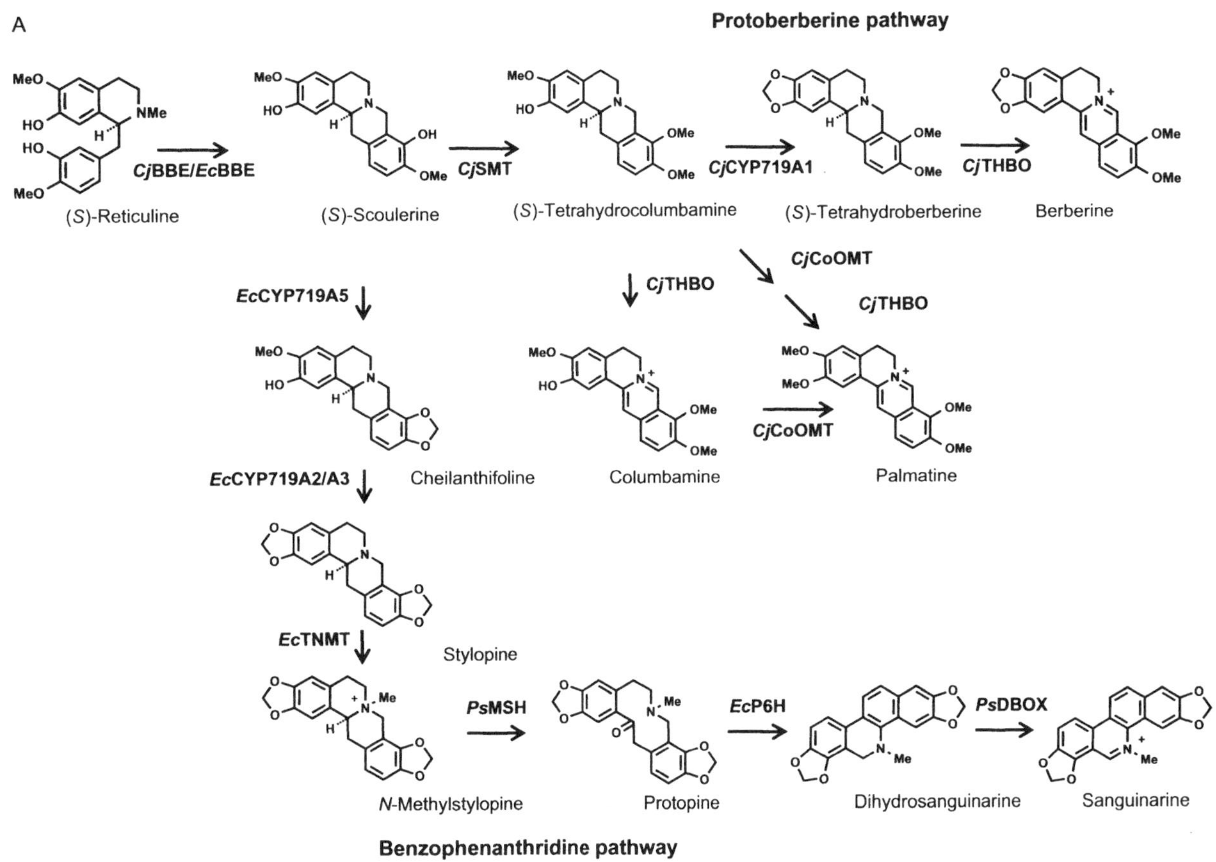

Protoberberine pathway
(S)-Reticuline
CjBBE/EcBBE
(S)-Scoulerine
CjSMT
(S)-Tetrahydrocolumbamine
CjCYP719A1
(S)-Tetrahydroberberine
CjTHBO
Berberine
CjCoOMT
CjTHBO
CjTHBO
EcCYP719A5
Cheilanthifoline
Columbamine
CjCoOMT
Palmatine
EcCYP719A2/A3
Stylopine
EcTNMT
N-Methylstylopine
PsMSH
Protopine
EcP6H
Dihydrosanguinarine
PsDBOX
Sanguinarine
Benzophenanthridine pathway

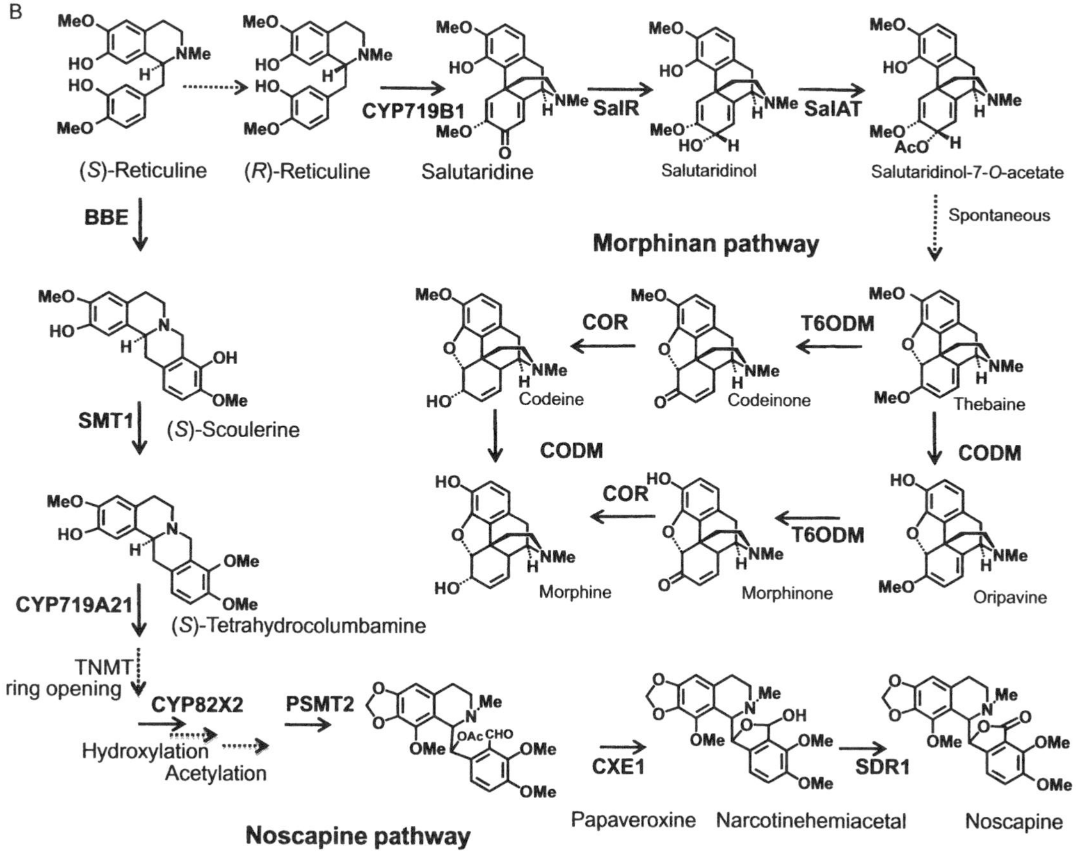

Figure 6.2 Extended pathways in isoquinoline alkaloid biosynthesis from reticuline to protoberberine, benzophenanthridine, morphinan and noscarpine alkaloids. (A) Protoberberine and benzophenanthridine alkaloid biosynthesis has mainly been characterized in *Coptis japonica* (*Cj*) and *Eschscholzia californica* (*Es*) cells, while some genes have been isolated from *Papaver somniferum* (*Ps*). (B) Morphinan and noscarpine biosynthesis were characterized in *Papaver somniferum*. The names of the biosynthetic enzymes for which cDNA has been isolated and characterized are shown. Abbreviations for biosynthetic enzymes are defined in the text, except for CoOMT (columbamine *O*-methyltransferase) and 7OMT (reticuline 7-*O*-methyltransferase). (For colour version of this figure, the reader is referred to the online version of this chapter.)

and a short-chain dehydrogenase/reductase (PSSDR1, Winzer et al., 2012)) in noscapine biosynthesis and those of *O*-methyltransferases in emetine biosynthesis (IpeOMT1/2/3, CiOMT, Cheong, Takemura, Yoshimatsu, & Sato, 2011; Nomura & Kutchan, 2010).

While many homologues are annotated after cloning of the first enzyme gene, we should pay close attention to the properties of each enzyme, as pathways in secondary metabolism are often diversified and connected in a complicated manner, based on the different substrate and reaction specificity of the enzyme acquired during the evolution of each species. For example, while *Coptis japonica* columbamine *O*-methyltransferase (*Cj*CoOMT) can use both tetrahydrocolumbamine and columbamine in palmatine biosynthesis, columbamine was once thought to be the exclusive substrate of CoOMT in *Berberis* plants (Morishige, Dubouzet, Choi, Yazaki, & Sato, 2002). Similarly, CNMT of *Coptis* and *Berberis* showed different substrate specificities; that is, the *Coptis* enzyme could *N*-methylate norlaudanosoline, while the *Berberis* enzyme could not. In this sense, the knowledge needed for both metabolic engineering and synthetic biology is still incomplete, and such information is greatly needed. We should be careful in the annotation

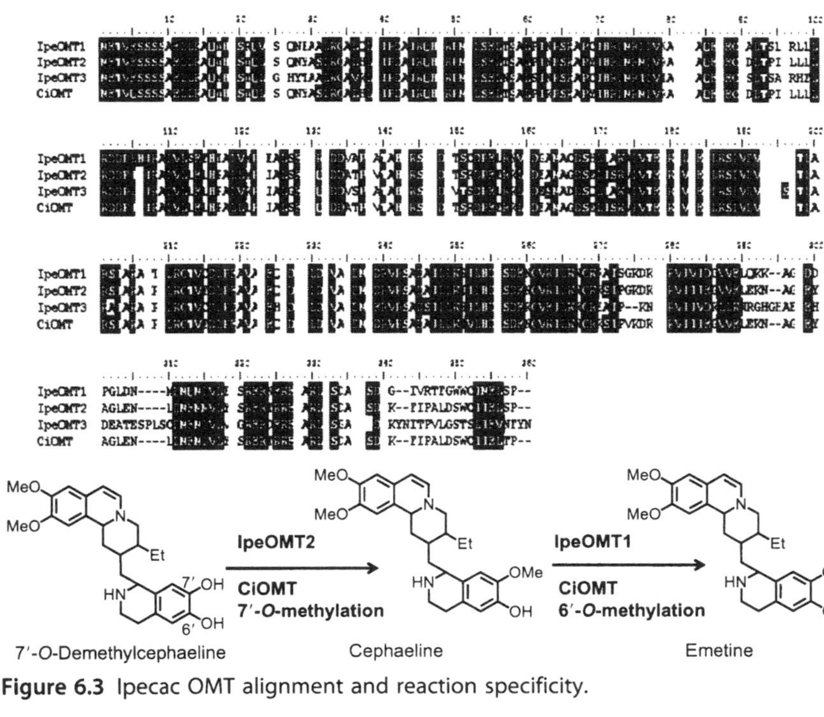

Figure 6.3 Ipecac OMT alignment and reaction specificity.

of an enzyme gene simply based on homology, as only a few changes in amino acids can modify reaction specificities; for example, Ipecac OMTs in emetine biosynthesis (Fig. 6.3; Cheong et al., 2011; Nomura & Kutchan, 2010).

3. METABOLIC PATHWAY ENGINEERING

3.1. Pathway engineering with overexpression of the rate-limiting enzyme and transcription factors in alkaloid biosynthesis

With an increase in the molecular tools available in metabolic engineering, we can now modulate the metabolic flow of plant cells. Among several strategies, overexpression of the gene for the enzyme in the rate-limiting step is commonly examined to improve the yield (Fig. 6.4). The overexpression of *C. japonica* SMT in the original cells was the first successful application of genetic engineering to IQA biosynthesis to increase berberine production (Sato et al., 2001), while 4′OMT overexpression was later successfully applied in intact plants (Inui et al., 2012). Allen et al. (2008) also reported that transgenic opium poppy with overexpression of *SalAT* had more than 30% greater total alkaloids than the control in three-independent trials over 3 years.

However, overexpression of an endogeneous gene can (often) induce the co-suppression of the gene and decrease the production of the end-metabolites (Jorgensen, 1995; Takemura, Chow, Todokoro, Okamoto, & Sato, 2010). Thus, heterologous expression of a homologue isolated from a different plant species is recommended. For example, *Cj*6OMT was expressed in California poppy (*Eschscholzia californica*) cells to overcome the rate-limiting step (Inui, Tamura, Fujii, Morishige, & Sato, 2007). Note that bottleneck step(s) may differ among plant species and cell lines due to variations in gene expression. In fact, the overexpression of *Cj*4′OMT was less effective for increasing alkaloid production than that of 6OMT in *E. californica* cells, whereas 4′OMT was effective in *C. japonica* plants (Inui et al., 2007, 2012). Changes in the bottleneck step may also occur after modification of the enzyme expression level through metabolic engineering.

As there are multiple bottleneck steps and these limiting steps can change with overexpression of the enzyme in the limiting step, overall regulation of the expression levels of biosynthetic genes in the pathway is the preferred strategy. Whereas information about the transcription factors in alkaloid biosynthesis is very limited, several transcription factors, such as the ERF,

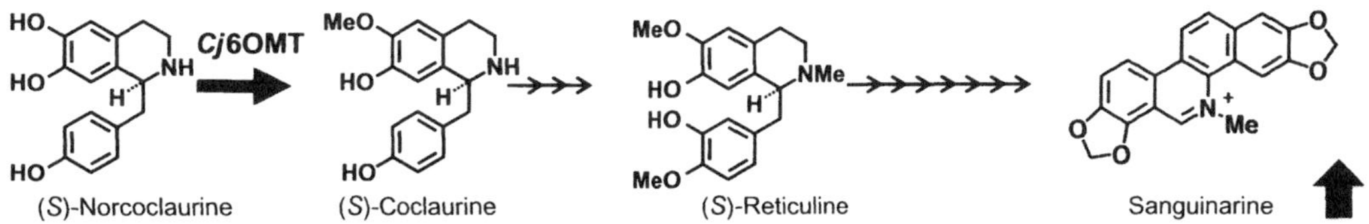

Figure 6.4 Yield improvement by the ectopic expression of *C. japonica* (*Cj*) 6OMT in California poppy cells.

WRKY and bHLH families in alkaloid biosynthesis, including IQA biosynthesis, have been isolated and characterized (Kato et al., 2007; Yamada, Kokabu, et al., 2011; Yamada, Koyama, & Sato, 2011; Yamada & Sato, 2013). While experimental results suggest that the activities of native transcription factors can be fine-tuned spatially and temporally (see Yamada & Sato, 2013), the successful application of the overexpression of heterologous *At*WRKY transcription factor in opium poppy to increase the morphinan alkaloid content (Apuya et al., 2008) suggests that transcription factors may be used to improve the yields of IQAs and other alkaloids.

3.2. Pathway engineering with trimming of pathways and the introduction of new branch pathways

While an increase in the metabolite yield is important for industrial application, modification of the metabolite profile by the introduction of a new metabolic branching point and/or the trimming of undesired pathways is another important approach in drugs development.

To increase metabolite diversity, we ectopically expressed *Cj*SMT, which is involved in berberine biosynthesis but not in benzophenanthridine alkaloid biosynthesis, in California poppy (*E. californica*) cells (Fig. 6.5). *Cj*SMT expression not only produced columbamine (oxidized product of tetrahydrocolumbamine) in transgenic *E. californica* cells (Sato et al., 2001) but also induced the accumulation of novel products, that is, allocryptopine and 10-hydroxychelerythrine, derived from *Cj*SMT reaction products by the endogenous enzyme reactions in *E. californica* cells, indicating that the alkaloid profile in transgenic cells was further diversified by the introduction of a branched pathway (Takemura, Ikezawa, Iwasa & Sato, 2010).

While an increase in metabolite diversity is preferable for the screening of novel products, the reduction of undesired metabolic diversity is also required to prepare a single purified chemical and increase the metabolic flow to the desired compound. Among the methods for gene silencing (e.g. antisense, RNA silencing, knockout with T-DNA insertion, mutation, etc.), RNA silencing with double-stranded RNA molecules (RNA interference; RNAi with short interfering (si) RNA or micro (mi) RNA, co-suppression of gene expression with overexpression of a target gene, or virus-induced gene silencing) is the most powerful and reliable, although transcription activator-like effector nucleases (TALEN) could be useful for preparing knockout mutant and novel mutants through homologous recombinants (Borgio, 2009; Desgagné-Penix & Facchini, 2012; Sato, 2005; Winzer et al., 2012; Zhang et al., 2013).

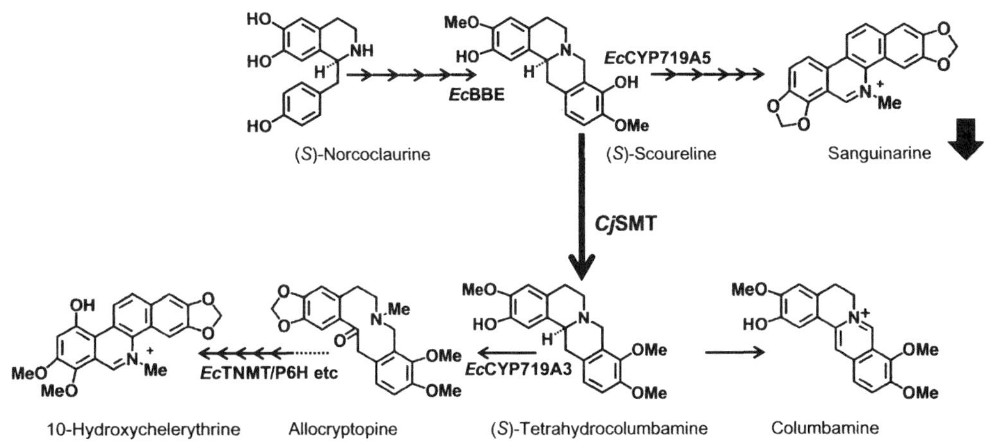

Figure 6.5 Metabolic diversification by the introduction of a branch pathway. The introduction of a novel branch pathway in California poppy induced the production of uncommon protoberberine-type alkaloids, whereas major products were further metabolized to dimethoxy-type benzophenanthridine alkaloids by endogenous enzymes.

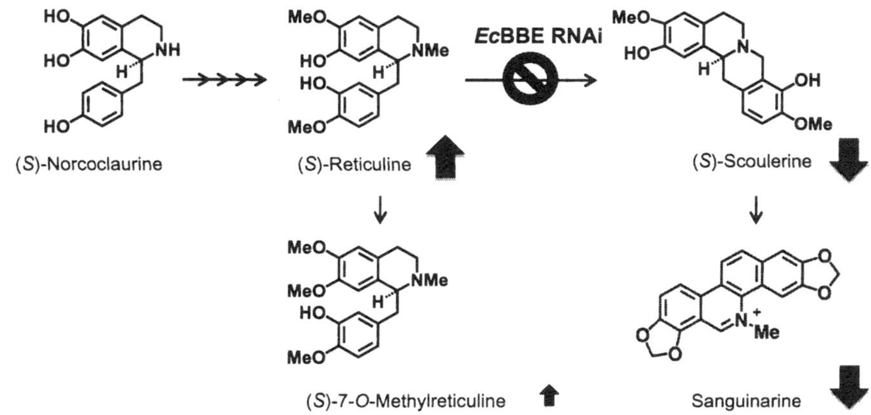

Figure 6.6 Quality modification by the trimming of a pathway. Knockdown of BBE expression with double-stranded RNA strongly reduced endogenous alkaloid accumulation and increased the accumulation of a precursor, reticuline and some metabolites such as 7-*O*-methylreticuline.

Successful RNAi of BBE in California poppy cells induced the accumulation of reticuline, a substrate of BBE (Fujii, Inui, Iwasa, Morishige, & Sato, 2007), whereas antisense RNA of BBE produced little accumulation of reticuline in transgenic cells (Park, Yu, & Facchini, 2002; Fig. 6.6). However, there is another pitfall in effective gene silencing; for example, an enhanced branch pathway from the accumulated reticuline to produce a byproduct, that is, 7-*O*-methylreticuline, which could scarcely be detected in control cells. In addition, suppression of a certain enzyme in a metabolic pathway may induce the unexpected disturbance of the pathway; for example, RNAi of COR genes in opium poppy induced the accumulation of a far-upstream precursor, (*S*)-reticuline (Allen et al., 2004). The disturbance of metabolism by the down-regulation of enzyme expression can be partly explained by the accumulation of an intermediate and feedback regulation of the pathway, or the disruption of an enzyme complex (metabolome) to catalyze sequential reactions, or off-target effects of RNAi on other unexpected gene targets (Allen et al., 2004, 2008).

While simple overexpression of the rate-limiting step enzyme/transcription factors, introduction of a branch pathway and trimming of a pathway through the use of gene silencing are currently applied in metabolic engineering in alkaloid biosynthesis, the combination of these strategies provides a more effective approach for improving both the quantity and quality of metabolites. For example, trimming of the endogenous pathway with the introduction of new branch was critical for drastic modification of the metabolite in blue rose production; that is, the accumulation of sufficient blue pigment, delphinidin, was only achieved by the combined expression of the gene for a novel blue pigment biosynthetic enzyme, that is, flavonoid 3′,5′-hydroxylase, and the down-regulation of endogenous dihydroflavonol 4-reductase by RNAi to shut down the undesired red pigment pathway (Katsumoto et al., 2007). Such a combined approach can be performed with a transcription factor and an enzyme for a rate-limiting step that is not under the control of the introduced transcription factor. This can provide greater flexibility in metabolic engineering.

4. PERSPECTIVES

While the biosynthetic pathways for many metabolites still require considerable characterization including of the involved enzymes and their genes, continued advances in molecular and genomic techniques will surely expand the production of useful natural products, especially alkaloids. Particularly, whole-genome sequence data bring considerable information on biosynthesis along with rapidly accumulating RNA sequence data. The combined analysis of transcriptomes, proteomes, metabolomes and genomes provides a new approach for identifying the candidate genes in uncharacterized pathways (Dang & Facchini, 2012; Desgagné-Penix, Farrow, Cram, Nowak, & Facchini, 2012; Desgagné-Penix et al., 2010; Farrow, Hagel, & Facchini, 2012; Takemura, Ikezawa, et al., 2010; Winzer et al., 2012). The establishment of transcriptome and proteome databases for the investigation of natural product metabolism in non-model plant systems has provided useful information for metabolic engineering. Desgagné-Penix et al. (2010) showed that profiling of the more abundant proteins in elicitor-treated opium poppy cell cultures revealed several uncharacterized enzymes that potentially catalyze steps in sanguinarine biosynthesis.

Structural biology also provides important cues for isolating key biosynthetic enzymes in pathways and their molecular functions, although structural information regarding the biosynthetic enzymes in secondary metabolism is very limited (Ilari et al., 2009; Wallner et al., 2012; Winkler et al., 2009).

Whereas we can use both plant and microbial cells as a platform for metabolite production (Chow & Sato, 2013; see also Chapter 7), we should be aware of their pros and cons. For instance, while plant cell or tissue cultures can produce more divergent chemicals than microbial systems, their complexity and instability are problematic for industrial production. Microbial systems enable alkaloid production on a relatively large scale within a short period but require more molecular information on the biosynthetic pathways and enzymes, whereas the emergence of synthetic biology offers the possibility of improving the production of useful products through the design and engineering of complex biological systems. While computer-assisted molecular design software is being developed to facilitate the synthesis of novel compounds with drug-like properties and desired activities, the use of information science for the reconstruction of metabolic pathways, that is, "retrobiosynthetic" systems will require more experimental data on the identification and analysis of biosynthetic enzymes and pathways. However, technological progress has been quite rapid and provides us with many options. The potentials of metabolic engineering in secondary metabolism are unlimited.

5. CONCLUSION

Metabolic engineering as well as synthetic biological approaches are at the forefront in natural product research. Our experiences in IQA biosynthesis provide new clues for the molecular characterization of biosynthetic pathways and novel production systems for natural products. Technological advances, including next-generation sequencing, offer a bright future for the production of not only IQAs but also other alkaloids. This high potential in natural product research will require more sophisticated techniques as we move from metabolites to drug discovery.

ACKNOWLEDGEMENT

This research was supported by the Ministry of Education, Culture, Sports, Science and Technology of Japan [Grant-in-Aid (No. 21248013 to F. S.)].

REFERENCES

Allen, R. S., Miller, J. A. C., Chitty, J. A., Fist, A. J., Gerlach, W. L., & Larkin, P. J. (2008). Metabolic engineering of morphinan alkaloids by overexpression and RNAi suppression of salutaridinol 7-O-acetyltransferase in opium poppy. *Plant Biotechnology Journal, 6*, 22–30.

Allen, R. S., Millgate, A. G., Chitty, J. A., Thistleton, J., Miller, J. A. C., Fist, A. J., et al. (2004). RNAi-mediated replacement of morphine with the non narcotic alkaloid reticuline in opium poppy. *Nature Biotechnology, 22*, 1559–1566.

Apuya, N. R., Park, J. H., Zhang, L., Ahyow, M., Davidow, P., Van Fleet, J., et al. (2008). Enhancement of alkaloid production in opium and California poppy by transactivation using heterologous regulatory factors. *Plant Biotechnology Journal, 6*, 160–175.

Beaudoin, G. A., & Facchini, P. J. (2013). Isolation and characterization of a cDNA encoding (*S*)-*cis*-*N*-methylstylopine 14-hydroxylase from opium poppy, a key enzyme in sanguinarine biosynthesis. *Biochemical and Biophysical Research Communications, 431*, 597–603.

Borgio, J. (2009). RNA interference (RNAi) technology: A promising tool for medicinal plant research. *Journal of Medicinal Plant Research, 3*, 1176–1183.

Cheong, B. E., Takemura, T., Yoshimatsu, K., & Sato, F. (2011). Molecular cloning of an *O*-methyltransferase from adventitious roots of *Carapichea ipecacuanha*. *Bioscience, Biotechnology, and Biochemistry, 75*, 107–113.

Choi, K. B., Morishige, T., Shitan, N., Yazaki, K., & Sato, F. (2002). Molecular cloning and characterization of coclaurine *N*-methyltransferase from cultured cells of *Coptis japonica*. *The Journal of Biological Chemistry, 277*, 830–835.

Chow, Y. L., & Sato, F. (2013). Metabolic engineering and synthetic biology for the production of isoquinoline alkaloids. In S. Chandra, H. Lata, & A. Varma (Eds.), *Biotechnology for medicinal plants: Micropropagation and improvement* (pp. 327–343). Berlin: Springer.

Croteau, R., Kutchan, T. M., & Lewis, N. G. (2000). Natural products (secondary metabolites). In B. B. Buchanan, W. Gruissem, & R. L. Jones (Eds.), *Biochemistry & molecular biology of plants* (pp. 1250–1318). Maryland: American Society of Plant Biologists.

Dang, T. T., & Facchini, P. J. (2012). Characterization of three O-methyltransferases involved in noscapine biosynthesis in opium poppy. *Plant Physiology, 159*, 618–631.

Desgagné-Penix, I., & Facchini, P. J. (2012). Systematic silencing of benzylisoquinoline alkaloid biosynthetic genes reveals the major route to papaverine in opium poppy. *The Plant Journal, 72*, 331–344.

Desgagné-Penix, I., Farrow, S. C., Cram, D., Nowak, J., & Facchini, P. J. (2012). Integration of deep transcript and targeted metabolite profiles for eight cultivars of opium poppy. *Plant Molecular Biology, 79*, 295–313.

Desgagné-Penix, I., Khan, M. F., Schriemer, D. C., Cram, D., Nowak, J., & Facchini, P. J. (2010). Integration of deep transcriptome and proteome analyses reveals the components of alkaloid metabolism in opium poppy cell cultures. *BMC Plant Biology, 10*, 252–268.

Dittrich, H., & Kutchan, T. M. (1991). Molecular cloning, expression, and induction of berberine bridge enzyme, an enzyme essential to the formation of benzophenanthridine alkaloids in the response of plants to pathogenic attack. *Proceedings of the National Academy of Sciences of the United States of America, 88*, 9969–9973.

Facchini, P. J., Huber-Allanach, K. L., & Tari, L. W. (2000). Plant aromatic L-amino acid decarboxylases: Evolution, biochemistry, regulation, and metabolic engineering applications. *Phytochemistry, 54*, 121–138.

Farrow, S. C., Hagel, J. M., & Facchini, P. J. (2012). Transcript and metabolite profiling in cell cultures of 18 plant species that produce benzylisoquinoline alkaloids. *Phytochemistry, 77*, 79–88.

Fujii, N., Inui, T., Iwasa, K., Morishige, T., & Sato, F. (2007). Knockdown of berberine bridge enzyme by RNAi accumulates (*S*)-reticuline and activates a silent pathway in cultured California poppy cells. *Transgenic Research, 16*, 363–375.

Gesell, A., Díaz Chávez, M. L., Kramell, R., Piotrowski, M., Macheroux, P., & Kutchan, T. M. (2011). Heterologous expression of two FAD-dependent oxidases with (*S*)-tetrahydroprotoberberine oxidase activity from *Argemone mexicana* and *Berberis wilsoniae* in insect cells. *Planta, 233*, 1185–1197.

Gesell, A., Rolf, M., Ziegler, J., Díaz Chávez, M. L., Huang, F.-C., & Kutchan, T. M. (2009). CYP719B1 is salutaridine synthase, the C-C phenol-coupling enzyme of morphine biosynthesis in opium poppy. *The Journal of Biological Chemistry, 284*, 24432–24442.

Grothe, T., Lenz, R., & Kutchan, T. M. (2001). Molecular characterization of the salutaridinol-7-O-acetyltransferase involved in morphine biosynthesis in opium poppy *Papaver somniferum*. *The Journal of Biological Chemistry*, *276*, 30717–30723.

Hagel, J. M., Beaudoin, G. A., Fossati, E., Ekins, A., Martin, V. J., & Facchini, P. J. (2012). Characterization of a flavoprotein oxidase from opium poppy catalyzing the final steps in sanguinarine and papaverine biosynthesis. *The Journal of Biological Chemistry*, *287*, 42972–42983.

Hagel, J. M., & Facchini, P. J. (2010). Dioxygenases catalyze the O-demethylation steps of morphine biosynthesis in opium poppy. *Nature Chemical Biology*, *6*, 273–275.

Hawkins, K., & Smolke, C. (2008). Production of benzylisoquinoline alkaloids in *Saccharomyces cerevisiae*. *Nature Chemical Biology*, *4*, 564–573.

Ikezawa, N., Iwasa, K., & Sato, F. (2007). Molecular cloning and characterization of methylenedioxy bridge-forming enzymes involved in stylopine biosynthesis in *Eschscholzia californica*. *The FEBS Journal*, *274*, 1019–1035.

Ikezawa, N., Iwasa, K., & Sato, F. (2008). Molecular cloning and characterization of CYP80G2, a cytochrome P450 that catalyzes an intramolecular C–C phenol coupling of (*S*)-reticuline in magnoflorine biosynthesis, from cultured *Coptis japonica* cells. *The Journal of Biological Chemistry*, *283*, 8810–8821.

Ikezawa, N., Iwasa, K., & Sato, F. (2009). CYP719A subfamily of cytochrome p450 oxygenase and isoquinoline alkaloid biosynthesis in *Eschscholzia californica*. *Plant Cell Reports*, *28*, 123–133.

Ikezawa, N., Tanaka, M., Nagayoshi, M., Shinkyo, R., Sakaki, T., Inouye, K., et al. (2003). Molecular cloning and characterization of CYP719, a methylenedioxy bridge-forming enzyme that belongs to a novel P450 family, from cultured *Coptis japonica* cells. *The Journal of Biological Chemistry*, *278*, 38557–38565.

Ilari, A., Franceschini, S., Bonamore, A., Arenghi, F., Botta, B., Macone, A., et al. (2009). Structural basis of enzymatic (S)-norcoclaurine biosynthesis. *The Journal of Biological Chemistry*, *284*, 897–904.

Inui, T., Kawano, N., Shitan, N., Yazaki, K., Kiuchi, F., Kawahara, N., et al. (2012). Improvement of benzylisoquinoline alkaloid productivity by overexpression of 3′-hydroxy-*N*-methylcoclaurine 4′-*O*-methyltransferase in transgenic *Coptis japonica* plants. *Biological & Pharmaceutical Bulletin*, *35*, 650–659.

Inui, T., Tamura, K., Fujii, N., Morishige, T., & Sato, F. (2007). Overexpression of *Coptis japonica* norcoclaurine 6-*O*-methyltransferase overcomes the rate-limiting step in benzylisoquinoline alkaloid biosynthesis in cultured *Eschscholzia californica*. *Plant & Cell Physiology*, *48*, 252–262.

Jorgensen, R. A. (1995). Cosuppression, flower color patterns, and metastable gene expression states. *Science*, *268*, 686–691.

Kato, N., Dubouzet, E., Kokabu, Y., Yoshida, S., Taniguchi, Y., Dubouzet, J. G., et al. (2007). Identification of a WRKY protein as a transcriptional regulator of benzylisoquinoline alkaloid biosynthesis in *Coptis japonica*. *Plant & Cell Physiology*, *48*, 8–18.

Katsumoto, Y., Fukuchi-Mizutani, M., Fukui, Y., Brugliera, F., Holton, T. A., Karan, M., et al. (2007). Engineering of the rose flavonoid biosynthetic pathway successfully generated blue-hued flowers accumulating delphinidin. *Plant & Cell Physiology*, *48*, 1589–1600.

Kraus, P. F., & Kutchan, T. M. (1995). Molecular-cloning and heterologous expression of a cDNA-encoding berbamunine synthase, a C-O-phenol-coupling cytochrome-P450 from the higher-plant *Berberis stolonifera*. *Proceedings of the National Academy of Sciences of the United States of America*, *92*, 2071–2075.

Lee, E. J., & Facchini, P. (2010). Norcoclaurine synthase is a member of the pathogenesis-related 10/Bet v1 protein family. *The Plant Cell*, *22*, 3489–3503.

Lee, E. J., & Facchini, P. J. (2011). Tyrosine aminotransferase contributes to benzylisoquinoline alkaloid biosynthesis in opium poppy. *Plant Physiology, 157*, 1067–1078.

Matsushima, Y., Minami, H., Hori, K., & Sato, F. (2012). Pathway engineering of benzylisoquinoline alkaloid biosynthesis in transgenic California poppy cells with ectopic expression of tetrahydroberberine oxidase from *Coptis japonica*. *Plant Biotechnology, 29*, 473–481.

Minami, H., Dubouzet, E., Iwasa, K., & Sato, F. (2007). Functional analysis of norcoclaurine synthase in *Coptis japonica*. *The Journal of Biological Chemistry, 282*, 6274–6282.

Minami, H., Kim, J.-S., Ikezawa, N., Takemura, T., Katayama, T., Kumagai, H., et al. (2008). Microbial production of plant benzylisoquinoline alkaloids. *Proceedings of the National Academy of Sciences of the United States of America, 105*, 7393–7398.

Morishige, T., Dubouzet, E., Choi, K.-B., Yazaki, K., & Sato, F. (2002). Molecular cloning of columbamine *O*-methyltransferase from cultured *Coptis japonica* cells. *European Journal of Biochemistry, 269*, 5659–5667.

Morishige, T., Tsujita, T., Yamada, Y., & Sato, F. (2000). Molecular characterization of the *S*-adenosyl-L-methionine: 3′-Hydroxy-*N*-methylcoclaurine 4′-*O*-methyltransferase involved in isoquinoline alkaloid biosynthesis in *Coptis japonica*. *The Journal of Biological Chemistry, 275*, 23398–23405.

Nakagawa, A., Minami, H., Kim, J. S., Koyanagi, T., Katayama, T., Sato, F., et al. (2011). A bacterial platform for fermentative production of plant alkaloids. *Nature Communications, 2*, http://dx.doi.org/10.1038/ncomms1327, Article number: 326.

Nomura, T., & Kutchan, T. M. (2010). Three new O-methyltransferases are sufficient for all *O*-methylation reactions of Ipecac alkaloid biosynthesis in root culture of *Psychotria ipecacuanha*. *The Journal of Biological Chemistry, 285*, 7722–7738.

Park, S., Yu, M., & Facchini, P. (2002). Antisense RNA-mediated suppression of benzophenanthridine alkaloid biosynthesis in transgenic cell cultures of California poppy. *Plant Physiology, 128*, 696–706.

Pauli, H., & Kutchan, T. M. (1998). Molecular cloning and functional heterologous expression of two alleles encoding (*S*)-*N*-methylcoclaurine 3′-hydroxylase (CYP80B1), a new methyl jasmonate-inducible cytochrome P-450-dependent mono-oxygenase of benzylisoquinoline alkaloid biosynthesis. *The Plant Journal, 13*, 793–801.

Samanani, N., Liscombe, D. K., & Facchini, P. J. (2004). Molecular cloning and characterization of norcoclaurine synthase, an enzyme catalyzing the first committed step in benzylisoquinoline alkaloid biosynthesis. *The Plant Journal, 40*, 302–313.

Sato, F. (2005). RNAi and functional genomics. *Plant Biotechnology, 22*, 431–442.

Sato, F., Hashimoto, T., Hachiya, A., Tamura, K.-I., Choi, K.-B., Morishige, T., et al. (2001). Metabolic engineering of plant alkaloid biosynthesis. *Proceedings of the National Academy of Sciences of the United States of America, 98*, 367–372.

Sato, F., Inai, K., & Hashimoto, T. (2007). Metabolic engineering in alkaloid biosynthesis: Case study in tyrosine- and putrescine-derived alkaloids. In R. Verpoorte, A. W. Alfermann, & T. S. Johnson (Eds.), *Applications of plant metabolic engineering* (pp. 145–173). Dordrecht: Springer.

Sato, F., Inui, T., & Takemura, T. (2007). Metabolic engineering in isoquinoline alkaloid biosynthesis. *Current Pharmaceutical Biotechnology, 8*, 211–218.

Sato, F., & Yamada, Y. (2008). Engineering formation of medicinal compounds in cell cultures. In H. J. Bohnert, H. Nguyen, & N. G. Lewis (Eds.), *Advances in plant biochemistry and molecular biology*: Vol. 1. (pp. 311–345). Amsterdam: Elsevier.

Takemura, T., Chow, Y.-L., Todokoro, T., Okamoto, T., & Sato, F. (2010). Overexpression of rate-limiting enzymes to improve alkaloid productivity. In J. M. Walker (Series Ed.), & A. G. Fett-Neto (Vol. Ed.), *Methods in molecular biology: Vol. 643. Plant secondary metabolism engineering: Methods and protocols* (pp. 33–45). New York: Springer.

Takemura, T., Ikezawa, N., Iwasa, K., & Sato, F. (2010). Metabolic diversification of benzylisoquinoline alkaloid biosynthesis through the introduction of a branch pathway in *Eschscholzia californica*. *Plant & Cell Physiology*, *51*, 949–959.

Takemura, T., Ikezawa, N., Iwasa, K., & Sato, F. (2013). Molecular cloning and characterization of a cytochrome P450 in sanguinarine biosynthesis from *Eschscholzia californica* cells. *Phytochemistry*, *91*, 100–108.

Takeshita, N., Fujiwara, H., Mimura, H., Fitchen, J., Yamada, Y., & Sato, F. (1995). Molecular cloning and characterization of *S*-adenosyl-L-methionine: Scoulerine-9-*O*-methyltransferase from cultured cells of *Coptis japonica*. *Plant & Cell Physiology*, *36*, 29–36.

Unterlinner, B., Lenz, R., & Kutchan, T. M. (1999). Molecular cloning and functional expression of codeinone reductase: The penultimate enzyme in morphine biosynthesis in the opium poppy *Papaver somniferum*. *The Plant Journal*, *18*(5), 465–475.

Verpoorte, R., van Der Heijden, R., van Gulik, W. M., & Ten Hoopen, H. J. G. (1991). Plant biotechnology for the production of alkaloids: Present status and prospects. In A. Brossi (Ed.), *The alkaloids: Chemistry and pharmacology*, Vol. 40, (pp. 1–187). New York: AcademicPress.

Vogel, M., Lawson, M., Sippl, W., Conarad, U., & Roos, W. (2010). Structure and mechanism of sanguinarine reductase, an enzyme of alkaloid detoxification. *The Journal of Biological Chemistry*, *285*, 18397–18406.

Wallner, S., Winkler, A., Riedl, S., Dully, C., Horvath, S., Gruber, K., et al. (2012). Catalytic and structural role of a conserved active site histidine in berberine bridge enzyme. *Biochemistry*, *51*, 6139–6147.

Winkler, A., Motz, K., Riedl, S., Puhl, M., Macheroux, P., & Gruber, K. (2009). Structural and mechanistic studies reveal the functional role of bicovalent flavinylation in berberine bridge enzyme. *The Journal of Biological Chemistry*, *284*, 19993–20001.

Winzer, T., Gazda, V., He, Z., Kaminski, F., Kern, M., Larson, T. R., et al. (2012). A *Papaver somniferum* 10-gene cluster for synthesis of the anticancer alkaloid noscapine. *Science*, *336*, 1704–1708.

Yamada, Y., Kokabu, Y., Chaki, K., Yoshimoto, T., Ohgaki, M., Yoshida, S., et al. (2011). Isoquinoline alkaloid biosynthesis is regulated by a unique bHLH-type transcription factor in *Coptis japonica*. *Plant & Cell Physiology*, *52*, 1131–1141.

Yamada, Y., Koyama, T., & Sato, F. (2011). Basic helix-loop-helix transcription factors and regulation of alkaloid biosynthesis. *Plant Signaling & Behavior*, *6*, 1627–1630.

Yamada, Y., & Sato, F. (2013). Transcription factors in alkaloid biosynthesis. *International Review of Cell and Molecular Biology*, *305*, 339–382.

Zhang, Y., Zhang, F., Li, X., Baller, J. A., Qi, Y., Starker, C. G., et al. (2013). Transcription activator-like effector nucleases enable efficient plant genome engineering. *Plant Physiology*, *161*, 20–27.

Ziegler, J., Voigtlaender, S., Schmidt, J., Kramell, R., Miersch, O., Ammer, C., et al. (2006). Comparative transcript and alkaloid profiling in Papaver species identifies a short chain dehydrogenase/reductase involved in morphine biosynthesis. *The Plant Journal*, *48*, 177–192.

CHAPTER SEVEN

Bioengineering of Isoquinoline Alkaloid Production in Microbial Systems

Akira Nakagawa*, Eitaro Matsumura*, Fumihiko Sato†, Hiromichi Minami*,1

*Research Institute for Bioresources and Biotechnology, Ishikawa Prefectural University, Suematsu, Nonoichi, Ishikawa, Japan
†Division of Integrated Life Science, Graduate School of Biostudies, Kyoto University, Oiwake-cho, Kitashirakawa, Sakyo-ku, Kyoto, Japan
[1]Corresponding author: e-mail address: minami@ishikawa-pu.ac.jp

Contents

1. Introduction 184
2. Design of IQA Biosynthetic Pathways and Their Enzymes in Bacteria 186
   2.1 Microbial reticuline production from dopamine 186
   2.2 Fermentative production of a key IQA biosynthesis intermediate, reticuline, from simple carbon sources 190
   2.3 Production of more complex IQAs using *Saccharomyces cerevisiae* 194
   2.4 Fermentative production of IQAs mediated by P450 oxidoreductases in *E. coli* 195
3. Perspectives 198
4. Conclusions 199
Acknowledgements 199
References 199

Abstract

Isoquinoline alkaloids (IQAs) are plant secondary metabolites that show diverse pharmaceutical activities. However, plants contain only low levels of IQAs, and the availability of different kinds of IQAs is limited. The biosynthetic pathways of IQAs have been studied extensively, and many of the genes related to IQA biosynthesis have been identified. This information has been used to create microbial production systems that are capable of producing high yields of IQAs in a short time. In fact, several microbial systems for production of IQAs have been established in the past 5 years, making mass production of IQAs possible. Here, we describe the current state of microbial production of IQAs including its strengths and weaknesses and discuss ways in which these systems can be developed to produce higher yields and a greater diversity of IQAs.

Advances in Botanical Research, Volume 68
ISSN 0065-2296
http://dx.doi.org/10.1016/B978-0-12-408061-4.00007-9

ABBREVIATIONS

4-HPAA 4-hydroxyphenylacetaldehyde
4′OMT 3′-hydroxy-*N*-methylcoclaurine 4′-*O*-methyltransferase
6OMT norcoclaurine 6-*O*-methyltransferase
BH_4 tetrahydrobiopterin
CM/PDH chorismate mutase/prephenate dehydrogenase
CNMT coclaurine *N*-methyltransferase
CPR NADPH-cytochrome oxidoreductase
DAHP 3-deoxy-D-arabino-heptulosonate-7-phosphate
DAHPS 3-deoxy-D-arabino-heptulosonate-7-phosphate synthase
DODC L-DOPA-specific decarboxylase
E4P erythrose-4-phosphate
fbr feedback-resistant
IPTG isopropyl-β-D-thiogalactoside
IQA isoquinoline alkaloid
L-DOPA L-dihydroxyphenylalanine
MAO monoamine oxidase
MH4 tetrahydromonapterine
NCS norcoclaurine synthase
PEP phosphoenolpyruvate
PEPS phosphoenolpyruvate synthase
PTS phosphotransferase system
SAM *S*-adenosyl-L-methionine
TH tyrosine hydroxylase
TyrAT tyrosine aminotransferase

1. INTRODUCTION

Isoquinoline alkaloids (IQAs) are plant secondary metabolites with diverse pharmaceutical activities. They are widely used as medicines; for example, morphine is prescribed for reducing intractable pain, and codeine is an antitussive included in most drugs used to treat the common cold. Although there is a high demand for IQAs, they are mainly obtained by extraction from plants, even though plants contain only low levels of these compounds. Three approaches have been attempted for large-scale IQA production: chemical synthesis, plant biotechnology and microbial production.

There are many examples of successful chemical syntheses of IQAs, for example, morphine and codeine (Gates & Tschudi, 1956) and (*S*)-reticuline

(Konda, Shioiri, & Yamada, 1975). Multi-step reactions are required for such syntheses because of the complex molecular structures of the products. For this reason, it has been difficult to establish practically feasible methods for their synthesis. In addition, because some chemical reactions need to be conducted under extreme conditions and/or require hazardous compounds, chemical syntheses have a high environmental load. Thus, to date, there has been no commercial production of IQAs by chemical synthesis methods.

In terms of plant biotechnology techniques, there have been some attempts to increase the amounts of alkaloids in plants (Allen et al., 2008, 2004; Frick, Kramell, & Kutchan, 2007; Fujii, Inui, Iwasa, Morishige, & Sato, 2007; Inui, Tamura, Fujii, Morishige, & Sato, 2007; Kempe, Higashi, Frick, Sabarna, & Kutchan, 2009; Larkin et al., 2007; Sato et al., 2001). However, plants have complex and often unknown regulation systems that control biosynthetic pathways and so it has proved difficult to increase the amounts of the desired products.

Recently, microbial production of plant secondary metabolites has attracted attention. Microbes can produce large amounts of compounds in a short time because of their rapid growth rate. The advantage of microbial production is that unlike plant production systems, microbial systems do not require a large space for culture. However, there are also disadvantages; because microbes do not produce IQAs naturally, all of the genes encoding IQA biosynthetic enzymes must be introduced. Therefore, it is desirable if all of the genes related to IQA synthesis are identified. If a gene has not been identified yet, an alternative enzyme with a similar function can be used in the pathway, or an alternative pathway can be constructed to bypass the reaction. Another disadvantage is that it is sometimes difficult to functionally express enzymes from other organisms in microbes. In such cases, the enzymes must be modified to facilitate their expressions.

We have studied IQA production using *Escherichia coli* (Minami et al., 2008) and successfully produced an important IQA intermediate, (*S*)-reticuline, from simple carbon sources (glucose or glycerol) (Nakagawa et al., 2011). In this chapter, we focus on (*S*)-reticuline production as a model of microbial IQA production and describe how to design the biosynthetic pathway and avoid the associated pitfalls. Because the major subfamilies of IQAs are produced through reactions mediated by P450 oxidoreductases, which are difficult to functionally express in *E. coli*, we also discuss the use of P450 enzymes for further development of microbial production of valuable IQAs.

2. DESIGN OF IQA BIOSYNTHETIC PATHWAYS AND THEIR ENZYMES IN BACTERIA

In plants, various types of IQAs are produced from L-tyrosine via (*S*)-norcoclaurine, which is synthesised from dopamine and 4-hydroxyphenylacetaldehyde (4-HPAA). (*S*)-reticuline is produced after three methylations and one hydroxylation of (*S*)-norcoclaurine (Fig. 7.1). Because almost all IQAs are synthesised via (*S*)-reticuline, we chose this compound as the first target for IQA production.

First, we attempted to produce reticuline and other IQAs from dopamine added to the culture medium. As a result, 11 mg/L reticuline was produced from 5 mM dopamine (Minami et al., 2008). Based on this success, we constructed a dopamine production pathway from a simple carbon source and combined it with the reticuline production pathway from dopamine. This resulted in a reticuline production pathway from glucose or glycerol (Nakagawa et al., 2011). From a synthetic biology viewpoint, there were some key points regarding the construction of this biosynthetic pathway, as described below.

2.1. Microbial reticuline production from dopamine

Expression of plant cytochrome P450 enzymes in bacteria, which lack the internal membrane-bound compartment for P450s, is another difficulty to overcome because the plant pathway for reticuline production includes a norcoclaurine 3′-hydroxylation step catalysed by P450 (CYP80B1) (Pauli & Kutchan, 1998; Fig. 7.1).

A synthetic biological approach can provide a 'shortcut pathway' using enzymes from other organisms. Conversion of dopamine to 3,4-dihydroxyphenylacetaldehyde by the monoamine oxidase (MAO) from *Micrococcus luteus* circumvents the hydroxylation step catalysed by CYP80B1 and produces norlaudanosoline via the Pictet–Spengler reaction, catalysed by norcoclaurine synthase (NCS). Because norlaudanosoline and its derivatives are potential substrates in IQA biosynthesis (Choi, Morishige, Shitan, Yazaki, & Sato, 2002; Sato, Tsujita, Katagiri, Yoshida, & Yamada, 1994), it was successfully converted into reticuline (Fig. 7.2). The shortcut pathway needs only dopamine as a substrate, unlike the plant pathway that requires both dopamine and 4-HPAA (Figs. 7.1 and 7.2). In the microbial production system, this pathway requires only five enzymes from dopamine,

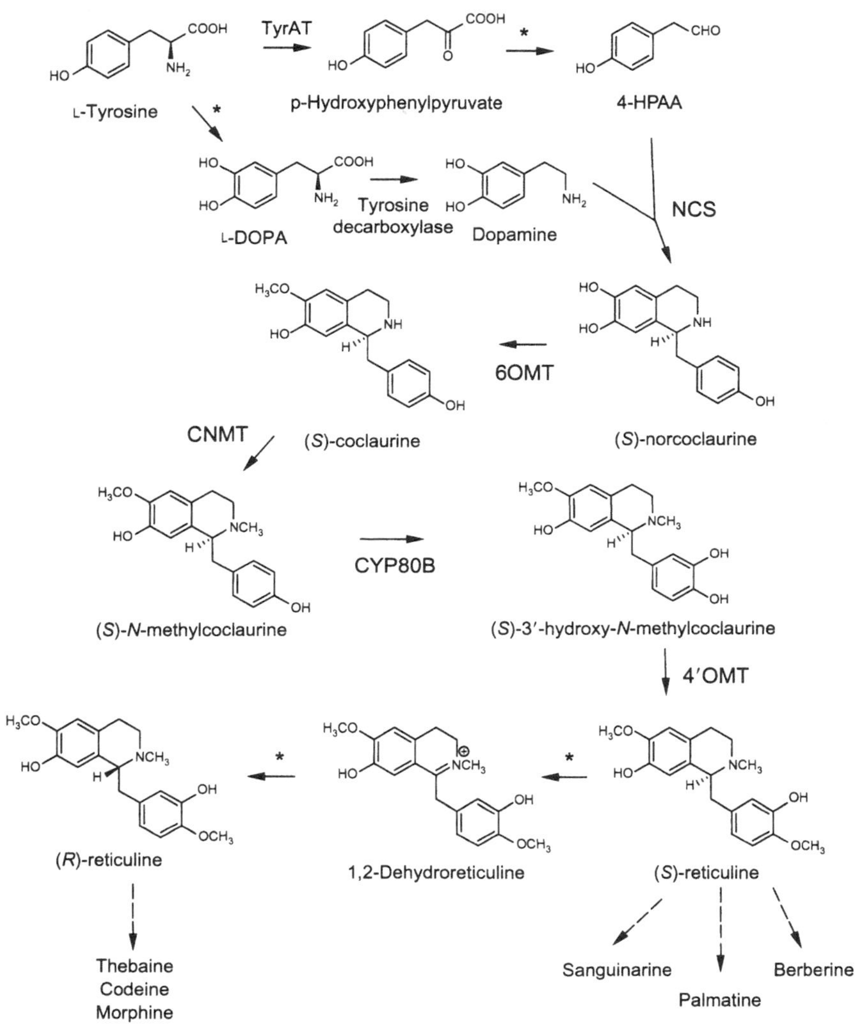

Figure 7.1 Natural biosynthetic pathway of IQAs in plants. Some enzymes have not been identified (asterisks). Arrows with dashed lines show IQAs derived from (*S*)- or (*R*)-reticuline.

compared with the nine-step reaction in the native (plant) pathway from tyrosine.

MAO is critical in this biosynthetic pathway. Because most eukaryotic MAOs are membrane proteins and, such as P450s, are difficult to express in *E. coli*, a soluble microbial MAO from *M. luteus* was expressed in *E. coli* (Roh et al., 2000). In the next step, an NCS from *Coptis japonica*

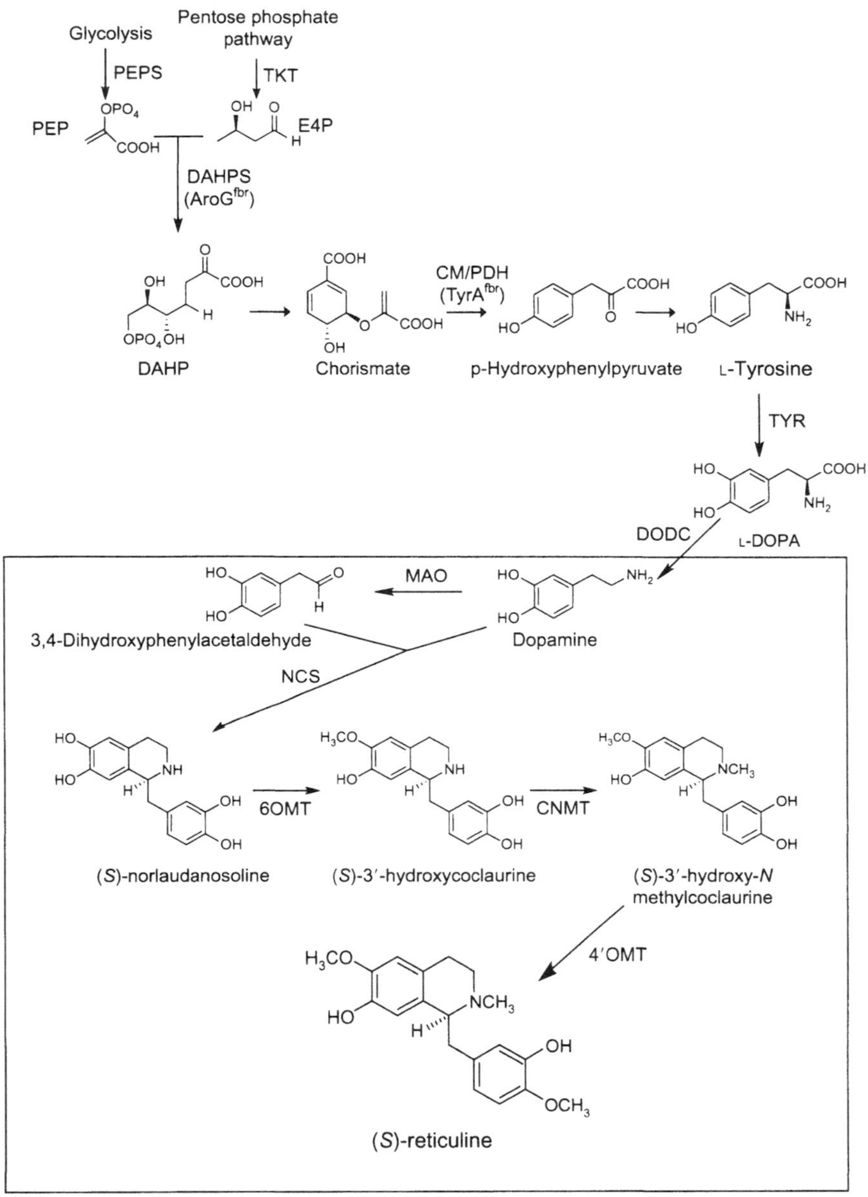

Figure 7.2 Constructed reticuline biosynthetic pathways in microbial production systems. Square indicates reticuline production pathway from dopamine.

(CjPR10A) was used because the active form of this enzyme is expressed at high levels in *E. coli*. In fact, two types of NCS (CjNCS1 and CjPR10A) were isolated from *C. japonica* cells (Minami, Dubouzet, Iwasa, & Sato, 2007). The signal peptide of CjPR10A was eliminated to achieve higher expression levels in *E. coli* (Minami et al., 2008).

In the following three steps catalysed by methyltransferases, norcoclaurine 6-*O*-methyltransferase (6OMT), coclaurine *N*-methyltransferase (CNMT) and 3′-hydroxy-*N*-methylcoclaurine 4′-*O*-methyltransferase (4′OMT) from *C. japonica* cells were used to produce reticuline from norlaudanosoline (Fig. 7.2). These enzymes were selected because they were the best-characterised methyltransferases. Whereas reticuline was produced from norlaudanosoline without optimising expressions of these three enzymes, removal of each enzyme showed the crucial importance of 6OMT to catalyse the initial reaction, whereas the other two enzymes functioned independently. Some intermediates were occasionally detected, especially 3'-hydroxycoclaurine and 3'-hydroxy-N-methylcoclaurine (Yusuke Shibutani, unpublished data), when enzyme activity and/or availability of co-factors were the rate-limiting steps of the reactions. However, the exogenous supply of *S*-adenosyl-L-methionine (SAM) as the methyl donor clearly improved reticuline production, indicating the importance of SAM supply in the reaction (Shibutani, unpublished data). In polyketide and biodiesel production processes requiring SAM, overexpression of SAM synthase increased production (Nawabi, Bauer, Kyrpides, & Lykidis, 2011; Wang, Boghigian, & Pfeifer, 2007). Similar to those results, enhanced expression of SAM synthetase markedly improved reticuline production from dopamine (Shibutani, unpublished data, see below).

The biosynthetic reticuline pathway in *E. coli* was reconstructed under the control of the T7 promoter, whereas the *MAO* gene from *M. luteus* was controlled by the *tac* promoter. In later experiments, codon optimisation enabled expression of *MAO* driven by the T7 promoter (Nakagawa et al., 2011). To increase metabolite production, enzyme expression should be improved by optimising codon usage, the promoter (type and strength), the plasmid copy number and so on. It was a promising starting point that *E. coli* BL21(DE3) cells containing pKK223-3-*NCS-MAO* and pACYC184-*6OMT-4′OMT-CNMT* produced (*R*,*S*)-reticuline at a yield of 11 mg/L medium from 5 mM dopamine without any optimisation. This overall yield of reticuline from dopamine (2.9%) was easily improved by increasing the supply of SAM to 14% (Shibutani, unpublished data). Importantly, dopamine and reticuline did not inhibit growth of *E. coli*. The low overall yield was also due to the instability of dopamine and reaction intermediates which were readily oxidised to form melanin-like pigments.

One unexpected result was that (*R*,*S*)-reticuline was produced, whereas NCS is (*S*)-stereospecific (Minami et al., 2007). This result indicated that the high rate of conversion of dopamine into its aldehyde by MAO induced a spontaneous condensation reaction to form norlaudanosoline. In fact, *E. coli* cells expressing all of the reticuline biosynthetic genes except *NCS* also produced racemic reticuline at the same level as that produced by a strain expressing *NCS*. Conversely, a crude enzyme preparation from transgenic *E. coli* cells produced stereospecific (*S*)-reticuline from dopamine with addition of SAM.

2.2. Fermentative production of a key IQA biosynthesis intermediate, reticuline, from simple carbon sources

The microbial synthesis of IQAs from a simple precursor such as dopamine or norlaudanosoline showed that microbes can produce IQAs but with low yields (Hawkins & Smolke, 2008; Minami et al., 2008). In addition, these substrates are expensive and not practically feasible for IQA production. The next challenges were to improve the yield and decrease the production costs by producing IQAs from much simpler substrates such as glucose. Here, we discuss the possibilities for producing IQAs from glucose. Several reactions have been examined for the fermentative production of reticuline, including the effective production of L-tyrosine, the conversion of L-tyrosine to L-dihydroxyphenylalanine (L-DOPA) and the conversion of L-DOPA to dopamine.

2.2.1 L-Tyrosine production from simple carbon sources

The production of L-tyrosine has been investigated using traditional and genetic modification techniques. Strains that overproduce L-tyrosine have been isolated from various microorganisms such as *Corynebacterium glutamicum*, *Arthrobacter globiformis* and *Brevibacterium lactofermentum*, using traditional mutagenesis methods (Hagino & Nakayama, 1973; Ito, Sakurai, Tanaka, Sato, & Enei, 1990; Roy, Murkhopadhyay, & Chatterjee, 1997). We constructed an L-tyrosine-over-producing *E. coli* strain based on our knowledge of the biochemical pathway and its regulation mechanism and genetic information (Chávez-Béjar et al., 2008; Lütke-Eversloh, Santos, & Stephanopoulos, 2007; Olson et al., 2007), as described below.

Briefly, the following key steps in L-tyrosine biosynthesis were modified: (1) condensation of phosphoenolpyruvate (PEP) and erythrose-4-phosphate

(E4P) by 3-deoxy-D-arabino-heptulosonate-7-phosphate synthase (DAHPS: *aroF*/*aroG*/*aroH*) to produce 3-deoxy-D-arabino-heptulosonate-7-phosphate (DAHP), (2) the supply of PEP from the glycolytic pathway and E4P from the pentose phosphate pathway, (3) conversion of the resultant 3-DAHP to chorismate through the shikimate pathway via seven reactions. TyrR, a 'global regulator', strictly regulates the shikimate pathway via the expression of *aroF*/*aroG*/*aroH* (DAHPS), *aroL* (shikimate kinase), chorismate mutase/prephenate dehydrogenase (CM/PDH: *tyrA*) and tyrosine aminotransferase (*tyrB*) (Pittard, Camakaris, & Yang, 2005). Prephenate is converted to *p*-hydroxyphenylpyruvate by a bifunctional enzyme CM/PDH (*tyrA*) and transformed into L-tyrosine by a tyrosine aminotransferase (*tyrB*).

First, the *tyrR* gene was disrupted to eliminate the repression of genes encoding shikimate pathway enzymes. Next, the feedback-resistant $AroG^{fbr}$ enzymes (i.e. Asp_{146} substituted with Asn; Kikuchi, Tsujimoto, & Kurahashi, 1997) and $TyrA^{fbr}$ (i.e. Met_{54} and Ala_{354} substituted with Ile and Val, respectively; Lütke-Eversloh & Stephanopoulos, 2007) were introduced into the *E. coli* Δ*tyrR* background to overcome feedback inhibition. These modifications resulted in an overall yield of $\leq$127 mg/L L-tyrosine (Lütke-Eversloh & Stephanopoulos, 2007). Next, an overexpression construct of phosphoenolpyruvate synthetase (PEPS: *ppsA*) and transketolase (TKT: *tktA*) was introduced into the above mutant to increase L-tyrosine production to 9.7 g/L with a yield of 0.1 g L-tyrosine/g glucose (Lütke-Eversloh et al., 2007). Mutants in which the PEP:sugar phosphotransferase system (PTS) is replaced by galactose permease and glucokinase to import glucose might be useful to improve yields and production capacities because PEP also serves as a phosphate donor in PTS (Flores, Xiao, Berry, Bolivar, & Valle, 1996; Gosset, Yong-Xiao, & Berry, 1996).

To produce reticuline from a simple carbon source, we introduced $aroG^{fbr}$, $tyrA^{fbr}$, *ppsA* and *tktA* driven by T7 promoters into the BL21 (DE3) *tyrR*::null strain of *E. coli* (Figs. 7.2 and 7.3). This L-tyrosine overproducing strain produced 4.4 g L-tyrosine from 15 g glycerol in a complex medium with a 29.5% molar ratio of the conversion efficiency (Nakagawa et al., 2011).

2.2.2 Conversion of L-tyrosine to dopamine

The full biosynthetic pathway of (*S*)-norcoclaurine from L-tyrosine (Fig. 7.1) has not been clarified yet (Lee & Facchini, 2011). Only tyrosine/DOPA decarboxylase and tyrosine aminotransferase (TyrAT) have

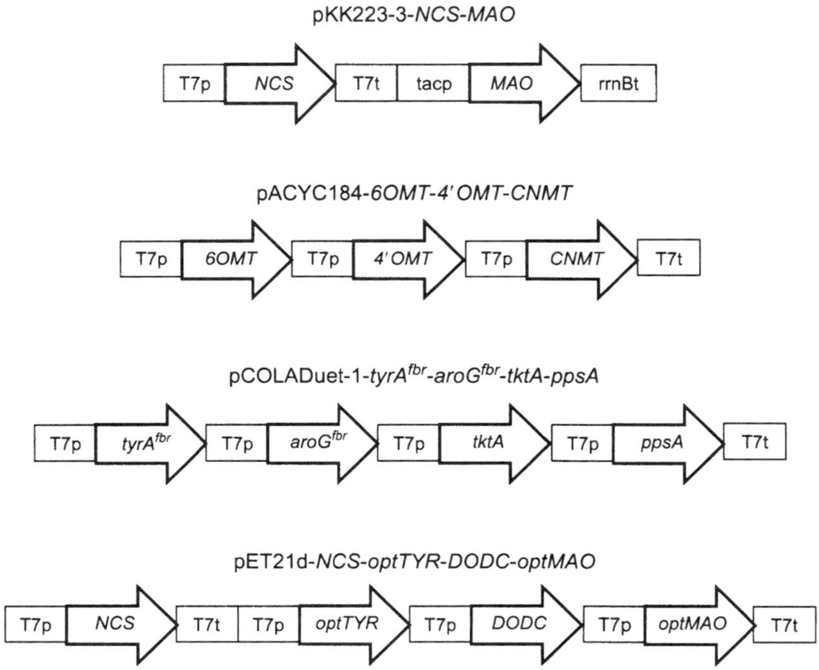

Figure 7.3 Plasmids used in microbial production of reticuline from dopamine or a simple carbon source. In reticuline production from dopamine, pKK223-3-*NCS-MAO* and pACYC184-*6OMT-4′OMT-CNMT* were introduced into *E. coli* BL21(DE3). In microbial production of reticuline from simple carbon sources, pCOLADuet-1-*tyrA*^{fbr}-*aroG*^{fbr}-*tktA-ppsA*, pET21d-*NCS-optTYR-DODC-optMAO* and pACYC184-*6OMT-4′OMT-CNMT* were introduced into the BL21(DE3) *tyrR*::null strain of *E. coli*. T7p, T7 promoter; T7t, T7 terminator; tacp, *tac* promoter; rrnBt, *rrnB* terminator; opt, codon usage optimised.

been isolated and characterised so far in the biosynthesis of dopamine and 4-HPAA (Facchini, Huber-Allanach, & Tari, 2000; Lee & Facchini, 2011).

There are several pathways from tyrosine to dopamine. Even when dopamine is produced via L-DOPA, there are three options. Plants and animals synthesise L-DOPA from L-tyrosine via tyrosine hydroxylase (TH). However, TH requires tetrahydrobiopterin (BH_4) as a cofactor, and *E. coli* does not contain BH_4. A recent study on hydroxytyrosol production suggested that tetrahydromonapterine (MH_4) could act as an alternative cofactor for the TH from mouse. Since *E. coli* can produce MH_4, Satoh, Tajima, Munekata, Keasling, & Lee, 2012 showed that TH was functional in *E. coli* over-expressing the MH_4-regenerating enzymes carbinolamine dehydratase and dihydropteridine reductase, both derived from human. They successfully produced L-DOPA from tyrosine and glucose using TH.

The second option was to overexpress 4-hydroxyphenylacetate 3-hydroxylase (HpaBC) in an L-tyrosine producing *E. coli* strain (Lee & Xun, 1998; Muñoz et al., 2011). HpaBC is a two-component FAD-dependent monooxygenase that oxidises a broad range of phenolic compounds (Xun & Sandvik, 2000). In a microbial production system using HpaBC, L-DOPA was produced at a yield of $\leq$320 mg/L from glucose in minimal medium (Muñoz et al., 2011).

As a third option, tyrosinase was used to hydroxylate L-tyrosine to form L-DOPA (Krishnaveni, Rathod, Thakur, & Neelgund, 2009). Tyrosinases, polyphenol oxidase family enzymes, catalyse the hydroxylation of monophenols to form *O*-diphenols (monophenolase activity) without requiring specific coenzymes such as BH_4 or FAD. Copper is required for tyrosinase activity, and it is simply added to the culture medium. Thus, we chose tyrosinase to convert L-tyrosine to L-DOPA. First, we selected tyrosinases from human, *Pholiota nameko* (mushroom), *Streptomyces castaneoglobisporus* (ScTYR) and verified their activity in *E. coli*. Only ScTYR was functionally expressed in *E. coli*, indicating that the bacterial enzyme showed the best expression in *E. coli*. *ScTYR* was transformed into the L-tyrosine overproducing strain described earlier, and L-DOPA was produced at a yield of 290 mg/L in the culture medium (Nakagawa et al., 2011).

In plants, dopamine is produced via the decarboxylation of L-DOPA by tyrosine/DOPA decarboxylases (Facchini & De Luca, 1995; Fig. 7.1). However, these enzymes also catalyse conversion of L-tyrosine into tyramine which is an undesirable substrate in the bacterial IQA synthetic pathway. Because tyramine can be recognised by MAO, a compound lacking the 3-OH group (norcoclaurine) would be produced, and the shortcut pathway via MAO would become problematic. Thus, an L-DOPA-specific decarboxylase (DODC) from the *Pseudomonas putida* strain KT2440 with more than 103-fold higher affinity for L-DOPA than for other aromatic amino acids (Koyanagi et al., 2012) was selected to produce dopamine from L-DOPA. This resulted in a dopamine yield of 280 mg/L with a conversion efficiency of 8.04% from L-tyrosine to dopamine (Nakagawa et al., 2011).

Tyrosinase also shows strong *O*-diphenolase activity towards *O*-quinones. This reaction forms undesirable by-products, L-dopaquinone and dopaminequinone, and results in the loss of substrate. In fact, L-DOPA, dopamine and norlaudanosoline were unstable in a culture of an ScTYR-expressing strain (Nakagawa et al., 2011). Therefore, ScTYR was replaced by the tyrosinase from *Ralstonia solanacearum* (RSc0337, RsTYR), which shows lower *O*-diphenolase activity than monophenolase activity, unlike all other known bacterial tyrosinases (García-Borrón & Solano, 2002). While the L-DOPA-producing strain expressing RsTYR yielded 290 mg/L L-DOPA, an identical yield to that produced by the strain expressing ScTYR, the dopamine producer expressing RsTYR produced 1.05 g/L dopamine, 3.7 times higher than that produced by the ScTYR-expressing strain (Nakagawa et al., 2011).

2.2.3 Reticuline production from a simple carbon source in E. coli

The biosynthesis of reticuline from simple carbon sources was reconstructed by combining the above three modified pathways using three plasmid vectors (Figs. 7.2 and 7.3; Nakagawa et al., 2011). All genes related to reticuline synthesis were transcribed by the T7 polymerase, controlled by the *lacUV5* promoter. Therefore, the reticuline synthetic pathway could be simultaneously activated by adding isopropyl-β-D-thiogalactoside (IPTG). The key parameters for reticuline production in this system were the concentrations of IPTG and dissolved oxygen. (*S*)-reticuline could be produced even without IPTG induction, while IPTG concentrations greater than 50 μM strongly inhibited production. These results indicated that strict control of IPTG induction is required for maximum production. Although oxygen is required for two enzymatic reactions in the pathway (those catalysed by tyrosinase and MAO), it is also consumed by the respiratory chain. Therefore, sufficient oxygen must be supplied but too much has a toxic effect on *E. coli* growth. Thus, we developed a jar fermenter culture system in which the concentration of dissolved oxygen could be strictly controlled. Finally, after optimising concentrations of IPTG and dissolved oxygen, reticuline production reached a maximum yield of 46.0 mg/L from glycerol in a fed-batch culture system (Nakagawa et al., 2011).

Whereas IQA fermentation was first achieved in a jar fermenter system, the shake flask culture method allowed for easier production. To bypass the aeration problem, a baffled shake flask was used for the fermentation culture. After optimising the media, IPTG concentration and the timing of its addition, agitation speed and the glycerol feed rate, reticuline was produced with a yield of 33.9 mg/L in 60 h in the small culture system (Nakagawa et al., 2012). Thus, it is possible to produce reticuline without a jar fermenter system.

2.3. Production of more complex IQAs using *Saccharomyces cerevisiae*

The reticuline fermentation system in *E. coli* provides the platform for production of more complex IQAs such as berberine, magnoflorine, morphine and so on (Fig. 7.1). These metabolites are produced via membrane-bound vesicles and membrane-bound P450 enzymes. Although the hydroxylation mediated by a P450 oxidoreductase was bypassed via the MAO shortcut pathway in the microbial production system (Fig. 7.2), P450-mediated reactions cannot be circumvented in the downstream pathways from (*S*)-reticuline. It is difficult to express membrane-bound enzymes in *E. coli* cells; therefore, one approach to use such enzymes is to express them in their unmodified form in an eukaryotic *Saccharomyces* system. The *ad hoc* reconstruction of the pathway in *S. cerevisiae* containing the berberine bridge enzyme or CYP80G2 with CNMT, and co-culture with *E. coli* producing reticuline from dopamine, resulted in successful production of magnoflorine or scoulerine (Minami et al., 2008; Fig. 7.4). These results provide another opportunity to produce more complex IQAs. Interestingly, the conversion efficiency of magnoflorine from reticuline was 65.5% and that of scoulerine from reticuline was 75.5%, whereas the yields of magnoflorine (7.2 mg/L culture in 72 h) and scoulerine (8.3 mg/L culture in 48 h) were not so high. Similar efficient production of (*S*)-scoulerine (65.4 mg/L), (*S*)-tetrahydrocolumbamine (68.2 mg/L), (*S*)-tetrahydroberberine (33.9 mg/L) or salutaridine (24.5 mg/L) from (*R*,*S*)-norlaudanosoline was reported in a single microbe, a transgenic *S. cerevisiae* strain (Hawkins & Smolke, 2008; Fig. 7.4). These successes suggest that there is great potential for the *Saccharomyces* system to produce various IQAs. There is still more potential to improve such systems because not all P450s and vesicle-bound enzymes such as tetrahydroberberine oxidase have been successfully expressed, even in yeast cells (Matsushima, Minami, Hori, & Sato, 2012).

2.4. Fermentative production of IQAs mediated by P450 oxidoreductases in *E. coli*

Because of energy and environmental concerns, there are great advantages in biosynthesising IQAs from low-cost, renewable resources such as glucose and glycerol. As mentioned earlier, the co-culture bioconversion system using *E. coli* and *S. cerevisiae* successfully produced IQAs from (*S*)-reticuline via a reaction involving P450 oxidoreductases. Although expressions of various P450 genes are possible in this system, it is hard to produce IQAs continuously because of the difficulty in balancing the growth of both strains. Thus, the development of an expression system for P450 genes in *E. coli* may be essential for fermentative biosynthesis of various IQAs because the production of (*S*)-reticuline from simple carbon sources has been established only in *E. coli*.

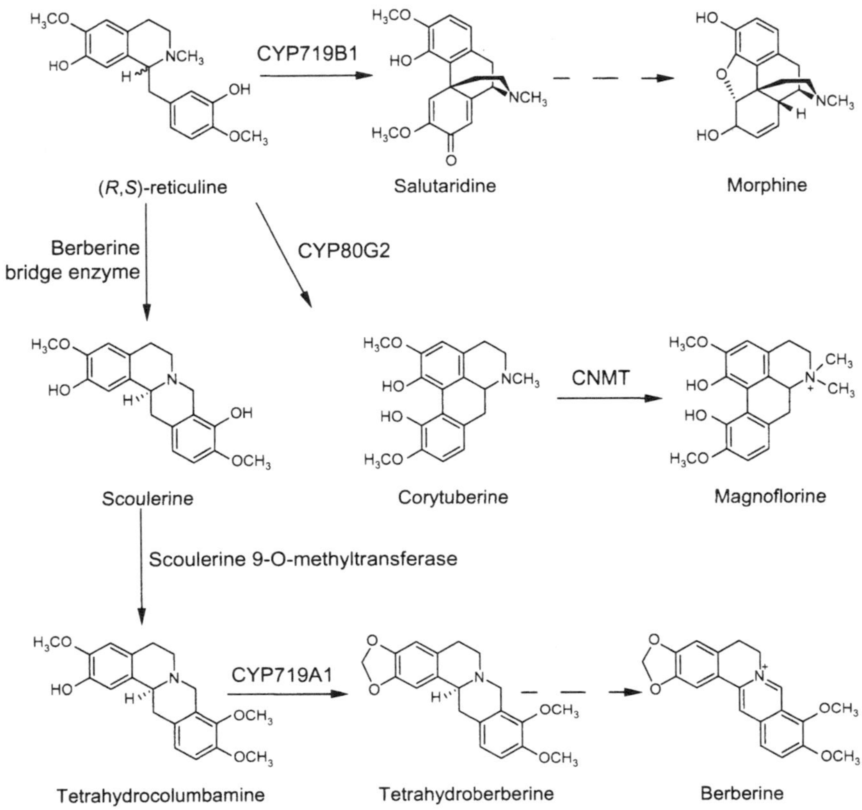

Figure 7.4 Reticuline-derived IQAs in *Saccharomyces* system. Arrows with dashed lines show final target products.

There have been several attempts to functionally express P450 genes in *E. coli*. Some strategies have resulted in expression of functional enzymes, but universal strategies that can be applied for most P450s have not been established yet. In fact, there are only a few examples of successful expression of plant P450 genes in *E. coli* (Quinlan, Jaradat, & Wurtzel, 2007). In the first successful heterologous expression of a P450 gene in *E. coli*, the first seven codons of CYP17A were modified; the second codon was replaced with GCT and four codons were optimised for bacterial expression (Barnes, Arlotto, & Waterman, 1991). Since then, there have been various attempts to functionally express heterologous P450 genes after altering their N-terminal region. The most common strategies have been to exchange the N-terminal sequence with the sequence 'MALLAVF', which was used in CYP17A expression (John, Hasler, He, & Halpert, 1994) and to eliminate the hydrophobic region from the N-terminal region (Iwata et al., 1998; von Wachenfeldt, Richardson, Cosme, & Johnson, 1997). In addition, processing can be optimised by fusing P450 genes with signal sequences from *E. coli*, such as the leader sequences *pelB* (pectate lyase B) and *ompA* (outer membrane protein A) because most eukaryotic P450 oxidoreductases exist as membrane-binding proteins in plant cells (Pritchard et al., 1997).

When engineering strains of *E. coli* express P450, an electron transfer component should be introduced along with the P450 gene. Although the catalytic cycle of a P450 oxidoreductase depends on transfer of electrons derived from NADPH through the FAD and FMN domains of redox partners (Jensen & Møller, 2010), the endogenous electron transfer system in *E. coli* is insufficient to support P450 activity (Jenkins & Waterman, 1994). Thus, to achieve sufficient P450 activity for bioconversion, two main strategies have been used to introduce appropriate electron transfer units.

The first strategy was to introduce various types of electron transfer components to activate P450 oxidoreductases (Harada et al., 2011; Iwata et al., 1998). In this system, it is easy to determine the appropriate redox transfer components because the expression method for each component has already been established, and the intended P450 genes can be co-expressed with various redox partners to test their activities. However, using membrane-binding-type P450 variants, the efficiency of electron transfer may be lower because the P450 enzymes are expressed in the periplasm while the redox partners are expressed in the cytosol.

The second strategy was to generate an artificial chimeric enzyme coupling a P450 oxidoreductase and an NADPH-cytochrome oxidoreductase (CPR) (Fisher, Shet, Caudle, Martin-Wixtrom, & Estabrook, 1992; Leonard & Koffas, 2007; Schückel, Rylott, Grogan, & Bruce, 2012). This system is attractive because of the highly efficient electron transfer from NADPH (Noble et al., 1999), but there were some difficulties in determining an appropriate redox partner and optimising the length of the linker sequence. Although plant, mammalian and bacterial CPRs have been used as redox partners in such fusion proteins, the most well-researched P450 fusion enzymes expressed in *E. coli* are naturally occurring bacterial fusion proteins, $P450_{BM3}$ (CYP102A1) from *Bacillus megaterium* and P450RhF (CYP116B2) from *Rhodococcus*. In particular, modified $P450_{BM3}$ enzymes have been developed that show various phenotypes, for example, short-chain alkane acceptable mutants (Meinhold, Peters, Chen, Takahashi, & Arnold, 2005), an NADH-dependent variant (Girvan et al., 2011) and high-coupling efficiency variants (Fasan, Chen, Crook, & Arnold, 2007). This information may help to develop new P450/CPR fusion proteins.

The reductase domain of $P450_{BM3}$ is very similar to eukaryotic CPRs (Jensen & Møller, 2010), and some fusion enzymes with eukaryotic P450 oxidoreductases have been expressed as active forms (Fairhead, Giannini, Gillam, & Gilardi, 2005; Helvig & Capdevila, 2000; Leonard & Koffas, 2007). In our laboratory, CYP80G2 modified using these strategies was used as a model of fermentative biosynthesis of an IQA mediated by a P450 oxidoreductase. The resulting strain has successfully produced corytuberine, an aporphine alkaloid, from glucose (Eitaro Matsumura, unpublished data).

3. PERSPECTIVES

Some IQAs have been produced successfully by microbial production systems. The IQAs produced to date are relatively easy to synthesise, so this will facilitate research on IQAs that has been hindered until now because of the low IQA content in plants. More complex IQAs that require more reaction steps could be produced in the near future. In addition, newly synthesised and unnatural IQAs could also be produced by combinations with enzymes acting on a broad range of substrates. Progress in these areas could increase the availability of IQAs for drug discovery.

While microbial production of IQAs is a very promising technique, there are still some disadvantages. For example, (*R*)-reticuline could not be produced from a simple carbon source, unlike (*R*,*S*)-forms from dopamine (Minami et al., 2008; Nakagawa et al., 2011). In plants, (*R*)-reticuline is synthesised from (*S*)-reticuline by two steps, chiral-specific reduction and oxidation; however, neither of the enzymes catalysing these steps have been characterised (Hirata, Poeaknapo, Schmidt, & Zenk, 2004). To produce IQAs derived from (*R*)-forms, such as codeine, thebaine and morphine, an (*R*)-form production system must be developed. As well as identifying important enzymes for these reactions, research should also focus on improving enzymes based on structural information, molecular mutagenesis and so on.

IQAs have been produced in complex media in microbial cultures. Although such media facilitate protein expression, they are very expensive. Pathways for biosynthesis of some IQAs have been established; therefore, one of the next challenges is to develop microbial production processes using synthetic media, which would drastically reduce production costs.

4. CONCLUSIONS

The important intermediate compound, (*S*)-reticuline, can be produced from a simple carbon source using a microbial production system with an artificial pathway. The microbial production platform is more suitable for reticuline production than plant extractions (Nakagawa et al., 2011). Furthermore, more complex IQAs could also be produced using the microbial production system, indicating that the system has the potential for production of many different kinds of IQAs.

Some disadvantages of microbial IQAs production can be avoided by modifying the pathway design or by choosing different enzymes. Some other disadvantages could be addressed by combining the microbial production system with other systems, for example, chemical synthesis and/or plant biotechnology.

These successes in producing IQAs show that a number of IQAs could be produced using this system. The further development and improvement of microbial IQA production systems will expand the availability of IQAs for drug discovery.

ACKNOWLEDGEMENTS

This work was supported by the Program for Promotion of Basic and Applied Research for Innovations in Bio-oriented Industry (BRAIN).

REFERENCES

Allen, R. S., Miller, J. A., Chitty, J. A., Fist, A. J., Gerlach, W. L., & Larkin, P. J. (2008). Metabolic engineering of morphinan alkaloids by over-expression and RNAi suppression of salutaridinol 7-*O*-acetyltransferase in opium poppy. *Plant Biotechnology Journal, 6*, 22–30.

Allen, R. S., Millgate, A. G., Chitty, J. A., Thisleton, J., Miller, J. A., Fist, A. J., et al. (2004). RNAi-mediated replacement of morphine with the nonnarcotic alkaloid reticuline in opium poppy. *Nature Biotechnology, 22*, 1559–1566.

Barnes, H. J., Arlotto, M. P., & Waterman, M. R. (1991). Expression and enzymatic activity of recombinant cytochrome P450 17 alpha-hydroxylase in *Escherichia coli. Proceedings of the National Academy of Sciences of the United States of America, 88*, 5597–5601.

Chávez-Béjar, M. I., Lara, A. R., Lopez, H., Hernández-Chávez, G., Martinez, A., Ramirez, O. T., et al. (2008). Metabolic engineering of *Escherichia coli* for L-tyrosine production by expression of genes coding for the chorismate mutase domain of the native chorismate mutase-prephenate dehydratase and a cyclohexadienyl dehydrogenase from *Zymomonas mobilis. Applied and Environmental Microbiology, 74*, 3284–3290.

Choi, K. B., Morishige, T., Shitan, N., Yazaki, K., & Sato, F. (2002). Molecular cloning and characterization of coclaurine *N*-methyltransferase from cultured cells of *Coptis japonica. The Journal of Biological Chemistry, 277*, 830–835.

Facchini, P. J., & De Luca, V. (1995). Phloem-specific expression of tyrosine/dopa decarboxylase genes and the biosynthesis of isoquinoline alkaloids in opium poppy. *The Plant Cell, 7*, 1811–1821.
Facchini, P. J., Huber-Allanach, K. L., & Tari, L. W. (2000). Plant aromatic L-amino acid decarboxylases: Evolution, biochemistry, regulation, and metabolic engineering applications. *Phytochemistry, 54*, 121–138.
Fairhead, M., Giannini, S., Gillam, E. M., & Gilardi, G. (2005). Functional characterisation of an engineered multidomain human P450 2E1 by molecular Lego. *Journal of Biological Inorganic Chemistry, 10*, 842–853.
Fasan, R., Chen, M. M., Crook, N. C., & Arnold, F. H. (2007). Engineered alkane-hydroxylating cytochrome P450(BM3) exhibiting native-like catalytic properties. *Angewandte Chemie (International Ed. in English), 46*, 8414–8418.
Fisher, C. W., Shet, M. S., Caudle, D. L., Martin-Wixtrom, C. A., & Estabrook, R. W. (1992). High-level expression in *Escherichia coli* of enzymatically active fusion proteins containing the domains of mammalian cytochromes P450 and NADPH-P450 reductase flavoprotein. *Proceedings of the National Academy of Sciences of the United States of America, 89*, 10817–10821.
Flores, N., Xiao, J., Berry, A., Bolivar, F., & Valle, F. (1996). Pathway engineering for the production of aromatic compounds in *Escherichia coli*. *Nature Biotechnology, 14*, 620–623.
Frick, S., Kramell, R., & Kutchan, T. M. (2007). Metabolic engineering with a morphine biosynthetic P450 in opium poppy surpasses breeding. *Metabolic Engineering, 9*, 169–176.
Fujii, N., Inui, T., Iwasa, K., Morishige, T., & Sato, F. (2007). Knockdown of berberine bridge enzyme by RNAi accumulates (*S*)-reticuline and activates a silent pathway in cultured California poppy cells. *Transgenic Research, 16*, 363–375.
García-Borrón, J. C., & Solano, F. (2002). Molecular anatomy of tyrosinase and its related proteins: Beyond the histidine-bound metal catalytic center. *Pigment Cell Research, 15*, 162–173.
Gates, M., & Tschudi, G. (1956). The synthesis of morphine. *Journal of the American Chemical Society, 78*, 1380–1393.
Girvan, H. M., Dunford, A. J., Neeli, R., Ekanem, I. S., Waltham, T. N., Joyce, M. G., et al. (2011). Flavocytochrome P450 BM3 mutant W1046A is a NADH-dependent fatty acid hydroxylase: Implications for the mechanism of electron transfer in the P450 BM3 dimer. *Archives of Biochemistry and Biophysics, 507*, 75–85.
Gosset, G., Yong-Xiao, J., & Berry, A. (1996). A direct comparison of approaches for increasing carbon flow to aromatic biosynthesis in *Escherichia coli*. *Journal of Industrial Microbiology, 17*, 47–52.
Hagino, H., & Nakayama, K. (1973). L-Tyrosine production by analog-resistant mutants derived from a phenylalanine auxotroph of *Corynebacterium glutamicum*. *Agricultural and Biological Chemistry, 37*, 2013–2023.
Harada, H., Shindo, K., Iki, K., Teraoka, A., Okamoto, S., Yu, F., et al. (2011). Efficient functional analysis system for cyanobacterial or plant cytochromes P450 involved in sesquiterpene biosynthesis. *Applied Microbiology and Biotechnology, 90*, 467–476.
Hawkins, K. M., & Smolke, C. D. (2008). Production of benzylisoquinoline alkaloids in *Saccharomyces cerevisiae*. *Nature Chemical Biology, 4*, 564–573.
Helvig, C., & Capdevila, J. H. (2000). Biochemical characterization of rat P450 2C11 fused to rat or bacterial NADPH-P450 reductase domains. *Biochemistry, 39*, 5196–5205.
Hirata, K., Poeaknapo, C., Schmidt, J., & Zenk, M. H. (2004). 1,2-Dehydroreticuline synthase, the branch point enzyme opening the morphinan biosynthetic pathway. *Phytochemistry, 65*, 1039–1046.
Inui, T., Tamura, K., Fujii, N., Morishige, T., & Sato, F. (2007). Overexpression of *Coptis japonica* norcoclaurine 6-*O*-methyltransferase overcomes the rate-limiting step in

benzylisoquinoline alkaloid biosynthesis in cultured *Eschscholzia californica*. *Plant & Cell Physiology*, *48*, 252–262.
Ito, H., Sakurai, S., Tanaka, T., Sato, K., & Enei, H. (1990). Genetic breeding of L-tyrosine producer from *Brevibacterium lactofermentum*. *Agricultural and Biological Chemistry*, *54*, 699–705.
Iwata, H., Fujita, K., Kushida, H., Suzuki, A., Konno, Y., Nakamura, K., et al. (1998). High catalytic activity of human cytochrome P450 co-expressed with human NADPH-cytochrome P450 reductase in *Escherichia coli*. *Biochemical Pharmacology*, *55*, 1315–1325.
Jenkins, C. M., & Waterman, M. R. (1994). Flavodoxin and NADPH-flavodoxin reductase from *Escherichia coli* support bovine cytochrome P450c17 hydroxylase activities. *The Journal of Biological Chemistry*, *269*, 27401–27408.
Jensen, K., & Møller, B. L. (2010). Plant NADPH-cytochrome P450 oxidoreductases. *Phytochemistry*, *71*, 132–141.
John, G. H., Hasler, J. A., He, Y. A., & Halpert, J. R. (1994). *Escherichia coli* expression and characterization of cytochromes P450 2B11, 2B1, and 2B5. *Archives of Biochemistry and Biophysics*, *314*, 367–375.
Kempe, K., Higashi, Y., Frick, S., Sabarna, K., & Kutchan, T. M. (2009). RNAi suppression of the morphine biosynthetic gene *salAT* and evidence of association of pathway enzymes. *Phytochemistry*, *70*, 579–589.
Kikuchi, Y., Tsujimoto, K., & Kurahashi, O. (1997). Mutational analysis of the feedback sites of phenylalanine-sensitive 3-deoxy-D-arabino-heptulosonate-7-phosphate synthase of *Escherichia coli*. *Applied and Environmental Microbiology*, *63*, 761–762.
Konda, M., Shioiri, T., & Yamada, S. (1975). Amino acids and peptides. XVIII. A biogenetic-type, asymmetric synthesis of (*S*)-reticuline from L-3-(3, 4-dihydroxyphenyl)-alanine by 1, 3-transfer of asymmetry. *Chemical & Pharmaceutical Bulletin*, *23*, 1063–1076.
Koyanagi, T., Nakagawa, A., Sakurama, H., Yamamoto, K., Sakurai, N., Takagi, Y., et al. (2012). Eukaryotic-type aromatic amino acid decarboxylase from the root colonizer *Pseudomonas putida* is highly specific for 3,4-dihydroxyphenyl-L-alanine, an allelochemical in the rhizosphere. *Microbiology*, *158*, 2965–2974.
Krishnaveni, R., Rathod, V., Thakur, M. S., & Neelgund, Y. F. (2009). Transformation of L-tyrosine to L-dopa by a novel fungus, *Acremonium rutilum*, under submerged fermentation. *Current Microbiology*, *58*, 122–128.
Larkin, P. J., Miller, J. A., Allen, R. S., Chitty, J. A., Gerlach, W. L., Frick, S., et al. (2007). Increasing morphinan alkaloid production by over-expressing codeinone reductase in transgenic *Papaver somniferum*. *Plant Biotechnology Journal*, *5*, 26–37.
Lee, E. J., & Facchini, P. J. (2011). Tyrosine aminotransferase contributes to benzylisoquinoline alkaloid biosynthesis in opium poppy. *Plant Physiology*, *157*, 1067–1078.
Lee, J.-., & Xun, L. (1998). Novel biological process for L-DOPA production from L-tyrosine by P-hydroxyphenylacetate 3-hydroxylase. *Biotechnology Letters*, *20*, 479–482.
Leonard, E., & Koffas, M. A. (2007). Engineering of artificial plant cytochrome P450 enzymes for synthesis of isoflavones by *Escherichia coli*. *Applied and Environmental Microbiology*, *73*, 7246–7251.
Lütke-Eversloh, T., Santos, C. N., & Stephanopoulos, G. (2007). Perspectives of biotechnological production of L-tyrosine and its applications. *Applied Microbiology and Biotechnology*, *77*, 751–762.
Lütke-Eversloh, T., & Stephanopoulos, G. (2007). L-Tyrosine production by deregulated strains of *Escherichia coli*. *Applied Microbiology and Biotechnology*, *75*, 103–110.
Matsushima, Y., Minami, H., Hori, K., & Sato, F. (2012). Pathway engineering of benzylisoquinoline alkaloid biosynthesis in transgenic California poppy cells with ectopic expression of tetrahydroberberine oxidase from *Coptis japonica*. *Plant Biotechnology*, *29*, 473–481.

Meinhold, P., Peters, M. W., Chen, M. M., Takahashi, K., & Arnold, F. H. (2005). Direct conversion of ethane to ethanol by engineered cytochrome P450 BM3. *Chembiochem: A European Journal of Chemical Biology*, *6*, 1765–1768.

Minami, H., Dubouzet, E., Iwasa, K., & Sato, F. (2007). Functional analysis of norcoclaurine synthase in *Coptis japonica*. *The Journal of Biological Chemistry*, *282*, 6274–6282.

Minami, H., Kim, J. S., Ikezawa, N., Takemura, T., Katayama, T., Kumagai, H., et al. (2008). Microbial production of plant benzylisoquinoline alkaloids. *Proceedings of the National Academy of Sciences of the United States of America*, *105*, 7393–7398.

Muñoz, A. J., Hernández-Chávez, G., de Anda, R., Martínez, A., Bolívar, F., & Gosset, G. (2011). Metabolic engineering of *Escherichia coli* for improving L-3,4-dihydroxyphenylalanine (L-DOPA) synthesis from glucose. *Journal of Industrial Microbiology & Biotechnology*, *38*, 1845–1852.

Nakagawa, A., Minami, H., Kim, J. S., Koyanagi, T., Katayama, T., Sato, F., et al. (2011). A bacterial platform for fermentative production of plant alkaloids. *Nature Communications*, *2*, 326.

Nakagawa, A., Minami, H., Kim, J. S., Koyanagi, T., Katayama, T., Sato, F., et al. (2012). Bench-top fermentative production of plant benzylisoquinoline alkaloids using a bacterial platform. *Bioengineered Bugs*, *3*, 49–53.

Nawabi, P., Bauer, S., Kyrpides, N., & Lykidis, A. (2011). Engineering *Escherichia coli* for biodiesel production utilizing a bacterial fatty acid methyltransferase. *Applied and Environmental Microbiology*, 77, 8052–8061.

Noble, M. A., Miles, C. S., Chapman, S. K., Lysek, D. A., MacKay, A. C., Reid, G. A., et al. (1999). Roles of key active-site residues in flavocytochrome P450 BM3. *The Biochemical Journal*, *339*(Pt 2), 371–379.

Olson, M. M., Templeton, L. J., Suh, W., Youderian, P., Sariaslani, F. S., Gatenby, A. A., et al. (2007). Production of tyrosine from sucrose or glucose achieved by rapid genetic changes to phenylalanine-producing *Escherichia coli* strains. *Applied Microbiology and Biotechnology*, *74*, 1031–1040.

Pauli, H. H., & Kutchan, T. M. (1998). Molecular cloning and functional heterologous expression of two alleles encoding (*S*)-*N*-methylcoclaurine 3'-hydroxylase (CYP80B1), a new methyl jasmonate-inducible cytochrome P-450-dependent mono-oxygenase of benzylisoquinoline alkaloid biosynthesis. *The Plant Journal*, *13*, 793–801.

Pittard, J., Camakaris, H., & Yang, J. (2005). The TyrR regulon. *Molecular Microbiology*, *55*, 16–26.

Pritchard, M. P., Ossetian, R., Li, D. N., Henderson, C. J., Burchell, B., Wolf, C. R., et al. (1997). A general strategy for the expression of recombinant human cytochrome P450s in *Escherichia coli* using bacterial signal peptides: Expression of CYP3A4, CYP2A6, and CYP2E1. *Archives of Biochemistry and Biophysics*, *345*, 342–354.

Quinlan, R. F., Jaradat, T. T., & Wurtzel, E. T. (2007). *Escherichia coli* as a platform for functional expression of plant P450 carotene hydroxylases. *Archives of Biochemistry and Biophysics*, *458*, 146–157.

Roh, J. H., Wouters, J., Depiereux, E., Yukawa, H., Inui, M., Minami, H., et al. (2000). Purification, cloning, and three-dimensional structure prediction of *Micrococcus luteus* FAD-containing tyramine oxidase. *Biochemical and Biophysical Research Communications*, *268*, 293–297.

Roy, A., Murkhopadhyay, S. K., & Chatterjee, S. P. (1997). Production of tyrosine by auxotrophic analogue resistant mutants of *Arthrobacter globiformis*. *Journal of Scientific and Industrial Research*, *56*, 727–733.

Sato, F., Hashimoto, T., Hachiya, A., Tamura, K., Choi, K. B., Morishige, T., et al. (2001). Metabolic engineering of plant alkaloid biosynthesis. *Proceedings of the National Academy of Sciences of the United States of America*, *98*, 367–372.

Sato, F., Tsujita, T., Katagiri, Y., Yoshida, S., & Yamada, Y. (1994). Purification and characterization of *S*-adenosyl-L-methionine: Norcoclaurine 6-*O*-methyltransferase from cultured *Coptis japonica* cells. *European Journal of Biochemistry/FEBS*, *225*, 125–131.

Satoh, Y., Tajima, K., Munekata, M., Keasling, J. D., & Lee, T. S. (2012). Engineering of L-tyrosine oxidation in *Escherichia coli* and microbial production of hydroxytyrosol. *Metabolic Engineering*, *14*, 603–610.

Schückel, J., Rylott, E. L., Grogan, G., & Bruce, N. C. (2012). A gene-fusion approach to enabling plant cytochromes p450 for biocatalysis. *Chembiochem*, *13*, 2758–2763.

von Wachenfeldt, C., Richardson, T. H., Cosme, J., & Johnson, E. F. (1997). Microsomal P450 2C3 is expressed as a soluble dimer in *Escherichia coli* following modification of its N-terminus. *Archives of Biochemistry and Biophysics*, *339*, 107–114.

Wang, Y., Boghigian, B. A., & Pfeifer, B. A. (2007). Improving heterologous polyketide production in *Escherichia coli* by overexpression of an *S*-adenosylmethionine synthetase gene. *Applied Microbiology and Biotechnology*, *77*, 367–373.

Xun, L., & Sandvik, E. R. (2000). Characterization of 4-hydroxyphenylacetate 3-hydroxylase (HpaB) of *Escherichia coli* as a reduced flavin adenine dinucleotide-utilizing monooxygenase. *Applied and Environmental Microbiology*, *66*, 481–486.

CHAPTER EIGHT

From Bioreactor to Entire Plants: Development of Production Systems for Secondary Metabolites

Thi Khieu Oanh Nguyen[*], Rebecca Dauwe[*], Frédéric Bourgaud[†,‡,§], Eric Gontier[*,§,1]

[*]Plant Biology & Innovation research Unit EA3900-UPJV, Université de Picardie Jules Verne, PRES UFECAP, Faculty of Sciences, Ilot des poulies, Amiens, France
[†]Université de Lorraine, Laboratoire Agronomie et Environnement, UMR 1121, ENSAIA, Vandœuvre Cedex, France
[‡]INRA, Laboratoire Agronomie et Environnement, UMR1121, ENSAIA, Vandœuvre Cedex, France
[§]Plant Advanced Technologies SA, Vandœuvre, France
[1]Corresponding author: e-mail address: eric.gontier@u-picardie.fr

Contents

1. Introduction 206
2. General Strategies to Produce Secondary Metabolites from Plant Material 208
3. Tropane Alkaloids Produced by Solanaceae 210
4. Furanocoumarins Produced by Rutaceae 211
5. Evaluating Plant Bioproduction and Selecting High-Producing Genotypes 212
   5.1 First step in chemical analysis 213
   5.2 Examples of plant genotype evaluation 214
6. First Steps in the Development of Agronomical Bioproduction 216
7. Biotechnological Approaches: From Callus to Organ *In Vitro* Culture 217
8. Process Engineering and Low-Cost 'Bioreactor' Design 218
9. Comparison Between Biotech (*In Vitro*) and Field Production 222
10. Hydroponics and Aeroponics as Open Bioreactors for the Production and Exploitation of High-Quality Biomass 223
11. Future Perspectives and Limitations; Concluding Remarks 226
Acknowledgements 228
References 228

Abstract

The production of secondary metabolites, and more specifically alkaloids, from medicinal plants is still an important objective for many research programs. When natural lead compounds have been discovered and when chemical synthesis cannot be easily performed, the extraction and purification of biomolecules from entire plants is generally the preferred solution. However, it is now established that plant cells and tissue cultures

Advances in Botanical Research, Volume 68
ISSN 0065-2296
http://dx.doi.org/10.1016/B978-0-12-408061-4.00008-0

in bioreactors can constitute an alternative solution to this agronomical approach. Our research programs devoted to the production of tropane alkaloids from *Datura innoxia* and furanocoumarins from *Ruta graveolens* have shown that hydroponics and aeroponics, techniques situated in-between field and fermentor scales, enable the entire plants to be used as efficient bioreactors. Revisiting scientific advances made in the past decades, the ethical, legal, biological and technological aspects are discussed in the light of the most recent literature, in order to establish a roadmap for further developments of plant secondary metabolite production systems.

ABBREVIATIONS

APS auto-priming syphon
FCs furanocoumarins
GC gas chromatography
HPLC high pressure liquid chromatography
HPTLC high pressure thin layer chromatography
MS mass spectrometry
PGPR plant growth-promoting Rhizobacteria
RI retention index
SMs secondary metabolites
TIS temporary immersion system
TLC thin layer chromatography

1. INTRODUCTION

Plants have always been the basis of food for humankind, but they have also been extensively used as non-food goods; for example, wood for building, heating and raw material. Other uses have consisted of preparing colours for art and types of body art (cosmetics), flavours for food and fragrances for perfumes. Additionally, they have been widely used in a variety of traditional medicines all over the world and through the ages. These uses, such as therapeutical preparations, decoction, extracts and cataplasms, were based on empirical knowledge accumulated for millennia, generation after generation. A large part of the biological activities found in medicinal plants are due to certain molecules synthesized by the plant as protective agents, many of which accumulate in response to stresses or biological aggressors. These molecules are only found in certain plant species, groups and families, and are classified as secondary metabolites (SMs).

An extensive range of such secondary compounds has been described in recent times and now more than 200,000 structures are known, grouped into different classes such as aliphatics, polyketides, carbohydrates, oxygen heterocycles, simple aromatics, benzofuranoids, benzopyranoids, flavonoids, tannins, lignans, polycylic aromatics, terpenoids, certain amino acids and peptides and alkaloids (Chapman & Hall, 1998 cited by Verpoorte, 2000). Within these groups of plant SMs, the three largest families are (i) phenolics, (ii) terpenes and steroids and (iii) alkaloids. Because phenolics are precursors to lignin synthesis, they are ubiquitously found in all higher plants. However, other compounds such as alkaloids are more sparsely distributed in the plant kingdom. They are more specific to defined plant genera and species. Like other secondary compounds, alkaloids are studied not only because of their biological properties but also for chemotaxonomy and chemical ecology.

Although plant SMs have been used for centuries in traditional medicine, they are now exploited for their value as pharmaceuticals, cosmetics (cosmeceuticals), fine chemicals or more recently nutraceuticals. In 1991, Payne et al. estimated that 25% of the molecules used by the pharmaceutical industry in western countries were of natural plant origin (Newman & Cragg, 2012; Payne, Bringi, Prince, & Shuler, 1994). As an example, aspirin (acetylsalicylate) is derived from salicylic acid. The natural molecule has been discovered in plants and could thus be isolated in large quantities from biomass. Nevertheless, because its chemical synthesis as an acetyl-derivative is easy and cheap, aspirin is produced in that way.

The complex nature of many secondary products, however, means that chemical synthesis is not cost-effective, and the production of plant SMs has, for a long time, been achieved through the field cultivation of medicinal plants. However, plants originating from particular biotopes can be hard to grow outside their local ecosystems. Moreover, some common plants cannot withstand large field cultures due to pathogen sensitivity. This has led to plant cell, tissue and organ cultures being considered an alternative way to produce the corresponding SMs (Bourgaud, Bouque, Gontier, & Guckert, 1997; Bourgaud, Gravot, Milesi, & Gontier, 2001). This approach is of especial value for those sources for which the collection of wild plants may endanger the species.

In this review, based on our own research activity, we will show how to explore the possibility of producing SMs from entire plants and from *in vitro* culture using bioreactors. Our examples will be based mainly on tropane alkaloids but other kinds of SMs will also be presented to illustrate the topic. Finally, it will be shown how the results and know-how obtained in the agronomical and biotechnological sectors can lead to the development of intermediate 'Agrobiotech' strategies using entire plants cultivated in controlled conditions using hydroponics and/or aeroponics. We will then conclude with a preview of future research and technology developments in the area of plant secondary metabolite production.

2. GENERAL STRATEGIES TO PRODUCE SECONDARY METABOLITES FROM PLANT MATERIAL

Identification of active compounds from plants is a prerequisite to production processes and will not be extensively developed in this review (see Ningthoujam, Talukdar, Potsangbam, & Choudhury, 2012). Briefly, the discovery process of a new active chemical entity from plants is illustrated by anticancer drug research. Although major plant anticancer drugs have been discovered by assessing cytotoxic activities, traditional medicines can reveal interesting drugs. Compared to the research of new biological activities through Blindfold strategies, the use of an ethnobotanical approach appears efficient (Ningthoujam et al., 2012). Nowadays, omics methods for assessing the toxicity of plant extracts can also be used for the targeted discovery of new activities (Efferth & Greten, 2012; Ouedraogo et al., 2012; Pelkonen et al., 2012).

The chemical synthesis of a given plant compound is the step following the discovery of a promising pharmacological property. In some cases, when the target molecule is complex and contains numerous asymmetric carbons, the chemical synthesis involves many steps with low yields. In these cases, production from natural plant resources appears to be the best option. The general scheme for the development of plant SM production may be represented as follows (Fig. 8.1).

Two main ways can be explored. The first uses the entire plant (Fig. 8.1, no. 1–3) and the second uses *in vitro* culture techniques (Fig. 8.1, no. 1, 2, 4–7) for the development of bioreactor cultures. However, before testing these methods, plant material must be collected in order to select the best ecotypes (Fig. 8.1, no. 1) on which chemical screening will be done. Plant collection in the wild is now submitted to international regulations. Consequently, a thorough knowledge of the legal aspects in relation to the *Convention on Biological Diversity* (http://www.cbd.int/convention/); signed by 150 government leaders at the 1992 Rio Earth summit, followed by the Convention on Biodiversity summit (Nagoya 2010) and of the specific rules

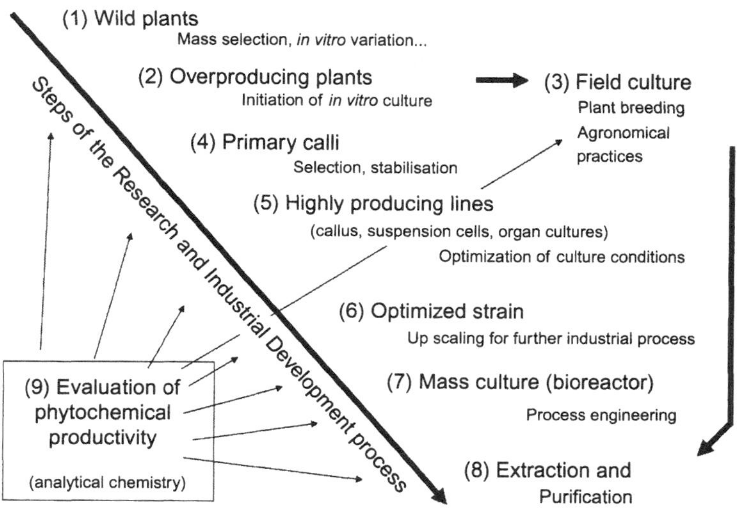

Figure 8.1 Scheme for the development of plant secondary metabolite production: from entire plant to bioreactor and use of entire plant as a bioreactor.

and laws of each country concerned is a minimum prerequisite. As far as only research is concerned and not commercial uses, worldwide networks of botanical gardens can be mobilized in order to access plant diversity. Once a certain set of genetic diversity is available, the potential of each entry (i.e. genotype or ecotype) can be evaluated. At this step and throughout the R&D program, there must be the capacity to evaluate the phytochemical levels and productivity. The aim of this first step is to select overproducing plants. This can be done by the direct mass selection of entire plants cultivated in greenhouses or experimental fields. The use of *in vitro* culture techniques can also serve to obtain somaclonal variations (Springer, 2012) and then high-producing plant lines.

Starting from these overproducing plants, field cultivation trials can be realized. Adaptation of agronomical practises and, later, genetic improvement by plant breeding must be done (Lidder & Sonnino, 2012). The goal is to establish field culture conditions in order to harvest the highest quality biomass for the extraction and purification of the targeted molecule.

Apart from such an agronomical strategy, it is possible to explore the potential of *in vitro* techniques for the development of plant SMs production. Most often, *in vitro* culture is established as callus culture from previously selected overproducing plants. The different callus lines are selected and stabilized in order to obtain high-producing lines that can be cultivated not only as calli but also as suspended cells or organ cultures. The optimization of culture conditions must then be performed in order to select optimized lines or strains that will be cultivated in a large bioreactor. The process engineering approach enables the set up of a production system to which the extraction and purification of SMs may be applied. At each step of this research program, the phytochemical state of the biological material must be controlled and a thorough knowledge of analytical chemistry techniques is essential.

From this short overview presenting the different ways of developing SMs production systems using entire plants in field or biotechnological methods, we will now illustrate the kind of results that can be obtained using our own experience on tropane alkaloid production in *Datura innoxia* species and on furanocoumarins (FCs) produced by *Ruta graveolens* species. A short introduction about these two plant-metabolite couples is proposed.

3. TROPANE ALKALOIDS PRODUCED BY SOLANACEAE

Tropane alkaloids, especially hyoscyamine and scopolamine, are widely used in medicine. They possess mydriatic, antispasmodic, anticholinergic, analgesic and sedative properties. For industrial purposes, they are extracted from various solanaceous plants belonging to the genera *Atropa*, *Duboisia*, *Datura* and *Hyoscyamus* (Mateus, Chekaoui, Christen, & Oksman-Caldentey, 2000; Oksman-Caldentey & Arroo, 2000). Hyoscyamine is usually the main alkaloid in plants such as *Hyoscyamus muticus* and *Atropa belladonna*, while scopolamine is only produced in small amounts (Mateus et al., 2000; Oksman-Caldentey & Arroo, 2000). Scopolamine is the more valuable compound, having higher physiological activity and possessing fewer side-effects; unfortunately, its yields are commonly lower. The world demand for this alkaloid has been estimated to be about 10 times greater than for hyoscyamine and its racemic form atropine (Hashimoto, Yun, & Yamada, 1993; Oksman-Caldentey & Arroo, 2000). A strong interest in boosting the scopolamine content of producing plants and their *in vitro* cultures has led to many research programs during recent decades. Furthermore, due to the relative complexity of their chemical structure and their rather long biosynthetic pathway (presented in Chapter 3), the synthetic production of these alkaloids is more expensive than their extraction from plant materials (Huang, Dai, Hu, Chen, & Zhu, 2005). Many authors assume that the future supply of tropane alkaloids will largely depend on biological methods of production (Dehghan, Häkkinen, Oksman-Caldentey, &

Ahmadi, 2012). It is thus important to improve tropane alkaloid production (especially scopolamine) to meet the expansion of industrial needs (Kai et al., 2007).

During the past three decades, different approaches based on callus and suspension cultures, somatic hybridization and root cultures have been applied for economical *in vitro* production of tropane alkaloids. As a promising system, hairy root cultures have been extensively studied by many research groups (Guillon, Trémouillaux-Guiller, Pati, Rideau, & Gantet, 2006a, 2006b; Jouhikainen et al., 1999; Mateus et al., 2000; Sevón, Biondi, Bagni, & Oksman-Caldentey, 2001; Sevón & Oksman-Caldentey, 2002; Zolala, Farsi, Gordan, & Mahmoodnia, 2007). In order to be commercially viable, hairy root-based technologies have to meet several criteria, one of which is to ensure that the yields of the products are high enough. In most cases, these yields are low and unsatisfactory. Therefore, integrated approaches for their improvement have to be developed, including genetic engineering (Lidder & Sonnino, 2012; Matsumoto & Gonsalves, 2012; Palazon et al., 2006; Rivera, Gómez-Lim, Fernández, & Loske, 2012), selection of highly productive lines (Zárate, 2010) and optimization of variables such as the nutrient medium, bioreactor design and environmental conditions. In the following sections, we will illustrate the results which were obtained in terms of not only tropane alkaloid production using entire plants but also *in vitro* cultures, especially of hairy roots. This historical perspective will present the potential of alternative ways using entire plants cultivated in soilless-controlled conditions, and new perspectives will then be discussed with regard to the most recent literature.

4. FURANOCOUMARINS PRODUCED BY RUTACEAE

Psoralen and its derivatives (Fig. 8.2), 5-methoxypsoralen (bergapten), 8-methoxypsoralen (xanthotoxin), 5,8-dimethoxypsoralen (isopimpinellin), belong to the furanocoumarin family. They have been used for the treatment of skin diseases (i.e. vitiligo and psoriasis) and several cancers (Edelson, 1988; Tioly-Bensoussan, Berreti, & Grupper, 1988). They have also been studied in relation to neurological illnesses like multiple sclerosis (Koppenhöfer, 1995). Nowadays, the pharmaceutical industry uses bergapten as a side-product from *Citrus bergamia* (bergamot tree) essential oil. However, this culture is declining and other sources need to be found. Plants belonging to the *Rutaceae* family are possible candidates for the production of such metabolites (McCloud,

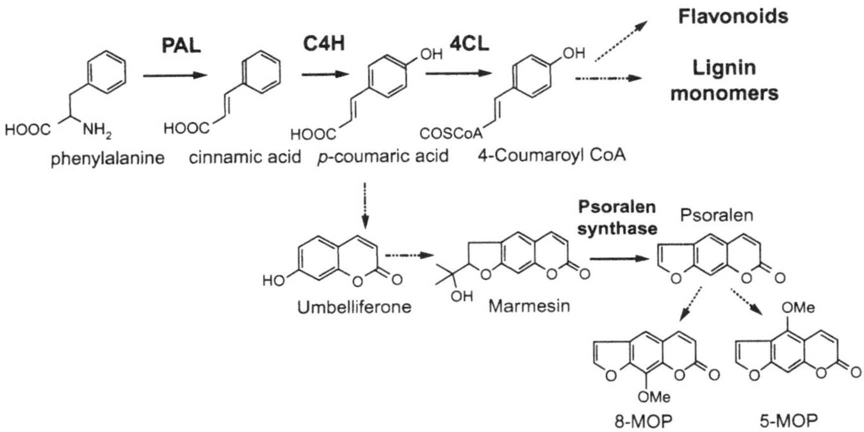

Figure 8.2 Simplified biosynthetic pathway of the main linear furocoumarins found in *R. graveolens* (psoralen, 8-MOP and 5-MOP). Dashed arrows represent multienzymatic reactions. Abbreviations: PAL, phenylalanine ammonia-lyase; C4H, *trans*-cinnamate 4 hydroxylase; 4CL, 4-coumarate:CoA ligase.

Berenbaum, & Tuveson, 1992). In *Rutaceae*, these compounds are involved as phytoalexins (Aliotta, Cafiero, DeFeo, & Sacchi, 1994).

This wide interest in FCs led us to investigate the possibility of their production by *in vitro* cultures of *R. graveolens*. Several types of plant material can be used to produce FCs with *R. graveolens*. Agronomical production has also been studied and will be described below as an illustration of the different key points to take into account.

5. EVALUATING PLANT BIOPRODUCTION AND SELECTING HIGH-PRODUCING GENOTYPES

When the initial plant material (seeds, plantlets, *in vitro* samples) has been obtained, the first challenge is to cultivate it in conditions that, at least, enable it to survive and, at best, provide maximum growth. An evaluation of the bioproducing capacities of such plant material must be done in parallel. Being able to extract, separate, purify, identify and quantify the molecules becomes the new priority. Techniques and methods for sampling (quenching in liquid nitrogen, freezing, drying in an oven), conditioning and preserving the samples (freeze-drying, low temperature freezing, storing in liquid nitrogen) must be applied. The extraction methods used may include decoction, simple solvent extraction, reflux extraction (Soxhlet), sonication or microwaves. Fractionation of the whole extract may eventually be needed prior to any other analysis: solid phase extraction, chromatography in open column, liquid–liquid purification, ultra- or nanofiltration, precipitation of salts and macromolecules with solvents or low temperatures.

5.1. First step in chemical analysis

Finally, chemical analysis can be performed. Thin layer chromatography (TLC) was widely used as a first step in characterizing phytochemical extracts. Such a simple and low-cost technique has been somewhat forgotten in favour of other more informative methods (see below). Nevertheless, using different sets of specific methods for the identification of individual compounds can be of great interest as a first step in the description of a complex mixture. Furthermore, with the development of high pressure thin layer chromatography (HPTLC), robust, repetitive and quantitative analyses can be performed quite easily.

Apart from TLC and HPTLC, other chromatographic methods can be used. High pressure liquid chromatography (HPLC) is most often applied to characterize complex phytochemical samples; it may be coupled to mass spectrometry (MS) and thus lead to information about the structure of the different metabolites contained within the original matrix. Access to high resolution mass spectra is needed to obtain useful information about the molecular weight and general formula of the metabolite. If only 'simple quad' technologies are of interest (easier to handle and cheaper), some limitations may rapidly block the identification processes.

Depending on the type of extraction used and the type of molecule extracted, other analytical techniques can be employed. Volatile or semi-volatile compounds may be analyzed using gas chromatography (GC), coupled if possible to MS. Compared to high pressure liquid chromatography coupled to mass spectrometry (HPLC–MS), GC–MS has the advantage of allowing the use of spectrum databases such as NIST or Wiley. Thus, potential identification may be achieved based on similarity between experimental mass spectra and mass spectra from the database (Kopka, 2006). In fact, the use of databases leads to the calculation of a probability. An absolute identification is commonly accepted if at least two independent characteristics of a molecule are comparable to the reference. This means that for identification, both the mass spectra and the retention time—calculated as Kovats retention index (RI)—must be identical (Kopka, 2006). The acceptable value of similarity is generally 0.8 but, for RI comparison, no statistics can be used to validate an identity. RI can thus be used essentially as a check to exclude potential MS hit if the difference between experimental and theoretical RI is too high. Finally, the injection of a pure molecule used as a standard is the only way to avoid any mistakes in terms of identification.

5.2. Examples of plant genotype evaluation

At the beginning of a research project aimed at studying and producing plant SMs, it is necessary to have a large variability of plant material in order to evaluate the different genotypes or ecotypes. In our research activity devoted to the production of FCs, mostly psoralen, from *Ruta* plants, we have collected 45 ecotypes belonging to four *Ruta* species (*R. graveolens*, *R. chalepensis*, *R. montana*, *R. angustifolia*). These ecotypes come from all around the Mediterranean sea and all this material was given mainly by the Botanical Garden of Nancy (http://www.cjbn.uhp-nancy.fr; France) which obtained access from other botanical gardens belonging to its own network. First of all, the seeds were sown in pots in our greenhouse and grown in order to characterize their growth capacities, their potential sensitivity to pests and their furanocoumarin profile. Similar trials were conducted in experimental fields in order to evaluate the behaviour of those ecotypes in the natural climatic conditions of the east of France (complete data in Milesi, Massot, Gontier, Bourgaud, & Guckert, 2001). Furanocoumarin content was determined in the aerial part of the different *Ruta* species using HPLC (Fig. 8.3). In those conditions, *R. graveolens* showed a higher growth rate and psoralen content so all of the further efforts to produce FCs have focused on one ecotype from this species (see results below, mainly from Milesi et al., 2001).

Another part of our research activity has been devoted to the study of tropane alkaloid production in *D. innoxia* Mill. The original plant material came from a collection of the Gif-sur-Yvette (France) research centre. Because *Datura* is autogamous, only a low genetic variability was available to start our research program. We thus attempted to create variability using androgenesis (Hérouart et al., 1991). We showed that such a technique could lead to an increase in the metabolic pattern (metabolite level) concerning tropane alkaloids. Some high-producing plants were obtained and their characteristics were stable enough to provide high-producing material (Hérouart et al., 1991). From this kind of new diversity (i.e. induced

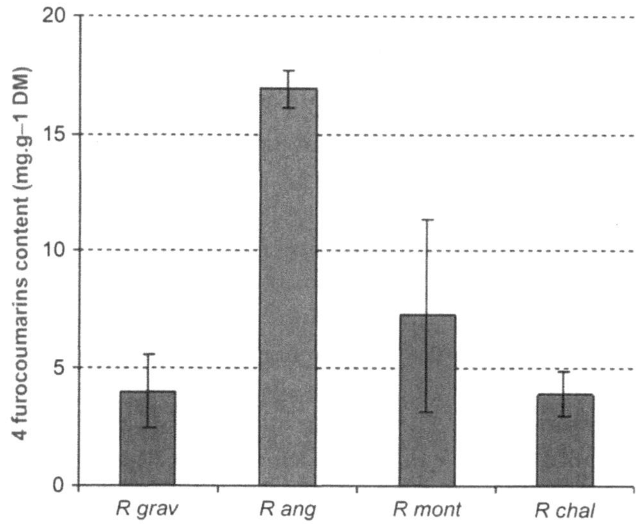

Figure 8.3 Furanocoumarin content of the leaves of *Ruta graveolens*, *R. angustifolia*, *R. montana* and *R. chalepensis* cultivated in an experimental field (+48° 44′ 24.75″, +6° 19′ 48.31″); the levels correspond to the sum of psoralen, 5-methoxy psoralen, 8-methoxy psoralen and 5,8-dimethoxy psoralen expressed in mg g^{-1} of dry weight. *Data from Milesi et al. (2001).* (For colour version of this figure, the reader is referred to the online version of this chapter.)

variability), we evaluated how a cloning method like micropropagation (through axillary bud culture) could be used to multiply high-producing plant lines (Gontier, Fliniaux, Barbotin, & Sangwan-Norreel, 1993, 1994). Thus, from one high-producing androgenic plant a population of plants was obtained with an alkaloid content higher than that of the two populations from the two different lower-producing androgenic *Datura* plants. This meant that micropropagation was a useful multiplication method for cloning plants producing high levels of tropane alkaloids. Moreover, a large variability was generated by this technique. In comparison to a population of plants obtained from seedlings, we showed that this variability was higher than that induced by sexual reproduction through self-fertilization. This specific result obtained in *D. innoxia* and devoted to tropane alkaloid biosynthetic potential is of much wider significance. Indeed, one could think that cloning (micropropagating) a high-quality genotype of a medicinal plant could be of interest in obtaining a population of high-value plants; nevertheless, such a cloning method should first be tested in case it leads to some kinds of undesirable variability.

6. FIRST STEPS IN THE DEVELOPMENT OF AGRONOMICAL BIOPRODUCTION

Field cultivation studies dedicated to plant SMs will be illustrated with the production of FCs by *R. graveolens* (FCs are not alkaloids but the way this work has been done can similarly be applied to alkaloid-producing plants). Plant growth and yields in FCs were first assessed in field trials amongst different ecotypes of *Ruta* obtained from botanical gardens and seed sellers. A 2-years experiment was established for which plants were sown in two different geographical locations which corresponded to contrasted soil and climatic conditions (Poutaraud, Bourgaud, Girardin, & Gontier, 2000). Two plots were situated at Illhaeusern in the Alsace plain (+48° 11′ 14.56″, +7° 26′ 4.03″, France), in clay–silt soil, at an altitude of 200 m. The third plot was at Luttenbach in the Vosges (+48° 2′ 10.73″, +7° 7′ 4.99″, France) in sandy soil at an altitude of 400 m. Three plots, each of 6 rows (70 cm between-row spacing) and 15 m in length were sown on 15 May 1997 on two sites at a seed density of 1.2 kg ha^{-1} and then rolled and irrigated. The plants were thinned on 16 June 1997 to a final density of 40,000 plants·ha^{-1}. The plots were fertilized with NPK at a ratio of 60:80:120 on 20 June 1997. Other complementary and larger experiments were also conducted in the Lorraine experimental field (Milesi et al., 2001) from March to May 1999 (La Bouzule-experimental farm-ENSAIA, Champenoux, France) and from 1997 to 2001 in a 1 ha production field in the centre of France (Soing-en-Sologne, +47° 25′ 13.52″, +1° 31′ 21.28″, France). Within all of these experiments, we could assess the growth potential of aerial parts, roots and fruits. We could determine the best sowing density (15,000, 30,000, 40,000 plants ha^{-1}), the best fertilization program and the best date for harvesting the biomass in order to extract FCs. These results thus led to an agronomical approach enabling the production of 4–10 kg of FCs per hectare and a maximum of 12 kg in the optimized conditions (Milesi et al., 2001; Poutaraud et al., 2000). This know-how for the production of specific FCs from *R. graveolens* species is of interest for industrial production. It can also serve as a model approach for the development of other kinds of metabolite (i.e. alkaloid) production but, as we will describe below, it constitutes a set of reference data for comparing the potential of entire plants to that of *in vitro* cultures.

7. BIOTECHNOLOGICAL APPROACHES: FROM CALLUS TO ORGAN *IN VITRO* CULTURE

As described above, the production of SMs from plant material can be performed through *in vitro* culture. In the case of tropane alkaloid production from *D. innoxia* Mill., callus cultures and then suspension cell lines have been established (Gontier, Boitel, Laberche, Barbotin, & Sangwan-Norreel, 1996). We have shown that these suspended cells grow fast and accumulate tropane alkaloids; nevertheless, independently to cultivation volumes (from 250-mL to 6-L Erlenmeyer flasks or in a 2.5-L batch bioreactor) these cells only accumulate low levels of hyoscyamine and scopolamine as compared to entire plants. It was also tried to immobilize the cells within calcium alginate beads and demonstrated that immobilization led to an improvement in alkaloid bioaccumulation in the cultures (Gontier, Sangwan, & Barbotin, 1994). This improvement was shown to be due essentially to the addition of calcium to the cell environment; furthermore, the increased hyoscyamine and scopolamine levels did not reach those measured in entire plants.

Based on the literature, hairy root cultures of *Datura* using *Agrobacterium rhizogenes* were initiated. This plant material grew well but not as fast as suspended cells. Hyoscyamine and scopolamine accumulated within the roots at levels equivalent to or higher than those previously measured in entire plants. The development of a specific bioreactor configuration (Boitel, Laberche, Gontier & Sangwan-Norreel, 1995) led to an improvement in growth (five times faster when expressed as productivity in g of dried biomass per day) and in tropane alkaloid levels (productivity expressed as µg per litre of culture per day: X5 for hyoscyamine and X2.5 for scopolamine).

Nevertheless, because of the specific structure of hairy roots, some problems were encountered. In comparison to hairy roots, plant cells can be homogeneously suspended in a culture system, whereas hairy roots are much more heterogeneous. Furthermore, it was shown that, after subculture, good growth could only be obtained if the root explants became entangled and formed a kind of raft that partially floated (Boitel, Laberche, et al., 1995). In order to favour this phenomenon, it was necessary to use a culture system corresponding to a fermentor in which most of the plunging parts were eliminated. Thus, the initial (expensive) bioreactor was finally used as a rather simple bubbling flask.

Furthermore, *D. innoxia* hairy roots accumulated hyoscyamine and scopolamine within the tissue and harvesting these compounds required destroying the biomass. Based on the literature and previous experimental background, a permeabilizing treatment using Tween20 as a detergent was developed (Boitel, Gontier, Laberche, Ducrocq, & Sangwan-Norreel, 1996; Boitel, Gontier, Laberche, Ducrocq, & Sangwan-Norreel, 1995). This allowed the leakage of most (or at least a large part) of the alkaloid initially contained within the root cells. It also led to a reactivation of the tropane alkaloid biosynthesis pathway and most probably stopped a downregulation presumably at the level of putrescine-*N*-methyl transferase (EC: 2.1.1.53). The supplementation of tropane alkaloid precursors in parallel with permeabilization then led to, at least, a partial biotransformation of these metabolites into the more interesting hyoscyamine and scopolamine. Such a process including permeabilization and the use of precursors for further biotransformation has been patented (Boitel, Gontier, Assaf, Laberche, & Sangwan, 1997).

Based on these previous results devoted to tropane alkaloid production in *D. innoxia*, the potential of *R. graveolens* for producing FCs using *in vitro* cultures was also explored. A comparison between callus, suspended cells and shoot cultures (Massot, Milesi, Gontier, Bourgaud, & Guckert, 2000; Milesi et al., 2000) showed that only the latter material is able to accumulate FCs at levels equivalent to or even higher than entire plants cultivated in fields (Fig. 8.4). However, *Ruta* shoot cultures do not grow as fast as suspended cells, cannot easily be elicited, need to aggregate for growth initiation and thus harbour quite similar properties to *Datura* hairy roots described above. *Ruta* shoots can be permeabilized using Tween20 but no efficient bioconversion of the precursor umbelliferone could be demonstrated.

8. PROCESS ENGINEERING AND LOW-COST 'BIOREACTOR' DESIGN

First, in order to prevent any misunderstanding, the term 'bioreactor' needs to be precisely defined. It was first used for large-scale vessels for plant biomass production. Usually, bioreactors are connected to (generally) computerized units controlling temperature, pH, aeration, stirring and various other devices. However, in some cases, *in vitro* culture vessels that differ from Erlenmeyer flasks, Petri dishes or culture boxes are also called 'bioreactors', when innovations and improvements in vessel design are introduced (Preil, 2005). Bioreactors have been primarily developed for culturing

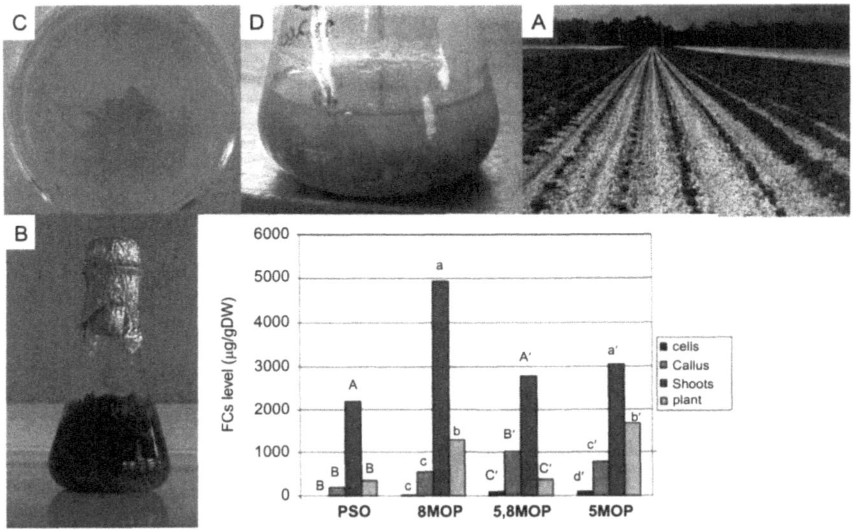

Figure 8.4 Furanocoumarin levels in *Ruta graveolens* entire plants (A) and *in vitro* cultures (shoots (B), callus (C), suspended cells (D)). PSO, psoralen; 8MOP, methoxy psoralen; 5MOP, 5methoxy psoralen; 5,8MOP, 5,8-dimethoxy psoralen. (See colour plate.)

microorganisms and later for plant cell suspensions (and organs like roots, shoots or embryos) to accumulate cell biomass for secondary metabolite production. Industrial cultivation of plant cells has been performed in some rather rare cases (Baque, Moh, Lee, Zhong, & Paek, 2012; Bourgaud et al., 2001). As an example, a large bioreactor cascade consisting of connected vessels of 75 L, 750 L, 7.5 m^3, 15 m^3 and 75 m^3 volumes was established and used for suspension cell cultures of *Echinacea purpurea*, *Rauwolfia serpentina* and some other species by the Diversa Company (Ahrensburg, Germany) during the late 1980s and early 1990s (Rittershaus, Ulrich, Weiss, & Westphal, 1989; Westphal, 1990). More generally, rather few industrial bioreactor applications have reached the market for the production of plant SMs (Bourgaud et al., 2001; Gontier, Clément, Bourgaud, & Guckert, 2001).

As indicated above, the aim of a bioreactor is to provide optimum growth conditions (by regulating chemical or physical parameters) to achieve the maximum yield and high quality of the biomass. In our own studies, in order to evaluate the potential of *Ruta* shoots when cultivated in culture systems larger than classic 250-mL Erlenmeyer flasks, we have based our strategy on the observation made on *D. innoxia* hairy root cultures.

In the latter case, it was necessary to simplify our fermentor configuration by eliminating most of the plunging parts (O_2 and pH probes, impeller). Thus, instead of buying an 'expensive' bioreactor and reducing its technological level, we decided to test the culture of *Ruta* shoots in a very simple bubbling flask culture system (Gontier et al., 2005). Simple glass vials were equipped with tubing, 0.2-μm HEPA (high efficiency particulate air) vent filters and compressed air was produced using (low-cost) aquarium pumps. Rather good growth was obtained in such conditions (Gontier et al., 2005). Because of the low price of such systems, many replicates could be carried out in parallel and thus the effects of the culture conditions (essentially the composition of the culture medium) could be studied (Gontier et al., 2005). The more sophisticated configuration developed is a temporary immersion system (TIS) including three compartments (Fig. 8.5). The plant material is cultivated in the first vessel (Fig. 8.5A). This can be weighed using a balance, which enables the real-time measurement of fresh weight without opening the system and thus risking microbial contamination. The weight is measured when the entire medium from vessel A (VA) has been transferred to vessel B (VB) (then the weight is the weight of the vessel plus that of the biomass). This transfer of the medium from VA to VB and the reverse is done

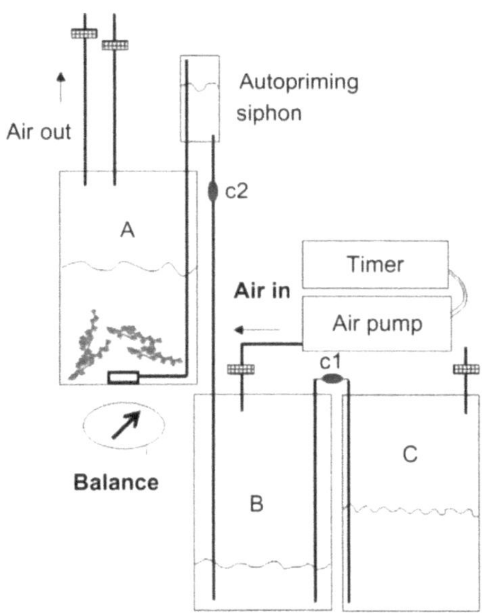

Figure 8.5 Configuration used for the culture of *R. graveolens* shoots in temporary immersion. (For colour version of this figure, the reader is referred to the online version of this chapter.)

automatically here. In fact, at the initial time, all the culture medium needed for the culture is contained in VB (further fresh medium is contained in VC and can be added to the culture in VB when needed without opening the system). When the air pump starts, the pressure in VB increases and the entire medium is transferred to VA through the auto-priming syphon (APS). When the air pump is stopped, part of the medium contained in the APS goes down into VB. This leads to a depressurization of the APS and thus the medium in VA goes back to VB through the APS. After that, commanded by the timer, the air pump starts again and VA is refilled. VC serves as a reserve of fresh medium added progressively throughout the culture. In these temporary immersion culture conditions, the shoots are exposed to oxygenated medium (because of the bubbling in VB) or to air (after the transfer of the medium from VA to VB). Compared to permanent immersion systems, this configuration provides a 40% better growth of the *Ruta* shoots (Gontier et al., 2005).

Furthermore, the addition of fresh medium produced a very high-density culture, but the quality of the tissue began to decrease and the FCs level was reduced by ~50% in some internal zones of the biomass pool. This limitation will be discussed in the following parts of this chapter.

These types of cultivation systems enabled the large-scale production of plant biomass (mostly shoots) and were described as really efficient for cloning millions of plants which were then sold to farmers. The use of simple Pyrex® vessels (Hempfling & Preil, 2005; Takayama & Akita, 2005) and also plastic bags (Savangikar, Savangikar, Daga, & Pathak, 2005) was very effective. In addition, Ziv (2005) presented a system using plastic bottles as bubbling flasks for the micropropagation of different plant species in liquid medium. This could be carried out industrially because a company nearby was able to sterilize her vessels using gamma irradiation.

As a partial conclusion, it is undoubtedly true that low-cost culture systems (bioreactors) can be developed and great success has already been achieved, at least for the production of biomass (micropropagation) in liquid medium (Hvoslef-Eide & Preil, 2005) and possibly in TISs and eventually in disposable culture bags. Currently, other kinds of very large and low-cost systems are being developed for the axenic or semi-axenic growth of microalgae in photoautotrophic or heterotrophic conditions (Perez-Garcia, Escalante, de-Bashan, & Bashan, 2011; Rawat, Ranjith Kumar, Mutanda, & Bux, 2013). The emergence of other companies specializing in the large-scale culture of plant material (Baque et al., 2012) to produce SMs is a further proof that process engineering and low-cost bioreactor technology are creating new industrial opportunities.

9. COMPARISON BETWEEN BIOTECH (*IN VITRO*) AND FIELD PRODUCTION

As illustrated by the examples of *Datura* and *Ruta*, suspension cells clearly grow well and fast compared to entire plants. Organ cultures, such as hairy roots and shoots, do not grow as fast but the quality of the biomass is high in terms of SMs content. Furthermore, it is possible to scale up the cultures in bioreactor devices specially adapted for the production of such kinds of biomass. Simple jars, potentially placing the culture in temporary immersion, most probably enable a better oxygenation of the tissues and thus growth can be optimized. In the examples of tropane alkaloids and FCs, most of the SMs are stored within the tissues and specialized cells. Permeabilization can thus be carried out and, with an appropriate treatment (detergent concentration, duration of the treatment), most of the biomass remains viable and can be used later for further production steps, again using permeabilization. In the case of *Datura*, the permeabilization process led to a reactivation of the tropane alkaloid biosynthesis pathway probably due to the decrease of a downregulation effect (this phenomenon was not observed in *R. graveolens*). Nevertheless, trying to increase the production potential of such organ cultures may lead to new problems, including the cost of the reactors, while simple low-cost devices can be developed. Inoculating large culture systems in axenic conditions is another problem to solve; because organs cannot be easily transferred by pumping, like suspended cells, inoculation of a large device becomes more difficult.

In contrast, in the case of *R. graveolens* plants, it was possible to cultivate the plant in open field surface of 1 ha. Good quality of biomass was produced containing relatively high levels of FCs, especially in fruits, and the canopy harvest led to the production of 4–12 kg ha^{-1} $year^{-1}$ of FCs, depending on the cultivation conditions.

However, all plant species cannot support field cultivation as growth can be very slow (*Panax vietnamensis* needs 6 years before the rhizomes can be harvested) and/or drastically affected by pests and diseases. Soilless cultivation conditions developed in greenhouses can thus constitute an alternative to field (or bioreactor) cultures.

In vitro, when cultivating roots or hairy roots, the aim is to produce a biomass that grows fast and also accumulates high levels of SMs. It is thus necessary to add sugar and vitamins to the culture medium and the sterility of the culture medium must be maintained. Alternatively, hydroponics and aeroponics can be used in non-sterile conditions to produce plants and give access to large quantities of root biomass. Such culture techniques were particularly developed to produce vegetables in the Pacific area during the Second World War (Morard, 1995). Plants cultivated with their roots immersed in (or watered with) a nutrient solution containing only minerals grow as fast as, and often faster than, plants in fields (agronomical production). Because the quality of the nutrient solution can be optimized and an artificial-controlled climatic environment can be applied, such cultures can be assimilated to a sort of opened bioreactor conducted under non-axenic conditions.

Based on such an analysis, the capacity of *D. innoxia* to grow in greenhouses, under hydroponic conditions, and to produce hyoscyamine and scopolamine within the roots and aerial parts, and to exude these alkaloids into the nutrient solution has been tested. The idea was to develop some kinds of large (indoor hydroponic) pools for the industrial production of biomass and metabolites.

10. HYDROPONICS AND AEROPONICS AS OPEN BIOREACTORS FOR THE PRODUCTION AND EXPLOITATION OF HIGH-QUALITY BIOMASS

In order to evaluate the potential of hydroponics for the growth of *D. innoxia* plants and for the production of tropane alkaloids, the first trials were carried out in individual bubbling flask systems (Gontier et al., 2002). Seedlings were transferred into homemade aquaponic devices. Their growth was measured through fresh and dry weight biomass. The hyoscyamine and scopolamine yields were measured and permeabilization was tested using Tween20 as a detergent. Parallel experiments were conducted in either hydroponic or aeroponic conditions. Plant growth occurred without any problems due to pests or disease. Most of the tropane alkaloids remained within the root and leaf tissues. An optimized permeabilizing treatment (Tween 3% for 24 h) liberated a large proportion of SMs into the nutrient solution. This treatment also reactivated the tropane alkaloid biosynthetic pathway and the precursors (phenylalanine and ornithine) added to the nutrient solution could be, at least partially, biotransformed into the more complex molecules scopolamine and hyoscyamine. These results are in accordance with those previously obtained in axenic conditions using hairy root cultures. The half-life of the molecule liberated into the culture medium was shown not to exceed 48 h, most probably due to enzymes from the plant root but also due to bacteria living in the liquid environment of the plant which could degrade alkaloids.

It was further demonstrated that permeabilizing treatment associated with feeding with precursors allows the possible harvest of alkaloids in the nutrient solution comparable to that obtained through extraction from the biomass of 1-month-old hydroponic plants (Fig. 8.6). However, with permeabilization, the plant remains viable and their roots can be rinsed before the plants are used again for further production steps, possibly using a treatment with Tween20 and precursors again. Furthermore, compared to soil culture, hydroponics can lead to plants having better quality in terms of SM content (Fig. 8.6). We also showed that placing the plant in conditions allowing a higher photosynthetic efficiency (slightly higher temperature, increase in light duration, Gontier et al., 2002) enabled the plant most probably to synthesize its own precursors so that the addition of exogenous molecules was no longer needed.

As an intermediate conclusion, it can be assumed that culture of *Datura* in soilless condition produces biomass in which roots can be permeabilized in a comparable way to that of hairy root cultures. Thus, because axenic culture conditions are not required in greenhouse conditions, it becomes possible to

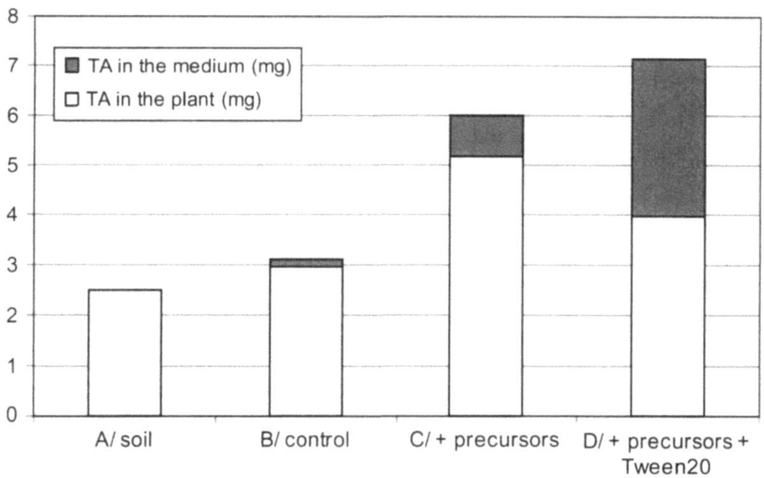

Figure 8.6 Tropane alkaloid amounts in the biomass and in the nutrient solution of plants cultivated in hydroponic conditions. (A) control plants cultivated in soil for 4 weeks; (B) plants cultivated in hydroponic conditions for 4 weeks; (C) plants as in (B) but then treated for 24 h with precursors added at 100 mg L^{-1} into the nutrient solution; (D) plants as in (C) but with addition of 3% Tween20 into the nutrient solution (see details of these results and experimental conditions in Gontier et al. (2002)).

imagine the culture of 1000 m^2 of plants and to produce and exploit large systems comparable in some aspects to 100 000-L bioreactors (Baque et al., 2012). Such a system established on a *Datura* model plant must nevertheless be tested and validated on a larger set of other plant species producing other kinds of SMs and including alkaloids. With this aim, the potential of hydroponics on *Ruta* producing FCs and *Taxus baccata* producing diterpenes (taxanes) has been tested. The results (Gontier et al., 2002) were not as positive as those obtained with the Solanaceae *Datura* but hydroponics was efficient in terms of plant growth and permeabilization could be successfully carried out.

Thus, a technique that has been applied on the model plant *Datura* could be transferred to at least two other plant species producing other kinds of SMs. From this result, we applied for a patent (Gontier et al., 2001) and we developed a business model (foundation of the company Plant Advanced Technologies in 2005; http://www.plantadvanced.com/). This business and technological model consists of choosing medicinal plants to produce specific metabolites, cultivating these plants in hydroponic or aeroponic conditions to obtain high-quality biomass (i.e. with elicitated levels of plant secondary compounds) and exploiting such a biomass through carefully adapted extraction processes, including permeabilization techniques. Therefore, based on this analysis of the potential interest of hydroponics and aeroponics, it can be assumed that it has the advantage of being relatively simple to setup and quite comparable to the agronomical approach as described in Fig. 8.1 (Points 1, 2, 3 and 8). Because the nutrient solution can be easily adapted in terms of mineral composition, pH, pO_2 and temperature adjustment, and because elicitation (Amdoun et al., 2009; 2010) can be imagined using elicitor or living microbes with properties such as 'plant growth-promoting Rhizobacteria' (PGPR) or other kinds of eliciting properties, hydroponics-aeroponics devices can be compared to simple non-axenic open bioreactors. In this kind of 'bioreactor' (i.e. opened cultivation system), the culture parameters can hardly be controlled as strictly as for axenic systems. Nevertheless, much larger biomass production can be achieved without any problems as compared to the same axenic cultivation systems. Soilless culture will neither replace field culture nor bioreactor-based approaches but it constitutes a novel opportunity to produce plant bioactive compounds and, in some specific cases, can lead to very efficient production systems which can be scaled up to industrial level. The use of genetically engineered plants is also easy to handle under the use of adapted confined conditions (i.e. genetically modified organism compliant greenhouses), avoiding problems associated to regulation rules of genetically modified organism uses and dissemination.

11. FUTURE PERSPECTIVES AND LIMITATIONS; CONCLUDING REMARKS

As demonstrated, the production of secondary compounds from plant material can be done through field culture, bioreactor culture or using intermediate systems that consist of cultivating entire plants in controlled conditions (non-axenic hydroponic–aeroponic cultures) under greenhouse conditions. Each of these techniques has their own advantages and limitations and it is still necessary to investigate potential ways to improve each of them.

Growth rate (Table 8.1) can be improved based on optimizing the medium and physical–chemical environment for *in vitro* culture (i.e. in a bioreactor). Concerning field production, plant breeding and agronomical practises (fertilization, crop protection) can be optimized also but environmental culture conditions cannot be highly easily modulated. Growth improvement in intermediate systems using plant soilless cultivation systems may be in-between the other two ways: because environmental optimization should be easier to get than in the field but, because of the lack of axenic character, potential improvement may be lower than in true *in vitro* culture. In non-axenic conditions, growth improvement can be obtained using viable bacteria such as PGPR that may lead to faster growth and sometimes higher SM levels in the plant biomass (Van de Mortel et al., 2012). Other kinds of non-PGPR bacteria may also be beneficial. As an example, addition of well-chosen *A. rhizogenes* rhizobacteria to the nutrient solution of *Datura* hydroponic cultures led to an improvement in both growth and tropane alkaloid levels. The maximum hyoscyamine and scopolamine yields were multiplied by 13 in the best conditions.

Table 8.1 Estimation of the possible levels for improving growth, secondary metabolite accumulation and harvest conditions by using biotechnological (*in vitro* culture, bioreactors), agronomical (field culture) and soilless (hydroponics and aeroponics-like systems) approaches

Possible improvement	*In vitro*	Hydroponics and aeroponics	Field culture
Growth rate	+++	++	+
SM levels	++++	+++	++
SM harvest	++++	+++	+

In each case, genetic manipulation (Matsumoto & Gonsalves, 2012) of the plant material can be employed in order to favour growth and/or specific secondary metabolite pathways (Palazon et al., 2006). Overexpression of transgenes (Gravot et al., 2004; Karamat et al., 2012; Larbat et al., 2007; Lièvre et al., 2005), multiple expression of transgenes (Bourgaud et al., 2006; Naqvi et al., 2010), gene silencing presenting an emphasis on RNA interference, transcription factors as a powerful tool for the engineering of biosynthetic pathways, the importance of novel gene promoters and optimization, together with the involvement of compartmentalization and transport are all various ways for modulating SMs synthesis (Zarate, 2010).

A great deal of work is being done in this area and the first results concerning the *de novo* creation of such metabolites (issuing from synthetic biology) are now available (Zurbriggen, Moor, & Weber, 2012). Nevertheless, for some applications, metabolic engineering may be excluded for legal reasons or due to the market which refuses genetically engineered products such as the cosmetics industry, for example. Apart from gene technologies, SMs levels can be improved through elicitation and/or feeding with precursors. While the first technique can be done *in vitro* and in the field, the second method (using precursors) cannot be so efficiently applied outdoor. However, in all cases, improvement of the initial plant material can be achieved for selecting high-producing plant material (Fig. 8.1, no. 2–5).

Among the key factors that may lead to improved SMs production, optimizing harvest conditions is one of the main ways in which innovative processes are reaching the market.

First of all, extraction and purification processes can now benefit from many new (or renewed) technologies. Microwave- and ultrasound-assisted extraction processes are widely studied and may improve extraction from biomass. Enzymatic hydrolysis can also be applied for some pre-extraction steps (mostly for primary metabolites such as lipids or for better extraction of fibres and polymers). New kinds of solvent, such as ionic solvents or even classic solvents (including water) placed in subcritical conditions may lead to improvement in extraction processes. Supercritical extractive processes using CO_2 have not yet yielded their full potential for industrial processes. Semi-preparative and preparative counter-current chromatography techniques are other useful tools for metabolite extraction–purification and new applications are regularly suggested. Membrane technologies such as micro-, ultra- and nanofiltration can also now be used as routine for concentrating extracts before further purification processes.

In contrast to the exploitation of 'sacrificed biomass' (i.e. frozen and/or dried, submitted to solvent extraction), the development of techniques enabling the survival of the plant material in continuous or semi-continuous mode can be applied to either bioreactor or hydroponic cultures. As shown above, permeabilization may be a technique of choice. The use of electric fields can also modify plant metabolism (Vallverdú-Queralt et al., 2013) and metabolite localization; it can be applied to the culture medium or directly onto entire plants between the canopy and the roots immersed in a conducting aqueous solution. This has been evaluated with at least partial success. Furthermore, the methodology using specific trapping of the low amounts of molecule being excreted by plant tissues or cells, in order to stimulate natural leakage and thus force the metabolic fluxes towards a higher efficiency, can be revisited using a specific adsorbent including molecular imprinted polymers.

Finally, bioreactor technology is also progressing and large culture systems are becoming more realistic for industrial applications based on plant cell and tissue culture. The boost associated with the development of microalgae culture projects for recycling CO_2 and the production of biodiesel have led to new ideas about a biosynthetic factory in which most, or all, the parts of the plant material can be exploited. In this development of a biorefinery concept, processes for the production of plant alkaloids and other SMs can find new biotechnological opportunities applied to either plants in bioreactors or entire plants used as bioreactors.

ACKNOWLEDGEMENTS

The personal results cited all along this chapter have been obtained with funds from the French ministry of higher education and research, ministry of Agriculture, region of Lorraine and region of Picardie. A part of the equipments used in those research activities have been obtained with European funds from FEDER.

REFERENCES

Aliotta, G., Cafiero, G., DeFeo, V., & Sacchi, R. (1994). Potential allelochemicals from *Ruta graveolens* and their action on radish seeds. *Journal of Chemical Ecology*, *20*, 2761–2775.

Amdoun, R., Khelifi, L., Khelifi-Slaoui, M., Amroune, S., Asch, M., Assaf-Ducrocq, C., et al. (2010). Optimization of the culture medium composition to improve the production of Hyoscyamine in elicited Datura stramonium L. Hairy roots using the response surface methodology (RSM). *International Journal of Molecular Sciences*, *11*, 4726–4740.

Amdoun, R., Khelifi, L., Khelifi-Slaoui, M., Amroune, S., Benyoussef, E.-H., Vu, T. D., et al. (2009). Influence of minerals and elicitation on *Datura stramonium* L. tropane alkaloid production: Modelization of the *in vitro* biochemical response. *Plant Science*, *177*, 81–87.

Baque, M. A., Moh, S. H., Lee, E. J., Zhong, J. J., & Paek, K. Y. (2012). Production of biomass and useful compounds from adventitious roots of high-value added medicinal plants using bioreactor. *Biotechnology Advances*, *30*, 1255–1267.

Boitel, M., Gontier, E., Assaf, C., Laberche, J. C., & Sangwan, B. S. (1997). Procédé de production de métabolites primaires et secondaires à partir d'organes végétaux cultivés *in vitro*. Patent FR 97 00641, 22 Jan 1997.

Boitel, M., Gontier, E., Laberche, J. C., Ducrocq, C., & Sangwan-Norreel, B. S. (1996). Inducer effect of Tween 20 permeabilization treatment used for release of stored tropane alkaloids in *Datura innoxia* Mill hairy root cultures. *Plant Cell Reports*, *16*, 241–244.

Boitel, M., Gontier, E., Laberche, J. C., Ducrocq, C., & Sangwan-Norreel, B. S. (1995). Permeabilization of *Datura innoxia* hairy roots for release of stored tropane alkaloids. *Planta Medica*, *61*, 287–290.

Boitel, M., Laberche, J. C., Gontier, E., & Sangwan-Norreel, B. (1995). Culture de racines transformées de *Datura innoxia* Mill. En bioréacteur: Comparaison de 3 types de procédés. *Compte Rendus de l'Académie des Sciences*, *318*, 593–598.

Bourgaud, F., Bouque, V., Gontier, E., & Guckert, A. (1997). Hairy root cultures for the production of secondary metabolites. *AgBiotech News and Informations*, *9*, 205–208.

Bourgaud, F., Gravot, A., Milesi, S., & Gontier, E. (2001). Production of plant secondary metabolites: A historical perspective. *Plant Sciences*, *161*, 839–851.

Bourgaud, F., Hehn, A., Larbat, R., Doerper, S., Gontier, E., Kellner, S., et al. (2006). Biosynthesis of coumarins in plants: A major pathway still to be unravelled for cytochrome P450 enzymes. *Phytochemistry Reviews*, *5*, 293–308.

Chapman & Hall (1998). Dictionary of natural products on CD-ROM, London, cited by R. Verpoorte (2000). In R. Verpoorte & A.W. Alfermann (Eds.), Metabolic engineering of plant secondary metabolism (pp. 286). Dortrecht: Kluwer Academic Publishers.

Dehghan, E., Häkkinen, S. T., Oksman-Caldentey, K. M., & Ahmadi, F. S. (2012). Production of tropane alkaloids in diploid and tetraploid plants and *in vitro* hairy root cultures of Egyptian henbane (Hyoscyamus muticus L.). *Plant Cell, Tissue and Organ Culture*, *110*, 35–44.

Edelson, R. (1988). Les médicaments photoactivés. *Pour la science*, *132*, 64–72.

Efferth, T., & Greten, H. J. (2012). Potential of 'Omics' technologies for implementation in research on phytotherapeutical toxicology. *Advances in Botanical Research*, *62*, 343–363.

Gontier, E., Boitel, M., Laberche, J. C., Barbotin, J. N., & Sangwan-Norreel, B. S. (1996). Evolution of tropane alkaloid level in a *Datura innoxia* Mill. cell suspension during successive numerous subcultures; partial restoration of the biosynthetic potentialities. *Revue de Cytologie et de Biologie végétales—le botaniste*, *18*, 89–96.

Gontier, E., Clément, A., Bourgaud, A., & Guckert, A. (2001). Method for producing metabolites from plants cultivated in soil-less medium. Patent WO 01/33942 A1.

Gontier, E., Clément, A., Tran, T. L. M., Lièvre, K., Gravot, A., Guckert, A., et al. (2002). Hydroponic combined with natural or forced root permeabilization: A promising technique for plant secondary metabolites production. *Plant Sciences*, *163*, 723–732.

Gontier, E., Fliniaux, M. A., Barbotin, J. N., & Sangwan-Norreel, B. S. (1993). Tropane alkaloid levels in the leaves of micropropagated *Datura innoxia* Mill. plants. *Planta Medica*, *59*, 432–435.

Gontier, E., Fliniaux, M. A., Barbotin, J. N., & Sangwan-Norreel, B. S. (1994). Contenu alcaloïdique foliaire de *Datura innoxia* Mill. Androgénétiques multipliés in vitro par culture de bourgeons axillaires. *Revue de Cytologie et de Biologie végétales—le botaniste*, *17*, 11–16.

Gontier, E., Piutti, S., Gravot, A., Milesi, S., Grabner, A., Massot, B., et al. (2005). Development and validation of an efficient low cost bioreactor for furanocoumarin production with *Ruta graveolens* shoot cultures. In A. K. Hvoslef-Eide & W. Preil (Eds.), *Liquid culture systems for in vitro plant propagation* (pp. 509–524). Dordrecht: Springer.

Gontier, E., Sangwan, B. S., & Barbotin, J. N. (1994). Effects of calcium, alginate and calcium-alginate immobilization on growth and tropane alkaloid levels in a stable suspension cell line of *Datura innoxia* Mill. *Plant Cell Reports, 13*, 533–536.

Gravot, A., Larbat, R., Hehn, A., Lièvre, K., Gontier, E., Goergen, J. L., et al. (2004). Cinnamic acid 4-Hydroxylase mechanism-based inactivation by psoralen derivatives: Cloning and characterisation of a C4H from a psoralen producing plant—*Ruta graveolens*—Exhibiting low sensitivity to psoralen inactivation. *Archives of Biochemistry and Biophysics, 422*, 71–80.

Guillon, S., Trémouillaux-Guiller, J., Pati, P. K., Rideau, M., & Gantet, P. (2006a). Hairy root research: Recent scenario and exciting prospects. *Current Opinion in Plant Biology, 9*, 341–346.

Guillon, S., Trémouillaux-Guiller, J., Pati, P. K., Rideau, M., & Gantet, P. (2006b). Harnessing the potential of hairy roots: Dawn of a new era. *Trends in Biotechnology, 24*, 403–409.

Hashimoto, T., Yun, D. J., & Yamada, Y. (1993). Production of tropane alkaloids in genetically engineered root cultures. *Phytochemistry, 32*, 713–718.

Hempfling, T., & Preil, W. (2005). Application of a temporary immersion system in mass propagation of Phalaenopsis. In A. K. Hvoslef-Eide & W. Preil (Eds.), *Liquid culture systems for in vitro plant propagation* (pp. 231–242). Dordrecht: Springer.

Herouart, D., Gontier, E., Sangwan, R. S., & Sangwan-Norreel, B. S. (1991). Analysis of the potential use of androgenic *Datura innoxia* for the development of stable cell cultures producing high amounts of tropane alkaloids. *Journal of Experimental Botany, 42*, 1073–1076.

Huang, F., Dai, X. D., Hu, Y. L., Chen, C. Y., & Zhu, G. Z. (2005). Progress in synthesis of tropane alkaloids. *Chemical Reagent, 27*, 141–144.

Hvoslef-Eide, A. K., & Preil, W. (2005). *Liquid culture systems for in vitro plant propagation.* Dordrecht: Springer.

Jouhikainen, K., Lindgren, L., Jokelainen, T., Hiltunen, R. H., Teeri, T., & Oksman-Caldentey, K. M. (1999). Enhancement of scopolamine production in Hyoscyamus muticus L. hairy root cultures by genetic engineering. *Planta, 208*, 545–551.

Kai, G., Chen, J., Li, L., Zhou, G., Zhou, L., Zhang, L., et al. (2007). Molecular cloning and characterization of a new cDNA encoding hyoscyamine 6b-hydroxylase from roots of Anisodus acutangulus. *Journal of Biochemistry and Molecular Biology, 40*, 715–722.

Karamat, F., Olry, A., Doerper, S., Vialart, G., Ullmann, P., Werck-Reichhart, D., et al. (2012). CYP98A22, a phenolic ester 3'-hydroxylase specialized in the synthesis of chlorogenic acid, as a new tool for enhancing the furanocoumarin concentration in Ruta graveolens. *BMC Plant Biology, 12*, 152.

Kopka, J. (2006). Current challenges and developments in GC–MS based metabolite profiling technology. *Journal of Biotechnology, 124*, 312–322.

Koppenhöfer, E. (1995). Kaliumkanalblocker in der Neurologie: 5MOP zur behandlung MS-bedingter Funktions de fizite. *TW Neurologie Psychiatrie, 9*, 585–591.

Larbat, R., Kellner, S., Specker, S., Hehn, A., Gontier, E., Bourgaud, F., et al. (2007). Molecular cloning and functional characterization of psoralen synthase, the first committed monooxygenase of furanocoumarin biosynthesis. *Journal of Biological Chemistry, 282*, 542–554.

Lidder, P., & Sonnino, A. (2012). Biotechnologies for the management of genetic resources for food and agriculture. *Advances in Genetics, 78*, 1–167.

Lièvre, K., Hehn, A., Tran, T. L. M., Gravot, A., Thomasset, B., Bourgaud, F., et al. (2005). Genetic transformation of the medicinal plant Ruta graveolens L. by an *Agrobacterium tumefaciens*-mediated method. *Plant Science, 168*, 883–888.

Massot, B., Milesi, S., Gontier, E., Bourgaud, F., & Guckert, A. (2000). Optimized culture conditions for the production of furanocoumarins by micropropagated shoots of *Ruta* gravolens. *Plant Cell, Tissue and Organ Culture, 62*, 11–19.

Mateus, L., Chekaoui, S., Christen, P., & Oksman-Caldentey, K. M. (2000). Simultaneous determination of scopolamine, hyoscyamine and littorine in plants and different hairy root clones of Hyoscyamus muticus by micellar electrokinetic chromatography. *Phytochemistry, 54*, 517–523.

Matsumoto, T. K., & Gonsalves D. (2012). Biolistic and other non-Agrobacterium technologies of plant transformation. Plant Biotechnology and Agriculture, Prospects for the 21st century, pp. 117–129.

McCloud, E. S., Berenbaum, M. R., & Tuveson, R. W. (1992). Furanocomarin content and phototoxicity of rough lemon (*Citrus jambhiri*) foliage exposed to enhanced utraviolet-B (UVB) irradiation. *Journal of Chemical Ecology, 18*, 1125–1137.

Milesi, S., Massot, B., Gontier, E., Bourgaud, F., & Guckert, A. (2001). *Ruta graveolens* L.: A promising species for the production of furanocoumarins. *Plant Science, 161*, 189–199.

Milesi, S., Massot, B., Piutti, S., Goergen, J. L., Murano, E., Toffanin, R., et al. (2000). *Ruta graveolens* cell cultures for psoralens production: Culture in bioreactor, effect of elicitation and calcium alginate immobilization. *International Journal of Biochromatography, 4*, 281–296.

Morard, P. (1995). *Les cultures végétales hors sol.* France: Publications agricoles Agen.

Naqvi, S., Farré, G., Sanahuja, G., Capell, T., Zhu, C., & Christou, P. (2010). When more is better: Multigene engineering in plants. *Trends in Plant Science, 15*, 48–56.

Newman, D. J., & Cragg, G. M. (2012). Natural products as sources of new drugs over the 30 years from 1981 to 2010. *Journal of Natural Products, 75*, 311–335.

Ningthoujam, S. S., Talukdar, A. D., Potsangbam, K. S., & Choudhury, M. D. (2012). Challenges in developing medicinal plant databases for sharing ethnopharmacological knowledge. *Journal of Ethnopharmacology, 141*, 9–32.

Oksman-Caldentey, K. M., & Arroo, R. (2000). Regulation of tropane alkaloid metabolism in plants and plant cell cultures. In R. Verpoorte & A. W. Alfermann (Eds.), *Metabolic engineering of plant secondary metabolism* (pp. 253–281). Dortrecht: Kluwer Academic Publishers.

Ouedraogo, M., Baudoux, T., Stévigny, C., Nortier, J., Colet, J. M., Efferth, T., et al. (2012). Review of current and "omics" methods for assessing the toxicity (genotoxicity, teratogenicity and nephrotoxicity) of herbal medicines and mushrooms. *Journal of Ethnopharmacology, 140*, 492–512.

Palazon, J., Moyano, E., Bonfill, M., Cusido, R. M., Pinol, M. T., & Teixeira da Silva, J. A. (2006). Tropane alkaloids in plants and genetic engineering of their biosynthesis. *Floriculture, Ornamental and Plant Biotechnology, II*, 209–221.

Payne, G. F., Bringi, V., Prince, C., & Shuler, M. L. (1992). The quest for commercial production of chemicals from plant cell culture. In G. F. Payne, V. Bringi, C. Prince, & M. L. Shuler (Eds.), *Plant cell and tissue culture in liquid systems* (pp. 1–10). Oxford: Oxford University Press.

Pelkonen, O., Pasanen, M., Lindon, J. C., Chan, K., Zhao, L., Deal, G., et al. (2012). Omics and its potential impact on R&D and regulation of complex herbal products. *Journal of Ethnopharmacology, 140*, 587–593.

Perez-Garcia, O., Escalante, F. M. E., de-Bashan, L. E., & Bashan, Y. (2011). Heterotrophic cultures of microalgae: Metabolism and potential products. *Water Research, 45*, 11–36.

Poutaraud, A., Bourgaud, F., Girardin, P., & Gontier, E. (2000). Cultivation of rue (*Ruta graveolens*) for the production of furanocoumarins of therapeutic value. *Canadian Journal of Botany, 78*, 1326–1335.

Preil, W. (2005). General introduction: A personal reflection on the use of liquid media for in vitro culture. In A. K. Hvoslef-Eide & W. Preil (Eds.), *Liquid culture systems for in vitro plant propagation* (pp. 1–18). Dordrecht: Springer.

Rawat, I., Ranjith Kumar, R., Mutanda, T., & Bux, F. (2013). Biodiesel from microalgae: A critical evaluation from laboratory to large scale production. *Applied Energy, 103*, 444–467.

Rittershaus, E., Ulrich, J., Weiss, A., & Westphal, K. (1989). Large scale industrial fermentation of plant cells: Experiences in cultivation of plant cells in a fermentation cascade up to a volume of 75000 litres. *Bioengineering, 5*, 28–34.

Rivera, A. L., Gómez-Lim, M., Fernández, F., & Loske, A. M. (2012). Physical methods for genetic plant transformation. *Physics of Life Reviews, 9*, 308–345.

Savangikar, V. A., Savangikar, C., Daga, R. S., & Pathak, S. (2005). Potentials for cost reduction in a new model of commercial micropropagation. In A. K. Hvoslef-Eide & W. Preil (Eds.), *Liquid culture systems for in vitro plant propagation* (pp. 403–414). Dordrecht: Springer.

Sevón, N., Biondi, S., Bagni, N., & Oksman-Caldentey, K. M. (2001). Transgenic Hyoscyamus muticus. In Y. P. S. Bajaj (Ed.), *Biotechnology in agriculture and forest* (pp. 171–200). Berlin: Springer.

Sevón, N., & Oksman-Caldentey, K. M. (2002). Agrobacterium rhizogenes mediated transformation: Root cultures as a source of alkaloids. *Planta Medica, 68*, 859–868.

Springer, N. M. (2012). Epigenetics and crop improvement. *Trends in Genetics, 29*(4), 241–247. http://dx.doi.org/10.1016/j.tig.2012.10.009.

Takayama, S., & Akita, M. (2005). Practical aspects of bioreactor application in mass propagation of plants. In A. K. Hvoslef-Eide & W. Preil (Eds.), *Liquid culture systems for in vitro plant propagation* (pp. 63–78). Dordrecht: Springer.

Tioly-Bensoussan, D., Berreti, B., & Grupper, C. (1988). PUVA and rePUVA in the treatment of mycosis fungoides: 58 Cases of mycosis with 710, 10 years follow up. In T. B. Fitzpatrick, P. Forlot, M. A. Pathak, & F. Urbach (Eds.), *Psoralens: Past, present and future of photochemoprotection and other biological activities* (pp. 293–300). Paris: John Libbey.

Vallverdú-Queralt, A., Oms-Oliu, G., Odriozola-Serrano, I., Lamuela-Raventós, R. M., Martín-Belloso, O., & Elez-Martínez, P. (2013). Metabolite profiling of phenolic and carotenoid contents in tomatoes after moderate-intensity pulsed electric field treatments. *Food Chemistry, 136*, 199–205.

Van de Mortel, J. E., De Vos, R. C., Dekkers, E., Pineda, A., Guillod, L., Bouwmeester, K., et al. (2012). Metabolic and transcriptomic changes induced in Arabidopsis by the rhizobacterium Pseudomonas fluorescens SS101. *Plant Physiology, 160*, 2173–2188.

Verpoorte, R. (2000). Secondary metabolism. In R. Verpoorte & A. W. Alfermann (Eds.), *Metabolic engineering of plant secondary metabolism* (pp. 1–29). Dordrecht: Kluwer Academic Publishers.

Westphal, K. (1990). Large-scale production of new biologically active compounds in plant-cell culture. In H. J. J. Nijk, L. H. W. van der Plas, & J. van Aartijk (Eds.), *Progress in plant cellular and molecular biology* (pp. 601–608). Dordrecht: Kluwer Academic Publishers.

Zarate, R. (2010). Plant secondary metabolism engineering. Methods, strategies, advances, and omics. *Comprehensive Natural Products II, 3*, 629–668.

Ziv, M. (2005). Simple bioreactors for mass propagation of plants. *Plant Cell, Tissue and Organ Culture, 81*, 277–285.

Zolala, J., Farsi, M., Gordan, H. R., & Mahmoodnia, M. (2007). Production a high scopolamine hairy root clone in Hyoscyamus muticus through transformation by Agrobacterium rhizogenes. *Journal of Agricultural Science and Technology, 9*, 327–339.

Zurbriggen, M. D., Moor, A., & Weber, W. (2012). Plant and bacterial systems biology as platform for plant synthetic bio(techno)logy. *Journal of Biotechnology, 160*, 80–90.

CHAPTER NINE

New Methods of Analysis and Investigation of Terpenoid Indole Alkaloids

Guitele Dalia Goldhaber-Pasillas, Young Hae Choi[1], Robert Verpoorte
Natural Products Laboratory, Institute of Biology Leiden, Leiden University, Leiden, The Netherlands
[1]Corresponding author: e-mail address: y.choi@chem.leidenuniv.nl

Contents

1. Introduction 235
2. Extraction and Purification Methods 236
   2.1 Solvent extraction 237
   2.2 Ionic liquids 239
   2.3 Molecular imprinted polymers 239
   2.4 Supercritical CO_2 240
3. Analytical Methods 242
   3.1 HPLC and hyphenated techniques 242
   3.2 Ultra high-pressure LC–MS 243
   3.3 Gas chromatography coupled to mass spectroscopy 248
   3.4 High-speed counter-current chromatography 251
   3.5 Capillary electrophoresis 253
   3.6 Quantitative nuclear magnetic resonance 259
4. Applications in Fingerprint Analysis 260
   4.1 NMR and LC–MS-based metabolic fingerprinting on TIAs 260
5. Conclusions 262
References 263

Abstract

Terpenoid indole alkaloids are biologically active compounds that have been used as pharmaceuticals among others as anticancer, antimalarial, antihypertensive and hypoglycemic agents for more than 40 years. Many efforts have been focused on their extraction, isolation, separation and structural elucidation by means of different approaches based on methodologies already established since the 1960s. New methodologies in extraction and sample preparation from different matrices include environmentally

Advances in Botanical Research, Volume 68
ISSN 0065-2296
http://dx.doi.org/10.1016/B978-0-12-408061-4.00009-2

friendly techniques such as ionic liquids or supercritical fluid extraction as well as the synthesis of molecularly imprinted polymers. Chromatography combined with spectroscopy is still the preferred analytical tool for alkaloid analysis, but recent improvements in mass spectroscopy and nuclear magnetic resonance-based technologies have also been applied to areas of research such as toxicology, quality control, metabolic fingerprinting and metabolic profiling. This review is intended to provide a more general rather than an exhaustive overview on the new methods for terpenoid indole alkaloid analysis focusing on hyphenated high-performance liquid chromatography and mass spectrometry approaches. We will also discuss extraction methods and the strength and weaknesses of the different analytical tools for their application in targeted or non targeted approaches.

ABBREVIATIONS

APCI atmospheric pressure chemical ionization
CCC counter-current chromatography
CE capillary electrophoresis
CID collision-induced dissociation
CPC centrifugal partition chromatography
CZE capillary zone electrophoresis
DAD photodiode array detection
ESI electrospray ionization
FID flame ionization detection
GC–MS gas chromatography coupled to mass spectroscopy
HPCCC high-performance counter-current chromatography
HPLC high-performance liquid chromatography
HSCCC high-speed counter-current chromatography
IT ion trap
LC–MS liquid chromatography coupled to mass spectroscopy
LOD limit of detection
MEKC micellar electrokinetic chromatography
MIP molecularly imprinted polymer
MS mass spectrometry
MS^n multistage mass (tandem mass) spectroscopy
NACE non-aqueous capillary electrophoresis
NMR nuclear magnetic resonance
qNMR quantitative nuclear magnetic resonance
SFE supercritical fluid extraction
SIM single ion monitoring
TIAs terpenoid indole alkaloids
TOF time of flight
UHPLC ultra high-pressure liquid chromatography
UV ultraviolet

1. INTRODUCTION

Terpenoid indole alkaloids (TIAs) are a group of ca. 3000 natural products among which a large number of compounds are being applied as medicine in pharmacology such as vinblastine, vincristine, strychnine and reserpine. Consequently, over the years, a lot of work has been done in bioprospecting for novel drugs from this group, and to improve the production of these compounds in plants or genetically modified organisms. All these studies require analytical tools to identify and quantify these compounds in various matrices such as plants and biological fluids. These tools are mainly based on chromatography and spectroscopy and different combinations of these. Thin-layer chromatography (TLC), gas chromatography (GC) (both since the 1960s) and high-performance liquid chromatography (HPLC; since the 1970s) offer efficient methods for the separation and identification of alkaloids based on retention behaviour that is further supported by more or less specific detection methods. Specificity in detection for TLC is obtained by colour reactions and ultraviolet (UV) spectroscopy, for GC by the combination with mass spectrometry (MS) and for HPLC by UV, MS and nuclear magnetic resonance (NMR) spectroscopy. Ever since the introduction of these methods, they have been extensively applied for alkaloids; however, in terms of separation principle, not much has changed since the first decade of use of these chromatographic tools (Baerheim Svendsen & Verpoorte, 1983; Verpoorte & Baerheim Svendsen, 1984). Stationary phases have improved, but the principle of retention of alkaloids has not changed, for example, in HPLC this is mainly reversed-phase chromatography with slightly acidic solvents, sometimes containing modifiers to reduce tailing due to adsorption of alkaloids on residual silanol groups in the stationary phase, or using ion-pair chromatography to change the selectivity. Though it is always claimed that the a new stationary phase has many advantages and is much better than the previous ones, the major disadvantage is that a different chromatogram will be obtained, requiring extensive validation, and direct comparison with previous results is not possible anymore.

This is where spectroscopic methods without hyphenation with chromatography have an advantage, as they do not involve a separation step. In the case of UV, infrared and NMR, the detection mechanism is based on pure physical properties, in case of a pure compound showing spectra specific for the chemical structure, in case of mixtures the sum of the spectra of the compounds present. In case of NMR spectroscopy, two-dimensional (2D) NMR can be applied to deconvolute the spectra of the compounds present. NMR has one major advantage above all methods and that is in ^{1}H NMR, all signals can be directly compared. The total signal intensity of each proton is only dependent on the molar concentration of the compound it is part of, which means that one can, without the need for individual calibration curves, quantify the compounds present in a mixture. In the case of mass spectroscopy, a manipulation is required to ionize the compound(s) present, which is a weakness as the spectra obtained are dependent on the instrumental conditions, as well as on the matrix. The advantage is that the resolution, is on the level of the molecular weight and in high-resolution mode, even the molecular formula can be obtained. The possibility of measuring fragmentation of the compounds adds further information for the identity of the compound.

In recent years combining the spectroscopy with chromatography has become common practice; thus, combining the advantages of both is particularly of great value in the identification of unknown, possibly novel, compounds. However, for the quantitative analysis of known compounds, liquid chromatography coupled to UV (LC–UV) and gas chromatography coupled to flame ionization detector (GC–FID) in many cases are fitting the needs of a quantitative analysis, as they are robust and reproducible methods.

2. EXTRACTION AND PURIFICATION METHODS

An elemental aspect in any analytical methodology is the compatibility of the extraction method with the analytical system, which will be ultimately reflected in the quality and usefulness of the chromatograms or spectra. Conventionally, extraction procedures for alkaloids have been done by different mechanical, physical and chemical processes such as Soxhlet extraction, maceration or percolation using organic solvents and liquid–liquid partitioning including aqueous solutions, although nowadays non-conventional methods which are more friendly to the environment have been successfully applied to alkaloid extraction such as ultrasound, microwave, supercritical fluids and ionic liquids (ILs).

The extraction has several aspects that need to be considered: the stability of compounds, the solubility and dissolution rate and the suitability of the extract for the analytical method. The stability is of concern from the very first moment of sampling, for example, when a plant is harvested, immediately the stress metabolism will start which may affect the alkaloids levels, for example, catabolism, *de novo* biosynthesis and (enzymatic) hydrolysis. Quenching of all metabolic activities is thus required, which can be achieved by heating or even treating with microwaves or freezing followed by freeze-drying. Moreover, the extraction procedure itself can cause the formation of artefacts (Maltese, van der Kooy, & Verpoorte, 2009; Verpoorte, Choi, Mustafa, & Kim, 2008) through reactivity of solvents or contaminations in solvents or by decomposition by heat (e.g. Soxhlet). Solubility can be good, but the rate of solubilization can be slow. Thus, extraction time is an important factor, as is the temperature; a balance must be found between these two in connection with the stability of the compounds to be extracted. In HPLC, the sample must be injected into a solvent that is similar to the mobile phase, for example, a chloroform extract cannot be directly injected on a reversed-phase column. In GC, the solvent must be volatile; in TLC, dirty extracts are not a problem as long as they do not interfere with the separation as a plate is only used once. Therefore, in every extraction procedure, such considerations must be made and it is important to properly validate the extraction method applied in any analytical protocol.

2.1. Solvent extraction

Extraction and sample preparation are the first key steps in plant analysis. Any analysis requires the extraction of the desired product from a complex matrix, dissolve it into the appropriate solvent and remove all the undesired compounds that might interfere with the analysis. TIAs have a lipophilic character as free bases at high pH and can be extracted with organic solvents such as alcohols, chloroform or ethyl acetate after plant material is basified with ammonia or sodium carbonate. A too high pH should be avoided as in that case phenolic alkaloids might not be fully extracted. Alternatively, alkaloids can be extracted with polar solvents at low pH, for example, water acidified with phosphoric acid (Moreno, van der Heijden, & Verpoorte, 1993), acetic acid (de Castro et al., 2012; Girardot et al., 2012; Jenks, 2002; Tanaka et al., 2007), tartaric acid (Jenett-Siems, Weigl, Kaloga, Schulz, & Eich, 2003), trifluoroacetic acid (Silvestrini et al., 2002) or hydrochloric acid (Andrade et al., 2005; Verma, Laakso, Seppanen-Lakso, Huhtikangas, & Riekkola, 2007). TIAs have also been extracted without adding acid or base with methanol (Paranhos, Fragoso, da Silveira, Henriques, & Fett-Neto, 2009; Sun & Liu, 2008), dichloromethane (Kumar, Bulumulla, Wimalasiri, & Reisch, 1994), acetone (Zhang, Yu, Liu, & Liu, 2007), ethanol (Wang et al., 2005), methanol–water

(Wang et al., 2012), chloroform or petrol (Etse, Gray, Thomas, & Waterman, 1989).

TIAs can be further purified by liquid–liquid extraction after basification, extraction with an immiscible organic solvent, for example, 10% of ammonia, and then extracted with ethyl acetate (Cao et al., 2012) or from an organic solvent with an aqueous acidic solution. In such liquid–liquid partitioning methods, one should keep in mind that non-polar counter ions like chloride, acetate and trifluoroacetic acids may form ion pairs with alkaloids that are well soluble in the organic solvents (Hermans-Lokkerbol & Verpoorte, 1986). Such approaches obviously separate alkaloids from other compounds with either non-polar (e.g. lipids) or polar character (e.g. sugars) that lack this dual lipophilic/hydrophilic character. It is therefore a targeted analysis for alkaloids.

Solid-phase extraction (SPE) involves selective extraction of TIAs from liquid samples onto a solid support using adsorption or ion exchange materials. For adsorption of TIAs in their free neutral form, reversed-phase materials such as C_8 and C_{18} on silica are widely used. Cation exchangers can be used to selectively bind alkaloids from aqueous extracts (Sheludko, Gerasimenko, Unger, Kostenyuk, & Stoeckgit, 1999). The choice of a SPE method also depends on the type of extract required in the analysis. For GC analysis, the alkaloids should be in the basic form, and for LC, it can be in the basic form or more commonly in the protonated form as in reversed-phase chromatography usually weakly acidic systems are applied. Depending on the method of detection, chlorine- or fluorine-containing solvents should be avoided for GC and for MS, and all components need to be volatile.

Quaternary indole alkaloids can be isolated by precipitation. A crude extract is prepared with an acidic aqueous solution and then they are precipitated with Mayer's reagent at pH 5 (Penelle et al., 2001; Verpoorte & Baerheim Svendsen, 1984) or as Reineckate salts at pH 8 (Ghosal & Srivastava, 1974) or pH 3 (Hu, Zhu, Prewo, & Hesse, 1989). After collection, the precipitate is dissolved into a mixture of organic solvents like acetone–methanol–water 6:2:1 v/v/v (Perera, Samuelson, van Beek, & Verpoorte, 1983; Verpoorte & Baerheim Svendsen, 1984) and exchanged to chlorides by means of an anion exchanger. Nevertheless, because these salts pose serious health risks to humans and to the environment, their use should be discouraged. Quaternary alkaloids can be adsorbed from aqueous extracts using cation exchangers.

2.2. Ionic liquids

ILs have been successfully applied to the extraction of complicated samples such as plant complexes or low-accumulation constituents like TIAs. Since the rate of solubilization in the often viscous IL is a limiting step, they are frequently used in ultrasound- or microwave-assisted extraction. Further steps to make the extract suitable for the analysis include liquid–liquid extraction, liquid-phase microextraction, solid-phase microextraction and aqueous two-phase system extraction. Particularly, microwave-assisted extraction has been the preferred option as it is a rapid, effective and cheap technique for the extraction of TIAs from different matrices (Sparr & Bjorklund, 2000).

In an IL-based ultrasound-assisted approach for the extraction of TIAs from *Catharanthus roseus*, Yang et al. (2011) tested the extraction efficiency of 1-allyl-3-methylimidazolium bromide at a concentration of 0.5 M and a solid–liquid ratio of 1:10 for 2 h of maceration time showing higher extraction efficiency when compared to seven different conventional extraction methods. The same cation and anion combination along with microwave-assisted extraction proved to be effective for isolating alkaloids from *Camptotheca acuminata* (Ma et al., 2012; Wang et al., 2011).

2.3. Molecular imprinted polymers

Molecularly imprinted polymers (MIPs) are an emerging technique that uses polymer materials with high selectivity and affinity towards a particular molecule, the template. MIPs are prepared by crosslinking monomers in a complexation solution in the presence of the template molecules. After the polymerization reaction is finished, the template is removed from the polymer by solvent extraction, leaving behind an imprint with a cavity that sterically and chemically binds the template molecule with high affinity (Xie et al., 2001).

Highly selective MIPs were used in SPE for preparative and analytical separations, using, for example, HPLC to separate vinblastine (Zhu, Huang, Li, & Yin, 2010), vindoline and catharanthine (Lopez et al., 2011) from a commercial extract of *C. roseus*. Methacrylic acid was used as the functional monomer, ethylene glycol dimethacrylate as the crosslinker and toluene or acetone as the porogenic solvent. Thermal polymerization yielded an MIP on which extracts were loaded on a polypropylene SPE cartridge containing the MIP. Analysis of the eluents by HPLC–UV showed high recoveries for vinblastine (93.8%) and catharanthine (101%)

and capacities of 750 and 818 μg/g, respectively, whereas for vindoline, recovery was only 33%. The MIP cavity apparently specifically binds catharanthine as a monomer as well as this moiety as part of the dimeric alkaloids.

2.4. Supercritical CO_2

Supercritical fluid extraction (SFE) has been applied to the isolation of natural products since the late 1970s to a few number of cases like the decafeination of coffee beans and tea leaves (Kaiser, Rompp, & Schmidt, 2001), but since the early 2000s, this technique has experienced an enormous development particularly in the food, toxicological, pharmaceutical and environmental areas. Compared with conventional organic solvents, supercritical fluids, and particularly CO_2, are non-toxic, low cost, environmentally friendly and because of high diffusivity reduce mass-transfer rates. They can be compressed at constant temperature, and therefore, increasing in density and solvent capacity. Their low surface tension facilitates analyte extraction, and since it is gaseous at room temperature and under constant pressure, it can be easily extracted rendering solvent-free analytes (Herrero, Mendiola, Cifuentes, & Ibañez, 2010). The main supercritical solvent used is CO_2 with its low polarity as its main disadvantage, which can be solved with the addition of a polar modifier (co-solvent), for example, 1–10% methanol or ethanol to expand its extraction range to include more polar compounds, which in turns will also reduce the analyte–matrix interactions improving their quantitative extraction. This can be done in two ways, either by mixing the modifier to the CO_2 flow or by mixing it with the raw material in the extraction cell (Mendiola, Herrero, Cifuentes, & Ibañez, 2007).

Studies regarding SFE have been performed for the major TIA's from *Catharanthus*, *Tabernaemontana* and *Evodia* species (Table 9.1). Lee et al. (1992) extracted vindoline and catharanthine from the leaves of *C. roseus* without the addition of a modifier. About 67.2% (w/w) of vindoline was recovered using a CO_2 flow rate of 150 ml/min at 40 °C, whereas in the case of catharanthine, a higher flow rate of 400 ml/min at the same temperature had a yield of 52% (w/w). Song et al. (1992) compared the addition or absence of ethanol as the modifier for vindoline extraction. The largest amount (58 wt%) was obtained without the addition of the modifier showing that vindoline solubility is more sensitive to pressure than to temperature. However, a mixture of CO_2–methanol–triethylamine (80:12:8) proved to be more effective than methanolic extraction of vinblastine and

Table 9.1 Experimental conditions for supercritical fluid extraction of TIAs

Plant material	Matrix	Target compound	Aim	Extraction conditions	Analytical method	References
Catharanthus roseus	Leaves	Vindoline, catharanthine	To selectively extract vindoline	CO_2, 150 bar, 40 °C, 10 h	HPLC, LC–MS	Lee et al. (1992)
	Leaves	Vindoline	To compare extraction with and without co-solvent	CO_2 + 3% ethanol, 300 bar, 35 °C, 5 h	HPLC, LC–MS	Song et al. (1992)
	Aerial parts and roots	Vinblastine, vincristine	To apply a basified SFE solvent	CO_2–methanol–triethylamine (80:12:8), 340 bar, 80 °C	LC–MS	Choi et al. (2002)
	Leaves	Catharanthine, vindoline, 3′,4′-anhydrovinblastine	To optimize the method and to compare it to conventional extraction methods	CO_2 + 6.6% methanol, 200–300 bar, 80 °C, 40 min	HPLC, LC–MS	Verma et al. (2008)
Evodia rutaecarpa	Fruits	Evodiamine, rutaecarpine	To test the use of methanol as co-solvent	CO_2 + 50% methanol, 280 bar, 62 °C, 1.3 h	HPLC	Liu et al. (2010)
Tabernaemontana catharinensis	Aerial parts	Coronaridine, voacangine, voacristine, voacangine hydroxylindolenine, voacristine hydroxylindolenine, 3-hydroxylcoronaridine	To evaluate temperature, pressure and co-solvent	CO_2 + 4.6% ethanol, 250 bar, 45 °C, 2 h	GC–MS, GC–FID, ^{1}H and ^{13}C NMR	Pereira et al. (2004)
Uncaria tomentosa	Root bark	Isopteropodine, pteropodine, isomitraphylline, uncarine F, mitraphylline, speciophylline, rhynchophylline, isorhynchophylline	To compare extraction with and without co-solvent	CO_2 + 10% methanol, 253 bar, 60 °C, 30 min	HPLC–MS and GC–MS	Lopez-Avila, Benedicto, and Robaugh (1997)

Abbreviations: FID, flame ionization detector.

vincristine (Choi, Yoo, & Kim, 2002) and CO_2 with 6.6% methanol for catharanthine (Verma, Hartonen, & Riekkola, 2008).

Alkaloids from *Tabernaemontana catharinensis* were extracted using a mixture of supercritical CO_2 and 4.6% ethanol (Pereira et al., 2004). The flow of CO_2 was held at 4 bar/min with a depressurization step of 70 bar (12 bar/min). Liu, Guo, Chang, Jiang, and Wang (2010) extracted the main alkaloids from unripe fruits of *E. rutaecarpa* using 50% methanol as the modifier in a flow rate of 0.4 ml/min with a static extraction for 5 min and then dynamic up to 90 min.

3. ANALYTICAL METHODS

3.1. HPLC and hyphenated techniques

HPLC has been the method of choice for the analysis of alkaloids. It has been recognized since the 1970s as the most versatile and most widely applied technique for efficient separation and analysis of alkaloids (Verpoorte & Baerheim Svendsen, 1984). Most of the separations are done on reversed-phase materials such as C_8, C_{18} and phenyl-bonded phases on silica. McCalley (2002) extensively described the importance of the characteristics of the stationary phases for the separation of alkaloids. The most common eluents for the separation of alkaloids are methanol–water and acetonitrile–water buffered at pH 2–4 (buffer strength $>$ 25 mM) in order to keep alkaloids in their more polar protonated form to reduce tailing due to interaction of the basic nitrogen with the residual acidic silanol groups of the stationary phase (Kingston, 1979; Verpoorte & Baerheim Svendsen, 1984). Ion-pair chromatography with, for example, long alkyl chain sulfonic acids is also used in alkaloid separations. Various amines such as triethylamine are sometimes added to reduce tailing. For detection, UV absorption is the most widely used since indole alkaloids have strong and specific UV chromophores that can be easily used to identify them, for example, using HPLC with photodiode array detection (DAD). Nevertheless, MS has been a major tool in the identification and structure elucidation of alkaloids, as it not only allows determination of the chemical structure of known and unknown compounds but also offers high sensitivity, and hence the combination of DAD and mass spectroscopy coupled with liquid chromatography, liquid chromatography coupled to mass spectroscopy (LC–MS) is the most selective detection for alkaloids (Verpoorte & Niessen, 1994). Mobile phases for LC–MS systems using isocratic separations on reversed-phase silica gel type stationary phases are fully volatile acidic eluents containing e.g. formic acid, acetic acid, trifluoroacetic acid, ammonium carbonate or ammonium formate (Table 9.2).

Typically, mass spectroscopy data for TIA analysis are acquired in the positive mode and based on the combination of retention time, UV and mass spectra, known compounds are rapidly dereplicated and new structures are identified from different plant matrices such as crude extracts from intact plants and organs, hairy roots, cell suspension cultures and from biological matrices. In multistage MS detection experiments, the use of deuterium-labelled internal standards has proved to be sensitive enough for the accurate quantification of yohimbine from commercially available aphrodisiacs and bark from *Pausinystalia yohimbe* by using yohimbine-d_3 (Zanolari, Ndjoko, Ioset, Marston, & Hostettmann, 2003) or clonazepam-d_4 for ibogaine and noribogaine determination from human tissues (Chèze, Lenoan, Deveaux, & Pépin, 2008). Recent approaches using direct-injection electrospray ionization (ESI)–MS/MS (Chen, Zhang, Zhang, Chen, & Chen, 2013; Zhou, Tai, Sun, & Pan, 2005) or flow-injection ESI–MS/MS (Favretto, Piovan, Filippini, & Caniato, 2001) can omit the sample preparation step and then be used to confirm the presence of alkaloids in different matrices. In addition, these techniques can provide characteristic structural information such as precursor and product ion information, which is useful for multicomponent screening purposes.

3.2. Ultra high-pressure LC–MS

One of the latest developments in LC–MS has been the introduction of very pH-stable stationary phases, sub-2-μm particles and monolith columns. This requires high pressures (>400 bar) and is now known as ultra high-pressure liquid chromatography (UHPLC). May achieve up to 100,000 number of plates per time unit (N/t_0) and peak capacities of 900 and reducing the analysis time by a factor of 20. With the same column length, a three-fold efficiency improvement can be observed compared to 5 μm supports (Nguyen, Guillarme, Rudaz, & Veuthey, 2006).

UHPLC interfaces with high-resolution tandem mass spectrometers and NMR can greatly improve analysis in terms of resolution, speed, reproducibility, sensitivity and unequivocal identification of trace compounds providing confirmative information for studies in e.g. quality control, fingerprinting, authentication, standardization or identification of biomarkers. Other approaches include microfractionation bioactivity-based analysis (Hou et al., 2012), chromatographic profiling (Xu et al., 2012), monitoring alkaloid production in cell suspension cultures (He, Yang, Tan, Zhao, & Hu, 2011;

Table 9.2 LC systems for quantitative and qualitative analysis of alkaloids

Plant species	Target compound(s)	Matrix	LC conditions column/particle size (μm)/mobile phase A and B/internal standard (IS)	Analytical method	References
Catharanthus roseus	Profiling approach	Roots	Luna C_{18}/5/A: ACN and B: 1% acetic acid in water/No IS	HPLC–DAD–ESI–MS/MS	Ferreres et al. (2010)
	Vindoline, vindolidine, vincristine, vinblastine, catharanthine, 19*S*-vindolinine, vindolinine	Commercial extract	Zorbax Eclipse XDB-C_8/5/A: 0.1% triethylamine and B: methanol/No IS	HPLC–ESI–MS/MS	Zhou et al. (2005)
	Vincristine	Human plasma	Luna C_8/3/A: 1% acetic acid in water and B: ACN/vincristine and vinblastine	LC–MS/MS	Guilhaumou et al. (2010)
	Vinblastine, vindoline, ajmalicine, catharanthine, vinleurosine	Stems	DL C_{18}/5/A: ACN and B: 10 mM ammonium acetate/No IS	LC–MS/MS	Chen et al. (2013)
Claviceps sp.	Ergometrine, ergosine, ergotamine, ergocornine, ergocryptine, ergocristine	Cereal and cereal products	XBridge MS C_{18}/3.5/A: water–0.2 M ammonium bicarbonate–methanol (85:5:10 v/v/v) and B: water–0.2 M ammonium bicarbonate–methanol (5:5:90 v/v/v)/methylergometrine and dihydroergotamine	LC–MS/MS	Di Mavungu et al. (2012)

Evodiae fructus	Evodiamine, rutaecarpine	Human serum	Venusil XBP C_{18}/5/A: 5 mM ammonium formate–methanol–water (85:15 v/v)/evodiamine and rutaecarpine	LC–ESI–MS/MS and LC–APCI–MS/MS	Wen, Li, Liu, Liao, and Liu (2006)
Mitragyna inermis	Uncarine D	Leaves	Waters C_{18} Symmetry/5/A: phosphate–methanol and B: methanol/naphthalene	HPLC–DAD	Fiot et al. (2005)
M. speciosa	Mitragynine	Urine	Zorbax C_{18}/5/A: 10 mM ammonium formate in water and B: 0.1% acetic acid in ACN/No IS	LC–IT–MS	Philipp et al. (2009)
Pausinystalia yohimbe	Yohimbine	Bark and commercial aphrodisiacs	Nucleosil 100-5 C_{18} AB/5/A: 2 mM triethylamine in water and B: 2 mM in ACN/Yohimbine-d_3 (for MS) and codeine (for UV)	HPLC–UV; HPLC–APCI–MS; HPLC–ESI–MS	Zanolari et al. (2003)
Rauvolfia serpentina	Reserpine, ajmaline, ajmalicine	Roots	Chromolith Performance C_{18}/4.6/A: 0.01 M sodium phosphate and B: 0.5% acetic acid in ACN/reserpine, ajmaline and ajmalicine	HPLC–DAD	Srivastava, Tripathi, Pandey, Verma, and Gupta (2006)

Continued

Table 9.2 LC systems for quantitative and qualitative analysis of alkaloids—cont'd

Plant species	Target compound(s)	Matrix	LC conditions column/particle size (μm)/mobile phase A and B/internal standard (IS)	Analytical method	References
R. serpentina × *Rhazya stricta*	Screening approach	Hybrid cell cultures	Nucleosil 100-5 C_{18}/A: 39 mM sodium phosphate in ACN and B: 3 mM sodium phosphate-2.5 mM hexanesulfonic acid in ACN/No IS	HPLC	Stöckigt et al. (2002)
R. verticillata	Fingerprint analysis	Roots and rhizomes	Diamonsil C_{18}/5/A: water and B: 0.1% formic acid/No IS	LC–Q–TOF–MS	Hong, Cheng, Wu, and Zhao (2010)
Strychnos nux-vomica	Strychnine	Urine	Chrompack cyanopropyl/3/A: ACN and B: 1% acetic acid in water/nalorphine	LC–APCI–MS/MS	Van Eenoo, Deventer, Roels, and Delbeke (2006)
		Blood	Hypurity C18/5/A: ACN and B: 20 mM sodium dihydrogen phosphate/chloroquine	LC–DAD	Duverneuil, de la Grandmaison, de Mazancourt, and Alvarez (2004)

Tabernanthe iboga	Ibogaine, noribogaine	Human plasma and blood	Zorbax eclipse XD8 C_8/5/A: 0.02% trimethylamine in ACN and B: 2 mM ammonium formate/fluorescein	LC–ESI–MS	Kontrimaviciute, Breton, Mathieu, Mathieu-Daudé, and Bressollee (2006)
		Biological fluids and hair	ODB Uptisphere C_{18}/5/A: 20% ACN and B: 2 mM formate/clonazepam-d_4	LC–ESI–MS/MS	Chèze et al. (2008)
Uncaria tomentosa	Pteropodine, isopteropodine, speciophylline, uncarine, mytrapylline, isomytraphylline, ryncophyllin, isoryncophyllin, corynoxeine, isocorynoxeine	Bark and leaves	Lichrosorb C_{18}/5/A: 30 mM ammonium acetate and B: methanol–ACN (1:1 v/v)/tryptophol	HPLC–ESI–MS	Montoro, Carbone, Zuniga-Quiroz, De Simone, and Pizza (2004)
		Bark	Zorbax XDB C_{18}/5/A: 35 mM triethylammonium acetate and B: ACN/Mytraphylline	HPLC–ESI–MS	Bertol, Franco, and de Oliveira (2012)
Vinca minor	Vinblastine, desacetylvinblastine, vincristine	Human plasma	Ultrasphere C_{18}/5/A: 15 mM ammonium acetate in methanol or ACN and B: ACN or methanol/Vinorelbine	LC–APCI–MS	Ramírez, Ogan, and Ratainn (1997)

Abbreviations: ACN, acetonitrile; APCI, atmospheric pressure chemical ionization; ESI, electrospray ionization; FID, flame ionization; IT, ion trap; MS/MS, tandem mass spectroscopy; Q, quadrupole; TOF, time of flight.

He, Yang, Xiong, et al., 2011), metabonomic approaches (Wang et al., 2010), toxicological studies (Liu, Zhu, Li, Yan, & Lei, 2011) or in functional studies (Lorenz, Olsovska, Sulc, & Tudzynski, 2010) using UHPLC–MS where alkaloids are include identified by their retention time, UV spectra, fragmentation pattern data and high-resolution MS data and in some cases confirmed by NMR experiments.

3.3. Gas chromatography coupled to mass spectroscopy

Most of the TIAs are polar compounds and not volatile due to their indolyl and tertiary amino group that is not amenable for derivatization, although some of them have been successfully analysed by capillary GC using high temperatures of injection (200–300 °C) and temperature gradients from 100 to 250 °C (Verpoorte, 2005). The combination of GC with MS is an efficient tool in the preliminary or even complete identification of alkaloids. This approach is used in fingerprinting and bioactivity-guided approaches and even applying hyphenation with capillary electrophoresis (CE) (Table 9.3). For complete identification, fragmentation of the molecular ion is important, which can be achieved by tandem mass spectroscopy (MS^n). For quantitative analysis, GC–FID or the specific Nitrogen detector have an advantage over gas chromatography coupled to mass spectroscopy (GC–MS) in the detection and quantitation. In GC–MS, each compound will have a different detector response, which means that absolute quantitation requires calibration curves of each single compound, whereas in the other detection methods, the detector response is more or less similar for all compounds, thus allowing comparison of the peaks within a chromatogram without the need for calibration compounds. That thus allows the analysis of rare alkaloids of which not sufficient material is available for making calibration curves.

Dagnino, Schripsema, Peltenburg, and Verpoorte (1991) showed the feasibility of capillary GC for the analysis of a wide range of TIAs, mainly found in the genus *Tabernaemontana*. Gallagher et al. (1995) developed a derivatization method to estimate ibogaine levels in biological samples by GC–MS using ibogaine-d_3 as internal standard. Several derivatizing agents were compared, for example, trifluoroacetyl, heptafluorobutyric anhydride, and trifluoroacetic anhydride in order to determine the best derivatization conditions in terms of choice of chemical reagent, conditions and detection parameters for the reliable quantitation of this alkaloid in different tissues after oral administration. The derivatization of noribogaine and ibogaine by ethylation (Hearn, Pablo,

Table 9.3 Methods for the detection of alkaloids in different samples using GC–MS techniques

Compound	Matrix	Work-up	Internal standard	Derivatization	Stationary phase	Detection mode	References
Ibogaine	Brain	LLE (*n*-hexane)	Ibogaine-d_3	Trifluoroacetic anhydride	DB-5MS (30 m × 0.25 mm; 0.1 μm)	EI, MS	Gallagher et al. (1995)
Strychnine	Liver, lung, brain, spleen, skeletal muscle, bile, urine, blood	LLE (butyl chloride)	Methapyrilene	–	DB-5MS (15 m × 0.25 mm; 0.25 μm)	EI, MS/MS, full scan	Rosano, Hubbard, Meola, and Swift (2000)
	Blood, liver, kidney, small intestine, urine	LLE (toluene–heptane–isoamyl alcohol 67:20:4 v/v/v)	Papaverine	–	HP1 (12.5 m × 0.2 mm; 0.33 μm)	EI, SIM	Marques et al. (2000)
	Blood	SPE	Papaverine	–	Ultra 2 (12 m × 0.25 mm; 0.25 μm)	EI–MS, SIM	Barroso et al. (2005)
	Slimming foods	SPE	Leucomalachite green	–	VF-5MS (30 m × 0.25 mm; 0.25 μm)	EI–MS/MS, full scan, SIM	Li et al. (2012)

Continued

Table 9.3 Methods for the detection of alkaloids in different samples using GC–MS techniques—cont'd

Compound	Matrix	Work-up	Internal standard	Derivatization	Stationary phase	Detection mode	References
Uleine, demethoxyaspidormine	Bark of *Himatanthus lancifolius*	SLE (1% HCl)	–	–	HP (30 m × 0.25 mm; 0.25 μm)	MS	Baggio et al. (2005)
Affinisine, voachalotine	Root bark of *Tabernaemontana laeta* and *T. hystrix*	SLE	Isatin	–	DB1 (30 m × 25 mm; 0.3 μm)	EI, FID	Vieira et al. (2008)
Aspidospermidine, demethoxypalosine, aspidocarpine, aspidolimine, fendlerine, aspidolimidine	Stem bark of *Aspidosperma spruceaunm*	SLE (methanol)	–	–	DB5-MS (30 m × 0.25 mm; 0.25 μm)	EI–MS/MS	Aguiar et al. (2010)
Yohimbine	Bark of *Pausinystalia yohimbe*	SLE (methanol)	Diazepam	–	DB5-MS (30 m × 0.32 mm; 0.25 μm)	EI, MS, SIM	Chen et al. (2008)
Voacangine, dregamine	Root extract from *T. elegans*	SLE (ethanol)	–	–	DB5-MS (30 m × 32 μm; 0.25 mm)	MS, Scan mode	Pallant et al. (2012)

Abbreviations: EI, electron impact; FID, flame ionization detection; LLE, liquid–liquid extraction; SIM, selected ion monitoring; SLE, solid–liquid extraction.

Hime, & Mash, 1995) or silylation (Alburges, Foltz, & Moody, 1995) after sample clean-up procedures with SPE has also proved to be effective in the determination of these alkaloids from blood, plasma and urine samples.

In bioactivity-guided experiments, a number of alkaloids have been identified from plant extracts or fractions by means of GC–MS (Cardoso, Vilegas, & Honda, 1998) or GC–FID (Cardoso, Vilegas, & Pozetti, 1997). Particularly, alkaloids from different *Tabernaemontana* species (Andrade et al., 2005; Pallant, Cromarty, & Steenkamp, 2012; Vieira et al., 2008) and *Himatanthus lancifolius* (Baggio et al., 2005) are amenable for analysis without any derivatization step.

Adulterants such as strychnine along with other compounds present in different commercially available slimming products were determined by GC–MS/MS. Optimization of the method included a sample clean-up step. Since the target compounds are weak bases, the extraction included SPE using a strong cation exchange cartridge which was washed with 2% formic acid, 30% methanol–water and 2% ammoniated methanol (Li et al., 2012).

3.4. High-speed counter-current chromatography

High-speed counter-current chromatography (HSCCC) is a two-phase solvent system, without solid phases, instead with liquid stationery phase, to resolve target compounds relying on the different partitioning of solutes between two immiscible solvents which makes it a very effective tool for the preparative separation and purification of natural products (Zhao & He, 2006).

The preparative isolation of alkaloids can be achieved by means of HSCCC. Because of the ionic nature of alkaloids, systems with a controlled pH are preferred for their separation. For example, pH-zone-refining counter-current chromatography (CCC) has been quite successful in separating alkaloids based on the pK_a values, showing the typically characteristic rectangular peaks for the analytes as common in displacement chromatography (Ito & Ma, 1996). Improved efficiency can be obtained by using ion-pairing gradients, for example, solvent two-phase systems consisting on methanol–chloroform–aqueous phosphate or citrate buffer (pH 4) containing perchlorate, acetate or chloride as the ion-pairing agent (Fang, Liu, Yang, Wang, & Huang, 2011; van der Heijden Hermans-Lokkerbol, Verpoorte, & Baerheim Svendsen, 1987). Important was the observation that ion pairs of alkaloids with chloride and perchlorate are quite well soluble in chloroform–methanol, something which is important information to keep in mind in liquid–liquid partitioning procedures for the isolation of alkaloids, as at acidic pH considerable amounts of alkaloids may pass into an organic solvent due to ion pairing (van der Heijden et al., 1987).

There are few studies on the separation and isolation of TIAs using CCC. They report on the separation of alkaloids from *C. roseus* (Renault et al., 1999), *Strychnos guaianensis* (Quetin-Leclercq et al., 1995), *S. nux-vomica* (Miao Cai, Xiang, An, & Ito, 1998), *Hortia oreadica* (Severino et al., 2009), *Geissospermum vellosi* (Mbeunkui, Grace, & Lila, 2012) and *Tabernaemontana* (van der Heijden et al., 1987), and *T. catharinensis* (Gonçalves, Curcino, Oliveira, & Braz-Filho, 2011).

Alkaloids from a crude extract of *C. roseus* as well as an artificial mixture of vinblastine, vincristine and catharanthine were successfully separated as monomers and dimers by means of centrifugal partition chromatography (CPC), a variation of CCC, in the pH-zone refining mode (Renault et al., 1999). The solvent phases used were methyl *tert*-butyl ether–acetonitrile–water (4:1:5 v/v/v). The upper organic phase was basified with 8 mM of triethylamine and used as mobile phase (ascending mode) or with 10 mM when used as the stationary phase (descending mode). The lower aqueous phase was acidified with 10 mM HCl (as a retainer stationary phase) or 8 mM (as a displacer mobile phase).

Quetin-Leclercq et al. (1995) briefly mentioned the fractionation of a chloroform residue by HSCCC when they isolated for the first time guianensine, an alkaloid from the stem bark of *Strychnos guianensis* using a multilayer-coil separator–extractor and a solvent system of ethyl acetate–methanol–water (4:1:3 v/v/v) where the lower aqueous phase was used as a stationary phase and the upper organic phase was pumped from the bottom to the upper part of the column which was also applied for strychnine and brucine separation from seeds of *S. nux-vomica* using a two-phase solvent system consisting of chloroform and 0.07 M sodium phosphate in a buffer solution of 0.04 M citric acid (1:1 v/v) (Miao et al., 1998).

Ingkaninan, Hazekamp, Hoek, Balconi, & Verpoorte (2000), Ingkaninan, Hermans-Lokkerbol, & Verpoorte (1999), reported the use of CPC for the pre-separation of crude extracts for rapid dereplication of known biological active compounds in plant materials. This included the analysis of several TIAs producing *Tabernaemontana* plants, allowing, for example, the rapid identification of two active TIAs (tubotaiwine and apparicine).

Severino et al. (2009) demonstrated the advantage of HSCCC in the isolation of the alkaloids rutaecarpin and dictamine from dichloromethane extract of *H. oreadica* leaves. They used the two-phase solvent system composed of *n*-hexane–ethanol–acetonitrile–water (10:8:1:1 v/v/v/v), where the upper phase was used as the mobile phase and the lower phase was used as the stationary phase in a tail-to-head elution mode. Further conventional methods of column chromatography yielded rutaecarpin and dictamnine with excellent recoveries compared to the concentration of the compounds quantified simultaneously by LC–APCI–MS/MS analysis of the same extract (93.1% and 84.9%, respectively). In a similar study, the combination of high-performance counter-current chromatography (HPCCC) and LC–MS/MS was successfully established to isolate indole alkaloids from the methanol extract from the stem bark of *G. vellosi* (Mbeunkui et al., 2012). Extract separation was achieved with the solvent system ethyl acetate–butanol–water (2:3:5 v/v/v) in an elution–extrusion with the upper phase as stationary phase and the combination of flash column chromatography. Identification of five different indole alkaloids was carried out with ESI multistage mass spectrometry (MS^n) data and confirmed by NMR methods.

Voachalotine and 12-methoxy-N_b-methylvoachalotine were resolved from the methanolic extract from the roots of *T. catharinensisin* by HSCCC in 4 h with a solvent system consisting of chloroform–methanol–water (5:10:6 v/v/v) with a 95% and 97% purity, respectively, and their identity was confirmed by ^1H and ^{13}C NMR experiments (Gonçalves et al., 2011).

Five indole alkaloids from the stem bark of *G. vellosii* were isolated with a combination of HPCCC and flash chromatography. To further analyse them, ESI–IT–TOF–MS and NMR experiments were conducted (Mbeunkui et al., 2012). In order to study the fragmentation pattern of these alkaloids, multiple tandem mass spectrometric data were produced by CID of the protonated molecule ion based on the most abundant ions $[M+2H]^{2+}$ and $[M+H]^+$ and a fragmentation pathway geissolosimine, geissospermine, geissoschizoline, geissoschizone and vellosiminol was proposed.

3.5. Capillary electrophoresis

CE represents an attractive analytical technique for the rapid qualitative and quantitative analysis of molecules with a wide range of polarity and molecular weight, including small molecules such as drugs but also macromolecules such as proteins or nucleic acids (Unger, 2009). Because of its versatility and high separation efficiency, CE is an alternative to the widely used RP-HPLC. CE has gained much interest for the analysis of natural products in plant extracts, quality control of herbal medicines, pharmaceutical formulations and food supplements (Ganzera, 2008; Verardo, Gomez-Caravaca, Seura-Carretero, Caboni, & Fernández-Gutiérrez, 2011).

The relatively poor sensitivity of CE, resulting from the small loading volumes, can be circumvented by the incorporation of pre-concentration strategies, while the advantages of MS detection are embodied in the improvement of detection sensitivity as well as the capability of both determining the exact mass of analytes and providing structural information, including the possibility to identify and determine co-migrating species in overlapping peaks (Niessen, Tjaden, & van der Greef, 1993; Ramautar, Somsen, & de Jong, 2011).

The ideal candidates for CZE are permanently charged molecules such as quaternary alkaloids and electrokinetic chromatography, but in fact, all acidic, basic and neutral compounds can be analysed by CE (Gotti, 2011). For TIA analysis, non-aqueous capillary electrophoresis (NACE) has been the most widely used since electrolytes such as ammonium acetate and ammonium formate can be used, allowing the hyphenation of CE and MS (Scriba, 2007). Buffer systems and CE methods used for TIA analysis are listed in Table 9.4.

NACE was useful for the separation of 11 *Vinca* alkaloids from an artificial mixture. Results were compared to those of HPLC using UV traces of both methods at 214 nm showing that although HPLC is more sensitive than CE in terms of limit of detection (LOD) and limit of quantification (LOQ), CE can be a good alternative by reducing analysis time and giving better resolution (Barthe et al., 2002).

Posch, Martin, et al. (2012) described an NACE-MS method to screen the psychoactive alkaloids present in two commercial preparations from *Mitragyna speciosa* as a quality control for added active compounds like the opioid *O*-desmethyltramadol, which can be fatal for humans. The use of a non-aqueous buffer system allowed the separation of diastereomers of mytraginine. The same methodology proved to have a high resolving power for the separation of iboga alkaloids from *Voacanga africana* although the choice of detector was not enough to discriminate between analytes with similar masses and migration times. In a similar study with preparations from *M. speciosa*, a higher selectivity and resolution were observed when BGE was switched to ammonium formate (Posch, Müller, et al., 2012). For the analysis of indole alkaloids from the root bark of *P. yohimbe* by NACE and GC–MS, the latter proved to be more sensitive (LOD 0.6 and 1.0 μg/ml, respectively) in terms of identification (Fig. 9.1; Chen et al., 2008).

An aqueous CE system using α, β or γ cyclodextrins (CD) was tested for the enantiomeric separation of vincamine, vinpocetine and vincadifformine. The best separations were achieved with β-CD and γ-CD. The proposed structures for the inclusion complexes were based on rotating-frame nuclear

Table 9.4 Buffer systems and CE methods for the analysis of terpenoid indole alkaloids from different plant species

Plant species	Target compound	CE method	Electrolyte	Analytical method	References
Catharanthus roseus	Vinblastine, vincristine	CZE	0.2 M ammonium acetate, pH 6.2	UV	Chu, Bodnar, White, and Bowman (1996)
	Vinblastine, vindoline, catharanthine	CE	20 mM ammonium acetate in 1.5% acetic acid	MS	Chen, Li, Zhang, Chen, and Chen (2011)
Claviceps purpurea	Ergonovinine, ergonovine, ergocornine, ergocryptine, ergocornine, ergocristine, ergosine, ergocristinine, ergotamine	CE	20 mM β-CD, 8 mM γ-CD, 2 M urea, 0.3% PVA in phosphate buffer, pH 2.5	UV	Frach and Blaschke (1998)
Evodiae fructus	Evodiamine, rutaecarpine, carboxyevodiamine, 1-methyl-2-nonyl-4(1*H*)-quinolone, 1-mehtyl-2-[(*Z*)-6-undecenyl]-4(1*H*)-quinolone, 1-methyl-2-undecyl-4(1*H*)-quinolone, evocarpine, 1-methyl-2-[(6*Z*,9*Z*)-6,9-pentadecadienyl]-4(1*H*)-quinolone, dihydroevodiamine	CZE, MEKC	CZE: 40 mM sodium dihydrogen phosphate–ACN (9:1 v/v) MEKC: 20 mM phosphate, 40 mM SDS, 9 mM sodium borate, pH 7.31	UV	Lee, Chuang, and Sheuu (1996)

Continued

Table 9.4 Buffer systems and CE methods for the analysis of terpenoid indole alkaloids from different plant species—cont'd

Plant species	Target compound	CE method	Electrolyte	Analytical method	References
Mitragyna speciosa	Mitragynine, paynantheine, 7-hydroxy-mitragynine	NACE	60 mM ammonium formate, 5% acetic acid in ACN	qTOF–MS	Posch, Müller, Schulz, Pütz, and Huhn (2012)
	Mitragynine, speciogynine, speciociliatine, mitraciliatine	NACE	58 mM ammonium formate, 1 M acetic acid in ACN	qTOF–MS	Posch, Martin, Pütz, and Huhn (2012)
Pausinystalia yohimbe	Yohimbine	NACE	20 mM ammonium acetate in 0.5% acetic acid	UV	Chen et al. (2008)
Phellodendron wilsonii	Berberine, palmatine, jatrorrhizine, phellondendrine, tetrahydropalmatine, magnoflorine, thalphenine	CE	60 mM ammonium acetate in 40% methanol, pH 4.5	UV, MS	Henion, Mordehai, and Cai (1994)
Psilocybe semilanceata	Psilocybin, baeocystin	CZE	10 mM borate, 10 mM phosphate, 25 mM SDS, pH 11.5	UV	Pedersen-Bjergaard, Rasmussen, and Sannes (1998)
Rauvolfia serpentina, *Rauvolfia serpentina* × *Rhazya stricta*,	Gramine, tryptamine, serpentine, alstonine, β-methylajmaline,	CZE	100 mM ammonium acetate in ACN (1:1 v/v), pH 3.1	UV, MS	Stöckigt et al. (2002), Stöckigt, Unger, Belder, and Stöckigtt (1997) and

Aspidosperma quebracho-blanco	tabersonine, vinblastine, corynanthine, vincristine, raufloridine, ajmaline, yohimbic acid, deserpidine, reserpine, rescinnamine				Unger, Stöckigt, Belder, and Stöckigtt (1997)
Strychnos nux-vomica	Strychnine, brucine	MEKC	50 mM phosphate, 100 mM SDS in ACN (4:1 v/v), pH 2.0		Wang, Han, Wang, Zang, and Wu (2006)
		CZE	10 mM phosphate buffer in methanol (9:1 v/v), pH 2.5	UV	Zong and Che (1995)
		NACE	25 mM Tris boric acid, methanol–ACN (6:2 v/v), pH 4.0	UV	Gu, Li, Zhu, and Zou (2006)
			30 mM ammonium acetate, acetic acid–ACN (1:1.5 v/v) in methanol	UV	Li et al. (2006)
S. pierrian	Strychnine, brucine, novacine, icajine		80 mM ammonium acetate, 0.1% acetic acid in water–methanol (4:6 v/v)	UV, MS	Feng, Yuan, and Lii (2003)
Uncaria tomentosa	Oxindole alkaloids	CZE	20 mM phosphate buffer, pH 5.6	UV	Stuppner, Sturn, and Konwalinkaa (1992)

Continued

Table 9.4 Buffer systems and CE methods for the analysis of terpenoid indole alkaloids from different plant species—cont'd

Plant species	Target compound	CE method	Electrolyte	Analytical method	References
Vinca	Catharanthine, vinorelbine, anhydrovinblastine, vinflunine, vindoline, 4-*O*-deacetylvinorelbine, 4-*O*-deaceylvinflunine, vindesine, 4′-deoxy-20′,20′-difluorovinblastine, vincristine	NACE	50 mM ammonium acetate, 25 mM SDS, 0.6 M acetic acid in methanol–ACN (75:25 v/v), pH 7.7	HPLC–DAD	Barthe et al. (2002)
	Vincamine, vinpocetine, vincadiformine	CE–CD	0.25–50 mM CD, 15 mM NaOH, pH 2.5	UV, NMR	Sohajda et al. (2010)
Voacanga africana	Voacamine, ibogaine, voacangine, 3-oxovoacangin	NACE	58 mM ammonium formate, 1 M acetic acid in ACN	qTOF–MS	Posch, Martin, et al. (2012)

Abbreviations: ACN, acetonitrile; CD, cyclodextrin; CE, capillary electrophoresis; CZE, capillary zone electrophoresis; MEKC, micellar electrokinetic chromatography; NACE, non-aqueous capillary electrophoresis; PVA, polyvinyl alcohol; SDS, sodium dodecyl sulphate.

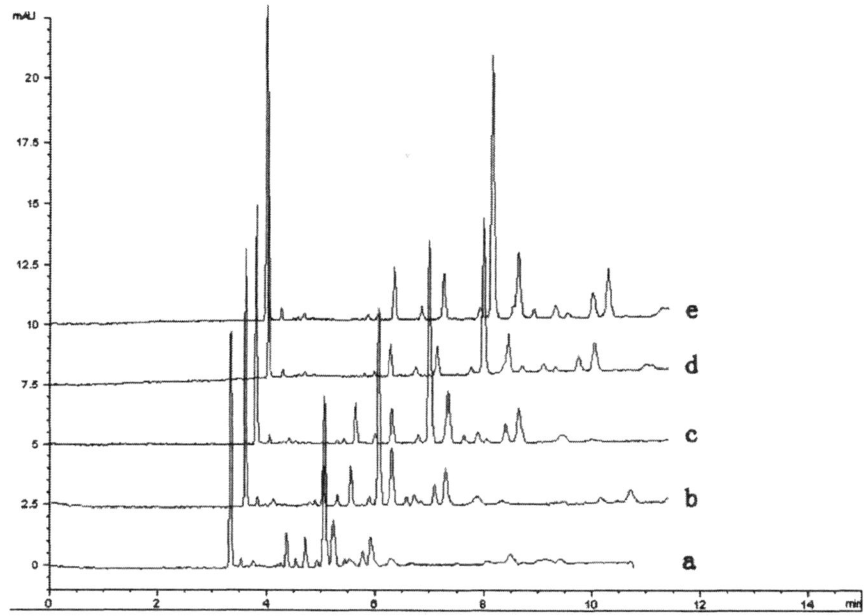

Figure 9.1 Electropherograms of several concentrations of ammonium acetate on the mobility of yohimbe, diazepam and berberine. Buffer solutions (a) 10 mM, (b) 15 mM, (c) 20 mM, (d) 25 mM, and (e) 30 mM ammonium acetate in 0.5% acetic acid. *Adapted with permission from Chen et al. (2008).*

Overhauser effect correlation spectroscopy (ROESY) experiments and their stability constants were determined by ^{1}H NMR chemical shift titrations for the three alkaloids (Sohajda et al., 2010). Using the same CDs, ergot alkaloids were successfully resolved in 12 min analysis time and 30-fold increased sensitivity when a laser-induced fluorescence detector was used (Frach & Blaschke, 1998).

3.6. Quantitative nuclear magnetic resonance

One of the major advantages of quantitative nuclear magnetic resonance (qNMR) is its primary analytical characteristic, because of which it can be applied in the quantitative estimation of purity of compounds without using any specific reference standard (Lindon & Nicholson, 2008). NMR-based metabolomics provides absolute and relative quantification of several metabolites in biological samples without separation of individual components in normal or modulated metabolism, so qNMR spectroscopy has been widely applied in environmental toxicity, drug toxicity, disease diagnosis, cancer metabolism, pathophysiology of disease, stress, nutrition, drug metabolism, plant metabolism, bacterial metabolism and cell–virus interactions (Bharti & Roy, 2012).

Camptothecin, 9-methoxycamptothecin, pumiloside and trigonelline were quantified by ^{1}H NMR analysis in root, stems and leaves from *Nothapodytes foetida* using DMSO-d_6 as solvent and 3,4,5-trimethoxybenzaldehyde as internal standard (Li, Lin, & Wu, 2005). The signals of H-7, H-10, H-19 and H-2 were selected as target signals for quantification of each alkaloid, respectively. Quantitation data were compared and confirmed to that of HPLC.

4. APPLICATIONS IN FINGERPRINT ANALYSIS

4.1. NMR and LC–MS-based metabolic fingerprinting on TIAs

NMR spectroscopy has a long-standing tradition to be applied to the characterization of pure compounds as it has been the case for the structural elucidation of TIAs. The continuous development of more and more sophisticated one-dimensional (1D) and 2D pulse sequences in NMR, various structure elucidation strategies have been developed and in the early 2000s, the unambiguous NMR-based structure elucidation of bisindoles is now inconceivable without using a 'holistic' NMR approach, that is, a full ^{1}H and ^{13}C NMR assignment in conjunction with the establishment of all spin–spin connectivities by a broad range of 2D-NMR methods, that is, homonuclear correlated spectroscopy (COSY), total correlated spectroscopy (TOCSY), heteronuclear single quantum coherence (HSQC), heteronuclear multiple bond correlation (HMBC), heteronuclear multiple quantum coherence (HMQC), nuclear Overhauser effect spectroscopy (NOESY) or ROESY experiments (Beni, Hada, Dubrovay, & Szantay, 2012).

There are excellent reviews covering the structural elucidation of the alkaloids in *Catharanthus* (Beni et al., 2012; Blaskó & Cordell, 1990; Dubrovay, Háda, Béni, & Szántay, 2012; Háda et al., 2012), *Aspidosperma* (Guimaraes, Braz-Filho, & Vieira, 2012), *Tabernaemontana* (Danieli & Palmisano, 1987; Nielsen, Hazell, Hazell, Ghia, & Torssell, 1994; Talapatra, Patra, & Talapatra, 1975), and *Strychnos* (Penelle et al., 2001, 2000; Rasoanaivo, Martin, Guittet, & Frappier, 2002) already published. However, the description of these experiments is beyond the scope of this discussion and interested readers are advised to consult the aforementioned publications.

Even though NMR is a crucial tool for identification and structure elucidation of pure samples, it can also make important contributions to the metabolic profiling by complementing MS-based approaches (Forseth & Schroeder, 2011). NMR-based metabolomics can be effectively applied to characterize and distinguish plants on species and genotype levels, different plant tissues within the same plant as well as for the detection of adulterants in foods and in health supplements for quality control (Holmes, Tang, Wang, & Seger, 2006). NMR can be integrated with chromatography to analyse herbal products to generate standardized 'metabolic fingerprints' which contains markers for activity (Heyman & Meyer, 2012).

One of the most common adulterants in *Strychnos* preparations is the 'false angostura bark', *Galipea officinalis*, whose bark closely resembles that of *S. nux-vomica*. An ^{1}H NMR method was developed for the quantitative analysis of strychnine and brucine in seeds and stem bark from *S. nux-vomica* (Frederich, Choi, & Verpoorte, 2003) along with a multivariate analysis which was useful for the metabolic profiling of *S. nux-vomica*, *S. ignatii* and *S. icaja* (Frederich et al., 2004). With this study, it was possible to discriminate the three species according to the composition in different organs, that is, seeds, leaves, stem bark and root bark. The compounds responsible for this discrimination were strychnine, brucine, loganin, fatty acids, icajine and sungucine. *Strychnos nux-vomica* and *S. icaja* stem bark could be distinguished by their content of brucine, but it was not possible to discriminate between the stem from *S. nux-vomica* and its adulterant arguing that the original material must have come from either stem bark or root bark.

Another interesting example concerns NMR-based profiling of *Cinchona* alkaloids in museum samples dating from 1850 to 1950 in order to determine the variation in contents from 117 different bark samples (Yilmaz, Nyberg, & Jaroszewski, 2012). An extraction system was developed using chloroform-d_1, methanol-d_4, D_2O and aqueous 70% perchloric acid (5:5:1:1 v/v/v/v). With an initial principal component analysis (PCA), it was possible to rule out four mislabelled samples that did not correspond at all to the *Cinchona* materials. With STOCSY-CA (statistical total correlation spectroscopy component analysis), it was possible to draw the conclusion that the variation methods in extracted alkaloids is not due to decomposition of quinine but an effect of the different cultivation methods of *Cinchona* trees over time.

Using 1D and 2D NMR, a comparison between the metabolic profile of healthy and phytoplasma-infected *C. roseus* plants was conducted along with multivariate data analysis in order to characterize and identify the metabolites responsible for the discrimination of the samples (Choi et al., 2004). Infected leaves showed an increase in signals of H-9 at δ 6.89 corresponding to vindoline showing a twofold increase than in healthy plants. The TIA precursors secologanin and loganic acid as well as chlorogenic acid and sugars, were four times higher than in healthy plants.

An MS-based fingerprint analysis was reported for yohimbe bark, and 18 different commercial dietary supplements, in order to determine the presence of yohimbine in the samples as well as to assess the quality of these supplements in the form of tablet, capsule or liquid (Sun & Chen, 2012). MS data were only used to confirm the identities of yohimbine, corynanthine and some other alkaloids, but the fingerprint analysis was conducted using the characteristic peaks in a chromatographic approach. In this case, all peaks are normalized against the area of yohimbine. With this method, the authors unambiguously demonstrated that 10 of the tested commercial preparations did not contain the amount of yohimbine claimed in the label of the product.

Mass spectroscopy does not only give the molecular weight, but each compound has also a characteristic fragmentation pattern, which is very useful for identification in GC-MS and LC-MS. Hesse (1974) has brought together all the information on mass spectroscopy of indole alkaloids, a very useful tool for identifying the identity of indole alkaloids.

5. CONCLUSIONS

The analysis of TIAs is a challenging task because of their complex chemical structures, usual low abundance and their difficult and time-consuming extraction procedures from different plant materials as well as from biological fluids. Only highly selective and sensitive methods will be suitable for such analyses. Sensitivity is the disadvantage of ^1H NMR when it comes to low contents of alkaloids in plant extracts, yet it is the only technique, which produces signals directly correlated with the amount of analytes in the sample. Even though CE offers some potential improvements in TIAs separation, it often faces sensitivity problems and in the case of GC, although a powerful tool, it is only suitable for a limited number of alkaloids which are volatile or amenable for derivatization. Consequently, liquid chromatographic or electrophoretic techniques in combination with different detectors have been mostly employed for TIAs analysis. Thus, HPLC in the reversed-phase mode has been and is the preferred separation technique for the analysis of TIAs.

The UV and DAD detection are robust detectors for targeted analysis. LC-MS offers further resolution and are of interest for more in depth analyses. Metabolomics as a novel approach is based on the different methods discussed here. But it requires a strict standardization to be able to store the results of the analyses with other laboratories. It should thus be based on standard protocols and public databases where the data are stored. Considering the extensive data presented in this review, it is clear that there is still a very long way to go to come to a chromatography based metabolomics in which all alkaloids can be analyzed. NMR despite its disadvantages of not being sensitive enough, seems closest to become a metabolomic platform in which also alkaloids can be analyzed.

REFERENCES

Aguiar, G. P., Wakabayashi, K. A. L., Luz, G. F., Oliveira, V. B., Mathias, L., Vieira, I. J. C., et al. (2010). Fragmentation of plumeran indole alkaloids from *Aspidosperma spruceanum* by electrospray ionization tandem mass spectrometry. *Rapid Communications in Mass Spectrometry*, *24*, 295–308.

Alburges, M. E., Foltz, R. L., & Moody, D. E. (1995). Determination of ibogaine and 12-hydroxy-ibogamine in plasma by gas chromatography-positive ion chemical ionization mass spectrometry. *Journal of Analytical Toxicology*, *19*, 381–386.

Andrade, M. T., Lima, J. A., Pinto, A. C., Rezende, C. M., Carvalho, M. P., & Epifanio, R. A. (2005). Indole alkaloids from *Tabernaemontana australis* (Müell. Arg) Miers that inhibit acetylcholinesterase enzyme. *Bioorganic & Medicinal Chemistry*, *13*, 4092–4095.

Baerheim Svendsen, A., & Verpoorte, R. (Eds.), (1983). *Chromatography of alkaloids, part I, thin-layer chromatography*. In *Chromatography library*, Vol. 23A, (pp. 1–534). Amsterdam: Elsevier.

Baggio, C. H., de Martini, G., de Souza, W., de Moraes, C. A., Brandao, L. M., Rieck, L., et al. (2005). Gastroprotective mechanisms of indole alkaloids from *Himatanthus lancifolius*. *Planta Medica*, *71*, 733–738.

Barroso, M., Gallardo, E., Margalho, C., Marques, E., Vieira, D. N., & López-Rivadulla, M. (2005). Determination of strychnine in human blood using solid-phase extraction and GC-EI-MS. *Journal of Analytical Toxicology*, *29*, 383–386.

Barthe, L., Ribet, J. P., Pelissou, M., Degude, M. J., Fahy, J., & Duflos, A. (2002). Optimization of the separation of *Vinca* alkaloids by nonaqueous capillary electrophoresis. *Journal of Chromatography. A*, *968*, 241–250.

Beni, Z., Hada, V., Dubrovay, Z., & Szantay, C., Jr. (2012). Structure elucidation of indole-indoline type alkaloids: A retrospective account from the point of view of current NMR and MS technology. *Journal of Pharmaceutical and Biomedical Analysis*, *69*, 106–124.

Bertol, G., Franco, L., & de Oliveira, B. H. (2012). HPLC analysis of oxindole alkaloids in *Uncaria tomentosa*: Sample preparation and analysis optimization by factorial design. *Phytochemical Analysis*, *23*, 143–151.

Bharti, S. K., & Roy, R. (2012). Quantitative ^{1}H NMR spectroscopy. *Trends in Analytical Chemistry*, *35*, 5–26.

Blaskó, G., & Cordell, G. A. (1990). Isolation, structure elucidation and biosynthesis of the bisindole alkaloids of Catharanthus. In B. Arnold, & S. Matthew (Eds.), *The alkaloids: Chemistry and pharmacology* (pp. 1–76). Orlando, FL: Academic Press.

Cao, M., Muganga, R., Nistor, I., Tits, M., Angenot, L., & Frederich, M. (2012). LC-SPE-NMR-MS analysis of *Strychnos usambarensis* fruits from Rwanda. *Phytochemistry Letters, 5*, 170–173.

Cardoso, C. A. L., Vilegas, W., & Honda, N. K. (1998). Qualitative determination of indole alkaloids, triterpenoids and steroids of *Tabernaemontana hilariana. Journal of Chromatography. A, 808*, 264–268.

Cardoso, C. A. L., Vilegas, W., & Pozetti, G. L. (1997). Gas chromatography analysis of indole alkaloids from *Tabernaemontana hilariana. Journal of Chromatography. A, 788*, 204–206.

Chen, Q., Li, N., Zhang, W., Chen, J., & Chen, Z. (2011). Simultaneous determination of vinblastine and its monomeric precursors vindoline and catharanthine in *Catharanthus roseus* by capillary electrophoresis-mass spectrometry. *Electrophoresis, 34*, 2885–2892.

Chen, Q., Li, P., Zhang, Z., Li, K., Liu, J., & Li, Q. (2008). Analysis of yohimbine alkaloid from *Pausinystalia yohimbe* by non-aqueous capillary electrophoresis and gas chromatography-mass spectrometry. *Journal of Separation Science, 31*, 2211–2218.

Chen, Q., Zhang, W., Zhang, Y., Chen, J., & Chen, Z. (2013). *Identification and quantification of active alkaloids in Catharanthus roseus by liquid chromatography-ion trap mass spectrometry. Food Chemistry, 139*, 845–852.

Chèze, M., Lenoan, A., Deveaux, M., & Pépin, G. (2008). Determination of ibogaine and noribogaine in biological fluids and hair by LC-MS/MS after *Tabernanthe iboga* abuse. Iboga alkaloids distribution in a drowning death case. *Forensic Science International, 176*, 58–66.

Choi, Y. H., Tapias, E. C., Kim, H. K., Lefeber, A. W. M., Erkelens, C., Verhoeven, J. T. J., et al. (2004). Metabolic discrimination of *Catharanthus roseus* leaves infected by phytoplasma using ^{1}H-NMR spectroscopy and multivariate data analysis. *Plant Physiology, 135*, 2398–2410.

Choi, Y. H., Yoo, K. P., & Kim, J. (2002). Supercritical fluid extraction and liquid chromatography-electrospray mass analysis of vinblastine from *Catharanthus roseus. Chemical and Pharmaceutical Bulletin, 50*, 1294–1296.

Chu, I., Bodnar, J. A., White, E. L., & Bowman, R. N. (1996). Quantification of vincristine and vinblastine in *Catharanthus roseus* plants by capillary zone electrophoresis. *Journal of Chromatography. A, 755*, 281–288.

Dagnino, D., Schripsema, J., Peltenburg, A., & Verpoorte, R. (1991). Capillary gas chromatographic analysis of indole alkaloids: Investigation of the indole alkaloids present in *Tabernaemontana divaricata* cell suspension culture. *Journal of Natural Products, 54*, 1558–1563.

Danieli, B., & Palmisano, G. (1987). Alkaloids from *Tabernaemontana*. In A. Brossi (Ed.), *The alkaloids: Chemistry and pharmacology* (pp. 1–130). New York: Elsevier Inc.

de Castro, A., Perez, T., de Carvalho, G. M., Pinto, L. L., da Silva, C. C., Tanaka, J. C., et al. (2012). Anti-leishmanial activity of alkaloid extracts obtained from different organs of *Aspidosperma ramiflorum. Phytomedicine, 19*, 413–417.

Di Mavungu, J. D., Malysheva, S. V., Sanders, M., Larionova, D., Robbens, J., Dubruel, P., et al. (2012). Development and validation of a new LC-MS/MS method for the simultaneous determination of six major ergot alkaloids and their corresponding epimers. Application to some food and feed commodities. *Food Chemistry, 135*, 292–303.

Dubrovay, Z., Háda, V., Béni, Z., & Szántay, C., Jr. (2012). *NMR and mass spectrometric characterization of vinblastine, vincristine and some new related impurities—Part I. Journal of Pharmaceutical and Biomedical Analysis*, http://dx.doi.org/10.1016/j.jpba.2012.08.019.

Duverneuil, C., de la Grandmaison, G. L., de Mazancourt, P., & Alvarez, J. C. (2004). Liquid chromatography/photodiode array detection for determination of strychnine in blood: A fatal case report. *Forensic Science International*, *141*, 17–21.

Etse, J. T., Gray, A. I., Thomas, D. W., & Waterman, P. G. (1989). Terpenoid and alkaloid compounds from the seeds of *Monodora brevipes*. *Phytochemistry*, *28*, 2489–2492.

Fang, L., Liu, Y., Yang, B., Wang, X., & Huang, L. (2011). Separation of alkaloids from herbs using high-speed counter-current chromatography. *Journal of Separation Science*, *34*, 2545–2558.

Favretto, D., Piovan, A., Filippini, R., & Caniato, R. (2001). Monitoring the production yields of vincristine and vinblastine in *Catharanthus roseus* from somatic embryogenesis. Semiquantitative determination by flow-injection electrospray ionization mass spectrometry. *Rapid Communications in Mass Spectrometry*, *15*, 364–369.

Feng, H. T., Yuan, L. L., & Li, S. F. Y. (2003). Analysis of Chinese medicine preparations by capillary electrophoresis-mass spectrometry. *Journal of Chromatography. A*, *1014*, 83–91.

Ferreres, F., Pereira, D. M., Valentao, P., Oliveira, J. M. A., Faria, J., Gaspar, L., et al. (2010). Simple and reproducible HPLC-DAD-ESI-MS/MS analysis of alkaloids in *Catharanthus roseus* roots. *Journal of Pharmaceutical and Biomedical Analysis*, *51*, 65–69.

Fiot, J., Baghdikian, B., Boyer, L., Mahiou, V., Azas, N., Gasquet, M., et al. (2005). HPLC quantification of uncarine D and the anti-plasmodial activity of alkaloids from leaves of *Mitragyna inermis* (Willd.) O. Kuntze. *Phytochemical Analysis*, *16*, 30–33.

Forseth, R. R., & Schroeder, F. C. (2011). NMR-spectroscopic analysis of mixtures: From structure to function. *Current Opinion in Chemical Biology*, *15*, 38–47.

Frach, K., & Blaschke, G. (1998). Separation of ergot alkaloids and their epimers and determination in sclerotia by capillary electrophoresis. *Journal of Chromatography. A*, *808*, 247–252.

Frederich, M., Choi, Y. H., Angenot, L., Harnischfeger, G., Lefeber, A. W. M., & Verpoorte, R. (2004). Metabolomic analysis of *Strychnos nux-vomica*, *Strychnos icaja* and *Strychnos ignatii* extracts by ^{1}H nuclear magnetic resonance spectrometry and multivariate analysis techniques. *Phytochemistry*, *65*, 1993–2001.

Frederich, M., Choi, Y. H., & Verpoorte, R. (2003). Quantitative analysis of strychnine and brucine in *Strychnos nux-vomica* using ^{1}H-NMR. *Planta Medica*, *69*, 1169–1171.

Gallagher, C. A., Hough, L. B., Keefner, S. M., Seyed-Mozaffari, A., Archer, S., & Glick, S. D. (1995). Identification and quantification of the indole alkaloid ibogaine in biological samples by gas chromatography-mass spectrometry. *Biochemical Pharmacology*, *49*, 73–79.

Ganzera, M. (2008). Quality control oh herbal medicines by capillary electrophoresis: Potential, requirements and applications. *Electrophoresis*, *29*, 3489–3503.

Ghosal, S., & Srivastava, R. S. (1974). Structure of erysophorine: A new quaternary alkaloid of *Erythrina arborescens*. *Phytochemistry*, *13*, 2603–2605.

Girardot, M., Deregnaucourt, C., Deville, A., Dubost, L., Joyeau, R., Allorge, L., et al. (2012). Indole alkaloids from *Muntafara sessilifolia* with antiplasmodial and cytotoxic activities. *Phytochemistry*, *73*, 65–73.

Gonçalves, M. S., Curcino, I. J., Oliveira, R. R., & Braz-Filho, R. (2011). Application of preparative high-speed counter-current chromatography for the separation of two alkaloids from the roots of *Tabernaemontana catharinensis* (Apocynaceae). *Molecules*, *16*, 7480–7487.

Gotti, R. (2011). Capillary electrophoresis of phytochemical substances in herbal drugs and medicinal plants. *Journal of Pharmaceutical and Biomedical Analysis*, *55*, 775–801.

Gu, X., Li, H., Zhu, R., & Zou, H. (2006). Determination of strychnine and brucine in *Strychnos nux-vomica* L. by nonaqueous capillary electrophoresis. *Chromatographia*, *63*, 289–292.

Guilhaumou, R., Solas, C., Rome, A., Giocanti, M., Andre, N., & Lacarelle, B. (2010). Validation of an electrospray ionization LC/MS/MS method for quantitative analysis of vincristine in human plasma samples. *Journal of Chromatography B, 878*, 423–427.

Guimaraes, H. A., Braz-Filho, R., & Vieira, I. J. (2012). ^{1}H and ^{13}C-NMR data of the simplest plumeran indole alkaloids isolated from *Aspidosperma* species. *Molecules, 17*, 3025–3043.

Háda, V., Dubrovay, Z., Lakó-Futó, A., Galambos, J., Gulyás, Z., & Aranyi, A. (2012). NMR and mass spectrometric characterization of vinblastine, vincristine and some new related impurities—Part II. *Journal of Pharmaceutical and Biomedical Analysis*, http://dx.doi.org/10.1016/j.jpba.2012.09.008.

He, L., Yang, L., Tan, R., Zhao, S., & Hu, Z. (2011). Enhancement of vindoline production in suspension culture of the *Catharanthus roseus* cell line C20hi by light and methyl jasmonate elicitation. *Analytical Sciences, 27*, 1243–1248.

He, L., Yang, L., Xiong, A., Zhao, S., Wang, Z., & Hu, Z. (2011). Simultaneous quantification of four indole alkaloids in *Catharanthus roseus* cell line C20hi by UPLC-MS. *Analytical Sciences, 27*, 433–438.

Hearn, W. L., Pablo, J., Hime, G. W., & Mash, D. C. (1995). Identification and quantitation of ibogaine and an o-demethylated metabolite in brain and biological fluids using gas chromatography-mass spectrometry. *Journal of Analytical Toxicology, 19*, 427–434.

Henion, J. D., Mordehai, A. V., & Cai, J. (1994). Quantitative capillary electrophoresis—Ion spray mass spectrometry on a benchtop ion trap for the determination of isoquinoline alkaloids. *Analytical Chemistry, 66*, 2103–2109.

Hermans-Lokkerbol, A., & Verpoorte, R. (1986). Droplet counter-current chromatography of alkaloids. The influence of pH-gradients and ion-pair formation on the retention of alkaloids. *Planta Medica, 4*, 299–302.

Herrero, M., Mendiola, J. A., Cifuentes, A., & Ibañez, E. (2010). Supercritical fluid extraction: Recent advances and applications. *Journal of Chromatography. A, 1217*, 2495–2511.

Hesse, M. (1974). Indolalkaoide. In H. Budzikiewicz (Ed.), *Progress in mass spectrometry. Teil 1: Text*, Vol. 1. Weinheim, Germany: Verlag Chemie GmbH.

Heyman, H. M., & Meyer, J. J. M. (2012). NMR-based metabolomics as a quality control tool for herbal products. *South African Journal of Botany, 82*, 21–32.

Holmes, E., Tang, H., Wang, Y., & Seger, C. (2006). The assessment of plant metabolite profiles by NMR-based methodologies. *Planta Medica, 72*, 771–785.

Hong, B., Cheng, W., Wu, C., & Zhao, C. (2010). Screening and identification of many of the compounds present in *Rauvolfia verticillata* by use of high-pressure LC and quadrupole TOF MS. *Chromatographia, 72*, 841–847.

Hou, Y., Cao, X., Wang, L., Cheng, B., Dong, L., Luo, X., et al. (2012). Microfractionation bioactivity-based ultra performance liquid chromatography/quadrupole time-of-flight mass spectrometry for the identification of nuclear factor-kB inhibitors and β2 adrenergic receptors agonists in an alkaloidal extract of the folk herb *Alstonia scholaris*. *Journal of Chromatography B, 908*, 98–104.

Hu, W. L., Zhu, J. P., Prewo, R., & Hesse, M. (1989). Alstogustine and 19-epialstogustine, quaternary indole alkaloids from *Alstonia angustifolia*. *Phytochemistry, 28*, 1963–1966.

Ingkaninan, K., Hazekamp, A., Hoek, A. C., Balconi, S., & Verpoorte, R. (2000). Application of centrifugal partition chromatography in a general separation and dereplication procedure for plant extracts. *Journal of Liquid Chromatography & Related Technologies, 23*, 2195–2208.

Ingkaninan, K., Hermans-Lokkerbol, A., & Verpoorte, R. (1999). Comparison of some centrifugal partition chromatography systems for a general separation of plant extracts. *Journal of Liquid Chromatography & Related Technologies, 22*, 885–896.

Ito, Y., & Ma, Y. (1996). pH-zone-refining counter-current chromatography. *Journal of Chromatography. A, 753*, 1–36.

Jenett-Siems, K., Weigl, R., Kaloga, M., Schulz, J., & Eich, E. (2003). Ipobscurines C and D: Macrolactam-type indole alkaloids from the seeds of *Ipomoea obscura*. *Phytochemistry, 62*, 1257–1263.

Jenks, C. (2002). Extraction studies of *Tabernanthe iboga* and *Voacanga africana*. *Natural Product Letters, 16*, 71–76.

Kaiser, C. S., Rompp, H., & Schmidt, P. C. (2001). Pharmaceutical applications of supercritical carbon dioxide. *Pharmazie, 56*, 907–926.

Kingston, D. G. I. (1979). High performance liquid chromatography of natural products. *Journal of Natural Products, 42*, 237–260.

Kontrimaviciute, V., Breton, H., Mathieu, O., Mathieu-Daudé, J. C., & Bressolle, F. M. M. (2006). Liquid chromatography-electrospray mass spectrometry determination of ibogaine and noribogaine in human plasma and whole blood. Application to a poisoning involving *Tabernanthe iboga* root. *Journal of Chromatography B, 843*, 131–141.

Kumar, V., Bulumulla, H. N. K., Wimalasiri, W. R., & Reisch, J. (1994). Coumarins and an indole alkaloid from *Pamburus missionis*. *Phytochemistry, 36*, 879–881.

Lee, M. C., Chuang, W. C., & Sheu, S. J. (1996). Determination of the alkaloids in *Evodiae fructus* by capillary electrophoresis. *Journal of Chromatography. A, 755*, 113–119.

Lee, H., Hong, W. H., Yoon, J. H., Song, K. M., Kwak, S. S., & Liu, J. R. (1992). Extraction of indole alkaloids from *Catharanthus roseus* by using supercritical carbon dioxide. *Biotechnology Techniques, 6*, 127–130.

Li, Y., He, X., Qi, A., Gao, W., Chen, X., & Hu, Z. (2006). Separation and determination of strychnine and brucine in *Strychnos nux-vomica* L. and its preparation by nonaqueous capillary electrophoresis. *Journal of Pharmaceutical and Biomedical Analysis, 41*, 400–407.

Li, C. Y., Lin, C. H., & Wu, T. S. (2005). Quantitative analysis of camptothecin derivatives in *Nothapodytes foetida* using ^{1}H-NMR method. *Chemical and Pharmaceutical Bulletin, 53*, 347–349.

Li, Y., Zhang, H., Hu, J. T., Xue, F., Li, Y. X., & Sun, C. J. (2012). A GC-EI-MS-MS method for simultaneous determination of seven adulterants in slimming functional foods. *Journal of Chromatographic Science, 50*, 928–933.

Lindon, J. C., & Nicholson, J. K. (2008). Analytical technologies for metabonomics and metabolomics, and multi-omic information recovery. *Trends in Analytical Chemistry, 27*, 194–204.

Liu, B., Guo, F., Chang, Y., Jiang, H., & Wang, Q. (2010). Optimization of extraction of evodiamine and rutaecarpine from fruit of *Evodia rutaecarpa* using modified supercritical CO_2. *Journal of Chromatography. A, 1217*, 7833–7839.

Liu, Y., Zhu, R., Li, H., Yan, M., & Lei, Y. (2011). Ultra-performance liquid chromatography-tandem mass spectrometric method for the determination of strychnine and brucine in mice plasma. *Journal of Chromatography B, 879*, 2714–2719.

Lopez, C., Claude, B., Morin, Ph., Max, J. P., Pena, R., & Ribet, J. P. (2011). Synthesis and study of a molecularly imprinted polymer for the specific extraction of indole alkaloids from *Catharanthus roseus* extracts. *Analytica Chimica Acta, 683*, 198–205.

Lopez-Avila, V., Benedicto, J., & Robaugh, D. (1997). Supercritical fluid extraction of oxindole alkaloids from Uncaria tormentosa. *Journal of High Resolution Chromatography, 20*, 231–236.

Lorenz, N., Olsovska, J., Sulc, M., & Tudzynski, P. (2010). Alkaloid cluster gene *ccsA* of ergot fungus *Claviceps purpurea* encodes chanoclavine I synthase, a flavin adenine dinucleotide-containing oxidoreductase mediating the transformation of N-methyl-dimethylallyltryptophan to chanocalvine I. *Applied and Environmental Microbiology, 76*, 1822–1830.

Ma, C. H., Wang, S. Y., Yang, L., Zu, Y. G., Yang, F. J., Zhao, C. J., et al. (2012). Ionic liquid-aqueous solution ultrasonic-assisted extraction of camptothecin and

10-hydroxtcamptothecin from *Camptotheca acuminata* samara. *Chemical Engineering and Processing: Process Intensification*, *57–58*, 59–64.
Maltese, F., van der Kooy, F., & Verpoorte, R. (2009). Solvent derived artifacts in natural products chemistry. *Natural Products Communications*, *4*, 447–454.
Marques, E. P., Gil, F., Proença, P., Monsanto, P., Oliveira, M. F., Castanheira, A., et al. (2000). Analytical method for the determination of strychnine in tissues by gas chromatography/mass spectrometry: Two case reports. *Forensic Science International*, *110*, 145–152.
Mbeunkui, F., Grace, M. H., & Lila, M. A. (2012). Isolation and structural elucidation of indole alkaloids from *Geissospermum vellosii* by mass spectrometry. *Journal of Chromatography B*, *885–886*, 83–89.
McCalley, D. V. (2002). Analysis of the *Cinchona* alkaloids by high-performance liquid chromatography and other separation techniques. *Journal of Chromatography. A*, *967*, 1–19.
Mendiola, J. A., Herrero, M., Cifuentes, A., & Ibañez, A. (2007). Use of compressed fluids for sample preparation: Food applications. *Journal of Chromatography. A*, *1152*, 234–246.
Miao, P., Cai, D. C., Xiang, B. R., An, D. K., & Ito, Y. (1998). Separation and purification of strychnine from crude extract of *Strychnos nux-vomica* L. By high-speed countercurrent chromatography. *Journal of Liquid Chromatography & Related Technologies*, *21*, 163–170.
Montoro, P., Carbone, V., Zuniga-Quiroz, J., De Simone, F., & Pizza, C. (2004). Identification and quantification of components in extracts of *Uncaria tomentosa* by HPLC-ES/MS. *Phytochemical Analysis*, *15*, 55–64.
Moreno, P. R. H., van der Heijden, R., & Verpoorte, R. (1993). Effect of terpenoid precursor feeding and elicitation on formation of indole alkaloids in cell suspension cultures of *Catharanthus roseus*. *Plant Cell Reports*, *12*, 702–705.
Nguyen, D. T. T., Guillarme, D., Rudaz, S., & Veuthey, J. L. (2006). Fast analysis in liquid chromatography using small particle size and high pressure. *Journal of Separation Science*, *29*, 1836–1848.
Nielsen, H. B., Hazell, A., Hazell, R., Ghia, F., & Torssell, K. B. G. (1994). Indole alkaloids and terpenoid from *Tabernaemontana markgrafiana*. *Phytochemistry*, *37*, 1729–1735.
Niessen, W. M. A., Tjaden, U. R., & van der Greef, J. (1993). Capillary electrophoresis-mass spectrometry. *Journal of Chromatography*, *636*, 3–19.
Pallant, C. A., Cromarty, A. D., & Steenkamp, V. (2012). Effect of an alkaloidal fraction of *Tabernaemontana elegans* (Stapf.) on selected micro-organisms. *Journal of Ethnopharmacology*, *140*, 398–404.
Paranhos, J. T., Fragoso, V., da Silveira, V. C., Henriques, A. T., & Fett-Neto, A. G. (2009). Organ-specific and environmental accumulation of psychollatine, a major indole alkaloid glucoside from *Psychotria umbellate*. *Biochemical Systematics and Ecology*, *37*, 707–715.
Pedersen-Bjergaard, S., Rasmussen, K. E., & Sannes, E. (1998). Strategies for the capillary electrophoretic separation of indole alkaloids in *Psilocybe semilanceata*. *Electrophoresis*, *19*, 27–30.
Penelle, J., Christen, P., Molgó, J., Tits, M., Brandt, V., Frederich, M., et al. (2001). 5′,6′-Dehydroguiachrysine and 5′,6′-dehydroguiaflavine, two curarizing quaternary indole alkaloids from the stem bark of *Strychnos guianensis*. *Phytochemistry*, *58*, 619–626.
Penelle, J., Tits, M., Christen, P., Molgo, J., Brandt, V., Frederich, M., et al. (2000). Quaternary indole alkaloids from the stem bark of *Strychnos guianensis*. *Phytochemistry*, *53*, 1057–1066.
Pereira, C. G., Marques, M. O. M., Barreto, A. S., Siani, A. C., Fernandes, E. C., & Meireles, M. A. M. (2004). Extraction of indole alkaloids from *Tabernaemontana catharinensis* using supercritical CO2+ethanol: An evaluation of the process variables and the raw material origin. *Journal of Supercritical Fluids*, *30*, 51–61.
Perera, P., Samuelson, G., van Beek, T. A., & Verpoorte, R. (1983). Tertiary indole alkaloids from leaves of *Tabernaemontana dichotoma*. *Planta Medica*, *47*, 148–150.

Philipp, A. A., Wissenbach, D. K., Zoerntlein, S. W., Klein, O. N., Kanogsunthornrat, J., & Maurer, H. H. (2009). Studies on the metabolism of mitragynine, the main alkaloid of the herbal drug Kratom, in rat and human urine using liquid chromatography-linear ion trap mass spectrometry. *Journal of Mass Spectrometry*, *44*, 1249–1261.

Posch, T. N., Martin, N., Pütz, M., & Huhn, C. (2012). Nonaqueous capillary electrophoresis-mass spectrometry: A versatile, straightforward tool for the analysis of alkaloids from psychoactive plant extracts. *Electrophoresis*, *33*, 1557–1566.

Posch, T. N., Müller, A., Schulz, W., Pütz, M., & Huhn, C. (2012). Implementation of a design of experiments to study the influence of the background electrolyte on separation and detection in non-aqueous capillary electrophoresis-mass spectrometry. *Electrophoresis*, *33*, 583–598.

Quetin-Leclercq, J., Llabres, G., Warin, R., Belem-Pinheiro, M. L., Mavar-Manga, H., & Angenot, L. (1995). Guianensine, a zwitterionic alkaloid from *Strychnos guianensis*. *Phytochemistry*, *40*, 1557–1560.

Ramautar, R., Somsen, G. W., & de Jong, G. J. (2011). CE-MS for metabolomics: Developments and applications in the period 2010-2012. *Electrophoresis*, *34*, 86–98.

Ramírez, J., Ogan, K., & Ratain, M. J. (1997). Determination of Vinca alkaloids in human plasma by liquid chromatography/atmospheric pressure chemical ionization mass spectrometry. *Cancer Chemotherapy and Pharmacology*, *39*, 286–290.

Rasoanaivo, P., Martin, M. T., Guittet, E., & Frappier, F. (2002). New contributions to the structure elucidation and pharmacology of *Strychnos* alkaloids. In Atta-ur-Rahman (Ed.), *Bioactive natural products: Studies in natural products chemistry*: Vol. 26, Part G, (pp. 1029–1072). Elsevier BV.

Renault, J. H., Nuzillard, J. M., Le Crouerour, G., Thepenier, P., Zeches-Hanrot, M., & Le Men-Olivier, L. (1999). Isolation of indole alkaloids from *Catharanthus roseus* by centrifugal partition chromatography in the pH-zone refining mode. *Journal of Chromatography. A*, *849*, 421–431.

Rosano, T. G., Hubbard, J. D., Meola, J. M., & Swift, T. A. (2000). Fatal strychnine poisoning: Application of gas chromatography and tandem mass spectrometry. *Journal of Analytical Toxicology*, *24*, 642–647.

Scriba, H. K. (2007). Non-aqueous capillary electrophoresis-mass spectrometry. *Journal of Chromatography. A*, *1159*, 28–41.

Severino, V. G. P., Cazal, C. M., Forim, M. R., Silva, M. G. F., Rodrigues-Filho, E. R., Fernandes, J. B., et al. (2009). Isolation of secondary metabolites from *Hortia oreadica* (Rutaceae) leaves through high-speed counter-current chromatography. *Journal of Chromatography. A*, *1216*, 4275–4281.

Sheludko, Y., Gerasimenko, I., Unger, M., Kostenyuk, I., & Stoeckgit, J. (1999). Induction of alkaloid diversity in hybrid plant cell cultures. *Plant Cell Reports*, *18*, 911–918.

Silvestrini, A., Pasqua, G., Botta, B., Monacelli, B., van der Heijden, R., & Verpoorte, R. (2002). Effects of alkaloid precursor feeding on a *Camptotheca acuminate* cell line. *Plant Physiology and Biochemistry*, *40*, 749–753.

Sohajda, T., Varga, E., Ivanyi, R., Fejos, I., Szente, L., Noszal, B., et al. (2010). Separation of vinca alkaloid enantiomers by capillary electrophoresis applying cyclodextrin derivatives and characterization of cyclodextrin complexes by nuclear magnetic resonance spectroscopy. *Journal of Pharmaceutical and Biomedical Analysis*, *53*, 1258–1266.

Song, K. M., Park, S. W., Hong, W. H., Lee, H., Kwak, S. S., & Liu, J. R. (1992). Isolation of vindoline from *Catharanthus roseus* by supercritical fluid extraction. *Biotechnology Progress*, *8*, 583–586.

Sparr, C., & Bjorklund, E. (2000). Analytical-scale microwave-assisted extraction. *Journal of Chromatography. A*, *902*, 227–250.

Srivastava, A., Tripathi, A. K., Pandey, R., Verma, R. K., & Gupta, M. M. (2006). Quantitative determination of reserpine, ajmaline, and ajmalicine in *Rauvolfia serpentina* by reversed-phase high-performance liquid chromatography. *Journal of Chromatographic Science, 44*, 557–560.

Stöckigt, J., Sheludko, Y., Unger, M., Gerasimenko, I., Warzecha, H., & Stöckigt, D. (2002). High-performance liquid chromatographic, capillary electrophoretic and capillary electrophoretic-electrospray ionisation mass spectrometric analysis of selected alkaloid groups. *Journal of Chromatography. A, 967*, 85–113.

Stöckigt, D., Unger, M., Belder, D., & Stöckigt, J. (1997). Analysis of *Rauwolfia* alkaloids employing capillary electrophoresis-mass spectrometry. *Natural Product Letters, 9*, 265–272.

Stuppner, H., Sturn, S., & Konwalinka, G. (1992). Capillary electrophoretic analysis of oxindole alkaloids from *Uncaria tomentosa*. *Journal of Chromatography, 609*, 375–380.

Sun, J., & Chen, P. (2012). Chromatographic fingerprint analysis of yohimbe bark and related dietary supplements using UHPLC/UV/MS. *Journal of Pharmaceutical and Biomedical Analysis, 61*, 142–149.

Sun, C., & Liu, H. (2008). Application of non-ionic surfactant in the microwave-assisted extraction of alkaloids from *Rhizoma Coptidis*. *Analytica Chimica Acta, 612*, 160–164.

Talapatra, B., Patra, A., & Talapatra, S. K. (1975). Terpenoids and alkaloids of the leaves of *Tabernaemontana coronaria*. *Phytochemistry, 14*, 1652–1653.

Tanaka, J. C. A., da Silva, C. C., Ferreira, I. C. P., Machado, G. M. C., Leon, L. L., & de Oliveira, A. J. B. (2007). Antileshmanial activity of indole alkaloids from *Aspidosperma*. *Phytomedicine, 14*, 377–380.

Unger, M. (2009). Capillary electrophoresis of natural products: Current applications and recent advances. *Planta Medica, 75*, 735–745.

Unger, M., Stöckigt, D., Belder, D., & Stöckigt, J. (1997). General approach for the analysis of various alkaloid classes using capillary electrophoresis and capillary electrophoresis-mass spectrometry. *Journal of Chromatography. A, 767*, 263–276.

van der Heijden, R., Hermans-Lokkerbol, A., Verpoorte, R., & Baerheim Svendsen, A. (1987). Pharmacognostical studies on *Tabernaemontana* XX. Ion-pair droplet counter-current chromatography of indole alkaloids from suspension cultures. *Journal of Chromatography, 396*, 410–415.

Van Eenoo, P., Deventer, K., Roels, K., & Delbeke, F. T. (2006). Quantitative LC-MS determination of strychnine in urine after ingestion of a *Strychnos nux-vomica* preparation and its consequences in doping control. *Forensic Science International, 164*, 159–163.

Verardo, V., Gomez-Caravaca, A. M., Seura-Carretero, A., Caboni, M. F., & Fernández-Gutiérrez, A. (2011). Development of a CE-ESI-micro TOF-MS method for a rapid identification of phenolic compounds in buckwheat. *Electrophoresis, 32*, 669–673.

Verma, A., Hartonen, K., & Riekkola, M. L. (2008). Optimization of supercritical fluid extraction of indole alkaloids from *Catharanthus roseus* using experimental design methodology—Comparison with other extraction techniques. *Phytochemical Analysis, 19*, 52–63.

Verma, A., Laakso, I., Seppanen-Lakso, T., Huhtikangas, A., & Riekkola, M. L. (2007). A simplified procedure for indole alkaloid extraction from *Catharanthus roseus* combined with a semi-synthetic production process for vinblastine. *Molecules, 12*, 1307–1315.

Verpoorte, R. (2005). Alkaloids. In P. Worsfold, A. Townshend, & C. Poole (Eds.), *Encyclopedia of analytical science* (pp. 56–61), 2nd ed. Elsevier.

Verpoorte, R., & Baerheim Svendsen, A. (Eds.), (1984). *Chromatography of alkaloids, part II, GLC and HPLC*. In *Chromatography library*, Vol. 23B, (pp. 1–467). Amsterdam: Elsevier.

Verpoorte, R., Choi, Y. H., Mustafa, R. N., & Kim, H. K. (2008). Metabolomics: Back to basics. *Phytochemistry Reviews, 7*, 525–538.

Verpoorte, R., & Niessen, W. M. A. (1994). Liquid chromatography coupled with mass spectrometry in the analysis of alkaloids. *Phytochemical Analysis, 5*, 217–232.

Vieira, I. J. C., Medeiros, W. L. B., Monnerat, C. S., Souza, J. J., Mathias, L., Braz-Filho, R., et al. (2008). Two fast screening methods (GC-MS and TLC-ChEI assay) for rapid evaluation of potential anticholinesterase indole alkaloids in complex mixtures. *Annals of the Brazilian Academy of Sciences*, *80*, 419–426.

Wang, C., Han, D., Wang, Z., Zang, X., & Wu, Q. (2006). Analysis of *Strychnos* alkaloids in traditional Chinese medicines with improved sensitivity by sweeping micellar electrokinetic chromatography. *Analytica Chimica Acta*, *572*, 190–196.

Wang, P., Sun, H., Lv, H., Sun, W., Yuan, Y., Han, Y., et al. (2010). Thyroxine and reserpine-induced changes in metabolic profiles of rat urine and therapeutic effect of Liu Wei Di Huang Wan detected by UPLC-HDMS. *Journal of Pharmacology and Biomedical Analysis*, *53*, 631–645.

Wang, C. H., Wang, G. C., Wang, Y., Zhang, X. Q., Huang, X. J., Zhang, D. M., et al. (2012). Cytotoxic dimeric indole alkaloids from *Catharanthus roseus*. *Fitoterapia*, *83*, 765–769.

Wang, S. Y., Yang, L., Zu, Y. G., Zhao, C. J., Sun, X. W., Zhang, L., et al. (2011). Design and performance evaluation of ionic-liquids-based microwave-assisted environmentally friendly extraction technique for camptothecin and 10-hydroxycamptothecin from Samara of *Camptotheca acuminata*. *Industrial and Engineering Chemistry Research*, *50*, 13620–13627.

Wang, J., Zheng, H., Efferth, T., Wang, R., Shen, Y., & Hao, X. (2005). Indole and carbazole alkaloids from *Glycosmis Montana* with weak anti-HIV and cytotoxic activities. *Phytochemistry*, *66*, 697–701.

Wen, D., Li, C., Liu, Y., Liao, Y., & Liu, H. (2006). Determination of evodiamine and rutecarpine in human serum by liquid chromatography-tandem mass spectrometry. *Analytical and Bioanalytical Chemistry*, *385*, 1075–1081.

Xie, J., Zhu, L., Luo, H., Zhou, L., Li, C., & Xu, X. (2001). Direct extraction of specific pharmacophoric flavonoids from gingko leaves using a molecularly imprinted polymer for quercetin. *Journal of Chromatography. A*, *934*, 1–11.

Xu, Y. J., Foubert, K., Dhooghe, L., Lemiere, F., Cimanga, K., Mesia, K., et al. (2012). Chromatographic profiling and identification of two new iridoid-indole alkaloids by UPLC-MS and HPLC-SPE-NMR analysis of an antimalarial extract from *Nauclea pobenguinii*. *Phytochemistry Letters*, *5*, 316–319.

Yang, L., Wang, H., Zu, Y. G., Zhao, C., Zhang, L., Chen, X., et al. (2011). Ultrasound-assisted extraction of the three terpenoid indole alkaloids vindoline, catharanthine and vinblastine from *Catharanthus roseus* using ionic liquid aqueous solutions. *Chemical Engineering Journal*, *172*, 705–712.

Yilmaz, A., Nyberg, N. T., & Jaroszewski, J. W. (2012). Extraction of alkaloids for NMR-based profiling: Exploratory analysis of an archaic *Cinchona* bark collection. *Planta Medica*, *78*, 1885–1890.

Zanolari, B., Ndjoko, K., Ioset, J. R., Marston, A., & Hostettmann, K. (2003). Qualitative and quantitative determination f yohimbine in authentic yohimbe bark and in commercial aphrodisiacs by HPLC-UV-API/MS methods. *Phytochemical Analysis*, *14*, 193–201.

Zhang, J., Yu, Y., Liu, D., & Liu, Z. (2007). Extraction and composition of three naturally occurring anti-cancer alkaloids in *Camptotheca acuminata* seed and leaf extracts. *Phytomedicine*, *14*, 50–56.

Zhao, C., & He, C. (2006). Preparative isolation and purification of atractylon and atractylenolide III from the Chinese medicinal plant *Atractylodes macrocephala* by high-speed counter-current chromatography. *Journal of Separation Science*, *29*, 1630–1636.

Zhou, H., Tai, Y., Sun, C., & Pan, Y. (2005). Rapid identification of vinca alkaloids by direct-injection electrospray ionization tandem mass spectrometry and confirmation

by high-performance liquid chromatography-mass spectrometry. *Phytochemical Analysis*, *16*, 328–333.
Zhu, Q. H., Huang, D. D., Li, L., & Yin, Y. G. (2010). Synthesis of molecularly imprinted polymers for the application of selective clean-up vinblastine from *Catharanthus roseus* extract. *Science China Chemistry*, *53*, 2587–2592.
Zong, Y. Y., & Che, C. T. (1995). Determination of strychnine and brucine by capillary zone electrophoresis. *Planta Medica*, *5*, 456–458.

CHAPTER TEN

Like Cures Like: Caffeine Immunizes Plants Against Biotic Stresses

Hiroshi Sano[*,†,1], **Yun-Soo Kim**[†,2], **Yong-Eui Choi**[†]
[*]Research and Education Center for Genetic Information, Nara Institute of Science and Technology, Nara, Japan
[†]Department of Forest Resources, Kangwon National University, Chuncheon, Republic of Korea
[1]Corresponding author: e-mail address: sano@gtc.naist.jp
[2]Present address: R&D Headquarters, Korea Ginseng Corporation, Daejeon, Korea

Contents

1. Introduction 274
2. Caffeine Biosynthesis 276
3. Caffeine-Producing Transgenic Plants 278
   3.1 Transgenic tobacco and chrysanthemum 278
   3.2 Caffeine production 279
4. Biotic Stress Tolerance 280
   4.1 Anti-herbivores 280
   4.2 Anti-pathogens 282
5. Activation of Defence System 283
   5.1 Defence-related genes and HR 283
   5.2 Induction of SA 284
6. Caffeine Targets 284
   6.1 Cyclic nucleotide PDE 285
   6.2 GABA receptor 286
   6.3 Adenosine receptors 287
7. Caffeine Signal Cascade 287
   7.1 Cyclic AMP 287
   7.2 Calcium 288
   7.3 Salicylic acid 289
   7.4 The overall route 289
   7.5 γ-Aminobutylic acid 291
8. Plant Immunization 291
9. Concluding Remarks 293
Acknowledgements 295
References 295

Advances in Botanical Research, Volume 68
ISSN 0065-2296
http://dx.doi.org/10.1016/B978-0-12-408061-4.00010-9

Abstract

Caffeine (1,3,7-trimethylxanthine) is a member of purine alkaloids and produced in over 80 plant species. It is one of the oldest and widely used secondary metabolites by mankind as stimulant and ingredient in drugs. Its physiological function in nature has not completely been elucidated but is thought to participate in the chemical defence against biotic attackers. To substantiate this idea, transgenic tobacco and chrysanthemum were constructed by expressing three distinct *N*-methyltranferases involved in the caffeine biosynthesis pathway. Resulting plants produced a low amount of caffeine (0.4–5 μg/g tissue) yet exhibited strong tolerance against herbivores and pathogens. Their self-defence system was autonomously activated without perceiving external stresses. This can be regarded as the priming of defence response, by which host plants become on standby to cope with a broad range of biotic stresses. The feature resembles mammalian immunization or vaccination, and it was proposed that plants can also be immunized by expressing a mildly toxic 'antigenic' chemical such as caffeine *in planta*. The caffeine signal was predicted to be successively transduced through phosphodiesterase, cyclic AMP, calcium flux and salicylic acid.

ABBREVIATIONS

BABA β-aminobutylic acid
CDPK calcium-dependent protein kinase
CNGC cyclic nucleotide-gated channel
GABA γ-aminobutylic acid
HR hypersensitive response
ICS isochorismic synthase
PAL phenylalanine ammonia lyase
PDE phosphodiesterase
SA salicylic acid
VOC volatile organic compound

1. INTRODUCTION

The first record of coffee utilization comes from Yemen in the sixth century. After the fifteenth century, coffee was widely drunk to shake off drowsiness in the Arabic society. In the seventeenth century, it was introduced into Europe and became one of the most popular beverages thereafter. In order to obtain coffee beans, international conflicts often happened during the eighteenth century. The unique component of the coffee plant is caffeine, which was first isolated by a German chemist Runge in 1819 (Weinberger & Bealer, 2002).

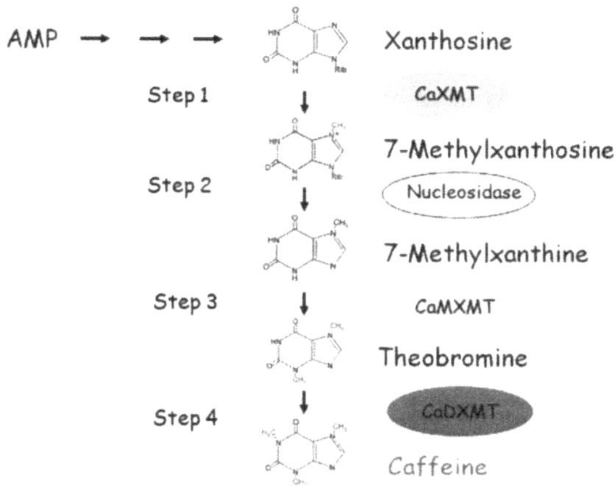

Figure 10.1 Caffeine biosynthesis in coffee plant. Caffeine is successively synthesized from the precursor, xanthosine through three methylation and one ribose removal steps. The first (1), third (3) and fourth (4) steps are N-methylation, and the second step (2) is ribose removal. The enzymes involved are *Coffea arabica* xanthosine methyltransferase (CaXMT) (step 1), 7-methylxanthosine nucleosidase and/or xanthosine methyltransferase (CaXMT) (step 2), 7-methylxanthine methyltransferase (CaMXMT) (step 3) and 3,7-dimethylxanthine (theobromine) methyltransferase (CaDXMT) (step 4). (See colour plate.)

Caffeine (1,3,7-trimethylxanthine) (Fig. 10.1) is a typical purine alkaloid and naturally produced in seeds, flowers and leaves of over 80 plant species. The best known producers include coffee (*Coffea arabica*), tea (*Camellia sinensis*), maté (*Ilex paraguariensis*), guaraná (*Paullinia cupana*), cola (*Cola nitida*) and cacao (*Theobroma cacao*) (Ashihara & Crozier, 1999). The average amount was estimated to be around 1–4% of dried tissues (Ashihara & Crozier, 2001), this roughly corresponds to ~10 mg/g fresh tissue of coffee leaves (Ogita, Uefuji, Yamaguchi, Koizumi, & Sano, 2003). Caffeine is also produced at lower level in a variety of plants such as the *Citrus* family (Kretschmar & Baumann, 1999; Stewart, 1985).

The physiological role of caffeine in nature is not completely determined yet, but ecochemical functions have so far been proposed (Ashihara & Crozier, 2001; Baumann, 2006). One is chemical defence against pathogens and herbivores. Exogenously applied caffeine was shown to be effective not only as a repellent and pesticide for moth larvae (Nathanson, 1984) but also to disturb the reproductive ability of several species of moths (Mathavan, Premalatha, & Christopher, 1985). Topical treatment of cabbage leaves with caffeine significantly reduced feeding by the slugs and was lethal to the snails

(Hollingsworth, Armstrong, & Campbell, 2002). Growth of bacterial and fungal pathogens was considerably reduced on the caffeine-containing culture medium (Kim & Sano, 2008). The other possible function of caffeine is allellopathic against competing plant species (Chou & Waller, 1980). Caffeine was shown to suppress germination of *Amaranthus spinosus* (Rizvi, Mukerji, & Mathur, 1981) and growth of *Arabidopsis* and tobacco seedlings (Mohanpuria & Yadav, 2009).

However, it is not clear whether naturally occurring caffeine functions as similarly as caffeine tested *in vitro*. One of the puzzling phenomena is that certain pathogens and herbivores normally attack caffeine-producing plants. For example, coffee plants greatly suffer from a fungal pathogen, coffee rust (*Hemileia vastatrix*), and tea plants are severely damaged by diverse insects including moth caterpillars and mites. Such pests are perhaps specialists, which particularly developed tolerant traits against caffeine. And yet these observations imply that caffeine may have other functions in addition to the direct toxicant against bio-attackers. In this chapter, we summarize the current understanding of biological functions of caffeine based on available literatures and experiments using transgenic plants which produced caffeine *in planta*.

2. CAFFEINE BIOSYNTHESIS

Caffeine is a derivative of xanthosine, an intermediate of purine catabolic pathways, and its biosynthetic pathway involves successive use of purine precursors such as AMP and GMP through multiple steps catalyzed by specified enzymes (Fig. 10.1; Ashihara, Sano, & Crozier, 2008). The first step of the final stage of caffeine synthesis is methylation of xanthosine by xanthosine methyltransferase (XMT), yielding 7-methylxanthosine (step 1). Its ribose moiety is then removed by 7-methylxanthosine nucleosidase (step 2). The resulting 7-methylxanthine is methylated at the 3-N position by 3-*N*-methylxanthine methyltransferase (MXMT or theobromine synthase), producing 3,7-dimethylxanthine (theobromine) (step 3). The 3,7-dimethylxanthine is further methylated at the 1-N position by 3,7-dimethylxanthine methyltransferase (DXMT or caffeine synthase) to give caffeine itself (1,3,7-trimethylxanthine) (step 4). All methylation reactions require *S*-adenosyl-L-methionine as a methyl donor.

Among four enzymes required for conversion of xanthosine into caffeine, methyltransferases are specific to the pathway, while nucleosidase appears to be common with broad substrate properties *in planta* (Uefuji, Ogita, Yamaguchi, Koizumi, & Sano, 2003). Identification and isolation of methyltransferases are accordingly critical to understand and reconstitute the caffeine biosynthesis pathway. To this end, genes encoding three methyltransferases were intensively searched from plants of the coffee family by PCR and library screening methods and finally successfully identified (Mizuno, Kato, et al., 2003; Mizuno, Okuda, et al., 2003; Ogawa, Herai, Koizumi, Kusano, & Sano, 2001; Uefuji et al., 2003). To date, seven cDNAs have been characterized from coffee plants; one for xanthosine methyltransferase (XMT/XRS), three for 7-methylxanthine methyltransferase or theobromine synthase (MXMT/CTS) and three for 3,7-dimethylxanthine methyltransferase or caffeine synthase (DXMT/CCS) (Ashihara et al., 2008; Kato & Mizuno, 2004).

A single gene for xanthosine methyltransferase (XMT/CmXRS) encodes a polypeptide consisting of 372 amino acids with an apparent molecular mass of 41.8 kDa. Notably, this enzyme was found to possess 7-methylxanthosine nucleosidase activity (McCarthy & McCarthy, 2007). It is expressed almost uniformly in aerial tissues of coffee plant (*C. arabica*), including leaves, floral buds and immature beans. In contrast, at least three genes encoding 7-methylxanthine methyltransferase (theobromine synthase) have been isolated. The number of amino acids in the putative polypeptides are 378 for MXMT1 (42.7 kDa) and 384 for MXMT2 and CTS2 (43.4 kDa). They differ by insertion or deletion of blocks of several residues in the C-terminal region. Their catalytic properties as judged from kinetic parameters, such as K_m values, are apparently distinct from each other. They are expressed in young leaves, floral buds and immature beans. Three genes were also identified for 3,7-dimethylxanthine methyltransferase (caffeine synthase); *DXMT*, *CCS1* and *CtCS7*, each encoding a 43-kDa polypeptide consisted of 384 amino acids. However, their kinetic properties differ with, for example, DXMT and CCS1 showing K_m values for theobromine of 1200 and 157 mM, respectively. Expression profiles are also distinct, *DXMT* being expressed exclusively in immature beans, while *CCS1* expression is ubiquitous, occurring in all tissues. The presence of isoforms of these enzymes with different properties suggests that caffeine is synthesized through multiple pathways depending on availability and concentration of the substrates (Mizuno, Kato et al., 2003; Mizuno, Okuda, et al., 2003; Mizuno, Tanaka, Kato, Ashihara, & Fujimura, 2001; Ogawa et al., 2001; Uefuji et al., 2003).

The deduced polypeptides of these enzymes have more than 82% similarity, and phylogenetic analysis indicates that they are more closely related to *C*-methyltransferases, including those for jasmonic acid, salicylic acid (SA) and benzoic acid, than to other *N*-methyltransferases. This suggests that coffee *N*-methyltransferases constitute a distinct sub-group within the plant methyltransferase family. Their cellular localization was determined by the green fluorescence protein fusion method, and this showed that all three enzymes are localized in the cytoplasm (Kodama, Shinya, & Sano, 2008; Ogawa et al., 2001). Caffeine, thus, appears to be synthesized in the cytoplasm and translocated to vacuoles via not yet identified mechanisms.

3. CAFFEINE-PRODUCING TRANSGENIC PLANTS

The physiological function of caffeine has been proposed to constitute a part of chemical defence systems against pathogens and herbivores (Ashihara & Crozier, 1999; Ashihara et al., 2008). Exogenously applied caffeine has been shown to markedly increase the resistance of food plants of several pests, affecting their growth and survival (Kim, Uefuji, Ogita, & Sano, 2006; Nathanson, 1984). Base on these observations, caffeine-producing transgenic plants, which originally do not synthesize the compound, were constructed and examined for tolerance traits against biotic attackers (Uefuji et al., 2005).

3.1. Transgenic tobacco and chrysanthemum

A multi-gene expression vector for the three coffee *N*-methyltransferase genes was first constructed (Uefuji et al., 2005). Practically, each of *CaXMT1*, *CaMXMT1/2* and *CaDXMT1* genes was initially independently introduced into pBI221, and then individual expression cassettes containing the CaMV *35S* promoter, cDNA clone and *NOS* terminator were removed and successively inserted into the multiple cloning site of pBluescript II SK (−). The three connected cassettes were finally replaced with the GUS coding sequence and *NOS* terminator of pIG121Hm (Hiei, Ohta, Komari, & Kumahsiro, 1994) and designated as pBIN-NMT777 (Fig. 10.2).

Tobacco (*Nicotiana tabacum* cv. Xanthi) leaf discs were transformed with pBIN-NMT777 by the *Agrobacterium* transformation method. After appropriate culture and selection, 23 kanamycin-resistant transgenic plantlets were obtained, among which 15 were confirmed by RT-PCR to express all three *N*-methyltransferase genes (Uefuji et al., 2005). The stature of mature plants was apparently normal (Fig. 10.2). Transgenic chrysanthemum plants (*Chrysanthemum morifolium* cv. Shinba) were similarly constructed by introducing

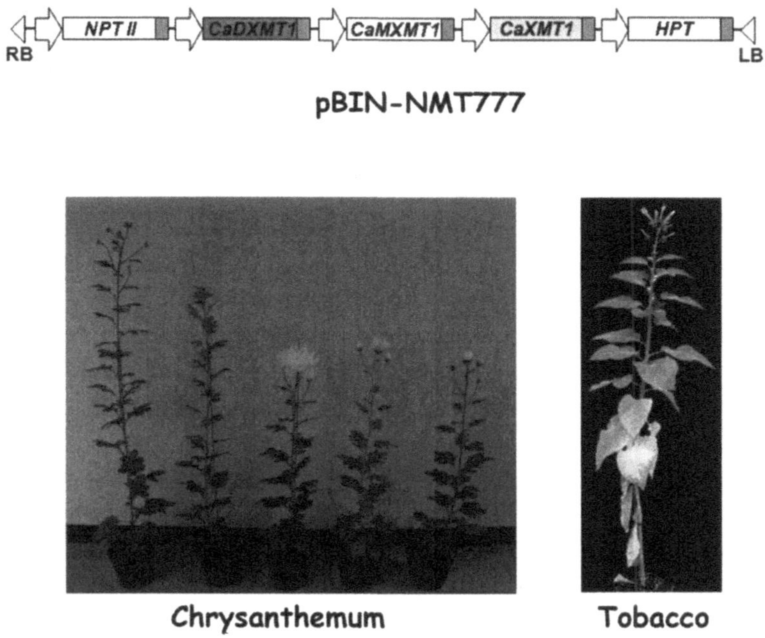

Figure 10.2 Transgenic plants-producing caffeine. A multi-gene transfer vector (pBIN-NMT777) was constructed using *CaXMT1*, *CaMXMT1* and *CaDXMT1*, all were driven by the *35S* (cauliflower mosaic virus *35S* RNA) promoter and *NOS* (nopaline synthase) terminator. The T-DNA region of the plasmid also contained genes for neomycin phosphotransferase (*NPT II*) and hygromycin phosphotransferase (*HPT*) as the antibiotic-resistance marker. The promoter is indicated by open arrows, and the terminator is indicated by shaded squares. The left and right borders are indicated by LB and RB, respectively (upper illustration) (Uefuji et al., 2005). The pBIN-NMT777 was introduced into chrysanthemum (bottom left) (Kim, Lim, Yoda, et al., 2011) and tobacco (bottom right) (Uefuji et al., 2005). Mature plants were apparently normal, but in chrysanthemum, dwarfism and early flowering were occasionally observed. (See colour plate.)

the multi-gene expression vector (pBIN-NMT777) into leaf discs via *Agrobacterium*-mediated transformation. After antibiotic selection, eight kanamycin-resistant transgenic plantlets were obtained, among which six were confirmed by the RT-PCR to express all three *N*-methyltransferase genes (Kim, Lim, Yoda, et al., 2011). A notable feature of the transgenic chrysanthemum was a phenotypic alteration at maturation, showing dwarfism and early flowering (Fig. 10.2).

3.2. Caffeine production

The selected lines were grown to maturity, and accumulation of purine alkaloids in leaves was examined by HPLC. Initial analysis using mature leaf samples of transgenic tobacco showed an efficient accumulation of caffeine and theobromine (Uefuji et al., 2005). Subsequent analysis indicated that, in immature leaves, caffeine was often undetectable whereas, in mature leaves, the average caffeine content was 0.2 μg/g fresh weight. When plants aged and entered the reproductive stage to form flower buds, caffeine content increased to over 5 μg/g fresh weight. Immature fruits contained caffeine at a rather low level, up to 1.3 μg/g fresh weight. No caffeine was detected in any parts of control plants. The results indicated that caffeine was indeed synthesized in transgenic tobacco leaves and that its content was higher in the older leaves of plants in the reproductive growth phase (Uefuji et al., 2005). Caffeine production was similarly examined in fully matured leaves of transgenic chrysanthemum by HPLC (Kim, Lim, Yoda, et al., 2011). In three transgenic lines, caffeine was accumulated at 3 μg/g fresh weight of the tissue, being comparable with values in transgenic tobacco plants (Kim & Sano, 2008; Uefuji et al., 2005). Results indicated that the introduced genes were actively transcribed, resulting in the efficient production of caffeine at the level of up to 5 μg/g fresh weight of the tissue.

4. BIOTIC STRESS TOLERANCE

Transgenic plants were tested for their tolerant traits against various biotic stresses including herbivore and pathogen attacks.

4.1. Anti-herbivores

As the representative herbivores, larvae of tobacco cutworm (*Spodoptera litura*) and beet armyworm (or small mottled willow moth; *Spodoptera exigua*) were selected. Both are known as severe pests for many crop plants in nature. Tobacco cutworm caterpillars at the third instar were starved for several hours and then allowed to select and feed on leaf discs prepared from transgenic or control tobacco plants. When fresh leaves producing caffeine at 5 μg/g tissue were subjected to a choice test together with non-caffeine-producing wild-type leaves, caterpillars positively avoided the transgenic sample, eating less than 0.02 cm^2 (1% of given leaves) (Fig. 10.3). In contrast, they preferentially ate up to 1.1 cm^2 of control wild-type leaves (50% of given leaves) (Fig. 10.3). The repellent effect was also observed with low caffeine content leaves (0.4 μg/g tissue), larvae eating only 4% of the transgenic, whereas up to 32% of the wild-type leaves (Uefuji et al., 2005). The similar result was obtained with caterpillars of beet armyworm feeding on

Figure 10.3 Biotic stress tolerance. *Resistance to insect larvae*: Tobacco cutworm (*S. litura*) larvae at the third instar were subjected to a choice test between three leaf discs from caffeine producing (5 μg/g fresh tissue) (TR) and three leaf discs from wild-type (WT) tobacco plants (Uefuji et al., 2005). *Resistance to aphids*: The third instar cotton aphids were subjected to a choice test between transgenic (3 μg/g fresh tissue) (TR) and wild-type (WT) chrysanthemum whole plants for a week, then leaf detached and photographed (Kim, Lim, Kang, et al., 2011). *Resistance to fungal pathogen*: Detached healthy mature leaves of transgenic (3 μg/g fresh tissue) (TR) and wild-type (WT) chrysanthemum plants were inoculated with spores of grey mould (*B. cinerea*), incubated at 20 °C for 15 days and photographed (Kim, Lim, Yoda, et al., 2011). *Resistance to bacterial pathogen*: Healthy leaves from transgenic (2.3 μg/g fresh tissue) (TR) and wild-type (WT) tobacco plants were inoculated with *P. syringae* pv. *glycinea* and necrotic lesion was observed 24 h later (Kim & Sano, 2008). *Resistance to viral pathogen*: Healthy leaves from transgenic (2.3 μg/g fresh tissue) (TR) and wild-type (WT) tobacco plants were inoculated with tobacco mosaic virus, kept at 30 °C for 2 days and then transferred to 20 °C (temperature shift). Lesions on whole leaf were observed 3 days after temperature shift (Kim & Sano, 2008). (See colour plate.)

transgenic chrysanthemum producing caffeine at 3 μg/g tissue. When the second instar caterpillars were subjected to a non-choice test, they vigorously fed on wild-type leaves eating up to 4.4 mm^2, while they positively avoided transgenic leaves, feeding less than 1.5 mm^2 (Kim, Lim, Kang, et al., 2011).

Aphids are another serious pest for agriculture and forestry, giving a severe damage by sucking sap. When the third instar cotton aphid (*Aphis gossypii*) was subjected to a choice test on wild-type and transgenic mature chrysanthemum plants, more than 80% of the insect preferred the wild-type leaves (27/33), whereas less than 20% gathered on the transgenic leaves (6/33) (Fig. 10.3; Kim, Lim, Kang, et al., 2011). These experiments clearly demonstrated that caffeine was effective in repelling the herbivore.

4.2. Anti-pathogens

More than 70% of known plant disease is caused by fungi, and 30% is caused by bacteria, viruses and other pathogens. The caffeine-producing transgenic tobacco and chrysanthemum plants were examined for response to each pathogen. Chrysanthemum plants were infected with a necrotrophic fungus grey mould (*Botrytis cinerea*), which causes death on flower, leaves, buds and fruits of many plant species (Kim, Lim, Yoda, et al., 2011). In the wild-type plant, lesions appeared 72 h after inoculation and rapidly developed from the infected site to outward leaves (Fig. 10.3). The lesion size exceeded 16 mm in diameter 5 days after infection. In the transgenic lines, the lesion appeared 90 h after inoculation, and the lesion size was smaller than the control, varying between 1 and 9 mm in diameter (Fig. 10.3).

Resistance against microbial pathogens was then examined (Kim & Sano, 2008). Healthy leaves of wild-type and transgenic tobacco plants were inoculated with *Pseudomonas syringae* pv. *glycinea*, which causes wild-fire disease in many plant species. In wild-type plants, distinct lesions were formed 24 h after infection and consistently developed into severe necrosis up to 48 h. In contrast, lesion development was remarkably inhibited in the transgenic line even 48 h after inoculation (Fig. 10.3). The number of propagated bacteria was lower in the transgenic line than in the control, showing 6.3×10^7 in wild type after 48 h infection, while 1.6×10^7 in the transgenic line, 1/4 that of the control.

Resistance against viral pathogens was also examined (Kim & Sano, 2008). Tobacco mosaic virus (TMV) has a wide host range over 120 plant species, and causes mottled patterns on leaves, ultimately leading to plant death. When healthy leaves of wild-type and transgenic tobacco plants were inoculated with TMV and kept at 30 °C, plants do not recognize infection and virus particles propagate. Upon shifting to 23 °C (temperature shift), the hypersensitive response (HR) takes place, and a series of defence system begins to operate. Physiologically, these responses can be visibly estimated by formation and development of necrotic lesions (Fig. 10.3). In wild-type plants, lesions appeared 48 h after temperature shift, further developing up to 48 h. In transgenic lines, lesions were similarly formed 48 h after temperature shift but did not develop further. In addition to slow lesion formation, their size in transgenic lines was much smaller than in the control (Fig. 10.3). The total number of lesions was also few in transgenic lines, showing only 15–30% of the control. These results pointed to the effectiveness of caffeine to confer tolerance against a wide range of pathogens.

5. ACTIVATION OF DEFENCE SYSTEM

Exogenously applied caffeine is toxic for diverse organisms at an average concentration between 0.01% and 0.3% (w/v solution) (Kim, Choi, & Sano, 2010). For example, caffeine up to 0.5% exhibited direct repelling effects on Lepidoptera caterpillars (Mathavan et al., 1985). Cabbage leaves sprayed with 0.1% caffeine solution were lethally toxic for snails and slugs (Hollingsworth et al., 2002). Caffeine concentration between 0.05% and 0.5% strongly inhibited the growth of pathogenic microbes, such as *Aspergillus ochraceus* (Tsubouchi, Terada, Yamamoto, Hisada, & Sakabe, 1985), cocoa pathogenic fungus, *Crinipellis perniciosa* (Aneja & Gianfagna, 2001) and bacterial pathogen, *P. syringae* (Kim & Sano, 2008).

The amount of caffeine produced in transgenic plants was at most 5 μg/g fresh weight. This value roughly corresponds to 5×10^{-4}%, being two to three orders of magnitude lower than those examined *in vitro*. And yet the transgenic plants showed almost equivalent tolerance against pests and pathogens to exogenously caffeine-treated plants. This apparent discrepancy in effective concentration raised a question whether or not caffeine in transgenic plants was directly toxic for organisms. It is rather conceivable that endogenously produced caffeine induced some chemical changes in leaves, thereby indirectly affecting plants' defence responses.

5.1. Defence-related genes and HR

Resistance against pathogens and herbivores is frequently associated with elevated expression of defence-related genes. One of such genes is *proteinase inhibitor-II* (*PI-II*) encoding a proteinous defence factor, which causes digestion dysfunction in larvae gut (Green & Ryan, 1972). Since it is expressed upon herbivore attack in many plant species, accumulation of its transcripts has been used as a hallmark of the onset of defence reaction (Green & Ryan, 1972). The status of *PI-II* expression was examined in tobacco leaves producing caffeine 2 μg/g tissue and found that its transcripts were constitutively accumulated regardless of the herbivore attack (Kim & Sano, 2008). Other representative for the defence onset is genes that encode pathogenesis-related (PR) proteins. Expression of *PR-1a* and *PR-2* is markedly induced upon pathogen infection (Ryals et al., 1996). Transcripts of *PR-1a* were constitutively accumulated in leaves of the transgenic tobacco without pathogen infection (Kim & Sano, 2008). Transcripts of *PR-2*, which encodes β-1,3-glucanase, were also constitutively accumulated in transgenic chrysanthemum under the non-stressed condition (Kim, Lim, Yoda, et al., 2011). Both genes were silent in the wild-type plants. These observations suggested that, in transgenic plants, a common self-defence system was autonomously activated in the absence of external stimuli.

5.2. Induction of SA

One of the hallmarks in plant defence response is SA, which simultaneously activates many defence-related genes including *PR-1a* and *PR-2* (Raskin, 1992). As transcripts of these *PR* genes were accumulated in transgenic lines, the status of SA level was then examined. In mature leaves of wild-type chrysanthemum, the amount of SA and its conjugate, salicylic acid glucoside (SAG), was 0.1 and 2.3 μg/g fresh weight, respectively, without pathogen attack. These values are comparable with those found in tobacco plants (Seo, Katou, Seto, Gomi, & Ohashi, 2007). In mature leaves of the transgenic chrysanthemum, levels of SA and SAG were 2.7 and 5.5 μg/g fresh weight, respectively, being constitutively 2.5-fold higher than that in the wild type (Kim, Lim, Yoda, et al., 2011). The increase in SA is also comparable with that found in tobacco leaves infected with TMV for 2 days (Seo, Seto, Koshino, Yoshida, & Ohashi, 2003). It is thus conceivable that endogenous caffeine stimulated production and/or deposition of SA and SAG, which possibly activated a series of defence reactions even under non-stressed conditions.

6. CAFFEINE TARGETS

The series of experiments showed that caffeine produced *in planta* stimulates the host defence against biotic stresses and that such a stress tolerance is, at least partly, mediated by an elevated level of SA. The molecular mechanism is currently not clear. The situation is figuratively comparable with a labyrinth, in which only the entrance (caffeine) and the exit (tolerant trait) are distinct. In order to find clues to predict the correct path(s), we screened documents describing diverse molecular events involved in stress signalling. In this section, we attempt to assemble such individual findings and propose a possible route connecting the entrance to the exit.

Caffeine and its derivatives antagonistically bind to multiple target molecules, thereby causing distinct physiological effects. Pharmacological analyses with mammalian cells have revealed three major molecular targets: cyclic nucleotide phosphodiesterases (PDE), γ-aminobutylic acid (GABA) receptors and adenosine receptors (Fredholm, Battig, Holmen, Nehlig, & Zvartau, 1999). PDEs hydrolyze cyclic nucleotides such as 3′,5′-cyclicAMP (cAMP), functioning as a molecular switch of cyclic nucleotide signalling pathway (Conti, 2000). GABA-receptors bind GABA, which plays an inhibitory role in nerve signal transduction (Goetz, Arslan, Wisden, & Wulff, 2007). Adenosine receptors bind adenosine and couple to G-proteins. They primarily control levels of cAMP, which serves as the second messenger in a broad range of signalling response (Ralevic & Burnstock, 1998). These caffeine targets have been established to be critical in neurotransmission in mammalian cells (Daly, 2007).

6.1. Cyclic nucleotide PDE

In mammals, PDE constitutes a superfamily, being classified into 11 groups. PDE1, PDE2, PDE3, PDE10 and PDE11 hydrolyze both cAMP and cGMP, whereas PDE4, PDE7, PDE8 specifically hydrolyze cAMP and PDE5, PDE6, PDE9 are specific to cGMP (Bender & Beavo, 2006). They negatively function in the cyclic nucleotide-mediated signalling pathways. Structurally, they are consisted of approximately 250 amino acids, with the N-terminal regulatory region and the C-terminal catalytic region (Francis, Blount, & Corbin, 2011). The catalytic site is well conserved among PDEs and contains the purine-binding pocket, to which methylxanthines are preferentially bound.

In contrast to mammalian PDEs, plant PDEs have been less characterized. Early studies with partially purified enzyme preparations have given controversial results in regard to substrate specificity, optimal pH and sensitivity to methylxanthines (Newton, Roef, Witters, & Van Onckelen, 1999; Newton & Smith, 2004). For example, samples from pea seedlings (Lin & Varner, 1972), bean seedlings (Dupon, Van Onckelen, & De Greef, 1987), soybean (Brewin & Northcote, 1973) and carrot (Niles & Mount, 1974) were insensitive to methylxanthine derivatives. This even raised a question whether or not plants possess a cyclic nucleotide-mediated signalling system

(Amrhein, 1977). However, further studies have indicated that plant PDEs are present in multiforms. PDEs isolated from spinach (Brown, Edwards, Newton, & Smith, 1979), common bean (*Phaseorus vulgaris*) (Brown, Alnajafi, & Newton, 1977), pea root (Chiatante, Newton, Crignola, Levi, & Brown, 1990) and black gram (*Vigna mungo*) (Lee & Abidin, 1989) exhibited a similar properties to mammalian enzymes, preferentially hydrolyzing 3′,5′-cAMP and being sensitive to methylxanthines. These findings suggest that some plant PDEs commonly function as mammalian counterparts (Brown & Newton, 1981) and are well targeted by caffeine and its derivatives.

6.2. GABA receptor

GABA is a non-protein amino acid and, in vertebrates, it serves as the major inhibitory neurotransmitter (Goetz et al., 2007). It is perceived through three major receptor proteins: $GABA_A$, $GABA_B$ and $GABA_C$, among which $GABA_A$ receptor is directly targeted by methylxanthines (Daly, 2007). The $GABA_A$ receptor is involved in Cl^- ion flux (Goetz et al., 2007). It is composed of several subunits, each with approximately 430 amino acids (50–60 kDa), and containing four transmembrane regions and regulatory sites (Olsen & Tobin, 1990). Xanthine was suggested to bind to the benzodiazepine (BZ)-positive modulatory site (Daly, 2007). The BZ site is distinct from the GABA-binding site, situated at the interface between α and γ subunits (Goetz et al., 2007). When positive antagonist such as BZ binds to the BZ site, the shape of receptor oligomers changes, and efficiency of GABA increases (Goetz et al., 2007). Caffeine was suggested to block the effect of benzodiazepine, resulting in decrease of the GABA efficiency (Shi, Padgett, & Daly, 2003).

In plants, GABA has been suggested to be one of the important factors in stress response (Bown & Shelp, 1997). It is involved in C/N balancing, cytosolic pH adjust, oxidative stress protection, defence against herbivores, osmoregulation and signalling (Bouche & Fromm, 2004). To date, however, no counter part of $GABA_A$ receptor has been found from plants. Using agonists and antagonists, the presence of $GABA_A$- and $GABA_B$-like receptors were circumstantially suggested in duckweed (*Lemna minor*), but no biochemical evidence was available (Kinnersley & Lin, 2000). A preliminary BLAST search indicated that the N-terminal regions of *Arabidopsis* glutamate receptor are related to $GABA_B$ receptors (Turano, Panta, Allard, & van Berkum, 2001). Quantum dots analysis also suggested the presence of $GABA_B$ receptor-like protein in tobacco pollen tube (Yu & Sun, 2007).

$GABA_B$ receptors are distinct from $GABA_A$ receptors having seven transmembrane regions and, upon coupling with G-proteins, they stimulate opening of K^+ channels (Bettler, Kaupmann, Mosbacher, & Gassmann, 2004). $GABA_B$ receptors are possibly not antagonized by methylxanthines (Parramon, Gonzalez, Herrero, & Oset-Gasque, 1995).

6.3. Adenosine receptors

In vertebrates, extracellular adenosine is a critical signalling molecule that mediates diverse metabolic pathways (Fredholm et al., 1994). It is perceived through adenosine receptors, which are grouped into four types, A1, A2a, A2b and A3. All of them function through coupling with G-proteins. They are composed of ~320–400 amino acids and contain seven transmembrane (TM) domains. TMI and TMVII form a barrel structure, which serves as the site of ligand binding (Fredholm et al., 1994). Adenosine receptors modulate adenylate cyclase activity, causing decrease (A1 and A3) or increase (A2a and A2b) in cAMP production (Ralevic & Burnstock, 1998). Xanthines and derivatives are nonselective antagonists, opposing adenosine receptor activation (Fredholm et al., 1999; Ribeiro & Sebastiao, 2010).

Plant adenosine receptors have so far not been documented. Far from that, adenosine itself has not been characterized as a signalling molecule in plants. Exogenously applied L-adenosine was shown to increase apoplastic ion concentration (Ries, Savithiry, Wert, & Widders, 1993), but, to our knowledge, no report is available describing the effect of naturally occurring D-adenosine.

7. CAFFEINE SIGNAL CASCADE

So far, the only target of caffeine known in plants is cyclic nucleotide PDE. Other two potential targets, $GABA_A$ receptors and adenosine receptors, are even not identified. Here, we discuss the possible caffeine signal cascade which starts from PDE.

7.1. Cyclic AMP

Involvement of PDE and its molecular substrate, cAMP, in plant stress response has repeatedly been reported. For example, when Arabidopsis was treated with isobutyl-1-methylxanthine (IBMX), a powerful inhibitor of PDE, the level of cAMP increased and onset of the HR was hastened (Ma et al., 2009). When cell suspension culture of alfalfa (*Medicago sativa*) was challenged by a glycoprotein elicitor derived from a pathogenic fungus, *Verticillium albo-atrum*, adenylate cyclase activity and eventually cAMP level increased within a few minute (Cooke, Smith, Walton, & Newton, 1994).

Cyclic AMP is considered to be the second messenger in plant signal transduction pathways (Assmann, 1995). It is involved in many physiological processes: ion fluxes, chloroplast development, pathogen response and gene transcription (Martinez-Atienza, van Ingelgem, Roef, & Maathuis, 2007). A critical role of cAMP in plant defence was substantially shown in different experimental systems. Generally, pathogen attack triggers the increase of cAMP, which stimulates production of diverse defence-related molecules. Treatment of *Arabidopsis* with a toxin, which was derived from a pathogenic fungus, *Verticillium dahliae*, resulted in increase of endogenous levels of cAMP and SA (Jiang, Fan, & Wu, 2005). When cultured cells of Mexican white cedar (*Cupressus lusitanica*) were elicited, cAMP increased up to five-fold followed by the increase of phytoalexins (Zhao, Guo, Fujita, & Sakai, 2004). Subsequently, cAMP was directly shown to control the defence activity. When alfalfa seedlings were treated with a cAMP analogue, dibutyryl cAMP, phenylalanine ammonia lyase (PAL) activity and phyto-alexin content increased (Cooke et al., 1994). An *Arabidopsis* mutant lacking a cAMP-gated cation channel did not display the HR upon pathogen infection (Clough et al., 2000).

7.2. Calcium

The downstream of cAMP in plants is not necessarily clear, but one of such components was suggested to be the cyclic nucleotide-gated channels (CNGCs). CNGCs are primary signalling molecules sensing extracellular stimuli (Talke, Blaudez, Maathuis, & Sanders, 2003). In particular, their role in controlling the Ca^{2+} flux and pathogen response is crucial (Ma & Berkowitz, 2011). Plant CNGC consists of ~700 amino acids with relative molecular mass of 80–85 kDa (Leng, Mercier, Yao, & Berkowitz, 1999). They contain six transmembrane structures and C-terminal cAMP-binding domain. When cAMP is bound, the transmembrane channel opens and Ca^{2+} flows into cytosol (Kaplan, Sherman, & Fromm, 2007; Talke et al., 2003). Cytosolic calcium is monitored by calcium sensors, typically represented by calcium-dependent protein kinases (CDPKs). Upon binding calcium, CDPK is enzymatically activated and switches on the phosphorylation signalling cascade, resulting in activation of diverse calcium-dependent reactions (Harmon, Gribskov, & Harper, 2000).

7.3. Salicylic acid

One of the key elements in defence reaction is SA. SA directly and indirectly regulates production of not only defence molecules but also many physiologically important components (Rivas-San Vicente & Plasencia, 2011). The intracellular level of SA fluctuates depending upon external stresses, and cAMP was shown to be a critical factor that controls SA level. For instance, when Arabidopsis seedlings were treated with an adenylate cyclase activator forskolin, the level of SA and expression of pathogenesis-related protein-1 (PR-1) increased, whereas treatment with adenylate cyclase inhibitor, 2′,5′-dideoxyadenosine, reduced SA levels (Jiang et al., 2005).

SA is synthesized via two distinct routes: isochorismic acid (IC) pathway and PAL pathway (Chen, Zheng, Huang, Lai, & Fan, 2009; Dempsey, Vlot, Wildermuth, & Klessig, 2011). Experiments using a PAL inhibitor (2-aminoindan-2-phosphonic acid; AIP) and PAL-silenced transgenic tobacco plants suggested that the PAL pathway was the major route of SA biosynthesis during pathogen response (Mauch-Mani & Slusarenko, 1996; Pallas, Paiva, Lamb, & Dixon, 1996). The activity of PAL increased upon phosphorylation, and a CDPK (AtCPK1) was shown to directly phosphorylate PAL (Cheng, Sheen, Gerrish, & Bolwell, 2001). The IC pathway was also suggested to be important in pathogen response (Dempsey et al., 2011; Wildermuth, Dewdney, Wu, & Ausubel, 2001). A fungal elicitor rapidly induced a CDPK (AtCPK1), and its overexpression accelerated isochorismic synthase (ICS) expression and SA synthesis (Coca & San Segundo, 2010).

7.4. The overall route

Taken together, the simplest model for the molecular pathway of caffeine action or caffeine signal pathway, in enhancing the defence response, can be summarized as followings (Fig. 10.4). First, caffeine directly blocks PDE, resulting in accumulation of cAMP. Second, increased cAMP activates CNGCs. Third, activated CNGC accelerates cytosolic Ca^{2+} flux. Fourth, high level of Ca^{2+} activates CDPKs, which directly phosphorylates PAL and ICS. Finally, enzymatically activated PAL and ICS upon phosphorylation enhance production of SA, which primes defence reactions.

A careful consideration is necessary as to the concentration, subcellular localization and expression of each components described earlier. *In vitro* assays with human materials indicated that caffeine concentrations to give a 50% inhibition were 20 μM for adenosine receptors, 300 μM for PDE and 800 μM for $GABA_A$ receptors (Fredholm et al., 1999). The caffeine

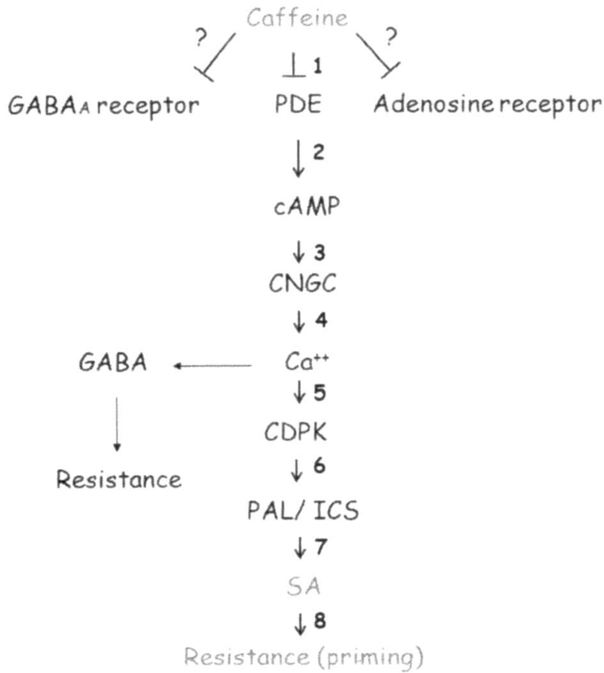

Figure 10.4 Caffeine signal cascade. The molecular cascade of caffeine signal is predicted based on available literatures and experimental data. *Step 1*: Caffeine directly blocks phosphodiesterase (PDE) (Francis et al., 2011). *Step 2*: Degradation of cAMP is inhibited and its level increases (Ma et al., 2009). *Step 3*: Increased cAMP activates cyclic nucleotide-gated channel (CNGC) (Talke et al., 2003). *Step 4*: Activated CNGC increases cytosolic Ca^{2+} level (Kaplan et al., 2007). *Step 5*: Increased Ca^{2+} activates calcium-dependent protein kinase (CDPK) activity (Harmon et al., 2000). *Step 6*: Activated CDPK directly phosphorylates phenylalanine ammonia lyase (PAL) and/or isochorismic synthase (ICS) (Cheng et al., 2001; Coca & San Segundo, 2010). *Step 7*: Activities of phosphorylated PAL and ICS increase, resulting in acceleration of salicylic acid (SA) production (Dempsey et al., 2011). *Step 8*: Increased level of SA primes defence response (Conrath, Pieterse, & Mauch-Mani, 2002). The $GABA_A$ receptors and adenosine receptors have not been identified in plants. However, increased Ca^{2+} induces GABA accumulation, which enhances defence reaction (Bown & Shelp, 1997). (See colour plate.)

concentration in transgenic tobacco and chrysanthemum was ~25 μM (5 μg/gram fresh weight) (Kim, Lim, Yoda, et al., 2011), being less than one tenth for effective inhibition of PDE. However, caffeine is synthesized in cytosol (Kodama et al., 2008), which occupies less than one tenth space in a plant cell. If newly synthesized caffeine remains in cytosol for a while, its concentration will be higher than 25 μM, perhaps being enough to interact with PDE.

Cyclic AMP localizes in cytosol, CNGC is on membrane and Ca^{2+} flows into and out of the cytosol. CDPK is a cytosolic calcium sensor and targets a variety of cytosolic proteins (Cheng, Willmann, Chen, & Sheen, 2002). Notably, AtCPK1 was shown to migrate from cytosol to lipid bodies and peroxisomes (Coca & San Segundo, 2010). PAL was shown to localize in microsomal and cytosolic fractions in tobacco (Achnine, Blancaflor, Rasmussen, & Dixon, 2004). This suggests SA to be synthesized in cytosol, although the exact cellular localization of SA biosynthesis is not completely determined. It was proposed to partly take place in plastid as ICS possesses the plastid localization signal at the N-terminal (Dempsey et al., 2011). And yet the majority of the players downstream of PDE have a chance to contact each other in cytosolic compartment. This allows caffeine to efficiently affect plant signalling procedures.

7.5. γ-Aminobutylic acid

In plants, GABA has been shown to accumulate in response to a variety of stresses. In soybean, GABA increased 20- to 40-fold upon cold and mechanical stresses (Wallace, Secor, & Schrader, 1984). Its role has not necessarily been clear yet, but generally thought to trigger primary defence reactions (Shelp, Brown, & Faure, 2006). In animals, GABA stimulates Cl^- influx through GABA-gated Cl^- channels (Goetz et al., 2007). Activity of this channel was shown to be decreased by treatments with cAMP, IBMX (PDE inhibitor) or forskolin (adenylate cyclase activator) (Heuschneider & Schwartz, 1989). In plants, no experimental data are available on its molecular mechanism (Park et al., 2010). However, GABA synthesis was found to greatly increase upon increase of cytosolic Ca^{2+} concentration (Bown & Shelp, 1997). This suggests a positive interaction between GABA and caffeine signal pathway (Fig. 10.4).

Note that adenosine receptors have so far not been identified from higher plants. In animal cells, adenosine receptors couple with G-proteins and modulate adenylate cyclase activity (Ralevic & Burnstock, 1998). Caffeine antagonizes the reaction. If a similar system is available in plants, caffeine would enhance the adenylate cyclase activity, thereby increasing the cAMP level. This converges on the main route of the predicted caffeine signal pathway (Fig. 10.4).

8. PLANT IMMUNIZATION

Caffeine was found to efficiently activate the defence system even in the absence of actual stresses. This resembles the 'priming' of the defence response, by which plants rapidly and efficiently cope with biotic and abiotic stresses (Conrath et al., 2006). Priming is typically seen in plants that had experienced challenges by pathogens or treatments with natural and synthetic compounds (Conrath et al., 2002). In particular, the effect of SA and its derivatives such as benzothiadiazole and β-aminobutylic acid (BABA) was distinct (Conrath et al., 2002). The underlying molecular mechanisms are currently not clear. Recent survey has suggested the involvement of protein kinases including mitogen-activated protein kinases (MAPKs) and CDPKs and chromatin modification (Conrath, 2011; Conrath et al., 2006; Pastor, Luna, Mauch-Mani, Ton, & Flors, 2013).

In regard to caffeine, priming by BABA appears to share a common feature. BABA is an isomer of GABA but rarely occurs in nature. Since over 40 years, BABA has been recognized to be a strong protector against a broad spectrum of plant disease caused by virus, bacteria, fungi and nematodes (Cohen, 2002; Jakab et al., 2001). When treated with BABA through the soil at 8 mg/l (~8 μM) 1 day before the inoculation with the oomycete *Peronospora parasitica*, *Arabidopsis* exhibited a strong resistance showing a complete inhibition of sporulation of the pathogen (Zimmerli, Jakab, Metraux, & Mauch-Mani, 2000). Similarly, when *Arabidopsis* seedlings were soil-drenched with 250 μM BABA, the number of diseased leaves caused by *P. syringae* was reduced to half that of the control (Ton et al., 2005). The molecular mechanism of BABA action in plants is not clear, but studies with *Xenopus* cells showed that BABA serves as an agonist of glycine, which, together with GABA, leads to hyperpolarization of the neural membrane through activating Cl^- influx (Schmieden & Betz, 1995).

As SA is a key component in defence reaction, its relationship to BABA has drawn much attention. In tobacco, BABA treatments enhanced virus resistance, which was strictly dependent on SA signaling (Siegrist, Orober, & Buchenauer, 2000). In potato, BABA-induced systemic resistance was indispensable of functional SA pathway (Eschen-Lippold, Altmann, & Rosahl, 2010). In contrast, BABA still protected SA-deficient *Arabidopsis* against oomycete pathogen, suggesting BABA functions downstream of SA (Zimmerli et al., 2000). It was concluded that requirement of SA for BABA action is diverse, depending on plant species and pathogens (Jakab et al., 2001). However, BABA treatment was found to be somehow detrimental for the host plant, showing stress-induced morphological response such as reduction of root and vegetative growth (Singh, Wu, & Zimmerli, 2010; Wu, Singh, Chen, & Zimmerli, 2010). BABA may induce a mild chronic stress, and this was referred as 'stress imprinting' (Singh et al., 2010). These observations are consistent with the case of caffeine. For example, exogenously applied caffeine conferred tobacco plants a strong resistance against *P. syringae* (Kim & Sano, 2008). Caffeine-producing chrysanthemum showed various morphological alterations including dwarfism and early flowering (Fig. 10.2). It is conceivable that these common features observed in plants treated with BABA and caffeine are mediated through SA, which plays a key role in not only stress response but also in developmental processes (Martinez, Pons, & Prats, 2004; Rivas-San Vicente & Plasencia, 2011).

Hence, activation of defence response by endogenously produced caffeine can be regarded as a kind of priming. However, the process differs from conventional priming, which takes place upon exogenous stimuli, such as biotic and abiotic stresses and chemical treatments. Caffeine is essentially toxic for organisms, and its production *in vivo*, even at low concentration, may result in a mild chronic stress or stress imprinting as suggested for BABA (Singh et al., 2010). In this context, the caffeine effect appears to be an analogy of immunization or vaccination in mammals. Vaccination is performed by administrating mild toxic 'antigenic substance,' thereby stimulating production of antibodies, which cope with invading pathogens. Plants can also be immunized by constitutively producing mildly toxic caffeine *in planta*, thereby stimulating defence system prior to receiving actual stresses (Kim et al., 2010).

9. CONCLUDING REMARKS

The available observation suggests the dual function of caffeine in nature: at high concentration, it is a direct toxicant for living organisms, whereas at low concentration, it serves as a priming agent to constitutively activate plants' defence system. A question then arises as to whether or not this is generally applicable to other plant secondary metabolites. Particularly, it is of interest to know whether or not they serve as priming agent, or more broadly, as 'antigen' for immunization.

Plants produce more than 50,000 secondary metabolites including alkaloids (12,000), terpenes (30,000) and phenolics (10,000) (Kennedy & Wightman, 2011). Their ecological role is essential to increase overall ability to survive and overcome environmental stresses, such as herbivore attack, pathogen infection or nutrient deprivation (Kennedy & Wightman, 2011; Wink, 2003). Physiologically, they function in direct host protection serving as *in planta* antibiotics, allelopathic defence against competitor plants, feeding deterrence and toxicity to herbivores and others (Bednarek, 2012; Kennedy & Wightman, 2011).

Little is known, however, about their potential role to activate or potentiate plants' innate defence system. Perhaps, the best characterized case is defence priming by volatile organic compounds (VOCs) (Beckers & Conrath, 2007). VOCs have been known to be rapidly emitted when plants are injured or attacked by herbivores and to stimulate nearby plants for defence (Engelberth, Albor, Schmelz, & Tumlinson, 2004; Ton et al., 2007). Upon clipping injury, sagebrush released a large amount of VOCs within 3 h. A gas chromatography analysis revealed the majority of compounds was terpenoids, such as 1,8-cineole, (*E*)-ocimene and *p*-cymene. Perceiving these compounds, a tobacco plants grown nearby clipped sagebrush showed enhanced defence activity against moth caterpillars (*Manduca sexta*). It was concluded that VOCs prime plants defence response, acting as the signalling factor between plants (Kessler, Halitschke, Diezel, & Baldwin, 2006). VOCs are clearly important mediators for plant–plant communication, but their capability as 'antigen' has not been tested.

An intriguing example of antigenic activity of the secondary metabolites is ginsenoside (Lee, Han, et al., 2012). Ginsenosides are triterpene saponins and classified into two groups, dammarane and oleanane types. When dammarenediol-II was produced in tobacco plants by introducing and expressing dammarenediol-II synthase gene from *Panax ginseng*, the mature plants exhibited a strong resistance against TMV infection, simultaneously showing the enhanced HR (Lee, Han, et al., 2012). This finding suggests that ginsenosides are potentially powerful priming agent or 'antigen' for immunization. The molecular mechanism is not clear, but, in mammalian cells, ginsenosides were shown to interact with $GABA_A$ receptors and affect the ion channels (Kimura et al., 1994; Lee, Choi, et al., 2012). $GABA_A$ receptors have not been identified in plants, and yet the possibility remains that GABA is involved in ginsenoside-induced priming of the defence response.

Considering the cases of VOCs, ginsenoside and caffeine, we speculate that plant secondary metabolites may commonly interact with the plant defence-related pathways and activate appropriate defence networks. The effect may usually be induced by exogenous application, naturally or artificially, such as VOCs, resulting in so-called priming. The effect is also induced by endogenously produced compounds as exemplified with caffeine and ginsenoside, this being equivalent to immunization. Thus, the answer to the question as to the priming or antigenic function of plants' secondary metabolites may be positive. Further analysis with other compounds will strengthen and generalize this idea.

Finally, it is worthy to mention that plant immunization with the secondary metabolites will open a new strategy to create plants with high tolerance against various environmental stresses. Contrary to the conventional transgenic plant, in which a particular stress-resistant trait is introduced, defence-primed plants are prepared by immunization, which generally gives a broad range of stress-resistance. The procedure is simple, and resulting plants are perhaps less loading on environment.

ACKNOWLEDGEMENTS

This work was supported by the World Class University (WCU) Project of the Ministry of Education, Science and Technology, Korea and by the Japan Society for the Promotion of Science, Japan.

REFERENCES

Achnine, L., Blancaflor, E. B., Rasmussen, S., & Dixon, R. A. (2004). Colocalization of L-phenylalanine ammonia-lyase and cinamate 4-hydroxylase for metabolic channeling in phenylpropanoid biosynthesis. *The Plant Cell*, *16*, 3098–3109.

Amrhein, N. (1977). The current status of cyclic AMP in higher plants. *Annual Review of Plant Physiology*, *28*, 123–132.

Aneja, M., & Gianfagna, T. (2001). Induction and accumulation of caffeine in young, actively growing leaves of cocoa (*Theobroma cacao* L.) by wounding or infection with *Crinipellis perniciosa*. *Physiological and Molecular Plant Pathology*, *59*, 13–16.

Ashihara, H., & Crozier, A. (1999). Biosynthesis and metabolism of caffeine and related purine alkaloids in plants. *Advances in Botanical Research*, *30*, 118–205.

Ashihara, H., & Crozier, A. (2001). Caffeine: A well known but little mentioned compound in plant science. *Trends in Plant Science*, *6*, 407–413.

Ashihara, H., Sano, H., & Crozier, A. (2008). Caffeine and related purine alkaloids: Biosynthesis, catabolism, function and genetic engineering. *Phytochemisty*, *69*, 841–856.

Assmann, S. M. (1995). Cyclic AMP as a second messenger in higher plants: Status and future prospects. *Plant Physiology*, *108*, 885–889.

Baumann, T. W. (2006). Some thoughts on the physiology of caffeine in coffee—And a glimpse of metabolite profiling. *Brazilian Journal of Plant Physiology*, *18*, 243–251.

Beckers, G. J. M., & Conrath, U. (2007). Priming for enhanced stress resistance: From the lab to the field. *Current Opinion in Plant Biology*, *10*, 425–431.

Bednarek, P. (2012). Chemical warfare or modulators of defence responses—The function of secondary metabolites in plant immunity. *Current Opinion in Plant Biology*, *15*, 407–417.

Bender, A. T., & Beavo, J. A. (2006). Cyclic nucleotide phosphodiesterases: Molecular regulation to clinical use. *Pharmacological Reviews*, *58*, 488–520.

Bettler, B., Kaupmann, K., Mosbacher, J., & Gassmann, M. (2004). Molecular structure and physiological functions of $GABA_B$ receptors. *Physiological Review*, *84*, 835–867.

Bouche, N., & Fromm, H. (2004). GABA in plants: Just a metabolite? *Trends in Plant Science*, *9*, 110–115.

Bown, A. W., & Shelp, B. J. (1997). The metabolism and functions of γ-aminobutyric acid. *Plant Physiology*, *115*, 1–5.

Brewin, N. J., & Northcote, D. H. (1973). Partial purification of a cyclic amp phosphodiesterase from soybean callus isolation of a non-dialysable inhibitor. *Biochimica et Biophysica Acta*, *320*, 104–122.
Brown, E. G., Alnajafi, T., & Newton, R. P. (1977). Cyclic nucleotide phosphodiesterase activity in *Phaseolus vulgaris*. *Phytochemistry*, *16*, 1333–1337.
Brown, E. G., Edwards, M. J., Newton, R. P., & Smith, C. J. (1979). Plurality of cyclic nucleotide phosphodiesterase in *Spinacea oleracea*, subcellular-distribution, partial-purification, and properties. *Phytochemistry*, *18*, 1943–1948.
Brown, E. G., & Newton, R. P. (1981). Cyclic AMP and higher plants. *Phytochemistry*, *20*, 2453–2463.
Chen, Z., Zheng, Z., Huang, J., Lai, Z., & Fan, B. (2009). Biosynthesis of salicylic acid in plants. *Plant Signaling & Behavior*, *4*, 493–496.
Cheng, S. H., Sheen, J., Gerrish, C., & Bolwell, G. P. (2001). Molecular identification of phenylalanine ammonia-lyase as a substrate of a specific constitutively active *Arabidopsis* CDPK expressed in maize protoplasts. *FEBS Letters*, *503*, 185–188.
Cheng, S. H., Willmann, M. R., Chen, H. C., & Sheen, J. (2002). Calcium signaling through protein kinases. The *Arabidopsis* calcium-dependent protein kinase gene family. *Plant Physiology*, *129*, 469–485.
Chiatante, D., Newton, R. P., Crignola, S., Levi, M., & Brown, E. G. (1990). The 3',5'-cyclic nucleotide phosphodiesterases of meristematic and differentiated tissue of pea roots. *Phytochemistry*, *29*, 2815–2820.
Chou, C. H., & Waller, G. R. (1980). Possible allelopathetic constituents of *Coffea Arabica*. *Journal of Chemical Ecology*, *6*, 643–654.
Clough, S. J., Fengler, K. A., Yu, I. C., Lippok, B., Smith, R. K., & Bent, A. F. (2000). The *Arabidopsis dnd1* "defense, no death" gene encodes a mutated cyclic nucleotide-gated ion channel. *Proceedings of the National Academy of Sciences of the United States of America*, *97*, 9323–9328.
Coca, M., & San Segundo, B. (2010). AtCPK1 calcium-dependent protein kinase mediates pathogen resistance in *Arabidopsis*. *The Plant Journal*, *63*, 526–540.
Cohen, Y. R. (2002). β-Aminobutyric acid-induced resistance against plant pathogens. *Plant Disease*, *86*, 448–457.
Conrath, U. (2011). Molecular aspects of defence priming. *Trends in Plant Science*, *16*, 524–553.
Conrath, U., Beckers, G. J. M., Flors, V., Garcia-Agustin, P., Jakab, G., Mauch, F., et al. (2006). Priming: Getting ready for battle. *Molecular Plant-Microbe Interactions*, *19*, 1062–1071.
Conrath, U., Pieterse, C. M. J., & Mauch-Mani, B. (2002). Priming in plant-pathogen interactions. *Trends in Plant Science*, *7*, 210–221.
Conti, M. (2000). Phosphodiesterases and cyclic nucleotide signaling in endocrine cells. *Molecular Endocrinology*, *14*(9), 1317–1327.
Cooke, C. J., Smith, C. J., Walton, T. J., & Newton, R. P. (1994). Evidence that cyclic AMP is involved in the hypersensitive response of *Medicago sativa* to a fungal elicitor. *Phytochemistry*, *35*, 889–895.
Daly, J. W. (2007). Caffeine analogs: Biomedical impact. *Cellular and Molecular Life Sciences*, *64*, 2153–2169.
Dempsey, D. A., Vlot, A. C., Wildermuth, M. C., & Klessig, D. F. (2011). Salicylic acid biosynthesis and metabolism. *The Arabidopsis Book*, *9*, e0156.
Dupon, M., Van Onckelen, H. A., & De Greef, J. A. (1987). Characterization of cyclic nucleotide phosphodiesterase activity in *Phaseolus vulgaris*. *Physiologia Plantarum*, *69*, 361–365.
Engelberth, J., Albor, H. T., Schmelz, E. A., & Tumlinson, J. H. (2004). Airborne signals prime plants against insect herbivore attack. *Proceedings of the National Academy of Sciences of the United States of America*, *101*, 1781–1785.

Eschen-Lippold, L., Altmann, S., & Rosahl, S. (2010). DL-β-aminobutyric acid-induced resistance of potato against *Phytophthora infestans* requires salicylic acid but not oxylipins. *Molecular Plant-Microbe Interactions*, *23*, 585–592.

Francis, S. H., Blount, M. A., & Corbin, J. D. (2011). Mammalian cyclic nucleotide phosphodiesterases: Molecular mechanisms and physiological functions. *Physiological Reviews*, *91*, 651–690.

Fredholm, B. B., Abbracchio, M. P., Burnstock, G., Daly, J. W., Harden, T. K., Jacobson, K. A., et al. (1994). Nomenclature and classification of purinoceptors. *Pharmacological Reviews*, *46*, 143–156.

Fredholm, B. B., Battig, K., Holmen, J., Nehlig, A., & Zvartau, E. (1999). Actions of caffeine in the brain with special reference to factors that contribute to its widespread use. *Pharmacological Reviews*, *51*, 83–153.

Goetz, T., Arslan, A., Wisden, W., & Wulff, P. (2007). GABA(A) receptors: Structure and function in the basal ganglia. *Progress in Brain Research*, *160*, 21–41.

Green, T. R., & Ryan, C. A. (1972). Wound-induced proteinase inhibitor in plant leaves: A possible defence mechanism against insects. *Science*, *175*, 776–777.

Harmon, A. C., Gribskov, M., & Harper, J. F. (2000). CDPKs—A kinase for every Ca^{2+} signal? *Trends in Plant Science*, *5*, 154–159.

Heuschneider, G., & Schwartz, R. D. (1989). cAMP and forskolin decrease γ-aminobutyric acid-gated chloride flux in rat brain synaptoneurosomes. *Proceedings of the National Academy of Sciences of the United States of America*, *86*, 2938–2942.

Hiei, Y., Ohta, S., Komari, T., & Kumahsiro, T. (1994). Efficient transformation of rice (*Oryza sativa* L.) mediated by *Agrobacterium* and sequence analysis of the boundaries of the T-DNA. *The Plant Journal*, *6*, 271–282.

Hollingsworth, R. G., Armstrong, J. W., & Campbell, E. (2002). Caffeine as a repellent for slugs and snails: At high concentrations this stimulant becomes a lethal neurotoxin to garden pests. *Nature*, *417*, 915–916.

Jakab, G., Cottier, V., Toquin, V., Rigoli, G., Zimmerli, L., Metraux, J. P., et al. (2001). β-Aminobutyric acid-induced resistance in plants. *European Journal of Plant Pathology*, *107*, 29–37.

Jiang, J., Fan, L. W., & Wu, W. H. (2005). Evidences for involvement of endogenous cAMP in *Arabidopsis* defense responses to *Verticillium* toxins. *Cell Research*, *15*, 585–592.

Kaplan, B., Sherman, T., & Fromm, H. (2007). Cyclic nucleotide-gated channels in plants. *FEBS Letters*, *581*, 2237–2246.

Kato, M., & Mizuno, K. (2004). Caffeine synthase and related methyltransferases in plants. *Frontiers in Bioscience*, *9*, 1833–1842.

Kennedy, D., & Wightman, E. L. (2011). Herbal extracts and phytochemicals: Plant secondary metabolites and the enhancement of human brain function. *Advances in Nutrition*, *2*, 32–50.

Kessler, A., Halitschke, R., Diezel, C., & Baldwin, I. T. (2006). Priming of plant defense responses in nature by airborne signaling between *Artemisia tridentata* and *Nicotiana attenuata*. *Oecologia*, *148*, 280–292.

Kim, Y. S., Choi, Y. E., & Sano, H. (2010). Plant vaccination: Stimulation of defense system by caffeine production in Planta. *Plant Signaling & Behavior*, *5*, 489–493.

Kim, Y. S., Lim, S., Kang, K. K., Jung, Y. J., Lee, Y. H., Choi, Y. E., et al. (2011). Resistance against beet armyworms and cotton aphids in caffeine-producing transgenic chrysanthemum. *Plant Biotechnology*, *28*, 393–395.

Kim, Y. S., Lim, S., Yoda, H., Choi, C. S., Choi, Y. E., & Sano, H. (2011). Simultaneous activation of salicylate production and fungal resistance in transgenic *Chrysanthemum* producing caffeine. *Plant Signaling & Behavior*, *6*, 409–412.

Kim, Y. S., & Sano, H. (2008). Pathogen resistance of transgenic tobacco plants producing caffeine. *Phytochemistry*, *69*, 882–888.

Kim, Y. S., Uefuji, H., Ogita, S., & Sano, H. (2006). Transgenic tobacco plants producing caffeine: A potential new strategy for insect pest control. *Transgenic Research*, *15*, 667–672.

Kimura, T., Saunders, P. A., Kim, H. S., Rheu, H. M., Oh, K. W., & Ho, I. K. (1994). Interactions of ginsenosides with ligand-bindings of GABAA and GABAB receptors. *General Pharmacology*, *25*, 193–199.

Kinnersley, A. M., & Lin, F. (2000). Receptor modifiers indicate that 4-aminobutyric acid (GABA) is a potential modulator of ion transport in plants. *Plant Growth Regulation*, *32*, 65–76.

Kodama, Y., Shinya, T., & Sano, H. (2008). Dimerization of *N*-methyltransferases involved in caffeine biosynthasis. *Biochimie*, *90*, 547–551.

Kretschmar, J. A., & Baumann, T. W. (1999). Caffeine in *Citrus* flowers. *Phytochemistry*, *52*, 19–23.

Lee, C. H., & Abidin, U. Z. (1989). Properties of a cyclic 3′,5′-nucleotide phosphodiesterase from *Vigna mungo*. *Biochemistry International*, *19*, 745–753.

Lee, B. H., Choi, S. H., Shin, T. J., Hwang, S. H., Kang, J., Kim, H. J., et al. (2012). Effects of ginsenoside metabolites on $GABA_A$ receptor-mediated ion currents. *Journal of Ginseng Research*, *36*, 55–60.

Lee, M. H., Han, J. Y., Kim, H. J., Kim, Y. S., Huh, G. H., & Choi, Y. E. (2012). Dammarenediol-II production confers TMV tolerance in transgenic tobacco expressing Panax ginseng dammarenediol-II synthase. *Plant & Cell Physiology*, *43*, 173–182.

Leng, Q., Mercier, R. W., Yao, W., & Berkowitz, G. A. (1999). Cloning and first functional characterization of a plant cyclic nucleotide-gated cation channel. *Plant Physiology*, *121*, 753–761.

Lin, P. P., & Varner, J. E. (1972). Cyclic nucleotide phosphodiesterase in pea seedlings. *Biochimica et Biophysica Acta*, *276*, 454–474.

Ma, W., & Berkowitz, G. A. (2011). Ca^{2+} conduction by plant cyclic nucleotide gated channels and associated signaling components in pathogen defense signal transduction cascade. *New Phytologist*, *190*, 566–572.

Ma, W., Qi, Z., Smigel, A., Walker, R. K., Verma, R., & Berkowitz, G. A. (2009). Ca^{2+}, cAMP, and transduction of non-self perception during plant immune responses. *Proceedings of the National Academy of Sciences of the United States of America*, *106*, 20995–21000.

Martinez, C., Pons, E., & Prats, G. (2004). Salicylic acid regulates flowering time and links defence responses and reproductive development. *The Plant Journal*, *37*, 209–217.

Martinez-Atienza, J., van Ingelgem, C., Roef, L., & Maathuis, F. J. M. (2007). Plant cyclic nucleotide signaling. *Plant Signaling & Behavior*, *2*, 540–543.

Mathavan, S., Premalatha, Y., & Christopher, M. S. M. (1985). Effects of caffeine and theophylline on the fecundity of four lepidopteran species. *Experimental Biology*, *44*, 133–138.

Mauch-Mani, B., & Slusarenko, A. J. (1996). Production of salicylic acid precursors is a major function of phenylalanine ammonia-lyase in the resistance of *Arabidopsis* to *Peronospora parasitica*. *The Plant Cell*, *8*, 203–212.

McCarthy, A. A., & McCarthy, J. G. (2007). The structure of two *N*-methyltransferases from the caffeine biosyntheyic pathway. *Plant Physiology*, *144*, 879–889.

Mizuno, K., Kato, M., Irino, F., Yoneyama, N., Fujimura, T., & Ashihara, H. (2003). The first committed step reaction of caffeine biosynthesis: 7-Methylxanthosine synthase is closely homologous to caffeine synthases in coffee (*Coffea arabica* L.). *FEBS Letters*, *547*, 56–60.

Mizuno, K., Okuda, A., Kato, M., Yoneyama, N., Tanaka, H., Ashihara, H., et al. (2003). Isolation of a new dual-functional caffeine synthase gene encoding an enzyme for the conversion of 7-methylxanthine to caffeine from coffee (*Coffea Arabica* L.). *FEBS Letters*, *534*, 75–81.

Mizuno, K., Tanaka, H., Kato, M., Ashihara, H., & Fujimura, T. (2001). cDNA cloning of caffeine (theobromine) synthase from coffee (*Coffea arabica* L.). In: *19th International scientific colloquium on coffee, ASIC, Paris* (pp. 815–818).

Mohanpuria, P., & Yadav, S. K. (2009). Retardation in seedling growth and induction of early scenscence in plants upon caffeine exposure is related to its negative effect on Rubisco. *Photosynthetica, 47*, 293–297.

Nathanson, J. A. (1984). Caffeine and related methylxanthine possible naturally occurring pesticides. *Science, 226*, 184–187.

Newton, R. P., Roef, L., Witters, E., & Van Onckelen, H. (1999). Cyclic nucleotides in higher plants: The enduring paradox. *New Phytologist, 143*, 427–455.

Newton, R. P., & Smith, C. J. (2004). Cyclic nucleotides. *Phytochemistry, 65*, 2423–2437.

Niles, R. M., & Mount, M. S. (1974). Cyclic nucleotide phosphodiesterase from carrot. *Phytochemistry, 13*, 2735–2740.

Ogawa, M., Herai, Y., Koizumi, N., Kusano, T., & Sano, H. (2001). 7-Methylxanthine methyltransferase of coffee plants: Gene isolation and enzymatic properties. *Journal of Biological Chemistry, 276*, 8213–8218.

Ogita, S., Uefuji, H., Yamaguchi, Y., Koizumi, N., & Sano, H. (2003). Production of decaffeinated coffee plants by genetic engineering. *Nature, 423*, 823.

Olsen, R. W., & Tobin, A. J. (1990). Molecular biology of GABAA receptors. *The FASEB Journal, 4*, 1469–1480.

Pallas, J. A., Paiva, N. L., Lamb, C., & Dixon, R. A. (1996). Tobacco plants epigenetically suppressed in phenylalanine ammonia-lyase expression do not develop systemic acquired resistance in response to infection by tobacco mosaic virus. *The Plant Journal, 10*, 281–293.

Park, D. H., Mirabella, R., Bronstein, P. A., Preston, G. M., Haring, M. A., Lim, C. K., et al. (2010). Mutations in gamma-aminobutyric acid (GABA) transaminase genes in plants or *Pseudomonas syringae* reduce bacterial virulence. *The Plant Journal, 64*, 318–330.

Parramon, M., Gonzalez, M. P., Herrero, M. T., & Oset-Gasque, M. J. (1995). $GABA_B$ receptors increase intracellular calcium concentrations in chromaffin cells through two different pathways: Their role in catecholamine secretion. *Journal of Neuroscience Research, 41*, 65–72.

Pastor, V., Luna, E., Mauch-Mani, B., Ton, J., & Flors, V. (2013). Primed plants do not forget. *Environmental and Experimental Botany*, (in press).

Ralevic, V., & Burnstock, G. (1998). Receptors for purines and pyrimidines. *Pharmacological Reviews, 50*, 413–492.

Raskin, I. (1992). Role of salicylic acid in plants. *Annual Review of Plant Physiology and Plant Molecular Biology, 43*, 439–463.

Ribeiro, J. A., & Sebastiao, A. M. (2010). Caffeine and adenosine. *Journal of Alzheimer's Disease, 20*, S3–S15.

Ries, S., Savithiry, S., Wert, V., & Widders, I. (1993). Rapid induction of ion pulses in tomato, cucumber, and maize plants following a foliar application of L(+)-adenosine. *Plant Physiology, 101*, 49–55.

Rivas-San Vicente, M., & Plasencia, J. (2011). Salicylic acid beyond defence: Its role in plant growth and development. *Journal of Experimental Botany, 62*, 3321–3338.

Rizvi, S. J. H., Mukerji, D., & Mathur, S. N. (1981). Selective phytotoxicity of 1,3,7-trimethylxanthine between *Phaseolus mungo* and some weeds. *Agricultural and Biological Chemistry, 45*, 1255–1256.

Ryals, J. A., Neuenschwander, U. H., Willits, M. G., Molina, A., Steiner, H. Y., & Hunt, M. D. (1996). Systemic acquired resistance. *The Plant Cell, 8*, 1809–1819.

Schmieden, V., & Betz, H. (1995). Pharmacology of the inhibitory glycine receptor: Agonist and antagonist actions of amino acids and piperidine carboxylic acid compounds. *Molecular Pharmacology, 48*, 919–927.

Seo, S., Katou, S., Seto, H., Gomi, K., & Ohashi, Y. (2007). The mitogen-activated protein kinases WIPK and SIPK regulate the levels of jasmonic and salicylic acids in wounded tobacco plants. *The Plant Journal, 49*, 899–909.

Seo, S., Seto, H., Koshino, H., Yoshida, S., & Ohashi, Y. (2003). A diterpene as an endogenous signal for the activation of defense response to infection with tobacco mosaic virus and wounding in tobacco. *The Plant Cell, 15*, 863–873.

Shelp, B. J., Brown, A. W., & Faure, D. (2006). Extracelluar γ-aminobutyrate mediates communication between plants and other organisms. *Plant Physiology, 142*, 1350–1352.

Shi, D., Padgett, W. L., & Daly, J. W. (2003). Caffeine analogs: Effects on ryanodine-sensitive calcium-release channels and GABAA receptors. *Cellular and Molecular Neurobiology, 23*, 331–347.

Siegrist, J., Orober, M., & Buchenauer, H. (2000). β-Aminobutyric acid-mediated enhancement of resistance in tobacco to tobacco mosaic virus depends on the accumulation of salicylic acid. *Physiological and Molecular Plant Pathology, 56*, 95–106.

Singh, P., Wu, C. C., & Zimmerli, L. (2010). β-Aminobutyric acid priming by stress imprinting. *Plant Signaling & Behavior, 5*, 878–880.

Stewart, I. (1985). Identification of caffeine in *Citrus* flowers and leaves. *Journal of Agricultural and Food Chemistry, 33*, 1163–1165.

Talke, I. N., Blaudez, D., Maathuis, F. J. M., & Sanders, D. (2003). CNGCs: Prime targets of plant cyclic nucleotide signaling? *Trends in Plant Science, 8*, 286–293.

Ton, J., D'Alessandro, M., Jourdie, V., Jakab, G., Karlen, D., Held, M., et al. (2007). Priming by airborne signals boosts direct and indirect resistance in maize. *The Plant Journal, 49*, 16–26.

Ton, J., Jakab, G., Toquin, V., Flors, V., Iavicoli, A., Maeder, M. N., et al. (2005). Dissecting the β-aminobutyric acid-induced priming phenomenon in *Arabidopsis*. *The Plant Cell, 17*, 987–999.

Tsubouchi, H., Terada, H., Yamamoto, K., Hisada, K., & Sakabe, Y. (1985). Caffeine degradation and increased ochratoxin A production by toxigenic strains of *Aspergillus ochraceus* isolated from green coffee beans. *Mycopathologia, 90*, 181–186.

Turano, F. J., Panta, G. R., Allard, M. W., & van Berkum, P. (2001). The putative glutamate receptors from plants are related to two superfamilies of animal neurotransmitter receptors via distinct evolutionary mechanisms. *Molecular Biology and Evolution, 18*, 1417–1420.

Uefuji, H., Ogita, S., Yamaguchi, Y., Koizumi, N., & Sano, H. (2003). Molecular cloning and functional characterization of three distinct *N*-methyltransferases involved in the caffeine biosynthetic pathway in coffee plants. *Plant Physiology, 132*, 372–380.

Uefuji, H., Tatsumi, Y., Morimoto, M., Kaothien-Nakayama, P., Ogita, S., & Sano, H. (2005). Caffeine production in tobacco plants by simultaneous expression of three coffee *N*-methyltransferases and its potential as a pest repellant. *Plant Molecular Biology, 59*, 221–227.

Wallace, W., Secor, J., & Schrader, L. E. (1984). Rapid accumulation of γ-aminobutylic acid and alanine in soybean leaves in response to an abrupt transfer to lower temperature, darkness, or mechanical stress. *Plant Physiology, 175*, 170–175.

Weinberger, B. A., & Bealer, B. K. (2002). *The world of caffeine*. NewYork: Routledge.

Wildermuth, M. C., Dewdney, J., Wu, G., & Ausubel, F. M. (2001). Isochorismate synthase is required to synthesize salicylic acid for plant defence. *Nature, 414*, 562–565.

Wink, M. (2003). Evolution of secondary metabolites from an ecological and molecular phylogenetic perspective. *Phytochemistry, 64*, 3–19.

Wu, C. C., Singh, P., Chen, M. C., & Zimmerli, L. (2010). L-Glutamine inhibits β-aminobutyric acid-induced stress resistance and priming in Arabidopsis. *Journal of Experimental Botany, 61*, 995–1002.

Yu, G. H., & Sun, M. X. (2007). Deciphering the possible mechanism of GABA in tobacco pollen tube growth and guidance. *Plant Signaling & Behavior, 2*, 393–395.

Zhao, J., Guo, Y., Fujita, K., & Sakai, K. (2004). Involvement of cAMP signaling in elicitor-induced phytoalexin accumulation in *Cupressus lusitanica* cell cultures. *New Phytologist, 161*, 723–733.

Zimmerli, L., Jakab, G., Metraux, J. P., & Mauch-Mani, B. (2000). Potentiation of pathogen-specific defense mechanisms in *Arabidopsis* by β-aminobutyric acid. *Proceedings of the National Academy of Sciences of the United States of America, 97*, 12920–12925.

CHAPTER ELEVEN

Alkaloids from Marine Bacteria

Sergey B. Zotchev[1]
Department of Biotechnology, Norwegian University of Science and Technology, Trondheim, Norway
[1]Corresponding author: e-mail address: sergey.zocthev@ntnu.no

Contents

1. Introduction 302
2. Occurrence and Biological Activities of Alkaloids from Marine Bacteria 302
   2.1 Alkaloids from marine actinomycete bacteria 302
   2.2 Alkaloids from marine cyanobacteria 310
   2.3 Alkaloids from other marine bacterial species 315
3. Biosynthesis of Alkaloids in Marine Bacteria 319
   3.1 Biosynthesis of violacein 319
   3.2 Biosynthesis of staurosporine and related bisinolde alkaloids 321
   3.3 Biosynthesis of diazepinomicin 322
   3.4 Biosynthesis of barbamide 323
   3.5 Biosynthesis of holomycin 324
   3.6 Biosynthesis of salinosporamide A 325
4. Conclusions 327
Acknowledgements 327
References 327

Abstract

Marine bacteria are rich and yet underexplored source of structurally diverse secondary metabolites, many of which possess unique biological activities. A large portion of these natural products are represented by compounds that can be classified as alkaloids. Marine actinomycete bacteria, especially those representing genus *Streptomyces*, have so far yielded most of the currently discovered marine alkaloids of bacterial origin, followed by cyanobacteria. This trend most likely reflects that fact that these bacteria are easier to culture compared with other, more rare and slow-growing species, suggesting that the development of new cultivation techniques can lead to more exciting discoveries. This chapter highlights some of the recent examples of alkaloids isolated from marine bacteria, providing brief information on their origin, structure and biological activities. Several examples of biosynthetic pathways for structurally diverse marine bacteria-derived alkaloids are also presented and discussed. Deciphering of the biosynthetic routes for alkaloid biosynthesis is an exciting research filed. It may not only provide new insights into the enzymology and chemistry of natural products but also deliver means for rational engineering of alkaloids to improve their drug-like properties.

Advances in Botanical Research, Volume 68
ISSN 0065-2296
http://dx.doi.org/10.1016/B978-0-12-408061-4.00011-0

1. INTRODUCTION

Alkaloids are structurally diverse compounds generally classified as such due to the basic character of the molecule (from Latin *alkali*) and a presence of at least one nitrogen atom, preferably in a heterocycle. Many of the naturally occurring alkaloids have biological activity, and some of them are being used as drugs in modern medicine (e.g. morphin, codeine, reserpine, etc.). Although originally discovered in plants and then in marine sponges, many alkaloids have recently been identified in cultivable microorganisms, which provide opportunities for their sustainable production.

A large number of biologically active alkaloids, many of which contain highly reactive halogen atom(s), have been isolated from marine sources, such as sponges, tunicates and other marine animals (Blunt, Copp, Keyzers, Munro, & Prinsep, 2013). However, a growing body of evidence suggests that these compounds are most likely produced not only by the macro-organisms themselves but also by their symbiotic or associated bacteria and fungi (Hentschel, Piel, Degnan, & Taylor, 2012). Challenges in cultivation of such microorganisms, however, make it difficult to unequivocally determine the true origin of some alkaloids. Metagenomics approach can be used in such cases to identify alkaloid biosynthesis genes, which can be traced to their microbial host (Freeman et al., 2012). This chapter reviews alkaloids recently identified in marine bacteria as well as biosyntheses and biological activities of some representative compounds.

2. OCCURRENCE AND BIOLOGICAL ACTIVITIES OF ALKALOIDS FROM MARINE BACTERIA

2.1. Alkaloids from marine actinomycete bacteria

Actinomycete bacteria represent a formidable source of chemically diverse biologically active molecules, some of which have found use as antibiotics (e.g. erythromycin, amphotericin B, vancomycin) and anti-cancer drugs (e.g. daunorubicin). Recent advances in isolation and characterization of actinomycetes from marine environment revealed their potential for producing many unique bioactive secondary metabolites (Hughes & Fenical, 2010), many of which can be classified as alkaloids.

Streptomyces is the most prolific genus of actinomycetes when it comes to the production of chemically diverse secondary metabolites. Over 50% of all antibiotics known today are produced by representative of this genus, which, until recently, have been preferentially isolated from terrestrial sources. Recent advances in cultivation of marine microbial species, including *Streptomyces* and other actinomycetes from deep-sea sediments and marine animals allowed isolation and characterization of novel metabolites with unique chemical structures and biological activities. It shall be noted that marine alkaloids typically isolated from marine sponges and tunicates may, in fact, be synthesized by associated marine bacteria, including actinomycetes (Kim & Dewapriya, 2012). Examples of diverse bioactive alkaloids isolated from marine actinomycete bacteria are given in Table 11.1, and their chemical structures are presented in Fig. 11.1A and B.

Altemicidin, a monoterpene alkaloid (Fig. 11.1A), has been one of the first bioactive alkaloids isolated from marine-derived *Streptomyces sioyaensis* SA-1758 (Takahashi et al., 1989). The latter bacterium was isolated from a sea mud collected at Guamo (Japan) and required addition of sea water to the media for both growth and altemicidin production. Altemicidin was shown to have acaricidal activity, killing 50% of newly hatched brine shrimp at a concentration of 3 μg/mL, as well as pronounced activity against carcinoma cell line (IC_{50}, 0.82 μg/mL). However, acute toxicity of altemicidin in mice model (LD_{50}, 0.3 mg/kg) apparently prevented this compound from being developed as a drug candidate.

Azamerone, a unique halogenated alkaloid, was isolated from a new *Streptomyces* sp. CNQ766 cultured from marine sediment collected near the Guam island in the tropical Pacific ocean (Jensen, Gontang, Mafnas, Mincer, & Fenical, 2005). This alkaloid has an atypical chloropyranophthalazinone core with a 3-chloro-6-hydroxy-2,2,6-trimethylcyclohexylmethyl side chain (Cho et al., 2006; Fig. 11.1A). As in the case of altemicidin, growth of the producing organism and biosynthesis of azamerone depended on the addition of sea water to the media. It is worth noting that azamerone represents the first reported natural compound with phthalazinone ring system. It has been later shown that azamerone biosynthesis includes novel rearrangement of the aryl diazoketone, where the aromatic ring is oxidatively cleaved and then re-aromatized with a dinitrogen group (where N-atoms are added in a sequential manner) to yield the phthalazinone core (Winter, Jansma, Handel, & Moore, 2009). Azamerone was shown to have a weak cytotoxic activity against mouse T-cells and macrophages, and no further reports on the biological activity of this alkaloid could be found.

Streptomyces sp. CNQ-583 isolated from a marine sediment produced several new pyrrolizidine alkaloids, including halogenated form of bohemamine, 5-chlorobohemamine C (Bugni et al., 2006; Fig. 11.1A).

Table 11.1 Selected alkaloids from marine actinomycete bacteria (see Fig. 11.1 for chemical structures)

Name	Sub-class	Source	References
Altemicidin	Monoterpene-alkaloid	*Streptomyces sioyaensis* from sea mud	Takahashi et al. (1989)
Ammosamide D	Pyrroloqui-noline	*Streptomyces variabilis* from marine sediment	Pan, Jamison, Yousufuddin, and MacMillan (2012)
Azamerone	Chloropyrano-phthalazinone	*Streptomyces* sp. from marine sediment	Cho, Kwon, Williams, Jensen, and Fenical (2006)
5-Chlorobo-hemamine	Pyrrolizidine	*Streptomyces* sp. from marine sediment	Bugni, Woolery, Kauffman, Jensen, and Fenical (2006)
Mansouramycins	Isoquinoli-nequinone	*Streptomyces* sp. from marine sediment	Hawas et al. (2009)
Marmycins	Anthraquinone/ angucycline	*Streptomyces* sp. from marine sediment	Martin et al. (2007)
Spiroindimicins	Bisindole	*Streptomyces* sp. from deep-sea marine sediment	Zhang et al. (2012)
Venezuelins	Phenoxazine	*Streptomyces venezuelae* from marine sediment	Ren et al. (2013)
Nitropyrrolins	Farnesyl-nitropyrrole	Unclassified *Streptomycetaceae* sp. from marine sediment	Kwon et al. (2010)
Caerulomycin I	Bipyridine	*Actinoalloteichus cyanogriseus* from marine sediment	Fu et al. (2011)
Diazepinomycin	Dibenzo-diazepine	*Micromonospora* sp. from marine ascidian	Charan et al. (2004)
TP-1161	Thiazolyl peptide	*Nocardiopsis* sp. from marine sediment	Engelhardt et al. (2010a)
Lynamycins A–E	Bisindole pyrrole	Marinispora sp.	McArthur et al. (2008)
Lodopyridone	Pyridone	*Saccharomonospora* sp. from marine sediment	Maloney et al. (2009)
Salinosporamide A	γ-Lactam-β-lactone	*Salinispora arenicola* from marine sediment	Feling et al. (2003)

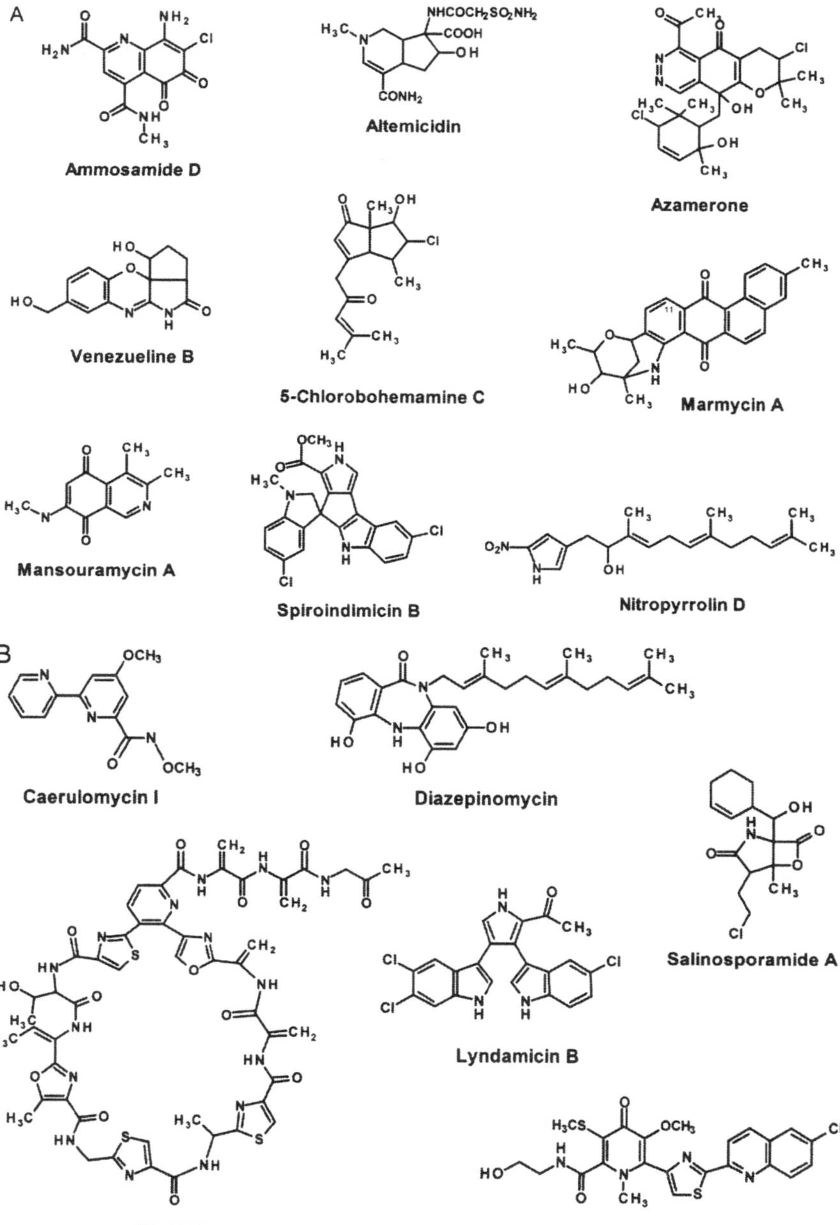

Figure 11.1 (A) Alkaloids from marine *Streptomyces* bacteria; (B) alkaloids from marine non-*Streptomyces* actinomycete bacteria.

Interestingly, none of these compounds demonstrated appreciable biological activity when tested for inhibition of the colon carcinoma cell line and antimicrobial activity.

Two anthraquinone/angucycline alkaloids designated as marmycins A and B were isolated from *Streptomyces* sp. CNH990 cultured from a marine sediment collected at the entrance to the Sea of Cortez, Mexico (Martin et al., 2007, Fig. 11.1A). Their biological activity was extensively tested *in vitro* using both cancer cell lines, bacteria and yeast. Marmycins A and B displayed no activity against human pathogens methicillin-resistant *Staphylococcus aureus*, vancomycin-resistant *Enterococcus faecium* and amphotericin-resistant *Candida albicans*. However, both compounds demonstrated activity against a number of tumour cell lines (Martin et al, 2007). Despite the presence of a chlorine atom at the C-11 position of the aromatic ring in marmycin B, this compound was shown to have considerably lower cytotoxic activity compared with its dehalogenated congener marmycin A (Fig. 11.1A), mean IC_{50} values for these compounds being 3.5 μM and 0.022 μM, respectively. Marmycin A seems to be of interest as a possible candidate for the development of a new anti-cancer drug, since synthetic routes to its derivatization have recently been designed (Zhang, Jiang, Ding, Yao, & Zhang, 2013).

Four cytotoxic isoquinoline quinone alkaloids mansouramycins A–D were isolated from *Streptomyces* sp. Mei37 derived from a sediment of Jade Bay on the southern German North Sea coast (Hawas et al., 2009). *In vitro* biological activity of mansouramycin A (Fig. 11.1A) was studied in some detail. This compound demonstrated moderate activity against both Gram-positive and Gram-negative bacteria, while being inactive against filamentous fungus and yeast. Its most profound activity was shown in assays against microalgae, *Chlorella* and *Scenedesmus*, and human tumour cell lines. In the latter assays, different mansouramycins displayed activity against cell lines representing solid tumours ranging from 0.089 to 13.44 μM (mean IC_{50} values). Remarkably, mansouramycins varied in their potency and selectivity towards different cell lines, which apparently depended on substitution pattern at positions C-3, C-4 and C-6. In the studies of human tumour xenograft in nude mice, mansouramycin B (carrying a chlorine atom substitution at C-6) demonstrated tumour-specific activity, suggesting that this compound may become a promising anti-cancer drug lead.

A series of novel bisindole alkaloids, spiroindimicins A–D, isolated from a culture broth of marine *Streptomyces* bacterium, were recently reported (Zhang et al., 2012). The producer, *Streptomyces* sp. SCSIO 03032, was isolated from a sediment collected at a depth of 3412 m from the Bay of Bengal in the Indian Ocean. Spiroindimicins have unprecedented skeletons featuring [5,6] or [5,5] spiro rings (structure for spiroindimicin B with [5,5] spiro ring is shown in Fig. 11.1B). As several other known bisindole alkaloids, most of the spiroindimicins displayed cytotoxic activity against several tumour cell lines, the spiroindimicin B being the most active with IC_{50} in the range of 4–12 μg/mL and spiroindimicin A showing no activity.

Phenoxazine alkaloids venezuelines were isolated from *Streptomyces venezuelae* KHG20-22 derived from the ocean sediment collected near the Guam island in the Pacific Ocean (Ren et al., 2013). Seven venezueline congeners were purified and characterized via structure elucidation and bioassays. The latter revealed that venezueline B (Fig. 11.1A) has moderate (IC_{50} range 5.74–9.56 μM) cytotoxicity towards five out of six tested tumour cell lines, showing no activity against human hepatoma cells. It was also shown that venezueline B induces the expression of orphan nuclear receptor Nur77, up-regulation of which triggers apoptosis in tumour cell lines (Liu et al., 2008).

Nitropyrrolins A–E, cytotoxic farnesyl–nitropyrrol alkaloids, were isolated from an unclassified bacterium of the family Streptopmycetaceae denoted as MAR4 strain CNQ-509 and isolated from a marine sediment collected from the Pacific Ocean off La Jolla, USA (Kwon et al., 2010). Nitropyrrolins represent hybrid polyketide-terpenoids apparently built from sesquiterpenoid and nitropyrrole moieties (Fig. 11.1A) and being the first secondary metabolites reported that possess such structural composition. All five nitropyrrolins were tested for biological activity against tumour cell lines and multi-resistant *S. aureus*. Nitropyrrolin D (Fig. 11.1A) was shown to be the most active compound against colon carcinoma cell line (IC_{50} 5.7 μM), while showing virtually no anti-bacterial activity.

Over the past 10 years, a small number of alkaloids have been isolated from actinomycetes of marine origin, which do not belong to the genus *Streptomyces*. This became possible mainly due to advances in culturing non-*Streptomyces* actinomycete bacteria and designing conditions favourable for the production of secondary metabolites (Jensen, Mincer, Williams, & Fenical, 2005). The first alkaloid isolated from a marine non-*Streptomyces* actinomycete bacterium was salinosporamide A (Fig. 11.1B; Feling et al., 2003). The obligate marine bacterium producing salinosporamide A and originally designated as *Salinospora* strain CNB-392 was isolated from a sediment collected from the tropical Pacific Ocean. Later, this bacterium was classified as *Salinispora arenicola* and proved to be a rich source of novel bioactive secondary metabolites (Freel, Nam, Fenical, & Jensen, 2011).

Salinosporamide A possesses a fused γ-lactam-β-lactone bicyclic ring similar to omuralide, a transformation product of the microbial metabolite lactacystin produced by a terrestrial *Streptomyces* sp. (Omura et al., 1991). Salinosporamide A displayed a potent and selective *in vitro* cytotoxicity against a National Cancer Institute panel of 60 cell lines, with a mean GI_{50} value (the concentration required to achieve 50% growth inhibition) of <10 nM. Soon after, it was shown that salinosporamide A specifically inhibits 20S proteasome involving β-lactone ring opening and irreversible binding triggered by the unique chloroethyl group (Groll, Huber, & Potts, 2006). This compound is currently undergoing Phase I clinical trials as an anti-cancer agent in patients with relapsed or refractory multiple myeloma (Gulder & Moore, 2010).

Diazepinomycin, an alkaloid of the dibenzodiazepine sub-class, was isolated from a *Micromonospora* sp. DPJ12, derived from the marine ascidian *Didemnum proliferum* collected at the Shishijima Island, Japan (Charan et al., 2004). This compound has a dibenzodiazepine core structure (Fig. 11.1B), which is extremely rare among natural products. In the initial study, diazepinomicin showed rather modest anti-microbial activity (minimal inhibitory concentration, MIC, ca. 32 μg/mL) against several Gram-positive bacteria. However, in 2009, Thallion Pharmaceuticals Inc. published a report on the discovery and characterization of TLN-4601, a natural compound structurally identical to deazepinomycin (Boufaied, Wioland, Falardeau, & Gourdeau, 2010). In this study, deazepinomycin demonstrated a broad cytotoxic activity and has shown *in vivo* anti-tumour activity in several xenograft models. The same compound was recently isolated from another marine actinomycete bacterium, *Micromonospora* strain RV115 isolated from the marine sponge *Aplysina aerophoba* collected near Rovinj, Croatia (Abdelmohsen et al., 2010). A different approach to the characterization of diazepinimycin revealed its anti-oxidant activity resulting in protection of cells from toxicity and DNA damage induced by hydrogen peroxide. Moreover, diazepinomycin was shown to inhibit protease cathepsin L implicated in the cancer progression and metastasis, thereby suggesting its possible mechanism of anti-cancer activity (Abdelmohsen et al., 2012).

Group of Fenical at Scripps Institute of Oceanography (USA) has reported a species of marine actinomycete, *Saccharomonospora* sp. CNQ490, isolated from a sediment collected at the mouth of the La Jolla Submarine Canyon, USA. This bacterium was shown to biosynthesize a novel alkaloid lodopyridone with unique carbon skeleton (Maloney et al., 2009; Fig. 11.1B) and required sea water for growth and production of this metabolite. Lodopyridone was shown to have significant activity against the human colon adenocarcinoma cell line (IC_{50} 3.6 μM) but did not exhibit any activity against methicillin-resistant *S. aureus*. Unfortunatelty, the yields of lodopyridone in fermentations were too low to continue with extensive biological characterization of this compound. Recently reported total synthesis of lodopyridone may alleviate this problem and supply enough material for the necessary tests (Burckhardt, Harms, & Koert, 2012).

A novel marine actinomycete bacterium, assigned to a proposed genus *Marinispora*, was isolated from a sediment sample collected at the Mission Bay in San Diego (USA). Its cultivation yielded several compounds showing activity against methicillin-resistant *S. aureus*. Subsequent purification and structure elucidation revealed that the bioactivity was due to five halogenated bisindole pyrroles, designated lynamicins A–E (McArthur et al., 2008). Lyndamycins were tested against a panel of Gram-positive and Gram-negative bacteria, showing broad-spectrum anti-microbial activity. Lyndamycin B (Fig. 11.1B) demonstrated the highest activity, with MIC values 1.8–18 μg/mL, with no distinction between antibiotic-resistant and -sensitive bacteria.

A marine *Nocardiopsis* sp. isolated from a sediment collected from the Trondheimfjord, Norway, yielded a thiopeptide antibiotic TP-1161 (Engelhardt et al., 2010a; Fig. 11.1B). As in the case of many other marine actinomycete bacteria, production of this compound was shown to be dependent on sea water. Structurally, TP-1161 belongs to the group of the *d* series of thiopeptide antibiotics characterized by the presence of a 2,3,6-trisubstituted pyridine unit located at the centre of a peptide macrocycle. Typical for the antibiotic of this class, TP-1161 demonstrated potent activity against a panel of Gram-positive bacteria, including vancomycin-resistant *Enterococci*, while showing no activity against Gram-negative strains. Soon after, using bioinformatics and mining of the *Nocardiopsis* draft genome sequence, the TP-1161 biosynthetic gene cluster was identified and characterized (Engelhardt et al., 2010b). This compound is produced first as a ribosomally synthesized pre-peptide, which is subsequently modified post-translationally by an unique enzymatic machinery.

Caerulomycins F–K are new bipyridine alkaloids produced, along with five known congeners, by the actinomycete bacterium *Actinoalloteichus cyanogriseus* WH1-2216-6 isolated from marine sediments collected from the seashore of Weihai, China (Fu et al., 2011; Fig. 11.1B). All these bipyridines were suggested to be biosynthesized by a hybrid pathway from amino acids and acyl units, although no experimental evidence was presented. Investigation of the biological activity of the new caerulomycins showed all of them, except for congener G, having cytotoxic effect on several tumour cell lines. Caerulomycin I (Fig. 11.1B) was shown to be the most active compound, displaying IC_{50} values in the range of 0.37–5.2 μM.

2.2. Alkaloids from marine cyanobacteria

Cyanobacteria, sometimes referred to as blue-green algae, represent another prominent group of marine bacteria known to produce a large variety of alkaloids. Most of these alkaloids are highly toxic to eukaryotic organisms and are likely to be produced by cyanobacteria as a defence warfare against predators. Around 300 alkaloids from marine cyanobacteria have been reported, majority of which are represented by peptides and polyketide–peptide hybrids biosynthesized by large multi-modular enzymes (see Section 3). The majority of these compounds were isolated from the filamentous order *Nostocales*, especially from the strains belonging to the genera *Lyngbya*, *Oscillatoria* and *Symploca* (Gerwick, Tan, & Sitachitta, 2001). Several cyanobacterial alkaloids have been exploited as starting points for the development of anti-cancer drug candidates, with several tubulin-targeting derivatives reaching clinical trials (Tan, 2007). Some of the cyanobacterial alkaloids, representing various sub-classes, are listed in Table 11.2, and their chemical structures are shown in Fig. 11.2.

Hectochlorin (Fig. 11.2) is a chlorinated lipopeptide isolated from a marine cyanobacterium *Lyngbya majuscula* collected from the Hector Bay, Jamaica (Marquez et al., 2002). Hectochlorin has a potent activity against eukaryotic cells, including yeasts and various tumour cell lines. In the tests against a panel of 60 different tumour cell lines at the National Cancer Institute, hectochlorin showed an average GI_{50} value of 5.1 μM, being the most active against cell lines in the colon, melanoma, ovarian and renal panels. Its cytotoxicity was shown to be due to promotion of actin polymerization and thus interfering with the normal progression of cell cycle.

Largazole, a macrocyclic depsipeptide, was isolated from cyanobacterium *Symploca* sp. collected from the Key Largo, Florida Keys, USA (Taori et al., 2008). Largazole (Fig. 11.2) was shown to inhibit the growth of highly invasive breast cancer cells (GI_{50} 7.7 nM), while being much less toxic to normal, non-transformed mammary epithelial cells (GI_{50} 122 nM). Such selectivity, very rarely observed among cytotoxic compounds, apparently triggered considerable interest to largazole, prompting its chemical derivatization and evaluation of analogues (Seiser, Kamena, & Cramer, 2008).

Table 11.2 Selected alkaloids from marine cyanobacteria (see Fig. 11.2 for chemical structures)

Name	Sub-class	Source	References
Hectochlorin	Lipopeptide	*Lyngbya majuscule*, marine isolate	Marquez et al. (2002)
Largazole	Macrocyclic depsipeptide	*Symploca* sp., marine isolate	Taori, Paul, and Luesch (2008)
Harman	β-carboline	*Geitlerinema* sp., marine isolate	Caicedo, Kumirska, Neumann, Stolte, and Thöming (2012)
Saxitoxin	3,4-Propinoperhydropurine	*Anabaena, Cylindrospermopsis, Aphanizomenon, Planktothrix* and *Lyngbya* spp. marine isolates	Martins, Pereira, Welker, Fastner, and Vasconcelos (2005)
Curacin A	N-containing lipid	*Lyngbya majuscule*, marine isolate	Gerwick et al. (1994)
Symplostatin 1	Peptolide	*Symploca hydnoides*, marine isolate	Harrigan et al. (1998)
Phormidinines	2-Alkylpyridine	*Phormidium* sp.	Teruya, Kobayashi, Suenaga, and Kigoshi (2005)
Lyngbyatoxin A	Diazoninoindol	*Lyngbya majuscula* Gomont	Cardellina, Marner, and Moore (1979)
Barbamide	Thiazolhexenamide	*Lyngbya majuscula*	Orjala and Gerwick (1996)
Bromoanaindolone	Indole	*Anabaena constricta*, marine isolate	Volk, Girreser, Al-Refai, and Laatsch (2009)

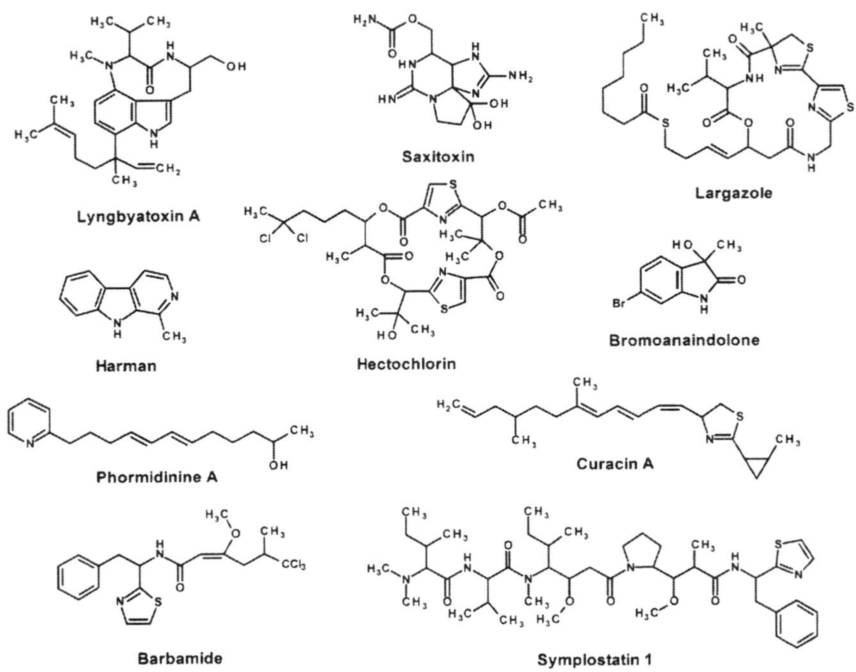

Figure 11.2 Alkaloids from marine cyanobacteria.

It has been shown that the cytotoxic activity of largazole is due to its being an inhibitor of the histone deacetylase enzymes involved in the proliferation of tumour cells (structural basis of the anti-proliferative activity of largazole, a depsipeptide inhibitor of the histone deacetylases; Cole, Dowling, Boone, Phillips, & Christianson, 2011). Recently, *in vivo* anti-tumour activity of largazole has been demonstrated in a human colon cancer xenograft mouse model, making this compound a promising lead for the development of anti-cancer drugs (Liu et al., 2010).

Harman (Fig. 11.2) is an indole alkaloid isolated from a marine cyanobacterium *Geitlerinema* sp. (Caicedo et al., 2012). It belongs to the group of compounds known as harmala alkaloids, many of which were shown to be inhibitors of monoamine oxidase (Herraiz & Chaparro, 2005). Harman was shown to bind to the benzodiazepine, imidazoline and serotonin receptors, thereby causing anti-depressant effects (Aricioglu & Altunbas, 2003; Baum, Hill, & Rommelspacher, 1996). However, its usefulness as a drug is questionable, since its close analogue, norharman, was shown to be a co-mutagen in the presence of aromatic amines, thereby increasing the carcinogenicity of the latter compounds (Totsuka et al., 1998).

Saxitoxin (Fig. 11.2) is probably the most famous neurotoxic alkaloid first discovered in dinoflagellates and later in the fresh-water and marine cyanobacteria (Wiese, D'Agostino, Mihali, Moffitt, & Neilan, 2010). Saxitoxin and its analogues are also known as the paralytic shellfish toxins, causative agents of shellfish poisoning resulting for the consumption of shellfish that have accumulated saxitoxin-producing organisms. Mechanism of neurotoxic action of saxitoxin is based on its reversible binding to the voltage-gated Na^+ channels selectively in the nerve, thereby blocking their normal function (Strichartz et al., 1986). Although the anaesthetic activity of saxitoxin has been known for many years (Adams, Blair, & Takman, 1976), its usefulness as a drug lead remains dubious due to the very high toxicity.

Another cyanobacterial alkaloid, curacin A (Fig. 11.2), received a considerable attention as a starting point for drug development. This compound, an unprecedented lipid linked to the 2-cyclopropyl-4-alkenyl substituted thiazoline unit, has been isolated from a marine cyanobacterium *L. majuscula* collected off the coast of Curaçao (Gerwick et al., 1994). Curacin A was shown to be a potent anti-mitotic agent that inhibits microtubule assembly, causing cell cycle arrest and apoptosis (Wipf, Reeves, & Day, 2004). Apparently, low water-solubility and chemical instability so far prevented the clinical development of curacin A, but synthetic analogues with improved solubility may pave the way for further development of this compound as an anti-cancer drug.

Marine cyanobacterium *Symploca hydnoides* isolated from a sample collected at the reef flat of the Pago Bay, Guam Island, was shown to produce a highly cytotoxic compound named symplostatin 1 (Harrigan et al., 1998; Fig. 11.2). This compound is a peptolide, apparently synthesized by a non-ribosomal peptide synthetase with subsequent modifications. Symplostatin 1 exhibited cytotoxicity against an epidermoid carcinoma line with an IC_{50} value of 0.3 ng/mL and is one of the most potent cytotoxic compounds known. The fact that it can be isolated from a cultivable bacterium makes it more valuable in terms of drug development, as opposed to its close homologue dolastatin 10, which can only be isolated in minute quantities from a mollusc *Dolabella auricularia* (Bai, Pettit, & Hamel, 1990). Cytotoxic activity of symplostatin 1 was shown to be due to its induction of the formation of abnormal mitotic spindles and accumulation of cells in metaphase, ultimately causing cell cycle arrest and apoptosis. It was also shown that symplostatin 1 inhibits the tubulin assembly strongly suggesting that tubulin is its intracellular target. *In vivo* evaluation of symplostatin 1 in murine model of mammary cancer demonstrated good activity, but apparent toxicity prevented this compound from being further developed.

Two rare 2-alkylpyridine alkaloids phormidinines A and B (Fig. 11.2) were isolated from a marine mat-forming cyanobacterium *Phormidium* sp. collected at Bise, Okinawa, Japan (Teruya et al., 2005). Unfortunately, biological activity of phormidinines was not reported, and these compounds remain, as of today, mere examples of chemical diversity of alkaloids produced by marine bacteria.

Lyngbyatoxin A, an inflammatory compound and causative agent of seaweed dermatitis ('swimmers itch') has been isolated from the extract of a marine cyanobacterium *L. majuscula* Gomont collected from the Hawaiian shallow-waters (Cardellina et al., 1979). Structurally, lynbyatoxin A belongs to the sub-class of diazoninoindole alkaloids (Fig. 11.2). This compound was shown to be a potent tumour promoter in mice (Fujiki et al., 1984), and this activity of lyngbyatoxin A is likely to be due to its competitive binding to protein kinase C (Jeffrey & Liskamp, 1986). Currently, no data exist on concentrations of lyngbyatoxin A in marine environments, and its health hazard to humans cannot be fully assessed.

The lipid extract from the cyanobacterium *L. majuscula* collected from Curaçao was shown to have several bioactivities, including brine shrimp toxicity, fish toxicity molluscicidal activity (Orjala & Gerwick, 1996). Further investigation revealed that the latter activity was due to a novel alkaloid of the thiazolehexenamide sub-class, named barbamide (Fig. 11.2). In the original study, purified barbamide exhibited molluscicidal activity with an LC_{100} (concentration causing 100% lethality) value of 100 μg/mL. Although further development of barbamide as a bioactive compound was not pursued, studies on its biosynthesis yielded discoveries of unprecedented enzymatic steps in natural product pathways. In particular, it has been shown that two halogenases, acting in tandem, provide a starting unit for barbamide biosynthesis via triple chlorination of leucine (Galonić, Vaillancourt, & Walsh, 2006).

Cyanobacterium *Anabaena constricta* of presumed marine origin from a Kuwait Culture Collection of Algae (Kuwait University) was shown to produce a brominated indole alkaloid named bromoanaindolone (Volk et al., 2009). Bromoanaindolone was shown to be a single constituent of the extracts from *A. constricta*, which is very rare for bacteria producing secondary metabolites that usually yield very complex extracts with tens of various metabolites.

In anti-microbial assay against *Bacillus cereus*, bromoanaindolone exhibited low activity with an MIC value of 128 μg/mL. Its activity against other cyanobacteria was somewhat higher, minimum toxic quantities determined in the specialized solid matrix assays ranging from 8 to 20 μg. No further studies on bromoanaindolone have been reported to date.

2.3. Alkaloids from other marine bacterial species

Although the absolute majority of alkaloids isolated from marine bacteria originate from actinomycetes and cyanobacteria, other bacterial species also were found to produce structurally unique bioactive alkaloids (Table 11.3).

Two highly brominated pyrrole alkaloids pentabromopseudulin and tetrabromobenzofuro(3,2-b) pyrrole (Fig. 11.3) were isolated from a strain of *Pseudoalteromonas* sp. associated with a surface of nudibranch collected from the Kaneohe Bay, Oahu (Fehér et al., 2010). Pentabromopseudulin has been previously isolated from *Pseudoalteromonas bromoutilis*, *Pseudoalteromonas luteoviolacea* and *Chromobacterium* and was shown to have a range of biological activities, including inhibition of both human lipoxygenases myosin-dependent processes (Fedorov et al., 2009; Ohri et al., 2005). Tetrabromobenzofuro(3,2-b)pyrrole was shown to be active against methicillin-resistant *S. aureus* (IC_{50} 1.93 mg/mL) and yeast *C. albicans* (no IC_{50} data provided).

Studies on the secondary metabolites from bacteria associated with the Caribbean ascidian *Stomozoa murrayi* (Kott, 1957) yielded a strain of *Acinetobacter* sp. producing a brominated indole alkaloid 6-bromoindole-3-carbaldehyde (Olguin-Uribe et al., 1997). The latter compound (Fig. 11.3) exhibited moderate anti-bacterial activity in agar diffusion assay and molluscicidal activity in settlement assay with *Balanus amphitrite* larvae. De-brominated analogue of 6-bromoindole-3-carbaldehyde was also isolated and tested, showing virtually no biological activity.

Violacein is perhaps the oldest known alkaloid first isolated from a terrestrial bacterium, *Chromobacterium violaceum*, and later confirmed present in marine isolates, such as *Collimonas* CT collected from the sea surface of the Trondheim Fjord, Norway (Hakvåg et al., 2009). Phylogenetic analysis of *vioA* and *vioB* violacein biosynthetic gene fragments from *Collimonas* CT demonstrated their closest relatedness to similar genes from violacein-producing *Duganella* sp. B2 isolated from a glacier in the People's Republic of China. At the same time, homology of these genes to the violacein biosynthesis genes from a soil-derived isolate, *C. violaceum*, was found to be

Table 11.3 Selected alkaloids from diverse marine bacteria (see Fig. 11.3 for chemical structures)

Name	Sub-class	Source	References
Pentabromopseudulin Tetrabromobenzofuro(3,2-b) pyrrole	Pyrrole	*Pseudoalteromonas* sp (CMMED 290) isolated from the surface of a nudibranch	Fehér, Barlow, McAtee, and Hemscheidt (2010)
6-Bromoindole-3-carbaldehyde	Indole	*Acinetobacter* sp. associated with an ascidian	Olguin-Uribe et al. (1997)
Violacein	Indolopyrrole	*Collimonas* sp. from sea surface	Hakvåg et al. (2009)
Thiomarinols A–G	Dithiolopyrrolone	*Alteromonas rava* from sea water	Shiozawa et al. (1993)
Marinoquinoline A	Pyrrolo[2,3-c] quinoline	marine gliding bacterium *Rapidithrix thailandica*	Sangnoi et al. (2008)
Aqabamycins A–G	Nitromaleimides	*Vibrio* sp. associated with corral	Al-Zereini, Fotso Fondja Yao, Laatsch, and Anke (2010)
Holomycin	Dithiopyrrolone	*Photobacterium halotolerans*	Wietz, Mansson, Gotfredsen, Larsen, and Gram (2010)
Bisucaberin	Cyclic dihydroxamate	*Alteromonas haloplanktis* from deep-sea mud	Takahashi, Kobayashi, et al. (1987) and Takahashi, Nakamura, et al. (1987)

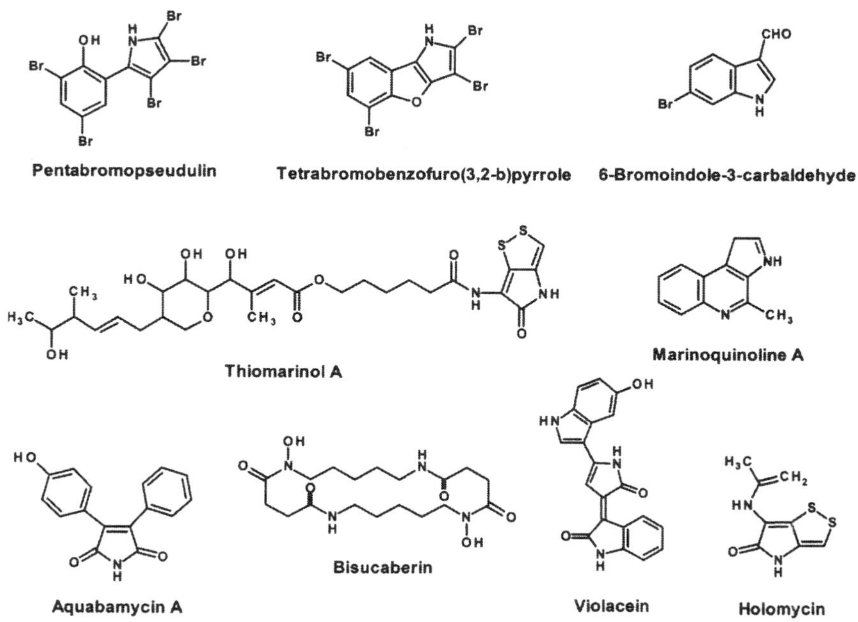

Figure 11.3 Alkaloids from diverse marine bacteria.

considerably lower. *Collimonas* CT and *Duganella* sp. B2 were also taxonomically close according to the 16S RNA-based phylogenetic analysis, suggesting vertical rather than horizontal transfer of violacein biosynthetic genes.

Violacein (Fig. 11.3) is an indolopyrrole alkaloid with a wide range of biological activities. In particular, violacein has been shown to cause apoptosis in myeloid leukaemia cells through the activation of genes for caspase 8, transcription of NF-κB and p38 MAPK (Durán et al., 2007). Violacein exhibited anti-microbial activity against Gram-positive bacteria, including *Mycobacterium tuberculosis*, but was inactive against Gram-negative species. However, violacein has a very low aqueous solubility, preventing its evaluation *in vivo* (i.e. in animal models) as a potential drug lead.

Nine novel alkaloids belonging to a dithiolopyrrolone sub-class were shown to be produced by a bacterium *Alteromonas rava* (later re-classified as *Pseudoalteromonas* sp.) isolated from sea water in Japan (Shiozawa et al., 1993). These compounds were designated thiomarinols A–G (Fig. 11.3), and their activity was tested against a panel of Gram-positive and Gram-negative bacteria. Thiomarinol A was shown to be the most active compound, inhibiting the growth of both types of bacteria with MIC values ranging from 0.1 to 25 μg/mL. Several thiomarinol derivatives were prepared by chemical synthesis, several of them shown good anti-microbial activity (Marion, Gao, Marcus, & Hall, 2009).

Interestingly, the biosynthesis of thiomarinols was shown to be governed by two gene clusters located on a plasmid, which yielded two components of thiomarinol: pseudomonic acid and pyrrothine. These components are then joined by an amide synthetase to provide the mature thiomarinol molecule (Fukuda et al., 2011). Recently, based on the knowledge of the biosynthetic pathway, a number of thiomarinol derivatives have been obtained via gene-specific mutagenesis and mutasynthesis (Murphy et al., 2011). Some of these analogues exhibited potent activity in anti-bacterial assays, hopefully paving a way for the development of thiomarinol analogues for anti-bacterial therapy.

The pyrroloquinoline alkaloid marinoquinoline A (Fig. 11.3) was isolated from a marine-gliding bacterium *Rapidithrix thailandica* collected from the Andaman Sea, Thailand (Srisukchayakul et al., 2007). Structurally, marinoquinoline A resembles tacrine, a potent inhibitor of human acetylcholinesterase (AChE). A shortage of acetylcholine is characteristic of patients with Alzheimer's disease, and the use of AChE inhibitors to slow down the hydrolysis of acetylcholine has been suggested as one of the strategies to treat such patients. In the AChE inhibition assay, the marinoquinoline A exhibited potent activity with an IC_{50} of 4.9 μM, while showing no appreciable cytotoxicity to the panel of cancer cell lines at 20 μM.

A novel *Vibrio* species isolated from the surface of the soft coral *Sinularia polydactyla* collected from the Red Sea, Japan, produced a series of nitrated maleimide alkaloids designated Aqabamycins A–G (Al-Zereini et al., 2010; Fig. 11.3). Aqabamycins were shown to have activity against Gram-positive bacteria and represent an interesting group of compounds for further biological evaluation and chemical derivatization due to the high degree of nitrosubstitution.

Holomycin (Fig. 11.3) is a dithiopyrrolone alkaloid produced predominantly by terrestrial bacteria but recently found in marine-derived *Photobacterium halotolerans*, a member of Vibrionaceae family (Wietz et al., 2010). This alkaloid was shown to be active against a wide variety of bacteria, with some exceptions (e.g. *Enterobacter cloacae*, *Morganella morganii* and *Pseudomonas aeruginosa*), but was inactive against fungi. Studies on holomycin mode of action suggested the inhibition of RNA chain elongation. However, only weak inhibition of *Escherichia coli* RNA polymerase in the *in vitro* assays was demonstrated, suggesting that holomycin might require metabolic conversion in the cell in order to inhibit RNA polymerase (Oliva, O'Neill, Wilson, O'Hanlon, & Chopra, 2001).

Another marine bacterium, *Alteromonas haloplanktis* SB-1123 isolated from the deep-sea mud, was shown to produce a novel siderophore designated as bisucaberin (Takahashi, Kobayashi, Asano, Yoshida, & Nakano, 1987; Takahashi, Nakamura, et al., 1987). Structurally, bisucaberin is a cyclic dihydroxamate (Fig. 11.3) and was found to bind ferric ions, thereby becoming soluble in water. This compound exhibited an interesting biological activity in that it selectively sensitized tumour cells to the cytolytic action of macrophages. Bisucaberin was also shown to act as a cytostatic on tumour cells, while displaying no cytocidal activity. In complex with iron, the bisucaberin became inactive, strongly suggesting that its ability to sensitize tumour cells is due to iron-chelating properties. Bisucaberin biosynthesis gene cluster has recently been cloned from a deep-sea metagenome, suggesting that this compound may play an important ecological role in marine bacterial communities, for example, by scavenging iron from sea water (Fujita, Kimura, Yokose, & Otsuka, 2012).

3. BIOSYNTHESIS OF ALKALOIDS IN MARINE BACTERIA

A considerable amount of work has been dedicated over the last decade to deciphering biosynthetic pathways for marine bacteria-derived alkaloids. These studies provided important new insights into the enzymology of natural product biosynthesis and pave the way for biosynthetic engineering that may yield new derivatives with improved pharmacological properties. Some of the most interesting examples of such deciphered pathways are presented below.

3.1. Biosynthesis of violacein

One of the first elucidations of biosynthetic pathways for alkaloids derived from marine environment was that for violacein, an antibiotic pigment produced by both terrestrial and marine bacteria (August et al., 2000). Initial biosynthesis studies have shown that all carbon and nitrogen atoms in violacein molecule originate from L-tryptophan, and oxygen atoms are derived from molecular oxygen (DeMoss & Evans, 1960; Hoshino, Kondo, Uchiyama, & Ogasawara, 1987; Hoshino, Takano, Hori, & Ogasawara, 1987). The four genes required for the biosynthesis of violacein, designated *vioA–D*, have been cloned from *C. violaceum* and functionally expressed in *E. coli* (Pemberton, Vincent, & Penfold, 1991). The violacein biosynthesis pathway has recently been revisited, adding another gene, *vioE*, to the pathway.

The route from L-tryptophan to violacein (Fig. 11.4) was found to be closely related to the biosynthesis of indolocarbazoles, a group of biologically active alkaloids frequently found in marine bacteria (Sánchez et al., 2006). First step in the violacein biosynthesis is accomplished by VioA, an L-amino acid oxidase, which converts L-tryptophan to indole-3-puruvic acid (IPA) alternating between its imine and keto forms. Next, VioB, an enzyme resembling a chromopyrrolic acid (CPA) synthase, assisted by VioE proposed to be involved in the intramolecular rearrangement of the indole ring converts IPA into deoxyproviolacein. It was suggested that VioE is essential for re-routing the pathway away from CPA, which would be the product of VioB alone, towards deoxyproviolacein. The latter compound can be a substrate for either of oxygenases VioC and VioD, which through intermediates proviolacein or deoxyviolacein complete the pathway, furnishing a violacein molecule. Clearly, the violacein biosynthesis pathway shares common initial steps with the biosynthesis of bisindole alkaloids such as staurosporine (see below).

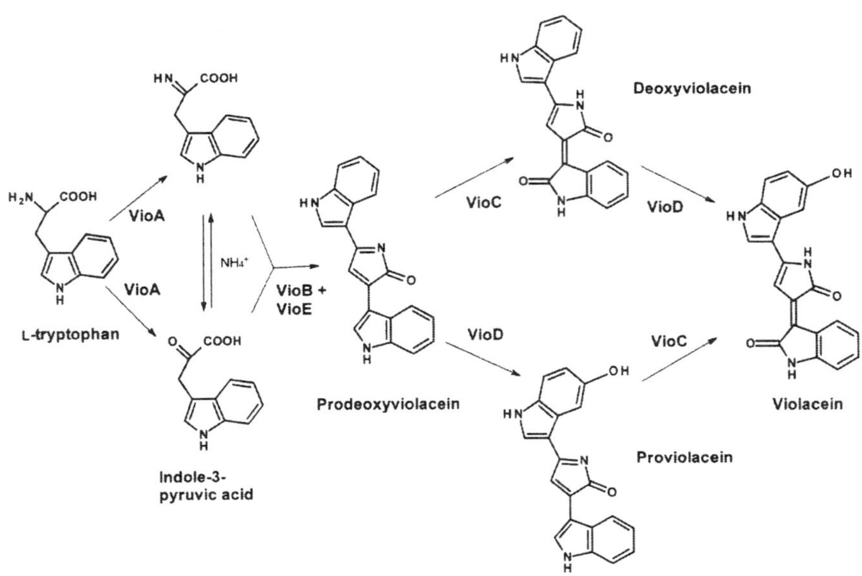

Figure 11.4 Proposed pathway for the biosynthesis of violacein.

3.2. Biosynthesis of staurosporine and related bisinolde alkaloids

Bisindole alkaloids, including marine actinomycete-derived lyndomycin B (Table 11.2, Fig. 11.2), are of considerable interest as exploratory compounds in search for anti-cancer drugs. Staurosporine, a bisindole alkaloid originally isolated from terrestrial actinomycete, and its analogues have received much attention as potent inhibitors of protein kinase C (Hoehn, Ghisalba, Moerker, & Peter, 1995; Takahashi, Kobayashi, et al., 1987; Takahashi, Nakamura, et al., 1987). The genes for biosynthesis of staurosporine have been reported (Onaka, Taniguchi, Igarashi, & Furumai, 2002), and functions of their products in the biosynthesis of this molecule are elucidated (Salas et al., 2005; Sánchez et al., 2005). Like in the violacein pathway, the biosynthesis of staurosporine and related bisindoles starts with the conversion of L-tryptophan to imine form of IPA by an oxidase StaO (Fig. 11.5).

Two molecules of IPA are then dimerized by an unusual heme-containing enzyme StaD. The product of StaD is believed to be an unstable intermediate, which is spontaneously converted into the CPA. The next step in the staurosporine biosynthesis is an oxidation of the CPA to staurosporinone, which is accomplished by enzymes cytochrome P450 StaP and oxidoreductase StaC encoded by the staurosporine cluster, and apparently assisted by non-specific flavodoxin reductase and ferredoxin. The biosynthesis is finalized by two enzymes, StaG and StaN, catalyzing coupling of the two indole nitrogens to an unusual sugar L-ristosamine.

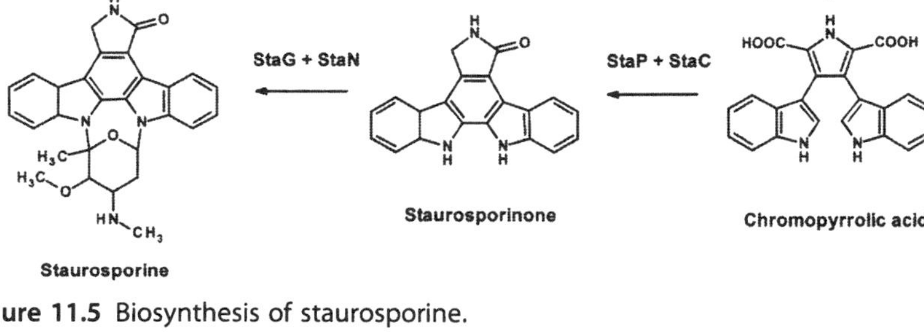

Figure 11.5 Biosynthesis of staurosporine.

3.3. Biosynthesis of diazepinomicin

Biosynthesis of the prenylated dibenzodiazepine alkaloid diazepinomycin produced by marine *Micromonospora* sp. (Table 11.2, Fig. 11.2) is an interesting example of a convergence of three different metabolic pathways to yield one particular molecule. From the biosynthetic point of view, diazepinomycin clearly represents a 'hybrid' molecule composed from three distinct moieties: benzoquinone, anthranilate and farnesyl diphosphate, each of which is synthesized separately. Deciphering of the diazepinomycin biosynthesis has become possible after cloning and analysis of its biosynthetic gene clusters, accompanied by feeding experiments (McAlpine et al., 2008).

The diazepinomycin biosynthesis gene cluster appears to contain 42 genes, and functions for many of them could be deduced from the bioinformatics analyses. According to the model proposed (Fig. 11.6), the 2-amino-6-hydroxybenzoquinone component is synthesized from phosphoenolpyruvate and erythrose-4-*O*-phosphate via a sequential action of enzymes homologous to 3-deoxy-D-arabinoheptulosonic acid 7-phosphate synthase, salicilate 1-monooxygenase, 2,4-dihydrohybenzoate monooxygenase and a multicopper oxidase. The other component of the molecule, 3-hydroxy-anthranilate (adenylated), is apparently formed from chorismic acid via the action of enzymes similar to anthranilate synthase,

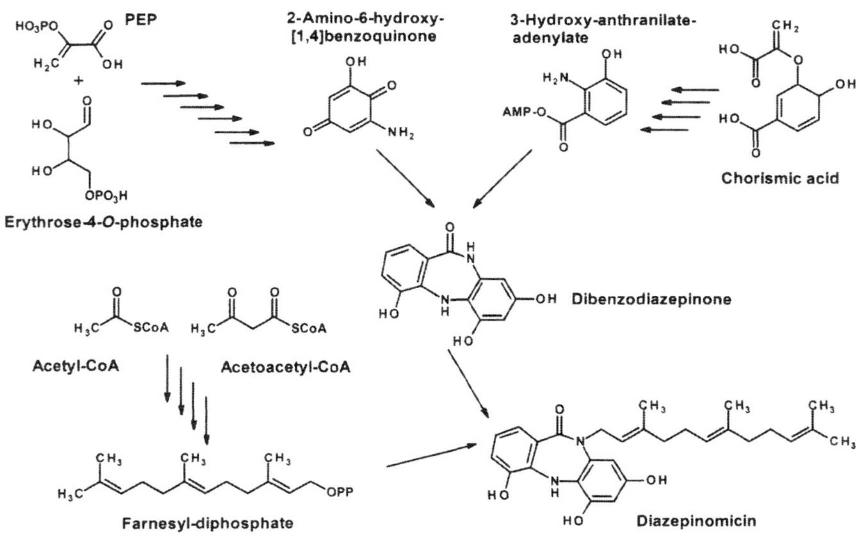

Figure 11.6 Proposed pathway for the biosynthesis of diazepinomicin in marine *Micromonospora* sp.

isochorismatase, 2,3-dihydro-2,3-dihydroxybenzoate dehydrogenase and adenylate ligase. The isoprenoid moiety is biosynthesized from acetyl-coenzyme A (CoA) and acetoacetyl-CoA by the enzymes of mevalonate pathway encoded in the diazepinomicin biosynthesis cluster. This part of the pathway starts with the condensation of acetoacetyl-CoA and acetyl-CoA by an enzyme similar to hydroxymethylglutaryl-CoA (HMG-CoA) synthase and is followed by a series of reactions involving HMG-CoA reductase, mevalonate- and phosphomevalonate kinases, diphosphomevalonate decarboxylase, isopentyl diphosphate isomerase and farnesyl diphosphate synthase. The biosynthetic pathway for the three parts described earlier converges in two steps. First, the 2-amino-6-hydroxybenzoquinone and 3-hydroxy-anthranilate-adenylate self-condense to form the dibenzo-diazepinone moiety. In a second and final step of molecule assembly, the farnesyl moiety is transferred to dibenzodiazepinone by a specific prenyltransferase to furnish diazepinomicin.

3.4. Biosynthesis of barbamide

Feeding studies on the biosynthesis of barbamide, a chlorinated thiazolhexenamide isolated from marine cyanobacterium *L. majuscule*, has shown it to be synthesized from L-leucine, L-cysteine, L-phenylalanine, acetate and *S*-adenosylmethionine (Sitachitta et al., 2000). Subsequent cloning and analysis of the barbamide biosynthetic gene cluster (Chang et al., 2002), followed by detailed analysis of a chlorination reaction converting L-leucine starter into trichlorleucin (Galonić et al., 2006) shed light on the biosynthesis of this unusual metabolite.

Barbamide biuosynthesis (Fig. 11.7) starts with the activation of L-leucine by a non-ribosomal peptide synthatase (NRPS)-like adenylation domain, represented by protein BarD, onto the peptide carrier protein (PCP) BarA. White tethered to BarA via a thioester bond, sequentially leucin is trichlorinated by the halogenases BarB2 and BarB1. Next, trichloroleucine is released from BarA by a thioesterase BarC and transferred to the hybrid NRPS–polyketide synthase protein BarE, where it is converted into a trichloroisovaleryl, presumably via the oxidative deamination by BarJ. The trichlorisovaleryl moiety is condensed with the malonyl-CoA-derived acetate and transferred to the acyl carrier protein (ACP) domain of the PKS BarF. While linked to BarF, the extended product undergoes an unusual reduction of the g-carbon resulting in the formation of a C-4–C-5 *E*-double bond by as-yet-unidentified mechanisms and

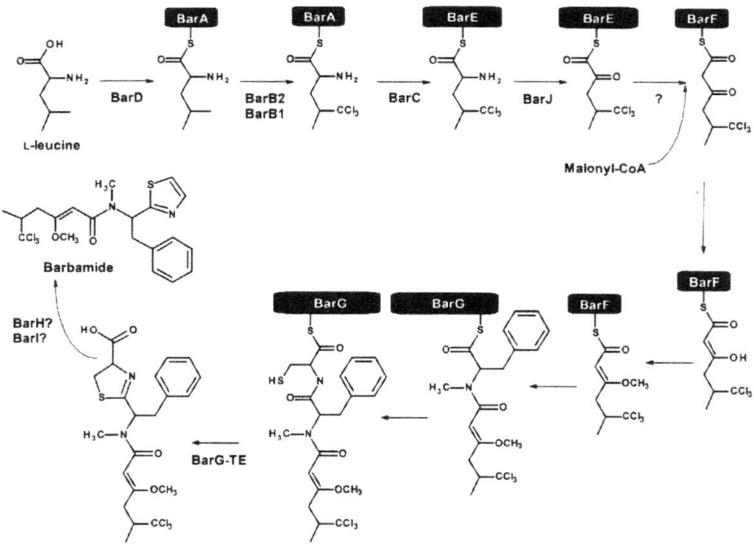

Figure 11.7 Proposed barbamide biosynthesis pathway in *Lyngbya majuscula*.

methylation of the resulting hydroxyl by the methyltransferase domain of BarF. The final steps of the barbamide biosynthesis comprise transfer of the BarF product onto the bi-modular NRPS BarG, where it is extended with L-phenylalanine and L-cysteine. It is not yet clear how the formation of the final thiazole moiety of barbamide is accomplished, since no enzymatic domains with such activity could be identified in BarG. It is possible that the formation of thiazole is catalyzed by the BarH and BarI proteins upon release of the aminoacyl carboxylic acid precursor from BarG by its thioesterase domain.

3.5. Biosynthesis of holomycin

Holomycin, the dithiopyrrolone alkaloid originally identified in a terrestrial *Streptomyces* bacterium, and later found to be produced by marine bacteria, is another example of a molecule synthesized by NRPS. The clue to its biosynthesis was provided after cloning and analysis of the biosynthetic gene clusters from *Streptomyces clavuligerus*, a well-known produced of clavulanic acid (Li & Walsh, 2010). The biosynthesis of holomycin is interesting also with respect to other marine bacteria-derived alkaloids containing dithiopyrrolone moiety, such as thiomarinols (Table 11.3, Fig. 11.3).

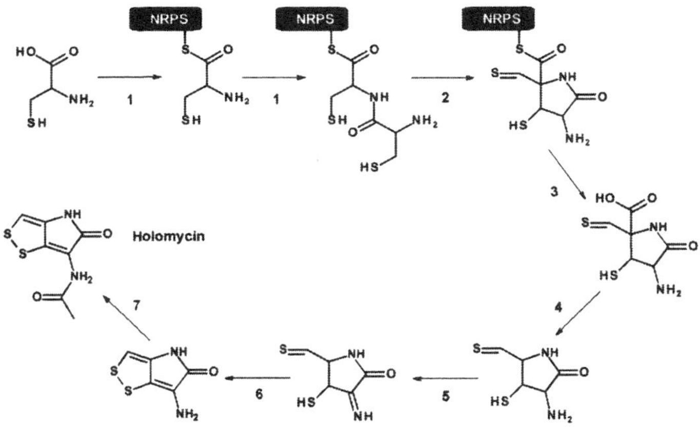

Figure 11.8 Proposed model for holomycin biosynthesis in *Streptomyes clavuligerus*. Numbers refer to various biosynthetic enzymes described in the text.

Two molecules of C-cysteine represent biosynthetic origin of holomycin (Fig. 11.8). Apparently, they are activated by the adenylation domain of the NRPS (1), and condensed on the same enzyme, probably assisted by a discrete NRPS condensation domain encoded by another gene in the cluster. According to the model, the dicysteine, while still tethered to the NRPS, is processed by an acyl-CoA dehydrogenase (2), yielding a product that spontaneously cyclizes. The next step in the biosynthesis is the release of the cyclized intermediate by a thioesterase (3). The released carboxylic acid intermediate is presumed to be a substrate to a decarboxylase (4), the product of which is converted to an imine intermediate by an oxidoreductase (5). The final step in the formation of holomycin dithiopyrrolone moiety is accomplished by a thioredoxin-disulphide reductase homologue (6), which converts the disulphide into dithiol. The last step in holomycin biosynthesis is acetylation of dithiopyrrolone by an acetyltransferase (7).

3.6. Biosynthesis of salinosporamide A

Alkaloid salinosporamide A from marine actinomycetes of the genus *Salinospora* received much attention both due to its potent anti-cancer activity and unusual biosynthetic pathway, which involves the formation of a specific chlorinated extender unit derived from *S*-adenosyl-L-methionine (SAM; Gulder & Moore, 2010). An enzyme SalL representing a novel class of enzymes and showing homology to fluorinases performs a displacement of the methionine in SAM with a simultaneous addition of a chlorine atom

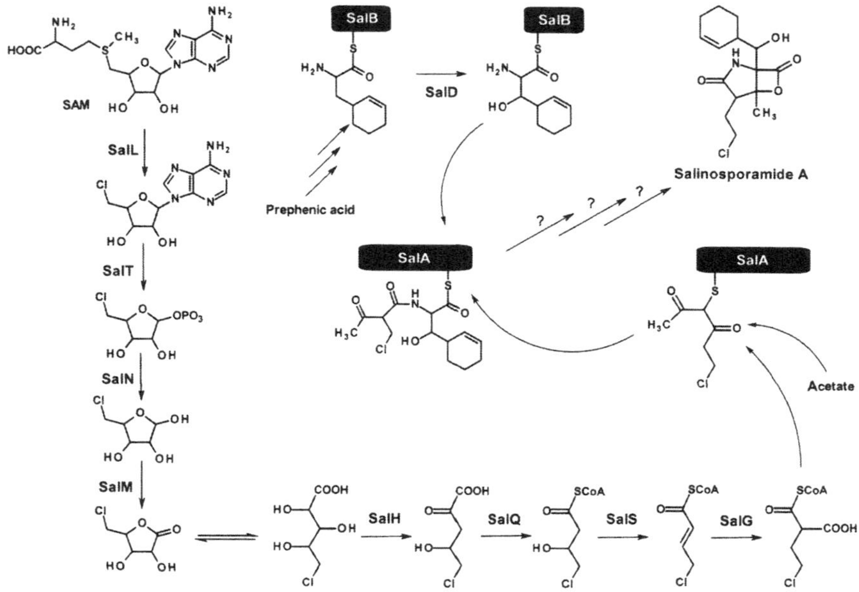

Figure 11.9 Pathway for the biosynthesis of salinosporamide A in *Salinispora* spp.

(Fig. 11.9). The resulting 5′-chloro-5′-deoxyadenosine is converted to 5-chloro-5-deoxy-D-ribose-1-phosphate by enzyme SalT. Phosphatase SalN modifies the product of SalT by removing a phosphate group, yielding 5-chloro-5-deoxy-D-ribose, which is then oxidized by SalM dehydrogenase first to 5-chloro-5-deoxy-D-ribono-1,4-lactone and then to 5-chlororibonate. The dehydratase SalH converts the latter intermediate to 5-chloro-3-hydroxybutyryl-2-oxopentanoate, which is then activated with CoA by SalQ enzyme via oxidative decarboxylation reaction and reduced by dehydratase SalS to give 4-chlorocrotoyl CoA. The final step in this unusual pathway is accomplished by a carboxylase/reductase SalG, which converts the latter substrate to the chloroethylmalonyl-CoA used as an extender unit for salinosporamide PKS SalA.

Two modular enzymes, hybrid PKS–NRPS SalA and NRPS SalB, are involved in the assembly of the salinosporamide A molecule. SalA is first loaded with an acetate and then performs its decarboxylative condensation with the chloroethylmalonyl-CoA extender biosynthesized as described earlier. SalB NRPS activates and binds a unique amino acid building block, presumably originating from the prephenic acid, an intermediate in the primary metabolic pathway for the biosynthesis of aromatic amino acids. The

SalB-bound intermediate is hydroxylated by the P450 monooxygenase SalD, and the resulting amino acid unit is fused with the chloroacyl precursor by the NRPS condensation domain of SalA. The process by which the resulting linear intermediate is converted to a bicyclic structure and release from SalA remains obscure, but it is suggested that type II thioesterase SalF may be involved (Gulder & Moore, 2010).

4. CONCLUSIONS

Marine environment is very diverse and represents a rich source of natural products that may be of interest for the development of new medicines. Over the past three decades, a number of structurally unique, biologically active alkaloids have been isolated from bacteria derived from various marine sources such as sea water, marine sediments and marine animals. These alkaloids are chemically diverse and often represent scaffolds, never before, or very rarely found in other natural products. The latter makes these compounds very interesting from the point of view of studying their biosynthesis. Alkaloids from marine bacteria possess a wide range of biological activities, encompassing anti-bacterial, anti-fungal, anti-tumour, molluscicidal, etc., thus positioning them for the drug discovery pipelines.

Importantly, bacteria can be sustainably grown in large quantities in the fermentors, thus providing material for purification of natural products and circumventing the need for collection of raw material from the sea. Although there still exist challenges for high cell density fermentations of most of the marine bacteria, which differ metabolically from their terrestrial counterparts, technologies such as cloning and heterologous expression of alkaloid biosynthesis gene clusters may provide an alternative to using the original producers. Metabolic engineering and synthetic biology will likely play major roles in the development of such technologies.

ACKNOWLEDGEMENTS

This work was supported by the Norwegian University of Science and Technology and the Research Council of Norway.

REFERENCES

Abdelmohsen, U. R., Pimentel-Elardo, S. M., Hanora, A., Radwan, M., Abou-El-Ela, S. H., Ahmed, S., et al. (2010). Isolation, phylogenetic analysis and anti-infective activity screening of marine sponge-associated actinomycetes. *Marine Drugs*, *8*, 399–412.

Abdelmohsen, U. R., Szesny, M., Othman, E. M., Schirmeister, T., Grond, S., Stopper, H., et al. (2012). Antioxidant and anti-protease activities of diazepinomicin from the sponge-associated *Micromonospora* strain RV115. *Marine Drugs*, *10*, 2208–2221.

Adams, H. J., Blair, M. R., Jr., & Takman, B. H. (1976). The local anesthetic activity of saxitoxin alone and with vasoconstrictor and local anesthetic agents. *Archives Internationales de Pharmacodynamie et de Thérapie*, *224*, 275–282.

Al-Zereini, W., Fotso Fondja Yao, C. B., Laatsch, H., & Anke, H. (2010). Aqabamycins A-G: Novel nitro maleimides from a marine *Vibrio* species. I. Taxonomy, fermentation, isolation and biological activities. *The Journal of Antibiotics*, *63*, 297–301.

Aricioglu, F., & Altunbas, H. (2003). Harmane induces anxiolysis and antidepressant-like effects in rats. *Annals of the New York Academy of Sciences*, *1009*, 196–200.

August, P. R., Grossman, T. H., Minor, C., Draper, M. P., MacNeil, I. A., Pemberton, J. M., et al. (2000). Sequence analysis and functional characterization of the violacein biosynthetic pathway from *Chromobacterium violaceum*. *Journal of Molecular Microbiology and Biotechnology*, *2*, 513–519.

Bai, R., Pettit, G. R., & Hamel, E. (1990). Binding of dolastatin 10 to tubulin at a distinct site for peptide antimitotic agents near the exchangeable nucleotide and vinca alkaloid sites. *Biochemical Pharmacology*, *39*, 1941–1949.

Baum, S. S., Hill, R., & Rommelspacher, H. (1996). Harman-induced changes of extracellular concentrations of neurotransmitters in the nucleus accumbens of rats. *European Journal of Pharmacology*, *314*, 75–82.

Blunt, J. W., Copp, B. R., Keyzers, R. A., Munro, M. H., & Prinsep, M. R. (2013). Marine natural products. *Natural Products Reports*, *30*, 237–323.

Boufaied, N., Wioland, M. A., Falardeau, P., & Gourdeau, H. (2010). TLN-4601, a novel anticancer agent, inhibits Ras signaling post Ras prenylation and before MEK activation. *Anti-Cancer Drugs*, *21*, 543–552.

Bugni, T. S., Woolery, M., Kauffman, C. A., Jensen, P. R., & Fenical, W. (2006). Bohemamines from a marine-derived *Streptomyces* sp. *Journal of Natural Products*, *69*, 1626–1628.

Burckhardt, T., Harms, K., & Koert, U. (2012). Total synthesis of lodopyridone. *Organic Letters*, *14*, 4674–4677.

Caicedo, N. H., Kumirska, J., Neumann, J., Stolte, S., & Thöming, J. (2012). Detection of bioactive exometabolites produced by the filamentous marine cyanobacterium *Geitlerinema* sp. *Marine Biotechnology (New York, N. Y.)*, *14*, 436–445.

Cardellina, J. H., II., Marner, F. J., & Moore, R. E. (1979). Seaweed dermatitis: Structure of lyngbyatoxin A. *Science*, *204*, 193–195.

Chang, Z., Flatt, P., Gerwick, W. H., Nguyen, V. A., Willis, C. L., & Sherman, D. H. (2002). The barbamide biosynthetic gene cluster: A novel marine cyanobacterial system of mixed polyketide synthase (PKS)-non-ribosomal peptide synthetase (NRPS) origin involving an unusual trichloroleucyl starter unit. *Gene*, *296*, 235–247.

Charan, R. D., Schlingmann, G., Janso, J., Bernan, V., Feng, X., & Carter, G. T. (2004). Diazepinomicin, a new antimicrobial alkaloid from a marine *Micromonospora* sp. *Journal of Natural Products*, *67*, 1431–1433.

Cho, J. Y., Kwon, H. C., Williams, P. G., Jensen, P. R., & Fenical, W. (2006). Azamerone, a terpenoid phthalazinone from a marine-derived bacterium related to the genus *Streptomyces* (*Actinomycetales*). *Organic Letters*, *8*, 2471–2474.

Cole, K. E., Dowling, D. P., Boone, M. A., Phillips, A. J., & Christianson, D. W. (2011). Structural basis of the antiproliferative activity of largazole, a depsipeptide inhibitor of the histone deacetylases. *Journal of the American Chemical Society*, *133*, 12474–12477.

DeMoss, R. D., & Evans, N. R. (1960). Incorporation of C14-labeled substrates into violacein. *Journal of Bacteriology*, *79*, 729–733.

Durán, N., Justo, G. Z., Ferreira, C. V., Melo, P. S., Cordi, L., & Martins, D. (2007). Violacein: Properties and biological activities. *Biotechnology and Applied Biochemistry*, *48*, 127–133.

Engelhardt, K., Degnes, K. F., Kemmler, M., Bredholt, H., Fjaervik, E., Klinkenberg, G., et al. (2010a). Production of a new thiopeptide antibiotic, TP-1161, by a marine Nocardiopsis species. *Applied and Environmental Microbiology*, *76*, 4969–4976.

Engelhardt, K., Degnes, K. F., & Zotchev, S. B. (2010b). Isolation and characterization of the gene cluster for biosynthesis of the thiopeptide antibiotic TP-1161. *Applied and Environmental Microbiology*, *76*, 7093–7101.

Fedorov, R., Böhl, M., Tsiavaliaris, G., Hartmann, F. K., Taft, M. H., Baruch, P., et al. (2009). The mechanism of pentabromopseudilin inhibition of myosin motor activity. *Nature Structural & Molecular Biology*, *16*, 80–88.

Fehér, D., Barlow, R., McAtee, J., & Hemscheidt, T. K. (2010). Highly brominated antimicrobial metabolites from a marine *Pseudoalteromonas* sp. *Journal of Natural Products*, *73*, 1963–1966.

Feling, R. H., Buchanan, G. O., Mincer, T. J., Kauffman, C. A., Jensen, P. R., & Fenical, W. (2003). Salinosporamide A: A highly cytotoxic proteasome inhibitor from a novel microbial source, a marine bacterium of the new genus *Salinospora*. *Angewandte Chemie (International Edition in English)*, *42*, 355–357.

Freel, K. C., Nam, S. J., Fenical, W., & Jensen, P. R. (2011). Evolution of secondary metabolite genes in three closely related marine actinomycete species. *Applied and Environmental Microbiology*, *77*, 7261–7270.

Freeman, M. F., Gurgui, C., Helf, M. J., Morinaka, B. I., Uria, A. R., Oldham, N. J., et al. (2012). Metagenome mining reveals polytheonamides as posttranslationally modified ribosomal peptides. *Science*, *338*, 387–390.

Fu, P., Wang, S., Hong, K., Li, X., Liu, P., Wang, Y., et al. (2011). Cytotoxic bipyridines from the marine-derived actinomycete Actinoalloteichus cyanogriseus WH1-2216-6. *Journal of Natural Products*, *74*, 1751–1756.

Fujiki, H., Suganuma, M., Hakii, H., Bartolini, G., Moore, R. E., Takayama, S., et al. (1984). A two-stage mouse skin carcinogenesis study of lyngbyatoxin A. *Journal of Cancer Research and Clinical Oncology*, *108*, 174–176.

Fujita, M. J., Kimura, N., Yokose, H., & Otsuka, M. (2012). Heterologous production of bisucaberin using a biosynthetic gene cluster cloned from a deep sea metagenome. *Molecular BioSystems*, *8*, 482–485.

Fukuda, D., Haines, A. S., Song, Z., Murphy, A. C., Hothersall, J., Stephens, E. R., et al. (2011). A natural plasmid uniquely encodes two biosynthetic pathways creating a potent anti-MRSA antibiotic. *PLoS One*, *6*, e18031.

Galonić, D. P., Vaillancourt, F. H., & Walsh, C. T. (2006). Halogenation of unactivated carbon centers in natural product biosynthesis: Trichlorination of leucine during barbamide biosynthesis. *Journal of the American Chemical Society*, *128*, 3900–3901.

Gerwick, W. H., Proteau, P. J., Nagle, D. G., Hamel, E., Blokhin, A., & Slate, D. L. (1994). Structure of Curacin A, a novel antimitotic, antiproliferative, and brine shrimp toxic natural product from the marine cyanobacterium *Lyngbya majuscula*. *The Journal of Organic Chemistry*, *59*, 1243–1245.

Gerwick, W. H., Tan, L. T., & Sitachitta, N. (2001). Nitrogen-containing metabolites from marine cyanobacteria. *The Alkaloids. Chemistry and Biology*, *57*, 75–184.

Groll, M., Huber, R., & Potts, B. C. (2006). Crystal structures of Salinosporamide A (NPI-0052) and B (NPI-0047) in complex with the 20S proteasome reveal important consequences of beta-lactone ring opening and a mechanism for irreversible binding. *Journal of the American Chemical Society*, *128*, 5136–5141.

Gulder, T. A. M., & Moore, B. S. (2010). Salinosporamide natural products: Potent 20S proteasome inhibitors as promising cancer chemotherapeutics. *Angewandte Chemie (International Edition in English)*, *49*, 9346–9367.

Hakvåg, S., Fjaervik, E., Klinkenberg, G., Borgos, S. E., Josefsen, K. D., Ellingsen, T. E., et al. (2009). Violacein-producing *Collimonas* sp. from the sea surface microlayer of costal waters in Trøndelag, Norway. *Marine Drugs*, 7, 576–588.

Harrigan, G. G., Luesch, H., Yoshida, W. Y., Moore, R. E., Nagle, D. G., Paul, V. J., et al. (1998). Symplostatin 1: A dolastatin 10 analogue from the marine cyanobacterium *Symploca hydnoides*. *Journal of Natural Products*, *61*, 1075–1077.

Hawas, U. W., Shaaban, M., Shaaban, K. A., Speitling, M., Maier, A., Kelter, G., et al. (2009). Mansouramycins A-D, cytotoxic isoquinolinequinones from a marine streptomycete. *Journal of Natural Products*, *72*, 2120–2124.

Hentschel, U., Piel, J., Degnan, S. M., & Taylor, M. W. (2012). Genomic insights into the marine sponge microbiome. *Nature Reviews. Microbiology*, *10*, 641–654.

Herraiz, T., & Chaparro, C. (2005). Human monoamine oxidase is inhibited by tobacco smoke: Beta-carboline alkaloids act as potent and reversible inhibitors. *Biochemical and Biophysical Research Communications*, *326*, 378–386.

Hoehn, P., Ghisalba, O., Moerker, T., & Peter, H. H. (1995). 3'-Demethoxy-3'-hydroxystaurosporine, a novel staurosporine analogue produced by a blocked mutant. *The Journal of Antibiotics*, *48*, 300–305.

Hoshino, T., Kondo, T., Uchiyama, T., & Ogasawara, N. (1987). Biosynthesis of violacein: A novel rearrangement in tryptophan metabolism with a 1,2-shift of the indole ring. *Agricultural and Biological Chemistry*, *51*, 965–968.

Hoshino, T., Takano, T., Hori, S., & Ogasawara, N. (1987). Biosynthesis of violacein: Origins of hydrogen, nitrogen and oxygen atoms in the 2-pyrrolidone nucleus. *Agricultural and Biological Chemistry*, *51*, 2733–2741.

Hughes, C. C., & Fenical, W. (2010). Antibacterials from the sea. *Chemistry*, *16*, 12512–12525.

Jeffrey, A. M., & Liskamp, R. M. (1986). Computer-assisted molecular modeling of tumor promoters: Rationale for the activity of phorbol esters, teleocidin B, and aplysiatoxin. *Proceedings of the National Academy of Sciences of the United States of America*, *83*, 241–245.

Jensen, P. R., Gontang, E., Mafnas, C., Mincer, T. J., & Fenical, W. (2005). Culturable marine actinomycete diversity from tropical Pacific Ocean sediments. *Environmental Microbiology*, 7, 1039–1048.

Jensen, P. R., Mincer, T. J., Williams, P. G., & Fenical, W. (2005). Marine actinomycete diversity and natural product discovery. *Antonie Van Leeuwenhoek*, *87*, 43–48.

Kim, S. K., & Dewapriya, P. (2012). Bioactive compounds from marine sponges and their symbiotic microbes: A potential source of nutraceuticals. *Advances in Food and Nutrition Research*, *65*, 137–151.

Kott, P. (1957). The sessile tunicate. *Scientific Reports from John Murray Expeditions, 1933–34*(10), 129–149.

Kwon, H. C., Espindola, A. P., Park, J. S., Prieto-Davó, A., Rose, M., Jensen, P. R., et al. (2010). Nitropyrrolins A-E, cytotoxic farnesyl-α-nitropyrroles from a marine-derived bacterium within the actinomycete family *Streptomycetaceae*. *Journal of Natural Products*, *73*, 2047–2052.

Li, B., & Walsh, C. T. (2010). Identification of the gene cluster for the dithiolopyrrolone antibiotic holomycin in *Streptomyces clavuligerus*. *Proceedings of the National Academy of Sciences of the United States of America*, *107*, 19731–19735.

Liu, Y., Salvador, L. A., Byeon, S., Ying, Y., Kwan, J. C., Law, B. K., et al. (2010). Anticolon cancer activity of largazole, a marine-derived tunable histone deacetylase inhibitor. *The Journal of Pharmacology and Experimental Therapeutics*, *335*, 351–361.

Liu, J., Zhou, W., Li, S. S., Sun, Z., Lin, B., Lang, Y. Y., et al. (2008). Modulation of orphan nuclear receptor Nur77-mediated apoptotic pathway by acetylshikonin and analogues. *Cancer Research*, *68*, 8871–8880.

Maloney, K. N., Macmillan, J. B., Kauffman, C. A., Jensen, P. R., Dipasquale, A. G., Rheingold, A. L., et al. (2009). Lodopyridone, a structurally unprecedented alkaloid from a marine actinomycete. *Organic Letters*, *11*, 5422–5424.

Marion, O., Gao, X., Marcus, S., & Hall, D. G. (2009). Synthesis and preliminary antibacterial evaluation of simplified thiomarinol analogs. *Bioorganic & Medicinal Chemistry*, *17*, 1006–1017.

Marquez, B. L., Watts, K. S., Yokochi, A., Roberts, M. A., Verdier-Pinard, P., Jimenez, J. I., et al. (2002). Structure and absolute stereochemistry of hectochlorin, a potent stimulator of actin assembly. *Journal of Natural Products*, *65*, 866–871.

Martin, G. D., Tan, L. T., Jensen, P. R., Dimayuga, R. E., Fairchild, C. R., Raventos-Suarez, C., et al. (2007). Marmycins A and B, cytotoxic pentacyclic C-glycosides from a marine sediment-derived actinomycete related to the genus *Streptomyces*. *Journal of Natural Products*, *70*, 1406–1409.

Martins, R., Pereira, P., Welker, M., Fastner, J., & Vasconcelos, V. M. (2005). Toxicity of culturable cyanobacteria strains isolated from the Portuguese coast. *Toxicon*, *46*, 454–464.

McAlpine, J. B., Banskota, A. H., Charan, R. D., Schlingmann, G., Zazopoulos, E., Piraee, M., et al. (2008). Biosynthesis of diazepinomicin/ECO-4601, a *Micromonospora* secondary metabolite with a novel ring system. *Journal of Natural Products*, *71*, 1585–1590.

McArthur, K. A., Mitchell, S. S., Tsueng, G., Rheingold, A., White, D. J., Grodberg, J., et al. (2008). Lynamicins A-E, chlorinated bisindole pyrrole antibiotics from a novel marine actinomycete. *Journal of Natural Products*, *71*, 1732–1737.

Murphy, A. C., Fukuda, D., Song, Z., Hothersall, J., Cox, R. J., Willis, C. L., et al. (2011). Engineered thiomarinol antibiotics active against MRSA are generated by mutagenesis and mutasynthesis of *Pseudoalteromonas* SANK73390. *Angewandte Chemie (International Edition in English)*, *50*, 3271–3274.

Ohri, R. V., Radosevich, A. T., Hrovat, K. J., Musich, C., Huang, D., Holman, T. R., et al. (2005). A Re(V)-catalyzed C-N bond-forming route to human lipoxygenase inhibitors. *Organic Letters*, *7*, 2501–2504.

Olguin-Uribe, G., Abou-Mansour, E., Boulander, A., Débard, H., Francisco, C., & Combaut, G. (1997). 6-Bromoindole-3-carbaldehyde, from an *Acinetobacter* sp. bacterium associated with the Ascidian *Stomozoa murrayi*. *Journal of Chemical Ecology*, *23*, 2507–2521.

Oliva, B., O'Neill, A., Wilson, J. M., O'Hanlon, P. J., & Chopra, I. (2001). Antimicrobial properties and mode of action of the pyrrothine holomycin. *Antimicrobial Agents and Chemotherapy*, *45*, 532–539.

Omura, S., Fujimoto, T., Otoguro, K., Matsuzaki, K., Moriguchi, R., Tanaka, H., et al. (1991). Lactacystin, a novel microbial metabolite, induces neuritogenesis of neuroblastoma cells. *The Journal of Antibiotics*, *44*, 113–116.

Onaka, H., Taniguchi, S., Igarashi, Y., & Furumai, T. (2002). Cloning of the staurosporine biosynthetic gene cluster from *Streptomyces* sp. TP-A0274 and its heterologous expression in *Streptomyces lividans*. *The Journal of Antibiotics*, *55*, 1063–1071.

Orjala, J., & Gerwick, W. H. (1996). Barbamide, a chlorinated metabolite with molluscicidal activity from the Caribbean cyanobacterium *Lyngbya majuscule*. *Journal of Natural Products*, *59*, 427–430.

Pan, E., Jamison, M., Yousufuddin, M., & MacMillan, J. B. (2012). Ammosamide D, an oxidatively ring opened ammosamide analog from a marine-derived *Streptomyces variabilis*. *Organic Letters*, *14*, 2390–2393.

Pemberton, J. M., Vincent, K. M., & Penfold, R. J. (1991). Cloning and heterologous expression of the violacein biosynthesis gene cluster from *Chromobacterium violaceum*. *Current Microbiology*, *22*, 355–358.

Ren, J., Liu, D., Tian, L., Wei, Y., Proksch, P., Zeng, J., et al. (2013). Venezuelines A-G, new phenoxazine-based alkaloids and aminophenols from *Streptomyces venezuelae* and the regulation of gene target Nur77. *Bioorganic & Medicinal Chemistry Letters*, *23*, 301–304.

Salas, A. P., Zhu, L., Sánchez, C., Braña, A. F., Rohr, J., Méndez, C., et al. (2005). Deciphering the late steps in the biosynthesis of the anti-tumour indolocarbazole staurosporine: Sugar donor substrate flexibility of the StaG glycosyltransferase. *Molecular Microbiology*, *58*, 17–27.

Sánchez, C., Braña, A. F., Méndez, C., & Salas, J. A. (2006). Reevaluation of the violacein biosynthetic pathway and its relationship to indolocarbazole biosynthesis. *ChemBioChem*, *7*, 1231–1240.

Sánchez, C., Zhu, L., Braña, A. F., Salas, A. P., Rohr, J., Méndez, C., et al. (2005). Combinatorial biosynthesis of antitumor indolocarbazole compounds. *Proceedings of the National Academy of Sciences of the United States of America*, *102*, 461–466.

Sangnoi, Y., Sakulkeo, O., Yuenyongsawad, S., Kanjana-opas, A., Ingkaninan, K., Plubrukarn, A., et al. (2008). Acetylcholinesterase-inhibiting activity of pyrrole derivatives from a novel marine gliding bacterium, *Rapidithrix thailandica*. *Marine Drugs*, *6*, 578–586.

Seiser, T., Kamena, F., & Cramer, N. (2008). Synthesis and biological activity of largazole and derivatives. *Angewandte Chemie (International Edition in English)*, *47*, 6483–6485.

Shiozawa, H., Kagasaki, T., Kinoshita, T., Haruyama, H., Domon, H., Utsui, Y., et al. (1993). Thiomarinol, a new hybrid antimicrobial antibiotic produced by a marine bacterium. Fermentation, isolation, structure, and antimicrobial activity. *The Journal of Antibiotics*, *46*, 1834–1842.

Sitachitta, N., Marquez, B. L., Williamson, R. T., Rossi, J., Robert, M. A., Gerwick, W. H., et al. (2000). Barbamide and dechlorobarbamide, molliscicidal agents from the marine cyanobacterium, *Lyngbya majuscula*: Biosynthetic pathway and origin of the chlorinated methyl group. *Tetrahedron*, *56*, 9130–9133.

Srisukchayakul, P., Suwannachart, C., Sangnoi, Y., Kanjana-opas, A., Hosoya, S., Yokota, A., et al. (2007). *Rapidithrix thailandica* gen. nov., sp. nov., a marine gliding bacterium isolated from samples collected from the Andaman sea, along the southern coastline of Thailand. *International Journal of Systematic and Evolutionary Microbiology*, *54*, 2275–2279.

Strichartz, G., Rando, T., Hall, S., Gitschier, J., Hall, L., Magnani, B., et al. (1986). On the mechanism by which saxitoxin binds to and blocks sodium channels. *Annals of the New York Academy of Sciences*, *479*, 96–112.

Takahashi, A., Ikeda, D., Nakamura, H., Naganawa, H., Kurasawa, S., Okami, Y., et al. (1989). Altemicidin, a new acaricidal and antitumor substance. II. Structure determination. *The Journal of Antibiotics*, *42*, 1562–1566.

Takahashi, I., Kobayashi, E., Asano, K., Yoshida, M., & Nakano, H. (1987). UCN-01, a selective inhibitor of protein kinase C from *Streptomyces*. *The Journal of Antibiotics*, *40*, 1782–1784.

Takahashi, A., Nakamura, H., Kameyama, T., Kurasawa, S., Naganawa, H., Okami, Y., et al. (1987). Bisucaberin, a new siderophore, sensitizing tumor cells to macrophage-mediated cytolysis. II. Physico-chemical properties and structure determination. *The Journal of Antibiotics*, *40*, 1671–1676.

Tan, L. T. (2007). Bioactive natural products from marine cyanobacteria for drug discovery. *Phytochemistry*, *68*, 954–979.

Taori, K., Paul, V. J., & Luesch, H. (2008). Structure and activity of largazole, a potent antiproliferative agent from the Floridian marine cyanobacterium *Symploca* sp. *Journal of the American Chemical Society*, *130*, 1806–1807.

Teruya, T., Kobayashi, K., Suenaga, K., & Kigoshi, H. (2005). Phormidinines A and B, novel 2-alkylpyridine alkaloids from the cyanobacterium *Phormidium* sp. *Tetrahedron Letters*, *46*, 4001–4003.

Totsuka, Y., Hada, N., Matsumoto, K., Kawahara, N., Murakami, Y., Yokoyama, Y., et al. (1998). Structural determination of a mutagenic minophenylnorharman produced by the co-mutagen norharman with aniline. *Carcinogenesis*, *19*, 1995–2000.

Volk, R. B., Girreser, U., Al-Refai, M., & Laatsch, H. (2009). Bromoanaindolone, a novel antimicrobial exometabolite from the cyanobacterium *Anabaena constricta*. *Natural Products Research*, *23*, 607–612.

Wiese, M., D'Agostino, P. M., Mihali, T. K., Moffitt, M. C., & Neilan, B. A. (2010). Neurotoxic alkaloids: Saxitoxin and its analogs. *Marine Drugs*, *8*, 2185–2211.

Wietz, M., Mansson, M., Gotfredsen, C. H., Larsen, T. O., & Gram, L. (2010). Antibacterial compounds from marine *Vibrionaceae* isolated on a global expedition. *Marine Drugs*, *8*, 2946–2960.

Winter, J. M., Jansma, A. L., Handel, T. M., & Moore, B. S. (2009). Formation of the pyridazine natural product azamerone by biosynthetic rearrangement of an aryl diazoketone. *Angewandte Chemie (International Edition in English)*, *48*, 767–770.

Wipf, P., Reeves, J. T., & Day, B. W. (2004). Chemistry and biology of curacin A. *Current Pharmaceutical Design*, *10*, 1417–1437.

Zhang, X., Jiang, X., Ding, C., Yao, Q., & Zhang, A. (2013). Unexpected N-glycosidation reaction of glycals with 1-amino-anthracene: Structure revision and application to the synthesis of new analogues of marmycin A. *Organic & Biomolecular Chemistry*, *11*, 1383–1389.

Zhang, W., Liu, Z., Li, S., Yang, T., Zhang, Q., Ma, L., et al. (2012). Spiroindimicins A-D: New bisindole alkaloids from a deep-sea-derived actinomycete. *Organic Letters*, *14*, 3364–3367.

AUTHOR INDEX

Note: Page numbers followed by "*f*" indicate figures, and "*t*" indicate tables.

A

Abbracchio, M.P., 287
Abdelmohsen, U.R., 308
Abidin, U.Z., 285–286
Abou-El-Ela, S.H., 308
Abou-Mansour, E., 315, 316*t*
Abraham, T.W., 55–56
Achnine, L., 291
Adams, H.J., 313
Aerts, R., 15–16, 18, 87–89
Aguiar, G.P., 249*t*
Aharonson, N., 100
Ahmad, A., 50–52
Ahmadi, F.S., 210–211
Ahmed, S., 308
Ahuja, P.S., 125–126, 129–131, 130*t*
Ahyow, M., 171–173
Aimi, N., 8, 12–13, 142, 144, 147–148, 149, 152–153
Aitken, W.M., 113–115
Aiyama, R., 142
Ajungla, L., 61*t*
Akita, M., 221
Alarco, A.M., 16–18, 86–87, 96
Albor, H.T., 294
Alburges, M.E., 248–251
Aliotta, G., 211–212
Allard, M.W., 286–287
Allen, R.S., 171, 175, 185
Allorge, L., 237–238
Almeida, I., 96, 98–99
Almeida, L., 4
Alnajafi, T., 285–286
Al-Refai, M., 311*t*, 314
Altmann, S., 292–293
Altunbas, H., 312
Alvarez, J.C., 244*t*
Alves, M.N., 43–44, 61*t*
Al-Zereini, W., 316*t*, 318
Amano, Y., 59
Amdoun, R., 61*t*, 225
Amini, A., 90
Ammer, C., 166–170
Amna, T., 153
Amrhein, N., 285–286
Amroune, S., 61*t*, 225
Anandalakshmi, R., 23–24
Andersen, A.H., 140–141
Anderson, J.C., 23–24
Anderson, L., 122
Andrabi, R., 153
Andrade, M.T., 237–238, 251
Andre, N., 244*t*
Andreu, F., 87, 89–90
Aneja, M., 283
Angelov, D., 47–48
Angenot, L., 238, 252, 261
Anke, H., 316*t*, 318
Ansarin, M., 58–59
Antúnez De Mayolo, G., 125–126
Apg, I.I.I., 78–79
Apostolides, Z., 128
Apuya, N.R., 171–173
Aranyi, A., 260–261
Arbain, D., 48–49
Archer, S., 248–251, 249*t*
Ardiles-Diaz, W., 9–10, 12, 15–16, 18, 21
Arenghi, F., 176
Aricioglu, F., 312
Aripova, S.F., 40, 46–47
Arisawa, M., 142
Arita, M., 145
Arlotto, M.P., 196–197
Armour, D., 92–93
Armstrong, J.W., 275–276, 283
Arnold, F.H., 197–198
Arroo, R., 210–211
Arslan, A., 285, 286, 291
Arvy, M.-P., 87
Asano, K., 316*t*, 319, 321
Asano, N., 47
Asano, T., 142, 149, 154
Asch, M., 225

Ashihara, H., 50–52, 112–115, 118, 119–120, 121*f*, 122, 123–126, 128–131, 133, 275–277, 278
Assaf, C., 218
Assaf-Ducrocq, C., 225
Assmann, S.M., 288
Atsumi, S.M., 4–5, 20, 23–24
Atta-Ur-Rahman, 43
August, P.R., 319–320
Ausubel, F.M., 289
Aynilian, G.H., 47–48
Ayora-Talavera, T., 24–26
Azas, N., 244*t*
Azemi, M.E., 52–54

B

Babbittet, P.C., 83–84
Bac, N.V., 4
Bacher, A., 6–8, 145
Bachmann, P., 58–59
Bacic, A., 83–84
Backenköhler, A., 82–83
Badawi, M.M., 142
Bae, H., 125–126
Baerheim Svendsen, A., 235, 238, 242–243
Baggio, C.H., 249*t*, 251
Baghdikian, B., 244*t*
Bagni, N., 211
Bai, R., 313–314
Bailey, B.A., 47–48, 125–126
Baker, D.D., 132
Baldwin, I.T., 294
Baller, J.A., 173
Balsevich, J., 16–18
Banskota, A.H., 322
Baque, M.A., 218–219, 221, 224–225
Barata, L.E.S., 44
Barber, A.E., 83–84
Barbosa-Filho, J.M., 40
Barbotin, J.N., 214–215, 217
Barleben, L., 12–13, 83–84, 85
Barlow, R., 315, 316*t*
Barmukh, R.B., 61*t*
Barnes, H.J., 196–197
Barres, M.L., 60–62, 61*t*
Barreto, A.S., 241*t*, 242
Barroso, M., 249*t*
Barthe, L., 254, 255*t*
Bartolini, G., 314
Barton, K., 96–98
Baruch, P., 315
Bashan, Y., 221
Battig, K., 285, 287, 289–290
Baudisch, B., 96–98
Baudoux, T., 208
Bauer, S., 187–189
Baulcombe, D.C., 23–24
Baum, S.S., 312
Baumann, T.W., 15–16, 87–89, 113–115, 114*t*, 118–119, 128–129, 275–276
Baxter, C., 10
Bealer, B.K., 274
Beasley, R.S., 142
Beaudoin, G.A., 166–170
Beavo, J.A., 285
Becker, A., 23–24
Beckers, G.J.M., 291–292, 294
Bednarek, P., 293–294
Beegum, A.S., 142
Behmanesh, M., 60–62
Belder, D., 255*t*
Belem-Pinheiro, M.L., 252
Bellimam, M.A., 40
Benabdelmouna, A., 91
Bender, A.T., 285
Benedetti, P., 154–155
Benedicto, J., 241*t*
Benhamron, S., 24–26
Béni, Z., 260–261
Bent, A.F., 288
Benyoussef, E.-H., 61*t*, 225
Berenbaum, M.R., 211–212
Bergey, D.R., 14–15, 96–98
Berkowitz, G.A., 287–288, 290*f*
Berlin, J., 83, 144
Bernan, V., 304*t*, 308
Bernasch, H., 48
Bernhardt, P., 26
Berreti, B., 211–212
Berry, A., 191
Bertol, G., 244*t*
Bettler, B., 286–287
Betz, H., 292
Beyreuther, K., 12–13, 147–148
Bharti, S.K., 259–260
Bhatt, A., 61*t*

Bhosale, B.B., 151
Bick, I.R.C., 48
Bienzle, D., 19–20, 21–22
Bieri, S., 47–48
Biget, L., 125
Biondi, S., 211
Bjorklund, E., 239
Bjorklund, J.A., 49–50, 54–56, 58
Bjornsti, M.-A., 142, 154–155
Blair, M.R. Jr., 313
Blanc, N., 14–15, 76–77, 96–98, 100
Blancaflor, E.B., 291
Blaschke, G., 254–259, 255*t*
Blaskó, G., 260–261
Blaudez, D., 288, 290*f*
Blokhin, A., 313
Blom, T.J.M., 13–15
Blount, M.A., 285, 290*f*
Blum, M.S., 44
Blunt, J.W., 302
Bodnar, J.A., 255*t*
Boger, D.L., 4
Boghigian, B.A., 187–189
Böhl, M., 315
Bohlmann, J., 4–5, 21–22, 27
Bohm, A., 46–47
Boisson, B., 24–26
Boitel, M., 217, 218
Boldt, R., 123–124
Bolívar, F., 191, 193
Bolwell, G.P., 289, 290*f*
Bonamore, A., 176
Bones, A.M., 99–100
Bonfill, M., 60, 211, 227
Boone, M.A., 310–312
Borgio, J., 173
Borgos, S.E., 315–317, 316*t*
Boronat, A., 5–8
Bot, E.S.M., 142
Botella, J.R., 92–93
Botta, B., 176, 237–238
Bouche, N., 286–287
Boufaied, N., 308
Boulander, A., 315, 316*t*
Bouque, V., 207
Bourgaud, A., 218–219, 225
Bourgaud, F., 207, 214, 215*f*, 216, 218–219, 227
Bouwmeester, K., 226
Bowers, J.H., 125–126
Bowman, R.N., 255*t*
Bown, A.W., 286–287, 290*f*, 291
Boyd, M.R., 142
Boyer, L., 244*t*
Brabant, A., 56–57
Brachet, A., 47–48
Braña, A.F., 320, 321
Branco, M.V.S.C., 40
Brandao, L.M., 249*t*, 251
Brandle, J.E., 19–20, 21–22
Brandt, V., 238, 260–261
Brandt, W., 52–54, 56–57
Braz-Filho, R., 249*t*, 251, 252, 253, 260–261
Bredholt, H., 304*t*, 309
Bremer, J., 87, 89–90
Bressolle, F.M.M., 244*t*
Breton, H., 244*t*
Brewin, N.J., 285–286
Bringi, V., 207
Brisson, N., 12–13, 144
Brock, A., 48, 56–57
Bronstein, P.A., 291
Brown, A.W., 291
Brown, E.G., 285–286
Brown, K.S. Jr., 44
Bruce, N.C., 197–198
Brugliera, F., 175–176
Buchanan, G.O., 304*t*, 307–308
Buchenauer, H., 292–293
Buell, C.R., 3*f*, 16, 80–82
Bugni, T.S., 303–306, 304*t*
Buitelaar, R.M., 151–152
Bulumulla, H.N.K., 237–238
Burchell, B., 196–197
Burch-Smith, T.M., 23–24
Burckhardt, T., 308–309
Burgener, L., 40, 43
Burgin, A.B., 154–155
Burlat, V., 8, 9, 10–12, 19–20, 76–77, 80–82, 83–84, 86–87, 92–93, 94–95, 96–98, 100, 146, 147
Burnett, R.J., 14–15, 96–98
Burnstock, G., 285, 287, 291
Burtin, D., 52
Busto, V.D., 62–63

Butler, M.S., 48
Bux, F., 221
Byeon, S., 310–312
Byrt, C., 83–84
Byun, S.Y., 152

C
Caboni, M.F., 253
Cafiero, G., 211–212
Cai, J., 255*t*
Cai, Z.Z., 62–63
Caicedo, N.H., 311*t*, 312
Caldwell, J., 83–84
Camakaris, H., 190–191
Campbell, E., 275–276, 283
Campos-Neves, A.S., 43
Campos-Neves, M.T., 43
Campos-Tamayo, F., 86–87
Cane, D.E., 143–144, 149
Canel, C., 24–26
Caniato, R., 243
Cao, M., 238
Cao, X., 243–248
Capdevila, J.H., 198
Capell, T., 227
Carbone, V., 244*t*
Cardellina, J.H. II., 142, 311*t*, 314
Cardillo, A.B., 62–63
Cardoso, C.A.L., 251
Carrillo-Pech, M., 87
Carroll, F.I., 47–48
Carte, B.K., 149
Carter, G.T., 304*t*, 308
Carvalho, A., 113–115
Carvalho, M.P., 237–238, 251
Casale, J.F., 47–48
Casey, P.J., 91–92
Castanheira, A., 249*t*
Caudle, D.L., 197–198
Cazal, C.M., 252–253
Cellarova, E., 155
Chahed, K., 6–8, 87
Chai, Y., 60
Chaki, K., 171–173
Chakravorty, D., 92–93
Champion, A., 15–16, 24–26, 87–89
Chan, F., 16–18
Chan, K., 208
Chandra, R., 59–60
Chandra, S., 59–60
Chandrashekar, A., 129
Chang, Y., 241*t*, 242
Chang, Z., 323
Chaparro, C., 312
Chapman, S.K., 197–198
Chappell, J., 6, 24–26
Charan, R.D., 304*t*, 308, 322
Chareonsap, P., 151, 152–153
Chase, M.W., 83–84
Chashmi, N.A., 62–63
Chatel, G., 9–10, 15–16, 87–89
Chatel, M.G., 15–16, 24–26
Chatterjee, D., 140–141
Chatterjee, S.P., 190
Chávez-Béjar, M.I., 190
Chayamarit, K., 142, 150–151, 154–155
Che, C.T., 255*t*
Chebbi, M., 90
Chekaoui, S., 210–211
Chen, C.Y., 210–211
Chen, H.C., 90, 291
Chen, J., 24–26, 52–54, 210–211, 243, 244*t*, 255*t*
Chen, M.C., 60, 292–293
Chen, M.M., 197–198
Chen, P., 262
Chen, Q., 243, 244*t*, 249*t*, 254, 255*t*, 259*f*
Chen, S., 26
Chen, W., 153–154
Chen, X., 239, 255*t*
Chen, Z., 243, 244*t*, 255*t*, 289
Cheng, B., 243–248
Cheng, K.D., 52–54
Cheng, S.H., 289, 290*f*, 291
Cheng, W., 244*t*
Chénieux, J.C., 6–8, 22–23, 87, 89–90
Cheong, B.E., 166–171
Chesters, N.C.J.E., 58–59
Chéze, M., 243, 244*t*
Chiang, V.L., 124–125
Chiatante, D., 285–286
Chiba, M., 149
Childs, K.L., 3*f*, 4–5, 16, 21, 80–82, 149
Chitnis, M., 142
Chitty, J.A., 171, 175, 185
Cho, J.Y., 303, 304*t*

Chockalingam, S., 100
Choi, C.S., 129–131, 130*t*, 278–280, 279*f*, 281*f*, 282, 283–284, 289–290
Choi, K.-B., 165–166, 170–171, 173, 185, 186–187
Choi, S.H., 294
Choi, Y.E., 129–131, 130*t*, 132, 278–281, 279*f*, 281*f*, 282, 283–284, 289–290, 293, 294
Choi, Y.H., 14, 236–237, 240–242, 241*t*, 261, 262
Chopra, I., 318–319
Chou, C.-H., 132, 275–276
Choudhury, M.D., 208
Chow, Y.-L., 164–165, 171, 176–177
Chrencik, J.E., 154–155
Christen, P., 47–48, 52, 210–211, 238, 260–261
Christiansen, K., 140–141
Christianson, D.W., 310–312
Christie, R., 47–48
Christopher, M.S.M., 275–276, 283
Christou, P., 227
Chu, I., 255*t*
Chuanasa, T., 142, 150–151, 154–155
Chuang, W.C., 255*t*
Chugh, A., 90–91
Cifuentes, A., 240
Cimanga, K., 243–248
Clastre, M., 6–8, 87, 91–92
Claude, B., 239–240
Claudel, P., 8, 9, 96–98, 146
Clément, A., 218–219, 223–225, 224*f*
Clifford, M.N., 113
Clough, S.J., 288
Cloutier, N., 15–16
Coca, M., 289, 290*f*, 291
Cohen, Y.R., 292
Colby, D.A., 4
Cole, K.E., 310–312
Colet, J.M., 208
Collu, G., 10, 146–148
Coltharp, C., 26
Combaut, G., 315, 316*t*
Compagnon, V., 16, 80–82
Conarad, U., 166–170
Conrath, U., 290*f*, 291–292, 294
Constabel, F., 87
Conti, M., 285
Contin, A., 8, 91
Cook, C.E., 140–141, 142
Cooke, C.J., 287–288
Copp, B.R., 302
Corbin, J.D., 285, 290*f*
Cordell, G.A., 142, 260–261
Cordi, L., 317
Coscia, C.J., 10–12
Cosme, J., 196–197
Costa, M.M., 4, 96, 98–99
Cottier, V., 292–293
Couladis, M.M., 55–56
Courdavault, V., 6–8, 9, 10–12, 14–15, 22–23, 76–77, 83–84, 86–89, 88*f*, 90, 91–93, 94–95, 96–98, 100, 146, 147
Courtois, M., 6–8, 9, 10–12, 19–20, 22–23, 87, 92–93
Covello, P.S., 4–5, 21–22, 27, 58–59
Cox, R.J., 318
Cragg, G.M., 207
Cram, D., 176
Cramer, N., 310–312
Crèche, J., 22–23, 87, 89–90
Crignola, S., 285–286
Cromarty, A.D., 249*t*, 251
Crook, N.C., 197–198
Croteau, R., 6, 9, 19–20, 164–165
Crouch, N.P., 10–12, 16, 19–20, 80–82, 83, 92, 94–95, 146–147
Crozier, A., 112–115, 118–120, 122, 123–125, 126, 128–131, 133, 275, 276–277, 278
Cui, L., 60
Cui, Y.Y., 46
Cuijpers, V., 52
Cunningham, D., 142
Curcino, I.J., 252, 253
Cusido, R.M., 211, 227
Cutler, A.J., 14–15, 16–18, 93–94, 96

D

da Silva, C.C., 237–238
da Silva, M.S., 40
da Silveira, V.C., 237–238
Daddona, P.E., 143–144, 149
Daga, R.S., 221
Dagnino, D., 248–251

D'Agostino, P.M., 313
Dai, J.R., 142
Dai, X.D., 210–211
D'Alessandro, M., 294
Daly, J.W., 285, 286, 287
Dang, T.T., 166–170, 176
Danieli, B., 260–261
Das, S., 59–60
Daughtry, C.S.T., 47–48
D'Auria, J.C., 44–46, 57, 58, 62–63
Davidow, P., 171–173
Davis, R.A., 48
Day, B.W., 313
de Anda, R., 193
De Budowski, J., 48–49
De Carolis, E.D., 16–18, 86–89, 96
de Carvalho, G.M., 237–238
de Castro, A., 237–238
de F.Agra, M., 40
De Greef, J.A., 285–286
de Groot, L., 52
de Jong, G.J., 254
de Koning, P., 13–15
de la Grandmaison, G.L., 244*t*
De Luca, V., 4–5, 8, 10, 12–13, 14–18, 19–20, 21–22, 23–26, 27, 80–82, 86–89, 93–95, 96, 144, 193
de Martini, G., 249*t*, 251
de Mayolo, G.A., 125–126
de Mazancourt, P., 244*t*
de Moraes, C.A., 249*t*, 251
de Oliveira, A.J.B., 237–238
de Oliveira, B.H., 244*t*
de Oliveira, S.L., 40
de Roetth, A., 46
De Simone, F., 244*t*
de Souza, W., 249*t*, 251
De Vos, R.C., 226
de Waal, A., 12–13, 19–20, 146–148
De Wet, H., 43
de Wolf, C.J.F., 15–16, 24–26
Deal, G., 208
Débard, H., 315, 316*t*
de-Bashan, L.E., 221
Debnath, S., 83–84
DeBrosse, C., 149
Décendit, A., 87, 89–90
De-Eknamkul, W., 85
DeFeo, V., 211–212
Degnan, S.M., 302
Degnes, K.F., 304*t*, 309
Degude, M.J., 254, 255*t*
Dehghan, E., 210–211
Dekker, N.H., 142
Dekkers, E., 226
Delauney, A.J., 50–52
Delazar, A., 48–49
Delbeke, F.T., 244*t*
DellaPenna, D., 3*f*, 16, 80–82
Delle Monache, F., 48–49
DeLong, D.C., 142, 150–151
DeMoss, R.D., 319–320
Dempsey, D.A., 289, 290*f*, 291
Denduangboripant, J., 142, 150–151, 154–155
Deng, B., 153–154
Deng, W.-W., 123–124, 126
Depiereux, E., 187–189
Deregnaucourt, C., 237–238
Desgagné-Penix, I., 173, 176
Dethier, M., 16–18
Deus-Neumann, B., 16–17
Devagupta, R., 12–13, 147–148
Deveaux, M., 243, 244*t*
Deventer, K., 244*t*
Deville, A., 237–238
Devoto, A., 87–89
deWaal, A., 12–13
Dewapriya, P., 302–303
Dewdney, J., 289
Dhooghe, L., 243–248
Di Mavungu, J.D., 244*t*
Di Stilio, V.S., 23–24
Díaz Chávez, M.L., 165–170
DiCosmo, F., 87
Diezel, C., 294
Dillehay, T.D., 40–41
Dimayuga, R.E., 304*t*, 306
Dinda, B., 83–84
Dinesh-Kumar, S.P., 23–24
Ding, C., 306
Ding, R., 60
Ding, X., 153–154
Dipasquale, A.G., 304*t*, 308–309
Dittrich, H., 165–166
Dixon, R.A., 289, 291

Do, Y.-Y., 146–147
Docimo, T., 52
Doerper, S., 227
Doireau, P., 22–23, 87–90, 91, 92–93
Domon, H., 316*t*, 317–318
Dong, L., 243–248
Douillard, H.Y., 142
Dowling, D.P., 310–312
Dräger, B., 48, 56–57
Draper, M.P., 319–320
Drea, S., 23–24
Duarte, P., 4, 96, 98–99
Dubost, L., 237–238
Dubouzet, E., 165–166, 170–173, 187–189
Dubouzet, J.G., 171–173
Dubrovay, Z., 260–261
Dubruel, P., 244*t*
Ducrocq, C., 218
Ducrot, P.H., 47
Ductocq, C., 218
Duflos, A., 254, 255*t*
Duke, J.A., 47–48
Duke, S.O., 47–48
Dully, C., 176
Dunford, A.J., 197–198
Dupéron, P., 91
Dupon, M., 285–286
Durán, N., 317
Duran-Patron, R., 55
Duverneuil, C., 244*t*
Dwyer, J.G., 10–12

E

Eagles, J., 55–56
Ebert, U., 46
Edelson, R., 211–212
Edwards, M.J., 285–286
Efferth, T., 208, 237–238
Eggleston, D., 149
Eich, E., 40, 237–238
Eilert, U., 17–18, 87, 93–94
Eisenreich, W., 6–8, 145
Ekanem, I.S., 197–198
Ekins, A., 166–170
El Bazaoui, A., 40
el Jaber-Vazdekis, N., 60–62, 61*t*
Elez-Martínez, P., 228
Ellingsen, T.E., 315–317, 316*t*
El-Sayed, M., 3*f*, 6–8, 7*f*, 9–10, 13–15, 87–89
Endo, T., 55, 56
Enei, H., 190
Eng, W.K., 142
Engelberth, J., 294
Engelhardt, K., 304*t*, 309
Epifanio, R.A., 237–238, 251
Erkelens, C., 262
Esaki, N., 10
Escalante, F.M.E., 221
Eschen-Lippold, L., 292–293
Espindola, A.P., 304*t*, 307
Estabrook, R.W., 197–198
Etse, J.T., 237–238
Evans, N.R., 319–320

F

Facchini, P.J., 2–5, 14, 17–18, 21–22, 23–26, 27, 50–54, 74–75, 76–77, 80–83, 85, 94–95, 96, 165–170, 167*f*, 173, 175, 176, 191–192, 193
Fahn, W., 16–17
Fahy, J., 254, 255*t*
Fairchild, C.R., 304*t*, 306
Fairhead, M., 198
Falardeau, P., 308
Fan, B., 289
Fan, L.W., 288, 289
Fang, L., 251–252
Faria, J., 244*t*
Farnswor, N.R., 47–48
Farnsworth, N.R., 142
Farré, G., 227
Farrow, S.C., 176
Farsi, M., 211
Fasan, R., 197–198
Fastner, J., 311*t*
Faure, D., 291
Favretto, D., 243
Fecker, L.F., 83
Fedewa, G., 4–5, 21, 149
Fedorov, R., 315
Fehér, D., 315, 316*t*
Fejos, I., 254–259, 255*t*
Feling, R.H., 304*t*, 307–308
Feng, H.T., 255*t*
Feng, X., 304*t*, 308

Fengler, K.A., 288
Fenical, W., 302, 303–306, 304*t*, 307–308
Fernandes, E.C., 241*t*, 242
Fernandes, J.B., 252–253
Fernández, F., 211
Fernández-Gutiérrez, A., 253
Ferreira, C.V., 317
Ferreira, I.C.P., 237–238
Ferreira, J.F.S., 47–48
Ferreres, F., 244*t*
Feth, F., 54–55
Fett-Neto, A.G., 26, 237–238
Filippini, R., 243
Fiot, J., 244*t*
Fischer, U., 151–152
Fisher, C.W., 197–198
Fisher, K.J., 27
Fist, A.J., 171, 175, 185
Fitchen, J., 165–166
Fjaervik, E., 304*t*, 309, 315–317, 316*t*
Flatt, P., 323
Fleet, G.W.J., 47
Fleury, V., 47
Fliniaux, M.A., 214–215
Flores, N., 191
Flors, V., 291–292
Foltz, R.L., 248–251
Forim, M.R., 252–253
Forseth, R.R., 261
Forster, P.I., 48
Fossati, E., 166–170
Fotso Fondja Yao, C.B., 316*t*, 318
Foubert, K., 243–248
Fox, J., 47–48
Frach, K., 254–259, 255*t*
Fragoso, V., 237–238
Franceschini, S., 176
Francis, S.H., 285, 290*f*
Francisco, C., 315, 316*t*
Franco, L., 244*t*
Franke-van Dijk, M.E.I., 13–15
Frappier, F., 260–261
Frédérich, M., 14, 238, 260–261
Fredholm, B.B., 113, 285, 287, 289–290
Freel, K.C., 307–308
Freeman, M.F., 302
Freitas, A.V.L., 44
Freud, S., 40–41
Frick, S., 185
Friedrich, A., 12–13
Fromm, H., 286–287, 288, 290*f*
Fu, P., 304*t*, 309–310
Fu, X., 60–62, 61*t*
Fujii, N., 149, 171, 175, 185
Fujiki, H., 314
Fujimori, A., 154–155
Fujimori, N., 118, 123–124, 125–126
Fujimoto, T., 307–308
Fujimura, T., 112–113, 118–119, 120, 124, 276–277
Fujita, K., 196–197, 288
Fujita, M.J., 319
Fujita, T., 10
Fujiwara, H., 165–166
Fukuchi-Mizutani, M., 175–176
Fukuda, D., 318
Fukui, T., 54–55
Fukui, Y., 175–176
Fukuyama, T., 4
Fulzele, D.P., 151–152
Furman, W., 142
Furmanowa, M., 151–152
Furumai, T., 321
Furze, J.M., 52

G

Galambos, J., 260–261
Galan, M.C., 26
Gallagher, C.A., 248–251, 249*t*
Gallardo, E., 249*t*
Galloway, G.L., 52
Galloway, M.P., 47–48
Galonić, D.P., 314, 323
Gantet, P., 9–10, 22–23, 87–89, 88*f*, 92–93, 211
Ganzera, M., 253
Gao, W., 255*t*
Gao, X., 152, 317–318
Garcia-Agustin, P., 291–292
García-Borrón, J.C., 193–194
Gaspar, L., 244*t*
Gaspar, T., 89–90
Gasquet, M., 244*t*
Gassmann, M., 286–287
Gatenby, A.A., 190
Gates, M., 184–185

Gazda, V., 166–170, 173, 176
Geerlings, A., 12–13, 14–15, 85, 96–98, 99–100
Geiger, L.Ph., 41–42, 49–50
Gelb, M.H., 91–92
Gentner, W.A., 47–48
Gerasimenko, I., 12–13, 85, 99–100, 238, 244*t*, 255*t*
Gerber, E., 91–92
Gerlach, W.L., 171, 175, 185
Gerrish, C., 289, 290*f*
Gershenzon, J., 44–46, 47–48, 52, 57, 62–63
Gerwick, W.H., 310, 311*t*, 313, 314, 323
Gesell, A., 165–170
Geu-Flores, F., 10–12, 83–84, 96–98, 147
Ghia, F., 260–261
Ghisalba, O., 321
Ghosal, S., 238
Gianfagna, T., 283
Giannini, S., 198
Gibbs, M., 122
Gibson, S.I., 24–26
Giddings, L.A., 3*f*, 12–13, 16, 80–82, 83–84
Giglioli-Guivarc'h, N., 24–26, 87, 88*f*, 91–93
Gil, F., 249*t*
Gilardi, G., 198
Gillam, E.M., 198
Gillard, J.W., 48
Gillies, F.M., 118–120, 126
Ginis, O., 90, 96–98
Giocanti, M., 244*t*
Girardin, P., 216
Girardot, M., 237–238
Girreser, U., 311*t*, 314
Girvan, H.M., 197–198
Gisi, D., 87–89
Gitschier, J., 313
Giulietti, A.M., 62–63
Glasl, S., 43
Gleissberg, S., 23–24
Glenn, W.S., 10–12, 83–84, 96–98, 147
Glévarec, G., 90, 96–98
Glick, S.D., 248–251, 249*t*
Gluck, M., 132
Goddijn, O.J.M., 147–148
Goergen, J.L., 218, 227
Goetz, T., 285, 286, 291
Goldmann, A., 47
Gomez-Caravaca, A.M., 253
Gómez-Lim, M., 211
Gomi, K., 284
Gonçalves, M.S., 252, 253
Gong, Y., 146
Góngora-Castillo, E., 4–5, 21, 149
Gonsalves D., 211, 227
Gontang, E., 303
Gontier, E., 207, 214–215, 215*f*, 216, 217, 218–221, 223–225, 224*f*, 227
Gonzalez, M.P., 286–287
Gooddijn, O.J.M., 12–13, 19–20
Gordan, H.R., 211
Gordon, H., 8, 10, 17–18, 19–20, 93–95
Gorman, E., 143–144
Gosset, G., 191, 193
Gotfredsen, C.H., 316*t*, 318–319
Gotti, R., 254
Gould, B., 23–24
Gourdeau, H., 308
Grabner, A., 219–221
Grace, M.H., 252–253
Graf, E., 43
Gram, L., 316*t*, 318–319
Gravot, A., 207, 218–221, 223–225, 224*f*, 227
Gray, A.I., 237–238
Green, T.R., 283–284
Greten, H.J., 208
Gribskov, M., 288, 290*f*
Griffin, W.J., 40
Grodberg, J., 304*t*, 309
Grogan, G., 197–198
Groll, M., 307–308
Gromova, I.I., 140–141
Grond, S., 308
Grosdemange-Billiard, C., 145
Gross, G.G., 56
Grosse, W., 83
Grossman, T.H., 319–320
Grothe, T., 166–170
Gruber, K., 176
Grupper, C., 211–212
Gu, W., 128
Gu, X., 255*t*
Guarnaccia, R., 10

Guckert, A., 207, 214, 215*f*, 216, 218–219, 223–225, 224*f*
Gueritte, F., 4
Guggisberg, A., 13
Guihur, A., 14–15, 17–18, 76–77, 96–99, 100
Guilet, D., 40, 43
Guilhaumou, R., 244*t*
Guillarme, D., 243
Guillod, L., 226
Guillon, S., 211
Guiltinan, M.J., 125–126
Guimaraes, H.A., 260–261
Guirimand, G., 8, 9, 10–12, 14–15, 17–18, 76–77, 90, 96–99, 100, 146
Guittet, E., 260–261
Guivarc'h, N., 6–8, 87
Gulati, A., 125–126
Gulder, T.A.M., 307–308, 325–327
Gulyás, Z., 260–261
Gunasekera, S.P., 142
Guo, F., 241*t*, 242
Guo, Y., 60, 288
Gupta, M.M., 21, 244*t*
Gurgui, C., 302
Gustowski, W., 151–152
Guymer, G.P., 48
Guzewska, J., 151–152

H

Hachiya, A., 171, 173, 185
Hada, N., 312
Háda, V., 260–261
Hagel, J.M., 24, 166–170, 176
Haginiwa, J., 142, 149
Hagino, H., 190
Haines, A.S., 318
Hakii, H., 314
Häkkinen, S.T., 210–211
Hakvåg, S., 315–317, 316*t*
Hale, T.I., 52
Halitschke, R., 294
Halkier, B.A., 99
Hall, D.G., 206–207, 317–318
Hall, L., 313
Hall, S., 313
Hallard, D., 2–4, 76–77, 80–82, 86–87
Halpert, J.R., 196–197
Hamaguchi, N., 55
Hamaguchni, N., 56
Hamdi, S., 6–8, 87
Hamel, E., 313–314
Hamilton, J.P., 3*f*, 4–5, 16, 21, 80–82, 149
Hamilton, J.T.G., 55
Hammerstone, J.F., 113–115
Hampp, N., 12–13, 147–148
Han, D., 255*t*
Han, J.Y., 294
Han, Y., 243–248
Han, Z., 140–141
Handel, T.M., 303
Hanora, A., 308
Hao, X., 237–238
Harada, H., 197
Harden, T.K., 287
Harigaya, Y., 83–84
Haring, M.A., 291
Harker, W.G., 154–155
Harkes, M.P., 151–152
Harmon, A.C., 288, 290*f*
Harms, K., 308–309
Harnischfeger, G., 261
Harper, J.F., 288, 290*f*
Harrigan, G.G., 311*t*, 313–314
Hartmann, F.K., 315
Hartmann, T., 44, 82–83
Hartonen, K., 240–242, 241*t*
Haruyama, H., 316*t*, 317–318
Hashimoto, K., 10
Hashimoto, T., 50–55, 56–57, 59, 164–165, 171, 173, 185, 210–211
Hashizume, K., 52
Hasler, J.A., 196–197
Hause, B., 56–57
Hawas, U.W., 304*t*, 306
Hawkins, K.M., 164–165, 190, 194–195
Hayashi, A., 59
Hazell, A., 260–261
Hazell, R., 260–261
He, C., 251
He, L., 243–248
He, X., 255*t*
He, Y.A., 196–197
He, Z., 166–170, 173, 176
Hearn, W.L., 248–251
Heble, M.R., 151–152

Hecht, S., 140–141
Heckendorf, A.H., 143–144, 149
Hedhili, S., 9–10, 87–89, 88*f*
Hehn, A., 227
Heim, W.G., 54–55
Held, M., 294
Helf, M.J., 302
Helvig, C., 198
Hemalal, K.D., 43
Hemling, M., 149
Hemmerlin, A., 145
Hempfling, T., 221
Hemscheidt, T.K., 13–15, 315, 316*t*
Henderson, C.J., 196–197
Henderson, V.E., 46
Henion, J.D., 255*t*
Henriques, A.T., 237–238
Henschke, H., 41–42, 49–50
Hentschel, U., 302
Herai, Y., 119, 125, 129, 276–277, 278
Hericourt, F., 17–18, 96–99
Hermans-Lokkerbol, A., 238
Hernández-Chávez, G., 190, 193
Hernandez-Vazquez, L., 59–60
Herouart, D., 214–215
Herraiz, T., 312
Herrero, M.T., 240, 286–287
Hertzberg, R., 140–141
Hervouet, N., 6–8, 9, 87–89, 94–95, 96–98
Herzfeld, T., 48
Hesse, M., 13, 15–16, 41–42, 49–50, 238, 262
Hesselink, P.G.M., 151–152
Heuschneider, G., 291
Heyman, H.M., 261
Hibi, N., 52–54
Hicks, M.A., 83–84
Hiei, Y., 278
Higashi, Y., 185
Higashiguchi, S., 52–54
Hildreth, S.B., 54–55
Hileman, L.C., 23–24
Hill, R., 312
Hilliou, F.A.O., 4, 15–16, 24–26, 96, 98–99
Hiltunen, R.H., 211
Hime, G.W., 248–251
Hirata, K., 198
Hirayama, C., 99–100
Hisada, K., 283
Ho, I.K., 294
Hoch, J.M., 142
Hoeffler, J.-F., 145
Hoehn, P., 321
Hofmann, A., 43
Hoge, J.H.C., 12–13, 19–20, 146–148
Hohnloser, W., 132
Hoki, U., 154–155
Hollander, C., 49–50
Hollingsworth, R.G., 275–276, 283
Holman, T.R., 315
Holmen, J., 285, 287, 289–290
Holmes, E., 261
Holton, T.A., 175–176
Honda, N.K., 251
Hong, B., 244*t*
Hong, K., 304*t*, 309–310
Hong, M.L.K., 61*t*
Hong, S.B., 24–26
Hong, W.H., 240–242, 241*t*
Honkavaara, P., 46
Hooykaas, P., 120–122
Hori, K., 165–166, 194–195
Hori, S., 319–320
Horvath, S., 176
Hoshino, H., 142, 149
Hoshino, T., 319–320
Hosoya, S., 318
Hostettmann, K., 243, 244*t*
Hothersall, J., 318
Hotze, M., 10–12, 19–20, 83, 92, 94–95, 146–147
Hou, R., 146
Hou, Y., 243–248
Hough, L.B., 248–251, 249*t*
Houghton, P.J., 142
Howe, G.A., 87–89
Hoye, T.R., 54–55
Hrovat, K.J., 315
Hsiang, Y.H., 140–141
Hsieh, M., 146
Hu, J.T., 249*t*, 251
Hu, W.L., 238
Hu, Y.L., 210–211
Hu, Z., 243–248, 255*t*
Huang, D.D., 239–240, 315
Huang, F.-C., 146–147, 166–170, 210–211

Huang, J., 289
Huang, L., 251–252
Huang, P.-L., 146–147
Huang, X.J., 237–238
Huang, Z., 146
Hubbard, J.D., 249*t*
Huber, R., 307–308
Huber-Allanach, K.L., 167*f*, 191–192
Hughes, C.C., 302
Hughes, E.H., 24–26
Huh, G.H., 294
Huhn, C., 254, 255*t*
Huhtikangas, A., 237–238
Hunnicutt, E.J., 43–44
Hunt, M.D., 283–284
Hur, B.K., 152
Husson, H.-P., 48–49
Hutchinson, C.R., 143–144, 149
Huysmans, S., 83–84
Hvoslef-Eide, A.K., 221
Hwang, I., 90
Hwang, S.H., 294

I

Iavicoli, A., 292
Ibañez, A., 240
Ibañez, E., 240
Ibanez, M., 12–13, 14–15
Ibanñez, M.M.-L., 85, 96–98, 99–100
Ichimaru, M., 43
Igarashi, Y., 321
Iijima, Y., 149
Ikeda, D., 303, 304*t*
Ikeda, H., 10
Ikenaga, T., 44
Ikezawa, N., 164–170, 173, 176, 185, 186, 187–189, 190, 194–195, 198
Iki, K., 197
Ilari, A., 176
Imagawa, H., 118, 123
Imanishi, S., 52
Imbault, N., 87–89, 91
Inai, K., 164–165
Ingkaninan, K., 252, 316*t*
Inoue, K., 10, 146–147
Inouye, H., 10
Inouye, K., 165–166
Inui, M., 187–189
Inui, T., 165, 171, 175, 185
Ioset, J.R., 243, 244*t*
Irie, M., 112–113, 118–119, 120
Irino, F., 118, 276–277
Irish, V.F., 23–24
Irmler, S., 10–12, 19–20, 83, 92, 94–95, 146–147
Ishida, M., 113–115, 124–125
Ishikawa, H., 4
Ito, E., 122, 126
Ito, H., 190
Ito, Y., 251–252
Ivanyi, R., 254–259, 255*t*
Iwama, M., 112–113, 118–119, 120
Iwanari, H., 59
Iwasa, K., 43, 165–170, 173, 175, 176, 185, 187–189
Iwata, H., 196–197

J

Jabos, D.J., 76–77, 80–82
Jacobs, D.I., 2–4
Jacobson, K.A., 287
Jain, A.K., 151
Jakab, G., 291–293, 294
James, R.D., 142
Jamison, M., 304*t*
Jander, G., 87–89
Jansma, A.L., 303
Janso, J., 304*t*, 308
Janssens, S.B., 83–84
Jaradat, T.T., 196–197
Jaroszewski, J.W., 261
Jeffrey, A.M., 314
Jelesko, J.G., 54–55
Jenett-Siems, K., 46–47, 237–238
Jenkins, C.M., 197
Jenks, C., 237–238
Jensen, K., 197, 198
Jensen, P.R., 303–306, 304*t*, 307–309
Jensen, S.R., 83
Jeong, M.J., 60
Jiang, H., 241*t*, 242
Jiang, J.-H., 59–60, 288, 289
Jiang, K., 146
Jiang, X., 306
Jiang, Y., 56–57
Jimenez, J.I., 310, 311*t*

Jirschitzka, J., 44–46, 57, 62–63
John, G.H., 196–197
Johns, S.R., 48–49
Johnson, E.F., 196–197
Johnson, R.K., 142
Jokelainen, T., 211
Jorgensen, C., 99–100
Jorgensen, K., 99–100
Jorgensen, R.A., 171
Josefsen, K.D., 315–317, 316*t*
Joshi, C.P., 124–125
Joshi, R., 125–126
Joshi, S.P., 151–152
Jouhikainen, K., 211
Jourdie, V., 294
Joyce, M.G., 197–198
Joyeau, R., 237–238
Jung, H.Y., 62–63
Jung, Y.J., 129–131, 130*t*, 280–281, 281*f*, 282
Justo, G.Z., 317

K

Kaczkowski, J., 55
Kagan, I.A., 145
Kagasaki, T., 316*t*, 317–318
Kai, G., 52–54, 56–57, 59–62, 61*t*, 210–211
Kai, M., 52
Kaiser, C.S., 240
Kaiser, H., 56–57
Kalicki, P., 48
Kallberg, Y., 56–57
Kaloga, M., 237–238
Kaltenbach, M., 16–17, 18
Kalt-Hadamowsky, M., 13
Kamena, F., 310–312
Kameyama, T., 316*t*, 319, 321
Kaminski, F., 166–170, 173, 176
Kanehara, T., 118–119
Kang, J., 294
Kang, K.K., 129–131, 130*t*, 280–281, 281*f*, 282
Kang, S.M., 62–63
Kang, Y.M., 60, 62–63
Kanjana-opas, A., 316*t*, 318
Kanogsunthornrat, J., 244*t*
Kantham, L., 43–44
Kaothien-Nakayama, P., 129–131, 130*t*, 278–281, 279*f*, 281*f*
Kaplan, B., 288, 290*f*
Karamat, F., 227
Karan, M., 175–176
Karasek, P., 142
Karathanasis, A., 40–41
Karimi, F., 60–63, 61*t*
Karlen, D., 294
Kashihara, E., 149
Kastell, A., 62–63
Katagiri, Y., 186–187
Katahira, R., 123–124
Katajamaa, M., 9–10, 12, 15–16, 18, 21
Katano, N., 10, 146–147
Katavic, P.L., 48
Katayama, T., 164–165, 185, 186, 187–190, 191, 193–195, 198, 199
Kato, A., 43, 48–49, 123, 128–129
Kato, H., 56–57
Kato, M., 112–115, 118–119, 120, 123–125, 128, 276–277
Kato, N., 171–173
Katoh, A., 54–55
Katou, S., 284
Katsir, L., 87–89
Katsumoto, Y., 175–176
Kauffman, C.A., 303–306, 304*t*, 307–309
Kaupmann, K., 286–287
Kavanagh, P., 47–48
Kawahara, N., 171, 312
Kawai, H., 21–22
Kawano, N., 171
Keasling, J.D., 192
Keefner, S.M., 248–251, 249*t*
Keiner, R., 56–57
Kellner, S., 227
Kelter, G., 304*t*, 306
Kemmler, M., 304*t*, 309
Kempe, K., 185
Keng, C.L., 61*t*
Kennedy, D., 293–294
Kern, M., 166–170, 173, 176
Kessler, A., 294
Kevers, C., 89–90
Keya, C.A., 122, 125–126
Keyzers, R.A., 302
Khajuria, R.K., 153

Khan, M.F., 176
Khanuja, S.P., 21
Khataee, E., 60–62, 61*t*
Khattak, K.F., 43
Khelifi, L., 61*t*, 225
Khelifi-Slaoui, M., 61*t*, 225
Khurana, J.P., 90–91
Khurana, P., 90–91
Kibble, N.A.J., 83–84
Kieber, J.J., 90
Kigoshi, H., 311*t*, 314
Kihlman, B.A., 113–115
Kijne, J.W., 15–16, 24–26, 87–89
Kikuchi, Y., 191
Kim, H.J., 294
Kim, H.K., 236–237, 262
Kim, H.S., 294
Kim, J.-S., 164–165, 185, 186, 187–190, 191, 193–195, 198, 199, 240–242, 241*t*
Kim, S.H., 55–56, 152
Kim, S.K., 302–303
Kim, W.S., 16–18, 19–20, 94–95
Kim, Y.D., 60, 62–63
Kim, Y.S., 129–131, 130*t*, 132, 275–276, 278–281, 279*f*, 281*f*, 282–284, 289–290, 292–293, 294
Kimura, N., 319
Kimura, T., 294
Kingston, D.G.I., 242–243
Kinnersley, A.M., 286–287
Kinoshita, T., 316*t*, 317–318
Kirch, W., 46
Kis, K., 6–8
Kiss, K., 145
Kissen, R., 99–100
Kitajima, M., 142, 145, 146–148, 149, 151–153
Kitamura, Y., 44
Kitao, N., 113–115, 124–125
Kiuchi, F., 171
Kizu, H., 47
Klapheck, S., 83
Klein, O.N., 244*t*
Klessig, D.F., 289, 290*f*, 291
Kletter, C., 43
Kliebenstein, D.J., 99–100
Klinkenberg, G., 304*t*, 309, 315–317, 316*t*
Klösgen, R.B., 96–98
Knorr, D., 62–63
Knox, E.B., 83–84
Kobayashi, E., 316*t*, 319, 321
Kobayashi, K., 149, 311*t*, 314
Koch, M., 48
Kodama, Y., 278, 289–290
Koelen, K.J., 56
Koepke, J., 12–13, 26, 83–84, 148
Koert, U., 308–309
Koffas, M.A., 197–198
Kohlhagen, G., 154–155
Kohno, J., 56, 59
Koizumi, N., 118, 119, 125, 129, 275, 276–277, 278
Kojima, H., 52
Kokabu, Y., 171–173
Koltun, D.O., 54–55
Komari, T., 278
Konda, M., 184–185
Kondo, T., 319–320
Kong, H., 47–48
Konno, K., 99–100
Konno, Y., 196–197
Konowalowa, R., 46–47
Kontrimaviciute, V., 244*t*
Konwalinka, G., 255*t*
Kopka, J., 213–214
Koppenhöfer, E., 211–212
Koscheski, P., 125
Koseki-Nakamura, M., 142, 147–148, 152–153
Koshiishi, C., 123, 128–129
Koshino, H., 284
Koshiro, Y., 123–124, 125–126, 128–129
Kostenyuk, I., 238
Koster, D.A., 142
Kosuth, J., 155
Kour, A., 153
Kowanko, N., 58–59
Koyama, T., 171–173
Koyama, Y., 112, 123–124, 128–129
Koyanagi, T., 164–165, 185, 186, 189–190, 191, 193–194, 198, 199
Kramell, R., 165–170, 185
Kramer, E.M., 23–24
Kraus, P.F., 166–170
Kretschmar, J.A., 113–115, 118–119, 275
Krishnamurthy, K.V., 151–152

Krishnaveni, R., 193
Kroymann, J., 99–100
Kubota, H., 119–120, 128–129
Kuboyama, T., 4
Kulkarni, D.K., 151–152
Kulshrestha, M., 9–10
Kumagai, H., 164–165, 185, 186, 187–189, 190, 194–195, 198
Kumahsiro, T., 278
Kumar, P.M., 153
Kumar, R.A., 23–24
Kumar, V., 125–126, 129–131, 130*t*, 237–238
Kumirska, J., 311*t*, 312
Kunert, G., 52
Kunze, G., 120–122
Kurahashi, O., 191
Kurasawa, S., 303, 304*t*, 316*t*, 319, 321
Kuroyanagi, M., 21–22
Kurz, W.G.W., 17–18, 87, 90
Kusano, T., 119, 125, 129, 276–277, 278
Kusari, S., 153–154, 155
Kushida, H., 196–197
Kutchan, T.M., 12–13, 78–79, 85, 147–148, 164–171, 185, 186
Kuzovkina, I.N., 83
Kuzovkov, A.D., 48
Kwak, S.S., 240–242, 241*t*
Kwan, J.C., 310–312
Kwon, H.C., 303, 304*t*, 307
Kyrpides, N., 187–189

L

Laakso, I., 237–238
Laatsch, H., 311*t*, 314, 316*t*, 318
Laberche, J.C., 217, 218
Lacarelle, B., 244*t*
Laflamme, P., 16–18, 19, 80–82, 86–87, 95
Lai, Z., 289
Lakó-Futó, A., 260–261
Lallemand, J.Y., 47
Lamb, C., 289
Lamberto, J.A., 48–49
Lammertyn, F., 9–10, 12, 15–16, 18, 21
Lamuela-Raventós, R.M., 228
Lan, X.Z., 60
Lang, Y.Y., 307
Lange, B.M., 6, 9, 19–20
Langlois, Y., 4
Lanoue, A., 10–12, 14–15, 76–77, 96–98, 100
Lara, A.R., 190
Larbat, R., 227
Larionova, D., 244*t*
Larkin, P.J., 171, 175, 185
Larsen, T.O., 316*t*, 318–319
Larson, T.R., 166–170, 173, 176
Larsson, S., 78–79, 79*f*, 81*f*
Last, R.L., 19–20
Laussermair, E., 16–17
Law, B.K., 310–312
Lawson, M., 166–170
Lazny, R., 48
Le Crouerour, G., 252
Le Men-Olivier, L., 252
Lean, M.E.J., 113
Lee, B.H., 294
Lee, C.H., 285–286
Lee, E.J., 165–166, 167*f*, 191–192, 218–219, 221, 224–225
Lee, H., 240–242, 241*t*
Lee, J., 193
Lee, M.C., 255*t*
Lee, M.H., 294
Lee, T.S., 192
Lee, Y.H., 129–131, 130*t*, 280–281, 281*f*, 282
Leech, M., 4, 96, 98–99
Leete, E., 49–56, 58–59
Lefeber, A.W.M., 8, 261, 262
Lefebre, A.W., 91
Lei, Y., 243–248
Lein, W., 120–122
Lemiere, F., 243–248
Leng, Q., 288
Lenoan, A., 243, 244*t*
Lenz, R., 166–170
Leo, E., 142
Leon, L.L., 237–238
Leonard, E., 197–198
Leow, H.M., 48
Lepingle, A., 47
Lerchl, J., 120–122
Levac, D., 4–5, 16–18, 19–20, 23–24, 94–95
Levi, M., 285–286

Levinson, H.Z., 43–44
Lewis, N.G., 164–165
Li, B., 324
Li, C.Y., 239, 244*t*, 260
Li, D.N., 196–197
Li, H., 243–248, 255*t*
Li, K., 249*t*, 254, 255*t*, 259*f*
Li, L., 52–54, 56–57, 60, 147–148, 210–211, 239–240
Li, M.Y., 60
Li, N., 255*t*
Li, P., 249*t*, 254, 255*t*, 259*f*
Li, Q., 249*t*, 254, 255*t*, 259*f*
Li, R., 23–24, 58–59
Li, S.F.Y., 142, 255*t*
Li, S.S., 304*t*, 306–307
Li, X., 173, 304*t*, 309–310
Li, Y.H., 125–126, 128, 146, 149, 249*t*, 251, 255*t*
Li, Y.X., 249*t*, 251
Liao, Y., 244*t*
Liao, Z.H., 60
Lidder, P., 209, 211
Liebisch, H.W., 48, 50–52, 55
Lièvre, K., 223–225, 224*f*, 227
Lila, M.A., 252–253
Lim, C.K., 291
Lim, S., 129–131, 130*t*, 278–281, 279*f*, 281*f*, 282, 283–284, 289–290
Lima, J.A., 237–238, 251
Lin, B., 307
Lin, C.H., 125, 260
Lin, F., 286–287
Lin, G.D., 40
Lin, P.P., 285–286
Lindgren, L., 211
Lindon, J.C., 208, 259–260
Ling, H., 146
Lingens, F., 132
Lippok, B., 288
Liscombe, D.K., 3*f*, 4–5, 14, 16–17, 21, 23–24, 50–52, 80–83, 85, 149, 165–166
Liskamp, R.M., 314
Litt, A., 23–24
Liu, B., 241*t*, 242
Liu, D., 87, 89–90, 237–238, 304*t*, 307
Liu, E.W., 23–24, 58–59
Liu, G., 87–89
Liu, H., 237–238, 244*t*
Liu, J.R., 240–242, 241*t*, 249*t*, 254, 255*t*, 259*f*, 307
Liu, K., 153–154
Liu, L.F., 140–141
Liu, P., 304*t*, 309–310
Liu, T., 52–54
Liu, X.Q., 60, 152
Liu, Y., 23–24, 243–248, 244*t*, 251–252, 310–312
Liu, Z., 237–238, 304*t*, 306–307
Llabres, G., 252
Loder, J.W., 48–49
Longevialle, P., 48–49
Lopes-Cardoso, M.I.L., 24–26, 146–148
Lopez, C., 239–240
Lopez, H., 190
Lopez-Avila, V., 241*t*
Lopez-Meyer, M., 12–13, 144
López-Rivadulla, M., 249*t*
Lorence, A., 80–82, 150, 152–153
Lorenz, N., 243–248
Loris, E.A., 12–13, 26, 83–84, 85, 148
Loske, A.M., 211
Lottspeich, F., 12–13, 147–148
Loukanina, N., 82–83
Lounasmaa, M., 40, 48
Loyola-Vargas, V.M., 24–26
Lozoya-Gloria, E., 24–26
Lu, B.-B., 59–62
Lu, H., 143–144
Lu, J.-L., 112
Lu, R.H., 54–55
Lu, X., 56–57
Lu, Y.-T., 147–148, 152
Lucena, H.F.S., 40
Lude, W., 43
Luesch, H., 310–312, 311*t*, 313–314
Luijendijk, T.J.C., 2–4, 12–13, 99–100, 101
Luna, E., 291–292
Luo, H., 146, 149, 239
Luo, X., 60–62, 61*t*, 243–248
Lütke-Eversloh, T., 190, 191
Luz, G.F., 249*t*
Lv, H., 243–248
Lydon, J., 47–48
Lykidis, A., 187–189
Lysek, D.A., 197–198

M

Ma, C.H., 239
Ma, J., 142
Ma, L., 304*t*, 306–307
Ma, W., 287–288, 290*f*
Ma, X.Y., 12–13, 26, 83–84, 85, 99–100, 148
Ma, Y., 251–252
Maathuis, F.J.M., 288, 290*f*
Machado, G.M.C., 237–238
Macheroux, P., 165–166, 176
MacKay, A.C., 197–198
MacLeod, B.P., 82–83
MacMillan, J.B., 304*t*, 308–309
MacNeil, I.A., 319–320
Macone, A., 176
Madhusoodanan, P.V., 142
Madyastha, K.M., 10–12
Maeder, M.N., 292
Mafnas, C., 303
Magallanes-Lundback, M., 4–5, 21, 149
Magnani, B., 313
Magnotta, M., 24–26
Mahadevan, R., 4–5, 21–22, 27
Mahiou, V., 244*t*
Mahlberg, P.G., 93–94
Mahmoodnia, M., 211
Mahroug, S., 6–8, 9, 14–15, 17–18, 76–77, 80–82, 86–89, 94–95, 96–99, 100
Maier, A., 304*t*, 306
Maille, M., 47
Malmberg, R.L., 52
Maloney, K.N., 304*t*, 308–309
Maltese, F., 236–237
Malysheva, S.V., 244*t*
Mandaokar, A., 87–89
Mann, J., 41–42
Mann, P., 46–47
Mano, Y., 83
Mansson, M., 316*t*, 318–319
Marathe, R., 23–24
Marchand, C., 142
Marcus, S., 317–318
Maresh, J.J., 2–4, 12–13, 14, 21, 22–23
Margalho, C., 249*t*
Marineau, C., 12–13, 144
Marini-Bettolo, G.B., 48–49
Marion, L., 50–52
Marion, O., 317–318
Marner, F.J., 311*t*, 314
Marques, E.P., 249*t*
Marques, M.O.M., 241*t*, 242
Marquez, B.L., 310, 311*t*, 323
Marston, A., 40, 43, 243, 244*t*
Martin, G.B., 23–24
Martin, G.D., 304*t*, 306
Martin, K.P., 142
Martin, M.T., 260–261
Martin, N., 254, 255*t*
Martin, V.J., 166–170
Martín-Belloso, O., 228
Martínez, A., 190, 193
Martinez, C., 292–293
Martinez-Atienza, J., 288
Martin-Hernandez, A.M., 23–24
Martino, G., 23–24
Martins, D., 317
Martins, R., 311*t*
Martin-Wixtrom, C.A., 197–198
Mash, D.C., 248–251
Massot, B., 214, 215*f*, 216, 218, 219–221
Mateus, L., 210–211
Mathavan, S., 275–276, 283
Mathias, L., 249*t*, 251
Mathieu, O., 244*t*
Mathieu-Daudé, J.C., 244*t*
Mathur, J., 96–98
Mathur, S.N., 275–276
Matsubayashi, Y., 52
Matsuda, J., 59
Matsui, K., 47
Matsumoto, K., 312
Matsumoto, T.K., 211, 227
Matsushima, Y., 165–166, 194–195
Matsuzaki, K., 307–308
Matsuzaki, R., 54–55
Mattern, D.J., 62–63
Mattern, M.R., 142
Mauch, F., 291–292
Mauch-Mani, B., 289, 290*f*, 291–293
Maurer, H.H., 244*t*
Mavar-Manga, H., 252
Max, J.P., 239–240

Mazars, C., 90–91
Mazzafera, P., 113–115, 114*t*, 118–119, 126, 128–129
Mbeunkui, F., 252–253
McAlpine, J.B., 322
McArthur, K.A., 304*t*, 309
McAtee, J., 315, 316*t*
McCalley, D.V., 242–243
McCarthy, A.A., 118, 125, 277
McCarthy, J.G., 118, 125, 277
McCarty, K., 23–24
McCloskey, J.A., 48–49
McCloud, E.S., 211–212
McCoy, E., 26
McKnight, T.D., 12–13, 14–15, 96–98, 143–144, 147–148
McPhail, A.T., 140–141, 142
Medeiros, W.L.B., 249*t*, 251
Media, D.H.G., 48–49
Medina-Bolivar, F., 152–153
Meehan, T.D., 10
Meier, A.C., 52–54
Meijer, A.H., 12–13, 146–148
Mein, H.F., 41–42, 49–50
Meinhold, P., 197–198
Meireles, M.A.M., 241*t*, 242
Meisner, J., 100
Melin, C., 96–98
Melo, P.S., 317
Melotto, M., 87–89
Memelink, J., 9–10, 12–13, 14–16, 18, 24–26, 85, 86–89, 96–98, 99–100, 146–148
Méndez, C., 320, 321
Mendiola, J.A., 240
Meng, C., 52–54
Menke, F.L., 15–16, 24–26, 87–89
Mentzer, M., 149
Meola, J.M., 249*t*
Mercier, R.W., 288
Merckx, V.S.F.T., 83–84
Mérillon, J.M., 87, 89–90
Mersey, B.G., 93–94
Merwe, C.F., 128
Mesia, K., 243–248
Message, B., 47
Mestichelli, L.J.J., 54–55
Metraux, J.P., 292–293
Meyer, J.J.M., 261
Meyer, O., 145
Michael, A.J., 52
Miersch, O., 166–170
Miettinen, K., 8, 9, 96–98, 146
Mihali, T.K., 313
Milat, M.L., 47
Miles, C.S., 197–198
Milesi, S., 207, 214, 215*f*, 216, 218–221
Miller, J.A.C., 171, 175, 185
Millgate, A.G., 175, 185
Mimura, H., 165–166
Min, J.Y., 60, 62–63
Minami, H., 164–166, 185, 186, 187–190, 191, 193–195, 198, 199
Mincer, T.J., 303, 304*t*, 307–308
Minero-García, Y., 87
Minor, C., 319–320
Mirabella, R., 291
Miranda-Ham, M.L., 87
Mirjalili, M.H., 59–60
Mirzaev, Y.R., 46–47
Misawa, M., 87, 90
Mitani, A., 54–55
Mitchell, S.S., 304*t*, 309
Mitchell-Olds, T., 99–100
Mitsuhara, I., 60–62
Miwa, A., 142, 149
Mizuno, K., 112–115, 118–119, 120, 124–125, 276–277
Mizusaki, S., 54–55
Mizutani, M., 10
Mochida, K., 149
Moerker, T., 321
Moffitt, M.C., 313
Moh, S.H., 218–219, 221, 224–225
Mohanpuria, P., 125–126, 129–131, 130*t*, 275–276
Molgó, J., 238, 260–261
Molina, A., 283–284
Mollenschott, C., 144
Møller, B.L., 99–100, 197, 198
Molyneux, R.J., 47
Monacelli, B., 95–96, 237–238
Monnerat, C.S., 249*t*, 251
Monsanto, P., 249*t*
Monteiro, A.M., 119–120
Montiel, G., 9–10, 15–16, 87–89

Montoro, P., 244*t*
Moody, D.E., 248–251
Moor, A., 227
Moore, B.S., 303, 307–308, 325–327
Moore, R.E., 311*t*, 313–314
Morant, A.V., 99–100
Morard, P., 222–223
Morath, P., 118
Mordehai, A.V., 255*t*
Moreno, P.R.H., 237–238
Moriguchi, R., 307–308
Morimoto, H., 119, 125
Morimoto, M., 129–131, 130*t*, 278–281, 279*f*, 281*f*
Morin, Ph., 239–240
Morinaka, B.I., 302
Morishige, T., 165–166, 170–171, 173, 175, 185, 186–187
Morita, M., 99
Moriyasu, M., 43
Mosbacher, J., 286–287
Mösli Waldhauser, S.S.M., 118–119, 128–129
Mothes, K., 55
Motz, K., 176
Moummou, H., 56–57
Mount, M.S., 285–286
Moyano, E., 60, 211, 227
Mueller, M.J., 15–16, 24–26, 87
Muganga, R., 238
Mukerji, D., 275–276
Müller, A., 254, 255*t*
Munekata, M., 192
Muñoz, A.J., 193
Munro, M.H., 302
Murakami, Y., 312
Murano, E., 218
Murata, J., 8, 10–12, 11*f*, 14–15, 16–20, 21–22, 24–26, 93–95, 101
Murkhopadhyay, S.K., 190
Murphy, A.C., 318
Musarrat, J., 153
Musich, C., 315
Mustafa, R.N., 236–237
Mustard, J.A., 132
Mutanda, T., 221
Mysore, K.S., 23–24

N

Nagai, C., 112, 123–124, 125–126, 128–129
Nagai, H., 142
Naganawa, H., 303, 304*t*, 316*t*, 319, 321
Nagayoshi, M., 165–166
Nagle, D.G., 311*t*, 313–314
Nahar, L., 48–49
Nahrstedt, A., 56–57
Nakabayashi, R., 149
Nakagawa, A., 164–165, 185, 186, 189–190, 191, 193–194, 198, 199
Nakai, S., 10
Nakajima, K., 52–54, 56–57
Nakakita, M., 52
Nakamura, H., 303, 304*t*, 316*t*, 319, 321
Nakamura, K., 196–197
Nakamura, M., 99–100, 151–152
Nakano, H., 316*t*, 319, 321
Nakayama, K., 190
Nam, S.J., 307–308
Namdeo, A.G., 151
Namjoyan, F., 52–54
Nandi, O.I., 82–83
Naqvi, S., 227
Naranjo, P., 39–72
Narantuya, S., 43
Narula, A., 59–60
Nash, R.J., 44, 47
Nasomjai, P., 58–59
Nathanson, J.A., 43–44, 275–276, 278
Naudascher, F., 87
Navarro, M., 142
Navarro-Ocana, A., 59–60
Nawabi, P., 187–189
Ndjoko, K., 243, 244*t*
Neelgund, Y.F., 193
Neeli, R., 197–198
Negishi, O., 118, 123
Nehlig, A., 285, 287, 289–290
Neilan, B.A., 313
Nelson, D., 83
Nelson, J.D., 142, 150–151
Nemoto, K., 83
Nessler, C.L., 12–13, 14–16, 80–82, 96–98, 144, 147–148, 150, 151, 152–153
Netherly, P.J., 40–41
Neuenschwander, U.H., 283–284

Neumann, J., 311*t*, 312
Newman, D.J., 207
Newman, J.D., 6
Newman, K.L., 27
Newmark, R.A., 58–59
Newton, R.P., 285–286, 287–288
Nguyen, D.T.T., 243
Nguyen, V.A., 323
Ni, X., 60–62, 61*t*
Nicholson, J.K., 259–260
Nielsen, H.B., 260–261
Niemann, A., 40–41, 47–48, 49–50
Niessen, W.M.A., 242–243
Nighat, F., 43
Nikam, T.D., 61*t*
Niki, T., 60–62
Niles, R.M., 285–286
Ningthoujam, S.S., 208
Nirala, N.K., 59–60
Nishimura, M., 142, 149
Nishitha, I.K., 142
Nishiyama, Y., 43
Nistor, I., 238
Niu, Y.Y., 46, 87–89, 146, 149
Noble, M.A., 197–198
Noé, W., 144
Noel, J.P., 125
Noguchi, M., 54–55
Nokata, K., 142
Nomura, T., 78–79, 85, 166–171
Northcote, D.H., 285–286
Nortier, J., 208
Noszal, B., 254–259, 255*t*
Nour-Eldin, H.H., 99
Nowak, J., 23–24, 58–59, 176
Nozomu, K., 129–131, 130*t*
Nurhayati, N., 82–83
Nuzillard, J.M., 252
Nyberg, N.T., 261

O

Ober, D., 82–83
O'Brien, J., 47–48
O'Connell, F.D., 113–115
O'Connor, S.E., 2–5, 14, 16–17, 21, 22–24, 26, 83–84, 148
Oda, J., 56–57
Oddone, A.M., 23–24
O'Donnell, C., 47–48
Odriozola-Serrano, I., 228
Oechslin, M., 115
Oertel, R., 46
Ogan, K., 244*t*
Ogasawara, N., 319–320
Ogawa, M., 119, 125, 129, 276–277, 278
Ogita, S., 113, 118, 125–126, 129–131, 130*t*, 275, 276–277, 278–281, 279*f*, 281*f*
Oh, K.W., 294
Ohagan, D., 58–59
O'Hagan, D., 55, 58–59
O'Hanlon, P.J., 318–319
Ohashi, Y., 60–62, 284
Ohgaki, M., 171–173
Ohri, R.V., 315
Ohsawa, M., 151–152
Ohta, S., 278
Ohtsubo, N., 60–62
Okada, S., 100
Okami, Y., 303, 304*t*, 316*t*, 319, 321
Okamoto, S., 197
Okamoto, T., 171
Oksman-Caldentey, K.-M., 60, 210–211
Okuda, A., 276–277
Oldham, N.J., 302
Olenski, T., 48
Olguin-Uribe, G., 315, 316*t*
Oliva, B., 318–319
Oliveira, J.M.A., 244*t*
Oliveira, M.F., 249*t*
Oliveira, R.R., 252, 253
Oliveira, S.L., 40
Oliveira, V.B., 249*t*
Olry, A., 227
Olsen, R.W., 286
Olson, M.M., 190
Olsovska, J., 243–248
Olsson, O., 118–119
Oms-Oliu, G., 228
Omura, S., 307–308
Onaka, H., 321
O'Neill, A., 318–319
Ongena, G., 56–57
Ooi, A., 10, 146–147
Orechoff, A., 46–47
Oresic, M., 9–10, 12, 15–16, 18, 21
Orjala, J., 311*t*, 314

Orober, M., 292–293
Oseki, K., 47
Oset-Gasque, M.J., 286–287
Ossetian, R., 196–197
Osswald, B., 132
Otalvaro, A.A.M., 62–63
Othman, E.M., 308
Otoguro, K., 307–308
Otsuka, M., 319
Ott, S.C., 46–47
Oudin, A., 6–8, 9, 10–12, 19–20, 22–23, 87–89, 90, 92–93, 94–95, 96–98
Ouedraogo, M., 208
Ouelhazi, L., 87, 89–90
Ouellet, M., 27
Ouwerkerk, P.B.F., 86–87
Ozaki, Y., 54–55
Ozawa, T., 118, 123

P

Pablo, J., 248–251
Padgett, W.L., 286
Paek, K.Y., 218–219, 221, 224–225
Paetz, C., 47–48
Page, J.E., 4–5, 21–22, 23–24, 27, 58–59
Paiva, N.L., 289
Palazon, J., 59–60, 211, 227
Palevitch, D., 100
Pallant, C.A., 249*t*, 251
Pallas, J.A., 289
Palle, K., 142
Palmer, K.H., 140–141, 142
Palmer, M.J., 132
Palmisano, G., 260–261
Pamboukdjian, N., 47
Pan, E., 304*t*
Pan, X.-W., 152
Pan, Y., 243, 244*t*
Panchuk, B.D., 17–18
Pandey, R., 244*t*
Panjikar, S., 12–13, 26, 80–82, 83–84, 85, 148
Panta, G.R., 286–287
Pantazis, P., 140–141
Papon, N., 22–23, 87, 89–90
Papper, E.M., 46
Paquette, S.M., 99–100
Paradise, E.M., 27
Paranhos, J.T., 237–238
Parchmann, S., 15–16, 24–26, 87
Pardo Torre, J.C., 47–48
Parello, J., 48–49
Park, D.H., 291
Park, D.J., 60, 62–63
Park, J.H., 171–173
Park, J.S., 304*t*, 307
Park, S.W., 175, 240–242, 241*t*
Parkerson, P.T., 40–41
Parr, A.J., 55–56, 58–59
Parramon, M., 286–287
Pasanen, M., 208
Paschke, R., 48
Pasqua, G., 95–96, 144, 237–238
Pasquali, G., 12–13, 19–20, 24–26, 147–148
Pastor, V., 291–292
Pathak, S., 221
Pati, P.K., 211
Patil, P.P., 61*t*
Patra, A., 260–261
Patra, B., 9–10
Pattanaik, S., 9–10
Paul, V.J., 310–312, 311*t*, 313–314
Pauli, H.H., 165–166, 186
Paulo Mde, Q., 40, 43
Pauw, B., 15–16, 24–26
Payne, G.F., 207
Peach, M.J.G., 58–59
Pedersen-Bjergaard, S., 255*t*
Peebles, C.A., 24–26
Peerless, A.C.J., 58
Peisker, K., 55
Pelcher, L.E., 23–24, 58–59
Pelissou, M., 254, 255*t*
Pelkonen, O., 208
Peltenburg, A., 248–251
Peltenburg-Looman, A.M.G., 10, 146–148
Pemberton, J.M., 319–320
Pena, R., 239–240
Penelle, J., 238, 260–261
Penfold, R.J., 319–320
Pépin, G., 243, 244*t*
Pereira, C.G., 241*t*, 242
Pereira, D.M., 244*t*
Pereira, L.G., 4, 96, 98–99
Pereira, P., 311*t*
Perera, P., 238

Perez, T., 237–238
Perez-Garcia, O., 221
Persson, B., 56–57
Peter, H.H., 321
Petermann, J., 115
Peters, M.W., 197–198
Petiard, V., 125
Pettit, G.R., 313–314
Pfeifer, B.A., 187–189
Philipp, A.A., 244*t*
Phillips, A.J., 310–312
Phillips, M.A., 96–98
Pi, Y., 146
Pichersky, E., 19–20, 125
Pichon, O., 90–91
Piel, J., 302
Pieterse, C.M.J., 290*f*, 291–292
Pimentel-Elardo, S.M., 308
Pineda, A., 226
Ping, N.S., 61*t*
Pinol, M.T., 211, 227
Pinto, A.C., 237–238, 251
Pinto, L.L., 237–238
Piotrowski, M., 165–166
Piovan, A., 243
Piraee, M., 322
Pittard, J., 190–191
Piutti, S., 218, 219–221
Pizza, C., 244*t*
Plasencia, J., 289, 292–293
Platonova, T.F., 48
Ploss, K., 10–12, 11*f*, 18–19, 20, 93–95, 101
Plowman, T., 40–41, 44
Plubrukarn, A., 316*t*
Poeaknapo, C., 198
Poehland, B., 149
Pol, B.B., 151–152
Pommier, Y., 140–141, 142, 154–155
Pons, E., 292–293
Porto, D.D., 26
Portsteffen, A., 56–57
Posch, T.N., 254, 255*t*
Potier, P., 4
Potsangbam, K.S., 208
Potts, B.C., 307–308
Pouchnik, D., 19–20
Pourquier, P., 154–155
Poutaraud, A., 216
Poutrain, P., 17–18, 90–91, 96–99
Power, E.F., 132
Power, R., 18
Pozetti, G.L., 251
Prats, G., 292–293
Pré, M., 9–10, 15–16, 87–89
Preil, W., 218–219, 221
Premalatha, Y., 275–276, 283
Presser, A., 43
Preston, G.M., 291
Prewo, R., 238
Prieto-Davó, A., 304*t*, 307
Prince, C., 207
Prinsep, M.R., 302
Pritchard, M.P., 196–197
Priti, V., 153
Priya, T., 151
Proença, P., 249*t*
Proiser, E., 122
Proksch, P., 304*t*, 307
Proteau, P.J., 313
Puhl, M., 176
Puri, S.C., 153
Pusset, J., 48–49
Pusset, M., 48–49
Pütz, M., 254, 255*t*
Pyykkö, I., 46

Q

Qazi, G.N., 153
Qi, A., 255*t*
Qi, Y., 173
Qi, Z., 287–288, 290*f*
Qian, Z., 147–148
Qiu, C., 146
Qu, X., 26
Queiroz, E.F., 40, 43
Quesada, A.L., 78–79, 85
Quesnel, A., 87
Quetin-Leclercq, J., 252
Quinlan, R.F., 196–197
Quinn, R.J., 48
Qureshi, A.A., 13, 14

R

Rabot, S., 58
Radom, L., 58–59
Radomski, K., 142

Radosevich, A.T., 315
Radwan, A.S., 55
Radwan, M., 308
Rahnama, H., 62–63
Rajam, M.V., 59–60
Ralevic, V., 285, 287, 291
Ramakrishna, A., 129
Ramautar, R., 254
Ramesha, B.T., 153
Ramin, H., 50–52
Ramírez, J., 244*t*
Ramirez, O.T., 190
Ramos-Valvidia, A.C., 8
Rana, J., 55
Rando, T., 313
Ranjith Kumar, R., 221
Rapisarda, A., 141–142
Raskin, I., 284
Rasmussen, K.E., 255*t*
Rasmussen, S., 291
Rasoanaivo, P., 260–261
Ratain, M.J., 244*t*
Ratcliff, F., 23–24
Rathod, V., 193
Rätsch, C., 43
Ravelo, A.G., 60–62, 61*t*
Raventos-Suarez, C., 304*t*, 306
Ravikanth, G., 153
Ravishankar, G.A., 129
Rawat, I., 221
Razzakov, N.A., 40
Reas, H.W., 46
Reed, D.W., 23–24, 58–59
Reeves, J.T., 313
Rehman, S., 153
Reichelt, M., 44–46, 47–48, 52, 57
Reid, G.A., 197–198
Reimann, A., 82–83
Reisch, J., 237–238
Ren, J., 304*t*, 307
Renault, J.H., 252
Renner, M.K., 54–55
Rezende, C.M., 237–238, 251
Reznicek, G., 43
Rheingold, A.L., 304*t*, 308–309
Rheu, H.M., 294
Rhodes, M.J.C., 52, 58
Ribeiro, J.A., 287
Ribet, J.P., 239–240, 254, 255*t*
Richardson, T.H., 196–197
Richter, U., 56–57
Rideau, M., 6–8, 9, 10–12, 19–20, 22–23, 87, 89–91, 211
Ridgway, J.E., 10–12
Rieck, L., 249*t*, 251
Riedl, S., 176
Riekkola, M.L., 237–238, 240–242, 241*t*
Ries, S., 287
Rigoli, G., 292–293
Rine, 91–92, 108
Rischer, H., 9–10, 12, 15–16, 18, 21
Rittershaus, E., 218–219
Rivas-San Vicente, M., 289, 292–293
Rivera, A.L., 211
Rivier, L., 40–41, 44
Rizvi, S.J.H., 275–276
Ro, D.K., 27
Robaugh, D., 241*t*
Robbens, J., 244*t*
Robbertse, H., 128
Roberts, M.A., 310, 311*t*, 323
Roberts, S.C., 155–156
Robins, R.J., 55–56, 58–59
Robinson, T., 58
Rodrigues-Filho, E.R., 252–253
Rodriguez, S., 16, 80–82
Rodríguez-Concepción, M., 6–8, 9, 87–89, 91–92, 94–95, 96–98
Rodriguez-Conception, A., 5–8
Rodriguez-Galindo, C., 142
Roeder, E., 151–152
Roef, L., 285–286, 288
Roels, K., 244*t*
Roepke, J., 10–12, 11*f*, 18–19, 20, 93–95, 101
Roepke, M.H., 46
Roepke, R., 8, 10, 17–18, 19–20
Roessner, C.A., 12–13, 147–148
Roessner, U., 83–84
Roewer, I., 15–16
Roh, J.H., 187–189
Rohdich, F., 6–8, 145
Rohmer, M., 5–8, 9
Rohr, J., 321
Roja, G., 151–152
Rolf, M., 166–170

Romanczyk, L.J., 113–115
Rome, A., 244*t*
Romero, M.A.V., 40
Rommelspacher, H., 312
Rompp, H., 240
Rookes, J.E., 92–93
Roos, W., 166–170
Rosahl, S., 292–293
Rosano, T.G., 249*t*
Rose, M., 304*t*, 307
Ross, J.R., 125
Rossen, J., 40–41
Rossi, J., 323
Rossiter, J.T., 99–100
Roth, A.D., 142
Roy, A., 190
Roy, R., 259–260
Roytrakul, S., 14
Rudaz, S., 243
Rügenhagen, C., 83
Runguphan, W., 21, 22–23, 26, 148
Runyon, S.P., 47–48
Ruppert, M., 12–13, 147–148
Russell, G.B., 48–49
Ryals, J.A., 283–284
Ryan, C.A., 283–284
Rylott, E.L., 197–198

S

Sabarna, K., 185
Sacchi, R., 211–212
Saito, K., 8, 12–13, 142, 144, 147–148, 149, 150–151, 152–153, 154–155
Sakabe, Y., 283
Sakai, H., 142, 149
Sakai, K., 288
Sakaki, T., 165–166
Sakulkeo, O., 316*t*
Sakurai, N., 193
Sakurai, S., 190
Sakurama, H., 193
Salas, A.P., 321
Salas, J.A., 320
Salim, V., 4–5, 10–12, 11*f*, 18–19, 20, 23–24, 93–95, 101
Salles, P., 23–24
Salvador, L.A., 310–312
Samanani, N., 85, 165–166
Samuelson, G., 238
San, K.Y., 24–26
San Segundo, B., 289, 290*f*, 291
Sanahuja, G., 227
Sánchez, C., 19–20, 320, 321
Sanchez-Perez, R., 99–100
Sandala, G.M., 58–59
Sandberg, G., 118–119, 128–129
Sanders, D., 288, 290*f*
Sanders, M., 244*t*
Sandmeier, E., 52
Sandonis, V., 15–16, 24–26
Sandvik, E.R., 193
Sangnoi, Y., 316*t*, 318
Sangwan, B.S., 217, 218
Sangwan, R.S., 214–215
Sangwan-Norreel, B.S., 214–215, 217, 218
Sannes, E., 255*t*
Sano, H., 113, 118, 119, 124, 125, 129–131, 130*t*, 132, 275–277, 278–281, 279*f*, 281*f*, 282–284, 289–290, 292–293
Santamaria, A.R., 95–96, 144
Santana, V.M., 142
Santos, C.N., 190
Sargent, M.V., 48–49
Sariaslani, F.S., 190
Sarker, S.D., 48–49
Sartoratto, A., 43–44, 61*t*
Sasaki, R., 149
Sasse, F., 83
Satdive, R.K., 151–152
Sato, F., 10, 164–173, 175, 176–177, 185, 186–190, 191, 193–195, 198, 199
Sato, H., 54–55
Sato, K., 190
Satoh, Y., 192
Satyanarayana, K.V., 129
Saunders, P.A., 294
Sausville, E.A., 141–142
Savangikar, C., 221
Savangikar, V.A., 221
Savithiry, S., 287
Savolainen, V., 83–84
Sawada, S., 142
Scavone, C., 43–44
Schaal, A., 56–57
Schafer, A., 15–16, 91–92
Schattat, M., 96–98

Schenk, P.M., 92–93
Schiff, M., 23–24
Schilmiller, A.L., 19–20
Schilperoort, R.A., 12–13, 19–20
Schipper, R.G., 52
Schirmeister, T., 308
Schlingmann, G., 304*t*, 308, 322
Schmelz, E.A., 294
Schmidt, E., 41–42, 49–50
Schmidt, G.W., 44–46, 47–48, 57
Schmidt, J., 10–12, 16–17, 18, 19–20, 83, 92, 94–95, 146–147, 166–170, 198
Schmidt, P.C., 240
Schmieden, V., 292
Schneider, B., 44–46, 47–48, 52, 57
Schnorrenberger, C.C., 41–42
Schoffin, I., 50–52
Scholz, A., 23–24
Schrader, L.E., 291
Schriemer, D.C., 176
Schripsema, J., 73–110, 248–251
Schröder, G., 10–12, 16–17, 18, 19–20, 83, 92, 94–95, 146–147
Schroeder, F.C., 261
Schubert, B.G., 113–115
Schückel, J., 197–198
Schultes, R.E., 41–42, 43
Schulthess, B.H., 118
Schulz, J., 237–238
Schulz, W., 254, 255*t*
Schütte, H.R., 48, 50–52, 55
Schwartz, H., 46
Schwartz, R.D., 291
Scott, A.I., 12–13, 14, 147–148
Scriba, H.K., 254
Scudiero, D.A., 141–142
Sebastiao, A.M., 287
Secor, J., 291
Seger, C., 261
Seiser, T., 310–312
Selby, M., 141–142
Sena-Filho, J.G., 40
Sensen, C.W., 19–20, 21–22
Senthil-Kumar, M., 23–24
Seo, S., 60–62, 284
Seppanen-Laakso, T., 9–10, 12, 15–16, 18, 21
Seppanen-Lakso, T., 237–238
Serenkov, G.P., 122
Seto, H., 284
Seura-Carretero, A., 253
Sevenet, T., 48
Severino, V.G.P., 252–253
Sevón, N., 211
Seyed-Mozaffari, A., 248–251, 249*t*
Shaaban, K.A., 304*t*, 306
Shaaban, M., 304*t*, 306
Shabbir, M., 43
Shanks, J.V., 24–26
Sharifi, M., 60–63
Sharma, J.P., 153
Shasany, A.K., 21
Shawl, A.S., 153
Sheen, J., 90, 289, 290*f*, 291
Shelp, B.J., 286–287, 290*f*, 291
Sheludko, Y., 12–13, 85, 99–100, 238, 244*t*, 255*t*
Shen, Y., 237–238
Sherden, N.H., 10–12, 83–84, 96–98, 147
Sherman, D.H., 323
Sherman, T., 288, 290*f*
Shet, M.S., 197–198
Sheu, S.J., 255*t*
Shi, D., 286
Shi, X.-G., 112
Shi, Y.-Y., 152
Shimizu, H., 118–119
Shimizu, M., 113–115, 114*t*
Shin, T.J., 294
Shindo, K., 197
Shinkyo, R., 165–166
Shinohara, C., 142
Shinya, T., 278, 289–290
Shioiri, T., 184–185
Shiozawa, H., 316*t*, 317–318
Shirley, N., 83–84
Shitan, N., 99, 165–166, 171, 186–187
Shoji, T., 54–55
Shrikhande, V.A., 151–152
Shukla, A.K., 21
Shuler, M.L., 207
Shweta, S., 153
Siani, A.C., 241*t*, 242
Siberil, Y., 24–26
Siegrist, J., 292–293
Sienkiewicz, M., 48

Siepmann, M., 46
Sierra, M., 13–15
Silva, M.G.F., 252–253
Silvarolla, M., 113–115, 114*t*
Silvestrini, A., 237–238
Sim, G.A., 140–141, 142
Simkin, A.D., 91–92
Simkin, A.J., 8, 9, 96–98, 146
Singh, A., 59–60
Singh, P., 292–293
Singh, S.K., 9–10, 47–48
Singla, B., 90–91
Sioumis, A.A., 48–49
Sippl, W., 166–170
Sirikantaramas, S., 154–155
Sitachitta, N., 310, 323
Slate, D.L., 313
Slater, L.A., 142
Slawin, A.M.Z., 58–59
Slusarenko, A.J., 15–16, 289
Smart, N.J., 90
Smetanska, I., 62–63
Smets, E.F., 83–84
Smigel, A., 287–288, 290*f*
Smith, C.J., 285–286, 287–288
Smith, D.M., 58–59
Smith, J.I., 90
Smith, R.K., 288
Smolke, C.D., 164–165, 190, 194–195
Snoeijer, W., 2–4, 76–77, 80–82
Soda, K., 10
Sohajda, T., 254–259, 255*t*
Sohani, M.M., 83–84
Solano, F., 193–194
Solas, C., 244*t*
Somsen, G.W., 254
Sonawane, K.B., 151–152
Song, H.J., 60
Song, J., 146, 149
Song, K.M., 240–242, 241*t*
Song, M.K., 46
Song, S.H., 152
Song, Z., 318
Sonnewald, U., 120–122, 123–124
Sonnino, A., 209, 211
Sonoda, M., 43
Soulaymani, A., 40
Souza, J.J., 249*t*, 251
Sparr, C., 239
Specker, S., 227
Speitling, M., 304*t*, 306
Spenser, I.D., 50–52, 54–55
Spiteller, M., 153–154, 155
Springer, N.M., 208–209
Srisukchayakul, P., 318
Srivastava, A., 244*t*
Srivastava, P.S., 59–60
Srivastava, R.S., 238
Stabler, D., 132
Staker, B.L., 154–155
Starker, C.G., 173
Stasolla, C., 123–124
Stauber, E.J., 19–20
Steenkamp, V., 249*t*, 251
Steiner, H.Y., 283–284
Stephanopoulos, G., 190, 191
Stephens, E.R., 318
Stevens, L.H., 12–13, 14–15
Stévigny, C., 208
Stewart, C.F., 142
Stewart, D.J., 142
Stewart, I., 113–115, 275
Stewart, L., 154–155
Stierle, A., 153
Stierle, D., 153
Stitt, M., 123–124
Stöckigt, D., 244*t*, 255*t*
Stöckigt, J., 12–13, 16–17, 26, 80–82, 83–84, 85, 99–100, 147–148, 244*t*, 255*t*
Stoeckgit, J., 238
Stolte, S., 311*t*, 312
Stopper, H., 308
St-Pierre, B., 9, 10–12, 14–18, 19–20, 76–77, 80–82, 83, 86–87, 92–93, 94–95, 96–99, 100, 146–147
Strack, D., 16–17, 18
Straughn, J.L., 143–144, 149
Streatfeild, D., 40–41
Strem, M.D., 125–126
Strichartz, G., 313
Strobel, G., 153
Stuppner, H., 255*t*
Sturn, S., 255*t*
Sudan, P., 153
Sudo, H., 8, 12–13, 142, 144, 145, 146–148, 149, 152–153, 154

Suenaga, K., 311*t*, 314
Suganuma, M., 314
Sugiyama, A., 99
Suh, W., 190
Sukrong, S., 151, 152–153
Sulc, M., 243–248
Sullivan, J.H., 47–48
Sultan, P., 153
Sun, C.J., 146, 149, 237–238, 243, 244*t*, 249*t*, 251
Sun, H., 243–248
Sun, J., 54–55, 262
Sun, M.X., 286–287
Sun, W., 243–248
Sun, X.W., 239
Sun, Y., 146, 149
Sun, Z., 307
Sundari, M.S.N., 100
Sung, P.-H., 146–147
Suttipanta, N., 9–10, 85
Suwannachart, C., 318
Suzuki, A., 196–197
Suzuki, K., 52–54
Suzuki, T., 118–120, 123–124, 125–126, 131
Svejstrup, J.Q., 140–141
Svoboda, G.H., 142, 150–151
Swift, T.A., 249*t*
Sykes, K.A., 54–55
Szabó, L.F., 13, 75–76, 77*f*, 78–79, 78*f*
Szántay, C. Jr., 260–261
Szente, L., 254–259, 255*t*
Szesny, M., 308

T

Taft, M.H., 315
Tafur, S., 142, 150–151
Tai, Y., 243, 244*t*
Tajima, K., 192
Takagi, Y., 193
Takahashi, A., 303, 304*t*, 316*t*, 319, 321
Takahashi, I., 316*t*, 319, 321
Takahashi, K., 197–198
Takano, T., 319–320
Takasawa, Y., 119–120, 126
Takayama, H., 142, 145, 146–148, 151–153
Takayama, S., 221, 314
Takemura, T., 164–165, 166–171, 173, 176, 185, 186, 187–189, 190, 194–195, 198
Takeshita, N., 165–166
Takman, B.H., 313
Talapatra, B., 260–261
Talapatra, S.K., 260–261
Talke, I.N., 288, 290*f*
Tallevi, S.G., 87
Talou, J.R., 62–63
Talukdar, A.D., 208
Tamaki, E., 54–55
Tamaki, K., 52–54
Tamminen, T., 40
Tamura, K.-I., 171, 173, 185
Tan, L.T., 304*t*, 306, 310
Tan, R., 243–248
Tanabe, Y., 54–55
Tanaka, H., 119, 276–277, 307–308
Tanaka, J.C.A., 237–238
Tanaka, M., 165–166
Tanaka, T., 190
Tang, H., 261
Tang, K.-X., 59–60
Taniguchi, S., 321
Taniguchi, Y., 171–173
Tanikawa, N., 113–115, 124–125
Tanizawa, K., 54–55
Tanksley, S.D., 125
Taori, K., 310–312, 311*t*
Tapias, E.C., 262
Tari, L.W., 73–110, 167*f*, 191–192
Tatsumi, Y., 129–131, 130*t*, 278–281, 279*f*, 281*f*
Tavares, J.F., 40
Taylor, M.W., 302
Teeri, T., 211
Teixeira da Silva, J.A., 211, 227
Templeton, L.J., 190
Ten Hoopen, H.J.G., 165
Tepfer, D., 47
Terada, H., 283
Teraoka, A., 197
Terrasaki, Y., 123–124
Teruya, T., 311*t*, 314
Teuber, M., 52–54
Thakur, M.S., 193

Thamm, A.M.K., 10–12, 11*f*, 18–19, 20, 93–95, 101
Thengane, S.R., 151–152
Thenmozhi, S., 100
Thepenier, P., 252
Thi, D.V., 61*t*
Thiersault, M., 15–16, 22–23, 24–26, 86–89, 90–91, 92–93
Thines, B., 87–89
Thio, M., 52
Thisleton, J., 185
Thistleton, J., 175
Thomas, D.W., 237–238
Thomasset, B., 227
Thöming, J., 311*t*, 312
Thorpe, T.A., 123–124
Tian, L., 304*t*, 307
Tillekeratne, L.M., 43
Tioly-Bensoussan, D., 211–212
Tits, M., 238, 260–261
Tjaden, U.R., 233–272
To, J.P., 90
Tobin, A.J., 286
Todokoro, T., 171
Tofern-Reblin, B., 46–47
Toffanin, R., 218
Tokuyama, H., 4
Tolkin, T.R., 23–24
Tomas-Barberan, F., 113
Tominaga, Y., 44
Tomioka, E., 47
Tomoda, Y., 123–124, 128
Ton, J., 291–292, 294
Tonfack, L.B., 56–57
Toquin, V., 292–293
Torssell, K.B.G., 260–261
Totsuka, Y., 312
Tozer, D.J., 58–59
Trainotti, L., 95–96, 144
Tran, T.L.M., 223–225, 224*f*, 227
Treimer, J., 12–13
Trémouillaux-Guiller, J., 211
Trigo, J.R., 43–44, 61*t*
Tripathi, A.K., 244*t*
Tritsch, D., 145
Trout, B.L., 12–13
Trusov, Y., 92–93
Tsai, T.H., 46
Tschudi, G., 184–185
Tsiavaliaris, G., 315
Tsubouchi, H., 283
Tsueng, G., 304*t*, 309
Tsujimoto, K., 191
Tsujita, T., 165–166, 186–187
Tudzynski, P., 243–248
Tumlinson, J.H., 294
Turano, F.J., 286–287
Turner, J.G., 87–89
Tuveson, R.W., 211–212
Tyler, R.T., 17–18

U

Uchiyama, T., 319–320
Udomson, N., 149
Uefuji, H., 118, 129–131, 130*t*, 275, 276–277, 278–281, 279*f*, 281*f*
Ueno, A., 21–22
Ueno, M., 151–152
Uesato, S., 10
Ugent, D., 40–41
Ullmann, P., 227
Ulrich, J., 218–219
Unger, M., 238, 244*t*, 253, 255*t*
Unterbusch, E., 16–17, 18
Unterlinner, B., 166–170
Unver, N., 10, 146–148
Uranchimeg, B., 141–142
Urano, A., 146–148
Urbanczyk-Lipkowska, Z., 48
Uria, A.R., 302
Usera, A.R., 14, 16–17
Usuda, S., 59
Utsui, Y., 316*t*, 317–318

V

Vaillancourt, F.H., 314, 323
Vakili, B., 60–62
Valentao, P., 244*t*
Valera, G.C., 48–49
Valle, F., 191
Valletta, A., 95–96, 144
Vallverdú-Queralt, A., 228
van Beek, T.A., 238
van Berkum, P., 286–287
van Breda, S.V., 128
Van de Mortel, J.E., 226

van der Fits, L., 9–10, 15–16, 18, 24–26, 87–89
van der Graaff, E., 120–122
van der Greef, J., 233–272
van der Heijden, R., 2–4, 8, 10, 12–13, 14–15, 24–26, 76–77, 80–82, 85, 91, 96–98, 99–100, 146–148, 165, 237–238, 251–252
van der Kooy, F., 236–237
van der Meijden, E., 99–100, 101
van der Rest, B., 56–57
Van Eenoo, P., 244*t*
Van Fleet, J., 171–173
van Gulik, W.M., 165
van Hengel, A.J., 151–152
van Ingelgem, C., 288
van Iren, F., 13–15
Van Onckelen, H.A., 285–286
van Soeren, J.H., 50–52
van Vliet, T.B., 13–15
Vandermeijden, E., 2–4
Vansiri, A., 22–23, 87, 89–90
Varga, E., 254–259, 255*t*
Varner, J.E., 285–286
Vasconcelos, V.M., 311*t*
Vásquez, V., 40–41
Vaughn, K.C., 47–48
Vázquez-Flota, F.A., 14–15, 16–18, 19–20, 86–87, 94–95, 96
Veau, B., 6–8, 87
Velasquez, L.M.E., 62–63
Verardo, V., 253
Verdier-Pinard, P., 310, 311*t*
Verhoeven, J.T.J., 262
Verhofstad, A.A.J., 52
Verma, A., 237–238, 240–242, 241*t*
Verma, D.P.S., 50–52
Verma, R., 287–288, 290*f*
Verma, R.K., 244*t*
Verma, V., 153
Verpoorte, A., 12–13, 19–20
Verpoorte, R., 2–4, 3*f*, 6–8, 7*f*, 9–10, 12–15, 26, 76–77, 80–82, 85, 86–89, 91, 96–98, 99–100, 101, 146–148, 165, 206–207, 235, 236–238, 242–243, 248–251, 261
Verpoorte, S., 14
Vetter, W., 48–49
Veuthey, J.L., 47–48, 243
Vialart, G., 227
Vieira, D.N., 249*t*
Vieira, I.J.C., 249*t*, 251, 260–261
Viglianti, G.A., 154–155
Vilegas, W., 251
Vincent, K.M., 319–320
Viraporn, V., 142, 150–151, 154–155
Vlot, A.C., 289, 290*f*, 291
Vogel, M., 166–170
Voigtlaender, S., 166–170
Volk, R.B., 311*t*, 314
von Wachenfeldt, C., 196–197
Voskuilen, J.T., 146–148
Vu, T.D., 225

W

Wagner, K.G., 54–55
Wakabayashi, K.A.L., 249*t*
Walker, R.K., 287–288, 290*f*
Wall, M.E., 140–141, 142
Wallace, W., 291
Waller, G.R., 123–124, 126, 131, 132, 275–276
Wallner, S., 176
Walsh, C.T., 314, 323, 324
Waltham, T.N., 197–198
Walton, N.J., 55–56, 58–59
Walton, T.J., 287–288
Wang, C.H., 237–238, 255*t*
Wang, D.-M., 112
Wang, G.C., 237–238
Wang, H., 46, 147–148, 239
Wang, J.C., 147–148, 154–155, 237–238
Wang, L., 243–248
Wang, M., 123–124, 125–126, 128–129
Wang, P., 243–248
Wang, Q., 146, 241*t*, 242
Wang, R., 237–238
Wang, S.Y., 239, 304*t*, 309–310
Wang, W., 147–148
Wang, X., 60–62, 61*t*, 251–252
Wang, Y., 142, 187–189, 237–238, 261, 304*t*, 309–310
Wang, Z., 60–62, 243–248, 255*t*
Wani, M.C., 140–141, 142
Wanner, H., 115
Warin, R., 252
Warzecha, H., 244*t*, 255*t*

Watase, K., 142
Waterman, M.R., 196–197
Waterman, P.G., 237–238
Watson, A.A., 44
Watson, M.B., 52
Watts, K.S., 310, 311*t*
Weber, W., 227
Wege, S., 23–24
Wei, Y., 304*t*, 307
Weigl, R., 46–47, 237–238
Weinberger, B.A., 274
Weiss, A., 218–219
Weissenberg, M., 100
Welker, M., 311*t*
Wen, D., 244*t*
Weng, S., 60–62, 61*t*
Werck-Reichhart, D., 83, 227
Werner, I., 43
Wert, V., 287
Wesnes, K.A., 46
Westergaard, O., 140–141
Westphal, K., 218–219
White, D.J., 304*t*, 309
White, E.L., 255*t*
Whitmer, S., 24–26
Wichers, H.J., 151–152
Widders, I., 287
Wiedenfeld, H., 151–152
Wiese, M., 313
Wietz, M., 316*t*, 318–319
Wightman, E.L., 293–294
Wigle, I.D., 54–55
Wijekoon, C.P., 23–24
Wikström, N., 83–84
Wildermuth, M.C., 289, 290*f*, 291
Wildung, M.R., 19–20
Willaman, J.J., 113–115
Williams, P.G., 303, 304*t*, 307–308
Williamson, R.T., 323
Willis, C.L., 318, 323
Willits, M.G., 283–284
Willmann, M.R., 291
Willstätter, R., 49–50
Wilson, J.M., 318–319
Wilson, S.A., 155–156
Wimalasiri, W.R., 237–238
Wingsle, G., 118–119
Wink, M., 40–42, 43–44, 128–129, 293–294
Winkler, A., 176
Winter, J.M., 303
Winzer, T., 166–170, 173, 176
Wioland, M.A., 308
Wipf, P., 313
Wiryani, R.D., 48–49
Wisden, W., 285, 286, 291
Wissenbach, D.K., 244*t*
Witte, L., 44
Witters, E., 285–286
Wodak, A., 52–54
Wolf, C.R., 196–197
Wong, C.W., 55
Woolery, M., 303–306, 304*t*
Woolley, J.G., 58–59
Wouters, J., 187–189
Wray, V., 54–55
Wright, G.A., 132
Wu, C.C., 10–12, 83–84, 96–98, 147, 244*t*, 292–293
Wu, G., 289
Wu, M., 10–12, 11*f*, 18–19, 20, 93–95, 101
Wu, Q., 255*t*
Wu, T.S., 260
Wu, W.H., 288, 289
Wulff, P., 285, 286, 291
Wurtzel, E.T., 196–197
Wyche, J., 140–141

X

Xavier, H.S., 40
Xia, Y., 60–62
Xiao, J., 60–62, 61*t*, 191
Xie, J., 239
Xiong, A., 243–248
Xu, H.-H., 152
Xu, X., 239
Xu, Y.J., 243–248
Xue, F., 249*t*, 251
Xun, L., 193

Y

Yadav, S.K., 125–126, 129–131, 130*t*, 275–276
Yahia, A., 89–90
Yama, S., 123, 128–129
Yamada, S., 184–185
Yamada, Y., 50–55, 56–57, 58, 59, 164–166, 171–173, 186–187, 210–211

Yamaguchi, Y., 118, 129–131, 130*t*, 275, 276–277
Yamakawa, T., 152–153
Yamamoto, H., 10, 146–147
Yamamoto, K., 193, 283
Yamazaki, M., 8, 12–13, 142, 144, 145, 146–148, 149, 150–151, 152–153, 154–155
Yamazaki, Y., 8, 12–13, 144, 145, 146–148
Yan, M., 243–248
Yan, X., 52–54, 56–57
Yang, B., 60–62, 251–252
Yang, C.X., 60
Yang, D.-P., 112
Yang, F.J., 239
Yang, J., 190–191
Yang, L., 239, 243–248
Yang, S., 60, 61*t*
Yang, T., 304*t*, 306–307
Yang, Y., 125
Yao, H., 146
Yao, Q., 306
Yao, W., 288
Yasui, H., 99–100
Ya-ut, P., 151, 152–153
Yazaki, K., 99, 165–166, 170–171, 186–187
Ye, C.-X., 112, 115, 119, 125
Ye, S., 128
Yi, Y., 142
Yilmaz, A., 261
Yin, Y.G., 239–240
Ying, Y., 310–312
Yoda, H., 129–131, 130*t*, 278–280, 279*f*, 281*f*, 282, 283–284, 289–290
Yoder, L.R., 93–94
Yokochi, A., 310, 311*t*
Yokose, H., 319
Yokoshima, S., 4
Yokota, A., 318
Yokoyama, Y., 312
Yoneyama, N., 118, 119, 125, 276–277
Yong-Xiao, J., 191
Yoo, K.P., 240–242, 241*t*
Yoon, J.H., 240–242, 241*t*
Yoshida, M., 316*t*, 319, 321
Yoshida, S., 171–173, 186–187, 284
Yoshida, W.Y., 311*t*, 313–314
Yoshimatsu, K., 166–171
Yoshimoto, T., 171–173
Yoshino, F., 149
Youderian, P., 190
Yousufuddin, M., 304*t*
Yu, F., 4–5, 20, 23–24, 197
Yu, G.H., 286–287
Yu, I.C., 288
Yu, M., 175
Yu, W., 52
Yu, Y., 237–238
Yuan, L.L., 9–10, 255*t*
Yuan, Y., 243–248
Yuenyongsawad, S., 316*t*
Yukawa, H., 187–189
Yukimune, Y., 50–52, 56
Yun, D.J., 210–211

Z

Zang, X., 255*t*
Zanolari, B., 40, 43, 243, 244*t*
Zárate, R., 26, 60–62, 61*t*, 211, 227
Zazopoulos, E., 322
Zeches-Hanrot, M., 252
Zelwer, C., 6–8, 9, 87–89, 94–95, 96–98
Zeng, J., 304*t*, 307
Zeng, L.J., 60
Zenk, M.H., 10, 12–15, 147–148, 198
Zhang, A., 60, 61*t*, 306
Zhang, C.L., 142
Zhang, D.M., 237–238
Zhang, F., 146, 173
Zhang, H.-M., 9–10, 59–60, 87–89, 91–92, 142, 249*t*, 251
Zhang, J., 237–238
Zhang, L., 59–62, 171–173, 210–211, 239
Zhang, Q., 304*t*, 306–307
Zhang, R., 52–54
Zhang, W.W., 46, 243, 244*t*, 255*t*, 304*t*, 306–307
Zhang, X.Q., 237–238, 306
Zhang, Y., 9–10, 52–54, 56–57, 61*t*, 87–89, 173, 243, 244*t*
Zhang, Z., 142, 249*t*, 254, 255*t*, 259*f*
Zhao, C.J., 239, 244*t*, 251
Zhao, J., 26, 288
Zhao, L., 208
Zhao, S., 243–248
Zhao, Y., 83
Zheng, H., 237–238

Zheng, X.-Q., 112, 115, 123–124, 125–126, 128–129
Zheng, Z., 289
Zhong, J.J., 218–219, 221, 224–225
Zhou, B.N., 142
Zhou, G., 210–211
Zhou, H., 243, 244*t*
Zhou, L., 210–211, 239
Zhou, W., 60, 307
Zhu, C., 227
Zhu, G.Z., 210–211
Zhu, H.X., 52–54
Zhu, J.P., 13, 238
Zhu, L., 46, 239, 321
Zhu, P., 52–54
Zhu, Q.H., 239–240
Zhu, R., 243–248, 255*t*
Ziegler, J., 2–5, 14, 26, 52–54, 74–75, 76–77, 80–82, 94–95, 96, 166–170
Zimmerli, L., 292–293
Ziv, M., 221
Zoerntlein, S.W., 244*t*
Zolala, J., 211
Zong, Y.Y., 255*t*
Zou, H., 255*t*
Zrenner, R., 121*f*, 123–124, 126
Zu, Y.G., 239
Zubieta, C., 125
Zuehlke, S., 153
Zuhlke, S., 153–154
Zulak, K.G., 50–52
Zuniga-Quiroz, J., 244*t*
Zuo, K., 146
Zurbriggen, M.D., 227
Zvartau, E., 285, 287, 289–290

SUBJECT INDEX

Note: Page numbers followed by "*f*" indicate figures, and "*t*" indicate tables.

A

Actinomycete bacteria
- altemicidin, 303
- anthraquinone/angucycline alkaloids, 306
- antibiotics, 302
- anti-cancer drugs, 302
- azamerone, 303
- bioactive alkaloids, 302–303, 304*t*, 305*f*
- bisindole alkaloids, 306–307
- caerulomycins F-K, 309–310
- CNQ-583, 303–306
- diazepinomycin, 308
- lodopyridone, 308–309
- lyndamicins, 309
- mansouramycins, 306
- nitropyrrolins A-E, 307
- phenoxazine alkaloids venezuelines, 307
- salinosporamide A, 307–308
- spiroindimicins, 306–307
- *Streptomyces*, 302–303
- TP-1161, thiopeptide antibiotic, 309

Adenosine receptors, 287

Agrobacterium rhizogenes, 152

Alkaloids
- bacterial species, 315–319
- biologically active, 302
- biosynthesis (*see* Marine bacteria)
- marine actinomycete bacteria, 302–310
- marine cyanobacteria, 310–315
- metagenomics approach, 302

Allantoic acid, 126

γ-Aminobutylic acid, 291

Angiosperm, MIE distribution
- camptothecin, 78–79, 79*f*
- Ipecac alkaloids, 78–79
- non-rearranged tryptamine moiety, 79
- primitive MIAs, 79
- 'quinoindole' alkaloids, 78–79
- vinblastine and vincristine, 76–77

Apocynaceae
- MIAs (*see* MIAs biosynthesis)
- PhytoMetaSyn project, 21–22
- putative iridoid biosynthetic genes, *L. japonica*, 22

Aqabamycins, 318

B

BABA. *See* β-aminobutylic acid (BABA)

β-aminobutylic acid (BABA), 291–292

Barbamide
- biosynthesis, 323–324
- molluscicidal activity, 314

BBE. *See* Berberine bridge enzyme (BBE)

Berberine bridge enzyme (BBE)
- antisense RNA, 175
- protoberberine and benzophenanthridine alkaloids, 167*f*
- RNAi, California poppy cells, 175

Biological activity
- alkaloids (*see* Alkaloids)
- anthraquinone/angucycline alkaloids, 306
- caerulomycins, 309–310
- mansouramycin A, 306
- nitropyrrolins, 307

Bioreactor
- cultivation systems, 221
- description, 218–219
- 'expensive', 219–221
- FCs, 221
- *in vitro* culture vessels, 218–219
- industrial cultivation, plant cells, 218–219
- large-scale culture, 221
- low-cost culture systems, 219–221
- micropropagation, 221
- plant material, 219–221
- Pyrex® vessels, 221
- *R. graveolens* shoots, 219–221, 220*f*
- suspension cell cultures, 218–219
- vessel A (VA) and vessel B (VB), 219–221

Biosynthesis
- barbamide, 323–324
- caffeine (*see* Caffeine)
- diazepinomicin, 322–323

Biosynthesis (*Continued*)
holomycin, 324–325
MIAs
developmental control, 86
environmental factors, 86–87
hormonal controls, 87–90
intracellular signalling, 90–93
spatial organisation (*see* Spatial organisation, MIA biosynthesis)
purine alkaloids (*see* Purine alkaloids)
salinosporamide A, 325–327
staurosporine and bisinolde alkaloids, 321
TAs
amino acids, 50–52
arginine decarboxylase (ADC), 52
cocaine, 58
D. stramonium, 58
hyoscyamine conversion, 59
immunolocalisation experiments, potatoes, 56–57
keto group reduction, 56
malonyl-CoA condensation, 55–56
MecgoR, 57
N-methylputrescine oxidase (MPO), 54–55
nitrogen-containing metabolite recruitment, 50–52
N-methyl putrescine formation, 52–54
ornithine decarboxylase (ODC), 52
putrescine *N*-methyltransferase (PMT), 52–54
radiolabelled feeding, 58–59
steps, 50–52
TRI and TRII, 56–57
virus-induced gene silencing techniques, 58–59
violacein, 319–320
Biosynthetic genes
early MIA biosynthesis
glucose moiety, strictosidine, 12–13
Pictet-Spengler condensation, 12–13
preakaummicine, 14
strictosidine aglycone formation, 13–14
early monoterpene biosynthesis
DXR and MECS, 6–8
IPP isomerisation, 8
mevalonate pathway, 6
plastidic MEP pathway, 6, 7*f*
terpenes, 5–6
iridoid pathway, 10
late MIA/vindoline pathway, *C. roseus*, 16–17
Bisucaberin, 319
6-Bromoindole-3-carbaldehyde, 315

C

Caffeine
biosynthesis
Caffeine synthase (*see* Caffeine, synthase)
and catabolism, 115
cellular localization, 278
in coffee plant, 275*f*
demethylases, 133
de novo purine biosynthesis, 120–122, 121*f*
ecological role, 113
enzymes and encoding genes, 116*t*
methyltransferases, 276–277
methyluric acids, 115
7-methylxanthine, 118
7-methylxanthosine synthase, 118
occurrence, plant kingdom, 114*t*
purified caffeine synthase, 118–119
recombinant enzymes, 119
regulation, 125–126
SAM route, 123
substrates, availability and concentration, 277
synthase, 117*f*
theobromine, 118–120
xanthosine, 276
cabbage leaves and, 275–276
catabolism, 126–128
ecochemical functions, 275–276
Hemileia vastatrix, coffee rust, 276
signal cascade
γ-aminobutylic acid, 291
calcium, 288
route, 289–291, 290*f*
salicylic acid, 289
synthase
crystallography analyses, 125
dual-functional, 119–120
expression, 128

genes, 112–113
localisation, 129
N-methyltransferases, 124
paraxanthine, 120
recombinant coffee, 119
RNAi constructs, 129–131
theobromine methylation, 119–120
young tea leaves, 118–119
targets
adenosine receptors, 287
GABA receptor, 286–287
PDE, cyclic nucleotide, 285–286
pharmacological analyses, 285
transgenic plants (*see* Transgenic plants)
xanthosine, 117*f*
Calcium, caffeine signal cascade, 288
Calcium-dependent protein kinases (CDPKs), 288
Camellia sinensis, 112, 126
Camptotheca acuminata
ASA genes, 143–144
bark and seeds, 150
callus and root cultures, 149, 151–152
CPT biosynthesis, 143–144
endophytic fungus, 153–154
MEP pathway, 146
3*(S)*-pumiloside, 149
TDC1 and TDC2 isoforms, 144
Camptothecin (CPT)
anti-cancer drugs, 155–156
biosynthetic pathway
non-mevalonate MEP, 143
post-strictosidine (*see* Post-strictosidine pathway)
shikimate, 143
tryptamine (*see* Tryptamine pathway)
biotechnological production
C. acuminata and *N. foetida*, 150
callus and cell suspension cultures, 150, 151–152
endophytic fungi, 150, 153–154
hairy roots, 150, 152–153
lactone ring, topotecan and irinotecan, 150
natural plant sources, 150–151
O. pumila and *C. acuminata*, 150
plants, 150
Chinese Happy tree ("Xi Shu"), 140–141
DNA synthesis, 140–141
hypoxia-inducible factor 1 (HIF-1), 141–142
metabolic engineering and synthetic biology, 155–156
producing plants, 142
resistance mechanisms
human topoisomerase I, 154–155
Met370Thr and Asn722Ser, 154–155
non-producing species, 155
O. pumila hairy root cultures, 154
prediction, 155
Saccharomyces cerevisiae, 154–155
target-based mutation, 155
topoisomerase I–drug interactions, 142
unique pentacyclic structure, 140–141, 141*f*
Capillary electrophoresis (CE)
aqueous system, 254–259
and CZE, 254
electropherograms, 254, 259*f*
Mitragyna speciosa, 254
and NACE, 254
NACE–MS method, 254
poor sensitivity, 254
rapid qualitative and quantitative analysis, 253
and ROESY, 254–259
TIAs, plant species, 254, 255*t*
Voacanga africana, 254
Capillary zone electrophoresis (CZE), 254, 255*t*
Catharanthus roseus
alkaloids, crude extract, 252
buffer systems and CE methods, 255*t*
catharanthine and vindoline, 4–5
cell culture, 86–87
cell suspensions, 147–148
'detonator' activity, 100
early MIA biosynthesis, 12–16
early monoterpene biosynthesis, 5–10
hairy root cell cultures, 148
healthy and phytoplasma-infected, 262
iridoid biosynthesis, 10–12
iridoid synthase, 147
late MIA biosynthesis pathway, 16–18
LC systems, alkaloids, 244*t*
MIA biosynthesis, 4
N-methyltransferase, 23–24

Catharanthus roseus (*Continued*)
secologanin synthase (SLS), 146
supercritical fluid extraction, TIAs, 241*t*
TDC, 144
TIA production, 146
vinblastine and vincristine, 2–4, 76
vindoline and catharanthine, 239–242
CCC. *See* Counter-current chromatography (CCC)
CDPKs. *See* Calcium-dependent protein kinases (CDPKs)
CE. *See* Capillary electrophoresis (CE)
Cellular localization, 278
Centrifugal partition chromatography (CPC), 252
Chemical defence
caffeine concentration, 283
related genes and HR
PI-II expression, 283–284
PR proteins, 283–284
SA induction, 284
*Cj*CoOMT. *See Coptis japonica* columbamine *O*-methyltransferase (*Cj*CoOMT)
CNGCs. *See* Cyclic nucleotide-gated channels (CNGCs)
CNMT. *See* Coclaurine *N*-methyltransferase (CNMT)
Cocaine
description, 47–48, 58
narcotic properties, 40–41
natural insecticide, 43–44
pure, 40–41
Coclaurine *N*-methyltransferase (CNMT), 165–166, 170–171, 188*f*, 189–190
Codeinone reductase (COR), 166–170, 175
Coffea arabica, 112
Compartmentalisation, 18–19, 24–26, 27, 28
Coptis japonica (Cj)
*Cj*6OMT, California poppy cells, 169*f*, 171, 172*f*
4'OMT, 171
protoberberine and benzophenanthridine alkaloid biosynthesis, 169*f*
SMT, 171
Coptis japonica columbamine *O*-methyltransferase (*Cj*CoOMT), 170–171
COR. *See* Codeinone reductase (COR)
Correlated spectroscopy (COSY), 260
COSY. *See* Correlated spectroscopy (COSY)
Counter-current chromatography (CCC)
and CPC, 252
and HSCCC (*see* High-speed counter-current chromatography (HSCCC))
pH-zone-refining, 251–252
and TIAs, 252
CPC. *See* Centrifugal partition chromatography (CPC)
Cyanobacteria
alkaloids, 311*t*, 312*f*
barbamide, 314
bromoanaindolone, 314
curacin A, 313
harman, 312
hectochlorin, 310
largazole, 310–312
lyngbyatoxin A, 314
saxitoxin, 313
symplostatin 1, 313–314
Cyclic nucleotide-gated channels (CNGCs), 288
CZE. *See* Capillary zone electrophoresis (CZE)

D

DAHPS. *See* 3-Deoxy-D-arabino-heptulosonate-7-phosphate synthase (DAHPS)
Datura innoxia
growth, 223
hairy roots, 218, 219–221
hydroponics, 223–224
tropane alkaloid production, 214–215, 217, 218
DBOX. *See* Dihydrobenzophenanthridine oxidase (DBOX)
3-Deoxy-D-arabino-heptulosonate-7-phosphate synthase (DAHPS), 190–191
1-Deoxy-D-xylulose-5-phosphate synthase (DXS), 145*f*, 146
Diazepinomicin
anti-microbial activity, 308
biosynthesis

gene cluster, 322–323
Micromonospora sp., 322
proposed model, 322–323, 322f
Dihydrobenzophenanthridine oxidase (DBOX), 166–170
L-DOPA-specific decarboxylase (DODC), 192f, 193
DXS. See 1-Deoxy-D-xylulose-5-phosphate synthase (DXS)

E

Encoding genes, 116t
Endophytic fungi, 150, 153–154
Environmental factors, MIA biosynthesis regulation
biotic stress, 87
light, 86–87
Enzymes
caffeine
biosynthesis, 116t, 117f
demethylases, 133
synthase, 117f
7-methylxanthosine synthase, 117f
N-methylnucleosidase, 117f
purine alkaloids, biosynthesis, 115
recombinant, 119
subcellular localisation, 96
tea leaves, detection, 123
theobromine synthase, 117f
Escherichia coli
biosynthetic reticuline pathway, 189–190
BL21 (DE3) *tyrR*::null strain, 191
IQA production, 185
MAO, 187–189
P450 fusion enzymes expression, 197–198
P450 genes, 195, 196–197
reticuline fermentation system, 194–195
reticuline production (*see* Reticuline production)
ScTYR, 193
tetrahydrobiopterin (BH4), 192
tetrahydromonapterine (MH4), 192
L-tyrosine-over-producing, 190
Eschscholzia californica
endogenous enzyme reactions, 173
protoberberine and benzophenanthridine alkaloid biosynthesis, 169f, 173

F

FCs production. *See* Furanocoumarins (FCs) production
Fermentative production, IQAs
artificial chimeric enzyme, 197–198
CYP17A expression, 196–197
E. coli (*see Escherichia coli*)
electron transfer components, 197
energy and environmental concerns, 195
engineering strains, 197
'MALLAVF' sequence, 196–197
$P450_{BM3}$ domain, 198
Furanocoumarins (FCs) production
agronomical, 212
application, 211–212
neurological illnesses, 211–212
psoralen and derivatives, 211–212, 212f
Rutaceae family, 211–212
Fusarium solani, 153, 155

G

GABA receptor, 286–287
Gas chromatography (GC)
alkaloids, 238
application, 248
GC–MS (*see* Gas chromatography to mass spectroscopy (GC–MS))
sensitivity problems, 262–263
Gas chromatography coupled to flame ionization detector (GC–FID), 236, 248, 251
Gas chromatography to mass spectroscopy (GC–MS)
adulterants, 251
advantages, 248
alkaloids detection, 248, 249t
application, 248–251
bioactivity-guided experiments, 251
detector response, 248
GC. *See* Gas chromatography (GC)
GC–FID. *See* Gas chromatography coupled to flame ionization detector (GC–FID)
GC–MS. *See* Gas chromatography to mass spectroscopy (GC–MS)

Gene discovery
  epidermis-enriched transcriptomic resources, 20
  orthology-based comparison, 21–22, 22*f*
  VIGS method, 28
Gene regulation
  early MIA biosynthesis, 15–16
  early monoterpene biosynthesis, 9–10
Geranyl diphosphate (GPP), 146
GPP. *See* Geranyl diphosphate (GPP)

H

Hairy roots
  CPT-producing plants, 152
  C. roseus, 148
  O. alata, 152–153
  O. pumila, 147–148, 149, 152–153
  polystyrene resin, 152–153
  transcriptome and metabolome analysis, 149
Herbivores
  biotic stress tolerance, 280–282
  caffeine treatment, 275–276
  defence-related genes, expression, 283–284
  GABA defence, 286–287
  pests, 280–281
  VOCs release, 294
Heteronuclear multiple ond correlation (HMBC), 260
Heteronuclear multiple quantum coherence (HMQC), 260
Heteronuclear single quantum coherence (HSQC), 260
High-performance liquid chromatography (HPLC)
  alkaloids retention, 235
  hyphenated techniques, 242–243
  and MIPs, 239–240
  mobile phase, 236–237
  reversed-phase mode, 262–263
  and SFE, 240–242, 241*t*
  and UHPLC, 243
  and UV, 239–240
High pressure thin layer chromatography (HPTLC), 213
High-speed counter-current chromatography (HSCCC)
  advantages, 252–253
  alkaloids preparative isolation, 251–252
  Catharanthus roseus crude extract, 252
  and CCC, 251–252
  and CPC, 252
  indole alkaloids, 253
  isolation, 252
  LC–APCI–MS/MS analysis, 252–253
  phase solvent system, 251
  voachalotine and 12-methoxy-N_b-methylvoachalotine, 253
HMBC. *See* Heteronuclear multiple ond correlation (HMBC)
HMBPP. *See* 1-Hydroxy-2-methyl-2 *(E)*-butenyl-4-diphosphate (HMBPP)
HMQC. *See* Heteronuclear multiple quantum coherence (HMQC)
Holomycin
  antibacterial activity, 318–319
  biosynthesis, 324–325
Hormonal controls, MIA biosynthesis regulation
  C. roseus, 87–90
  cytokinin signal transduction, 90
  gibberelic acid, 90
  jasmonate (JA), 87–89
4-HPAA. *See* 4-Hydroxyphenylacetaldehyde (4-HPAA)
HPLC. *See* High-performance liquid chromatography (HPLC)
HPP. *See* *p*-Hydroxyphenylpyruvate (HPP)
HPTLC. *See* High pressure thin layer chromatography (HPTLC)
HSCCC. *See* High-speed counter-current chromatography (HSCCC)
HSQC. *See* Heteronuclear single quantum coherence (HSQC)
1-Hydroxy-2-methyl-2 *(E)*-butenyl-4-diphosphate (HMBPP), 146
4-Hydroxyphenylacetaldehyde (4-HPAA), 186–187, 187*f*
4-Hydroxyphenylacetate 3-hydroxylase (HpaBC)
p-Hydroxyphenylpyruvate (HPP), 190–191

I

IBMX. *See* Isobutyl-1-methylxanthine (IBMX)
Immunization, 291–293
Internal phloem-assisted parenchyma (IPAP) cells, 9
Intracellular signalling, MIA biosynthesis regulation
  calcium, 90–91
  protein prenylation events
    CaaX-PTases, 92–93
    C. roseus, 92–93
    protein isoprenylation, 91–92
Ionic liquids (ILs), 239
IPAP cells. *See* Internal phloem-assisted parenchyma (IPAP) cells
IPP. *See* Isopentenyl diphosphate (IPP)
Iridoid biosynthesis
  biosynthetic genes, 10, 11*f*
  gene regulation, 12
  localisation, 10–12
Isobutyl-1-methylxanthine (IBMX), 287–288
Isopentenyl diphosphate (IPP)
  biosynthesis, 146
  formation, 145
  secologanin, 146–147
Isoquinoline alkaloids (IQAs)
  application, 184
  biosynthetic pathways and enzymes, bacteria
    dopamine/norlaudanosoline, 190
    fermentative production (*see* Fermentative production, IQAs)
    L-tyrosine production (*see* L-Tyrosine production)
    reticuline production (*see* Reticuline production)
    Saccharomyces cerevisiae, 194–195, 196*f*
  chemical syntheses, 184–185
  complex media, microbial cultures, 198
  disadvantages, 199
  Escherichia coli, 185
  (R)-forms, 198
  microbial production, 198
  microbial production, secondary metabolites, 185
  plant (*see* Plant isoquinoline alkaloids)
  plant biotechnology techniques, 185
  (S)-reticuline, 199
  synthesised and unnatural, 198

J

Jasmonate (JA), 87–89

L

LC–MS. *See* Liquid chromatography coupled to mass spectroscopy (LC–MS)
LC–UV. *See* Liquid chromatography coupled to UV (LC–UV)
Liquid chromatography coupled to mass spectroscopy (LC–MS)
  metabolic fingerprinting, TIAs, 260–262
  supercritical fluid extraction, TIAs, 240–242, 241*t*
  ultra high-pressure, 243–248
Liquid chromatography coupled to UV (LC–UV), 236

M

MAO. *See* Monoamine oxidase (MAO)
Marine bacteria
  alkaloids
    aqabamycins, 318
    bisucaberin, 319
    6-bromoindole-3-carbaldehyde, 315
    holomycin, 318–319
    pentabromopseudulin, 315
    pyrroloquinoline alkaloid marinoquinoline A, 318
    tetrabromobenzofuro(3,2-b) pyrrole, 315
    thiomarinols, 317–318
    violacein, 315–317
  barbamide biosynthesis, 323–324
  diazepinomicin biosynthesis, 322–323
  holomycin biosynthesis, 324–325
  salinosporamide A biosynthesis, 325–327
  staurosporine and related bisinolde alkaloids biosynthesis, 321
  violacein biosynthesis, 319–320
Mass spectrometry (MS)
  based fingerprint analysis, 262
  and CE, 254
  data libraries, 263

Mass spectrometry (MS) (*Continued*)
GC–MS (*see* Gas chromatography to mass spectroscopy (GC–MS))
LC–MS (*see* Liquid chromatography coupled to mass spectroscopy (LC–MS))
and NACE, 254
Medicinal plants, 41–42
MEP pathway. *See* 2C-Methyl-D-erythritol-4-phosphate (MEP) pathway
Metabolic pathway engineering
alkaloid biosynthesis
branch pathway, 173, 174*f*
*Cj*6OMT, 171
drugs development, 173
ectopic expression, *C. japonica*, 171, 172*f*
endogenous pathway, 175–176
Eschscholzia californica cells, 173
gene silencing, 173
metabolite diversity, 173, 174*f*
overexpression, endogeneous gene, 171
rate-limiting step enzyme/transcription factors, 175–176
RNAi, 175
transcription factors, 171–173
trimming, 175–176, 175*f*
MIA biosynthesis pathway
mutant STR, 26
non-natural MIAs, 26
and ORCA3, 24–26
Saccharomyces cerevisieae, 27
TAs
Agrobacterium rhizogenes, 59–60
biotic and abiotic elicitors, 60–62, 61*t*
multigene construct transformation, 60
plant defense responses induction, 60–62
salicylic acid (SA), 60–62
scaling-up, hairy root cultures, 62–63
scopolamine and atropine, 59–60
Scopolia parviflora, 60
2C-Methyl-D-erythritol-4-phosphate (MEP) pathway
IPP to secologanin, 146–147
isopentenyl diphosphate formation, 145–146
non-mevalonate, 143
Methyltransferases, 276–277
Mevalonate (MVA) pathway, 145
MIAs. *See* Monoterpene indole alkaloids (MIAs)
MIAs biosynthesis
C. roseus, 4
description, 2–4
early MIA biosynthesis
biosynthetic genes, 12–14
gene regulation, 15–16
localization, 14–15
early monoterpene biosynthesis
biosynthetic genes (*see* Biosynthetic genes, early monoterpene biosynthesis)
gene regulation, 9–10
localisation, 9
homology-based cloning approaches, 4–5
iridoid biosynthesis (*see* Iridoid biosynthesis)
large-scale genomic approaches, functional characterisation
candidate genes screening tools, 22–24
shared pathways, Apocynaceae family, 21–22
transcriptome sequences, 21
late biosynthesis pathway
biosynthetic genes, 16–17
vindoline biosynthesis, 17–18
medicinal plant genome projects, 4–5
metabolic engineering (*see* Metabolic pathway engineering)
organisation and spatial separation
epidermis, biosynthetic site, 19–20
epidermis-enriched transcriptomic resources, gene discovery, 20
multi-cell-type coordination, 19
Micrococcus luteus, 186–190
Mitragyna speciosa
buffer systems and CE methods, 255*t*
LC systems, alkaloids, 244*t*
NACE–MS method, 254
Molecularly imprinted polymers (MIPs)
emerging technique, 239
preparation, 239
Monoamine oxidase (MAO)
biosynthetic pathway, 187–189
M. luteus, 186–187, 189–190
tyramine, 193

Monoterpene indole alkaloids (MIAs)
biosynthesis (*see* MIAs biosynthesis)
biosynthetic origin
complex alkaloids, 79–80
C. roseus and *R. serpentina*, 80, 81*f*
strictosidine β-glucosidase (SGD), 80
tabersonine metabolism, 80–82, 82*f*
distribution, angiosperm (*see* Angiosperm, MIE distribution)
evolutionary origin
cytochrome P450 enzyme, 83
hydrolysis, strictosidine, 85
iridoids, 83–84
monophyletic and non-monophyletic origin, 82–83
phylogenetic analysis, 85
Pictet-Spengler-type condensations, 85
secologanin biosynthesis, 83–84
strictosidine synthase-like (SSL), 83–84
tryptophan decarboxylase (TDC), 83
structural diversity
'cyclovincosan' skeletons, 76, 77*f*
dimer formation, 76
tryptamine subunit, 75
vincosans, 76
MS. *See* Mass spectrometry (MS)
MVA pathway. *See* Mevalonate (MVA) pathway

N

NACE. *See* Non-aqueous capillary electrophoresis (NACE)
NADPH:cytochrome P450 reductases (CPRs), 145*f*, 146–147
"Natural Viagra," 43
NCS. *See* Norcoclaurine synthase (NCS)
NOESY. *See* Nuclear Overhauser effect spectroscopy (NOESY)
Non-aqueous capillary electrophoresis (NACE)
application, 254
buffer systems and CE methods, TIAs, 254, 255*t*
MS method, 254
Vinca alkaloids, 254
Non-ribosomal peptide synthatase (NRPS), 323–324, 325
Norcoclaurine 6-*O*-methyltransferase (6OMT), 189
Norcoclaurine synthase (NCS)
C. japonica, 187–189
*Cj*NCS1, 165–166
(R,S)-reticuline, 190
*Tf*NCS, 165–166
types, 187–189
Nothapodytes foetida
bark and seeds, 150
cell suspension cultures, 151–152
cotyledons, 151–152
CPT (*see* Camptothecin (CPT))
cytochrome P450 proteins, 146–147
endophytic, 153
NRPS. *See* Non-ribosomal peptide synthatase (NRPS)
Nuclear Overhauser effect spectroscopy (NOESY), 260
Nuclear time bomb
'detonator' activity, *Catharanthus* cells, 100
glycoside hydrolysis, 99–100
Ligustrum obtusifolium leaves, 99–100
strictosidine aglycone, 100

O

6OMT. *See* Norcoclaurine 6-*O*-methyltransferase (6OMT)
Ophiorrhiza
O. alata, 151, 152–153
O. japonica, 147–148, 155
O. mungos, 142, 150–151
O. pumila
callus culture, 152–153
cell suspension culture cells, 149
CPT, 152–153
cytochrome P450 proteins, 146–147
hairy root cultures, 147–148, 149, 154
stems and roots, 147–148
STR expression, 147–148
suspension culture, 151–152
TDC, 144

P

PAL. *See* Phenylalanine ammonia lyase (PAL)
Pathogens. *See* Stress tolerance

PDE. *See* Phosphodiesterase (PDE)
Pentabromopseudulin, 315
PEP. *See* Phosphoenolpyruvate (PEP)
PEPS. *See* Phosphoenolpyruvate synthetase (PEPS)
PGPR. *See* Plant growth-promoting Rhizobacteria (PGPR)
Phenylalanine ammonia lyase (PAL), 288
Phosphodiesterase (PDE), 285–286
Phosphoenolpyruvate (PEP), 190–191
Phosphoenolpyruvate synthetase (PEPS), 191
Phosphotransferase system (PTS), 191
Pictet-Spengler-type condensations, 85
Plant growth-promoting Rhizobacteria (PGPR), 225, 226
Plant isoquinoline alkaloids
  biosynthesis and pathway characterization
    Agrobacterium rhizogenes, 165
    benzophenanthridine, 166–170, 169*f*
    Coptis and *Berberis* enzyme, 170–171
    corytuberine synthase, 166–170, 167*f*, 169*f*
    enzymatic reaction, 165–166, 167*f*
    homologues, 170–171
    Ipecac OMT alignment, 170–171, 170*f*
    Lithospermum erythrorhizon, 165
    L-tyrosine, 165–166, 167*f*
    molecular biology, 165
    morphinan alkaloid, 166–170, 169*f*
  combined analysis, 176
  low-molecular-weight chemicals, 164–165
  metabolic pathway engineering (*see* Metabolic pathway engineering)
  and microbial cells, 176–177
  productivity and metabolic profiles, 164–165
  "retrobiosynthetic" systems, 176–177
  structural biology, 176
  transcriptome and proteome databases, 176
Plant material, SMs
  active compounds identification, 208
  agronomical strategy, 209–210
  chemical synthesis, 208
  Datura innoxia and *Ruta graveolens* species, 210
  entire plant, 208–209, 209*f*
  FCs production (*see* Furanocoumarins (FCs) production)
  in vitro culture techniques, 208–209
  omics methods, 208
  overproducing plants and field cultivation trials, 209
  process engineering approach, 209–210
  tropane alkaloids production (*see* Tropane alkaloids (TAs), production)
Post-strictosidine pathway
  CPT biosynthesis, 149
  synthesis, 147–148
PPA. *See* Prephenate (PPA)
Prephenate (PPA), 190–191
Priming. *See also* Immunization
  BABA and, 291–292
  caffeine concentration, 293
  defence response activation, 293
PTS. *See* Phosphotransferase system (PTS)
Purine alkaloids
  biosynthesis
    caffeine (*see* Caffeine)
    cellular purine nucleotide pools, 123
    N-methyltransferases, genes and molecular structure, 124–125
    radio-labelled precursors, activity estimation, 123–124
    SAM route, 123
    xanthine and uric acid skeletons, 115
    xanthosine, 115–120
  biotechnology, 129–131
  caffeine, catabolism, 126–128
  distribution
    subcellular, 128–129
    in tissues, 128
  in planta function, 131–132
  occurrence, plant kingdom, 113–115, 114*t*
  structures, 112*f*
Pyrroloquinoline alkaloid marinoquinoline A, 318

Q

qNMR. *See* Quantitative nuclear magnetic resonance (qNMR)
Quantitative nuclear magnetic resonance (qNMR), 259–260

R

Rauvolfia serpentina
STR, 148
TIA-producing plants, 147–148, 149
Reticuline production
biosynthetic pathways, 186–187, 188*f*
E. coli, 189–190
IQAs, 186
MAO, 187–189
natural biosynthetic pathway, 186–187, 187*f*
NCS, 190
norcoclaurine 6-*O*-methyltransferase (6OMT), 189
plant cytochrome P450 enzymes, 186
S-adenosyl-L-methionine (SAM), 189
simple carbon source, *E. coli*, 194
synthetic biological approach, 186–187
RNAi. *See* RNA interference (RNAi)
RNA interference (RNAi)
BBE, 175
short interfering (si), 173
ROESY. *See* Rotating-frame nuclear Overhauser effect correlation spectroscopy (ROESY)
Rotating-frame nuclear Overhauser effect correlation spectroscopy (ROESY), 254–259, 260
Ruta graveolens
cultivation, 222
furanocoumarin, 215*f*, 216, 219*f*
furocoumarins, 212*f*
growth rate and psoralen content, 214
in vitro cultures, 212
temporary immersion, 220*f*

S

SA. *See* Salicylic acid (SA)
Saccharomyces cerevisiae
co-culture bioconversion system, 195
complex IQAs, 194–195, 196*f*
S-adenosyl-L-methionine (SAM)
E. coli cells, 190
polyketide and biodiesel production processes, 189
Salicylic acid (SA)
caffeine signal cascade, 289
induction, 284
Salinosporamide A biosynthesis, 325–327
SAM. *See* *S*-adenosyl-L-methionine (SAM)
(S)-scoulerine 9-*O*-methyltransferase (SMT), 171, 173
Secologanin synthase (SLS), 146–147, 149, 153–154
Secondary metabolites (SMs)
agronomical bioproduction, 216
application, 206
bioproduction and selection
challenges, 212–213
chemical analysis, 213–214
fractionation, 212–213
plant genotype evaluation, 214–215
techniques and methods, 212–213
biosynthetic factory, 228
biotechnological approaches, 217–218
complex nature, 207
compounds, 206–207
de novo creation, 227
description, 206
electric fields, 228
engineering and 'bioreactor' design (*see* Bioreactor)
environmental optimization, 226
extraction and purification processes, 227
FCs production (*see* Furanocoumarins (FCs) production)
field and biotech production, 222–223
genetic manipulation, 227
greenhouse conditions, 226
hydroponics and aeroponics, 223–225
natural plant origin, 207
PGPR and non-PGPR bacteria, 226
plant material (*see* Plant material, SMs)
possible levels, growth, 226, 226*t*
research activity, 207–208
'sacrificed biomass', 228
supercritical extractive processes, 227
tropane alkaloids production (*see* Tropane alkaloids (TAs), production)
SFE. *See* Supercritical fluid extraction (SFE)
Shikimate pathway. *See* Tryptamine pathway
Signal cascade, caffeine
γ-aminobutylic acid, 291
calcium, 288

Signal cascade, caffeine (*Continued*)
route, 289–291, 290*f*
salicylic acid, 289
SLS. *See* Secologanin synthase (SLS)
SMs. *See* Secondary metabolites (SMs)
Solanaceae. *See* Tropane alkaloids (TAs), production
Solid-phase extraction (SPE)
application, 251
clean-up procedures, 248–251
and MIPs, 239–240
polypropylene, 239–240
selective extraction, TIAs, 238
Spatial organisation, MIA biosynthesis
biological function, 101
metabolites compartmentation, 93–94
nuclear time bomb, 99–101
specialised cells compartmentation
C. acuminata, 95–96
C. roseus, 81*f*, 94–95
TDC, STR and MAT transcripts, 95
subcellular organisation
D4H and DAT, 98–99
in silico analysis, biosynthetic enzyme sequences, 96
loganic acid methyltransferase, 96–98
plastids, 96–98
strictosidine synthesis, 96–98
tryptophan and geraniol, 96–98
SPE. *See* Solid-phase extraction (SPE)
Statistical total correlation spectroscopy component analysis (STOCSY–CA), 261
Staurosporine and bisinolde alkaloids biosynthesis, 321
STOCSY–CA. *See* Statistical total correlation spectroscopy component analysis (STOCSY–CA)
Streptomyces
alkaloids, 304*t*, 305*f*
holomycin, 324
mansouramycins A–D, 306
marmycins A& B, 306
secondary metabolites, production, 302–303
spiroindimicins A–D, 306–307
venezuelines, 307
Stress tolerance
anti-herbivores, 280–282
anti-pathogens
hypersensitive response, 282–283
lesion development, 282
necrotrophic fungus, 282
TMV, 282–283
viral pathogens, 282–283
aphids, 282
biotic, 281*f*
tobacco cutworm caterpillars, 280–281
Supercritical fluid extraction (SFE), 240–242

T

TDC. *See* Tryptophan decarboxylase (TDC)
Temporary immersion system (TIS), 219–221
Terpenoid indole alkaloids (TIAs)
advantages, 235–236
analytical methods
and CE, 253–259
and GC-MS, 248–251
HPLC and hyphenated techniques, 242–243
and HSCCC, 251–253
and qNMR, 259–260
ultra high-pressure LC–MS, 243–248
analytical system, 236
application, 235
common chromatographic/ electrophoretic techniques, 262–263
extraction, 236–237
ionic liquids, 239
and MIPs, 239–240
MS-data libraries, 263
natural products, 235
NMR and LC–MS-based metabolic fingerprinting, 235–236, 260–262
quantitative analysis, 236
reversed-phase, 235
separation and identification, 235
solubility, 236–237
solvent extraction, 237–238
stationary phase, 235
supercritical CO_2, 240–242
UV and DAD detectors, 262–263
Tetrabromobenzofuro(3,2-b) pyrrole, 315

Tetrahydroprotoberberine oxidase (THBO), 165–166, 194–195
TH. *See* Tyrosine hydroxylase (TH)
THBO. *See* Tetrahydroprotoberberine oxidase (THBO)
Theobroma cacao, 112
Theobromine synthase, 119
Thin-layer chromatography (TLC), 213, 235, 236–237
Thiomarinols, 317–318
TIS. *See* Temporary immersion system (TIS)
TLC. *See* Thin-layer chromatography (TLC)
TOCSY. *See* Total correlated spectroscopy (TOCSY)
Total correlated spectroscopy (TOCSY), 260
Transgenic plants
  caffeine, 129–131, 132
  caffeine production
    HPLC, examination, 279–280
    purine alkaloids, accumulation, 279–280
  caffeine synthase gene, 130*t*, 133
  transgenic tobacco and chrysanthemum
    Agrobacterium transformation method, 278–279
    kanamycin-resistant transgenic plantlets, 278–279
    multi-gene transfer vector, 279*f*
  types, 129–131
Tropane alkaloids (TAs)
  atropine, 41–42, 41*f*
  biosynthesis (*see* Biosynthesis, TAs)
  calystegines, 44
  Catuaba, 43
  core structure, 40, 40*f*
  Datura species, 41–42
  description, 40
  Erythroxylaceae plants, 42*f*, 43
  metabolic engineering (*see* Metabolic pathway engineering, TAs)
  plants
    Arabidopsis thaliana, 48
    Brugine, 48–49, 49*f*
    calystegines, 47
    cocaine, 47–48
    Convolvulaceae, 46–47
    dicotyledon lineages, 44–46
    pyranotropanes, 48
    Rhizophoraceae genera, 48–49
    scattered distribution, angiosperms, 44–46, 45*f*
    scopolamine, 48–49
    Solanaceae, 46
    truxillines, 47–48
  production
    biological methods, 210–211
    chemical structure and biosynthetic pathway, 210–211
    entire plants, 211
    in vitro production, 211
    industrial purposes, 210–211
    medicine, 210–211
    plants, 210–211
    promising system and hairy root cultures, 211
  scopolamine, 41–42, 43–44, 53*f*
  sequestration, 44
  Solanaceous plants, 41–42, 43
  TA cocaine, 40–41, 53*f*
Tryptamine pathway
  biosynthetic, 143, 143*f*
  C. acuminata, 143–144
  CPT, 143–144
  MEP pathway (*see* 2C-Methyl-D-erythritol-4-phosphate (MEP) pathway)
  TDC1 and TDC2 isoforms, 144
  TSB mRNA and protein levels, 144
Tryptophan decarboxylase (TDC)
  C. acuminata TDC1 and TDC2 isoforms, 144
  CPT biosynthesis, 153–154
  molecular masses and positive correlation, 149
Tyrosine hydroxylase (TH), 192
L-Tyrosine production
  conversion
    biosynthetic pathway, 191–192
    HpaBC, 193
    L-DOPA, 192
    plants, 193
    Pseudomonas putida, 193
    Ralstonia solanacearum, 193–194
    ScTYR, 193
    tyrosinase, 193–194

L-Tyrosine production (*Continued*)
simple carbon sources
biosynthesis, 190–191
isolation, 190
PEPS, 191
plasmids use, microbial production, 191, 192*f*
shikimate pathway, 190–191
tyrR gene, 191
via *(S)*-norcoclaurine, 186

U

UHPLC. *See* Ultra high-pressure liquid chromatography (UHPLC)
Ultra high-pressure liquid chromatography (UHPLC), 243–248

V

VIGS. *See* Virus-induced gene silencing (VIGS)
Vindoline biosynthesis
localisation, 17–18
modulation, MeJA and light, 18
Violacein
apoptosis, 317
biosynthesis
bisindole alkaloids and, 320
genes, 319–320
pathway, 319–320, 320*f*
Chromobacterium violaceum, 315–317
Virus-induced gene silencing (VIGS)
advantage, 23–24
description, 24
gene discovery process, 28
gene functional analysis, 23–24
technology, 24
VOCs. *See* Volatile organic compounds (VOCs)
Volatile organic compounds (VOCs), 294

X

Xanthosine, 276
Xanthosine monophosphate (XMP), 118
XMP. *See* Xanthosine monophosphate (XMP)

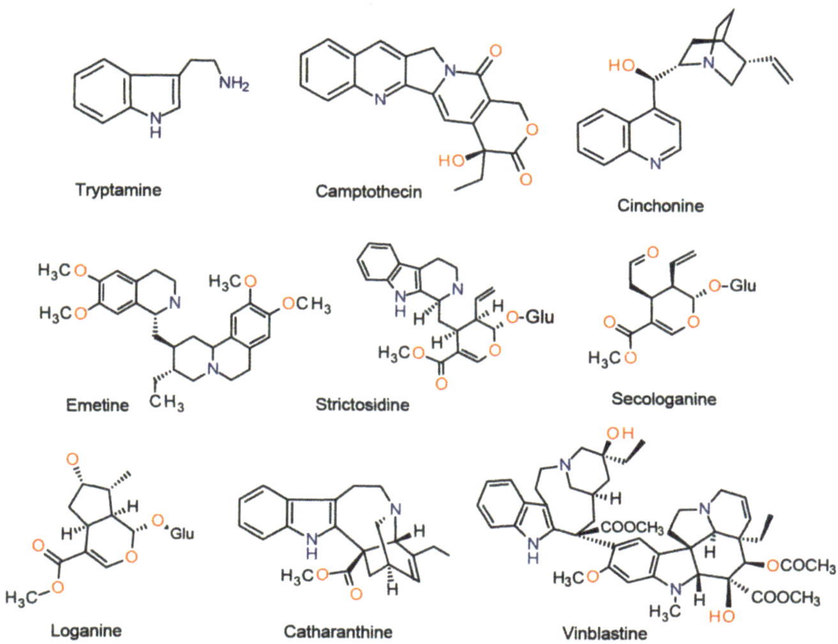

Benoit St-Pierre *et al.*, Figure 3.1 Selected monoterpenoid indole alkaloids and modified MIAs.

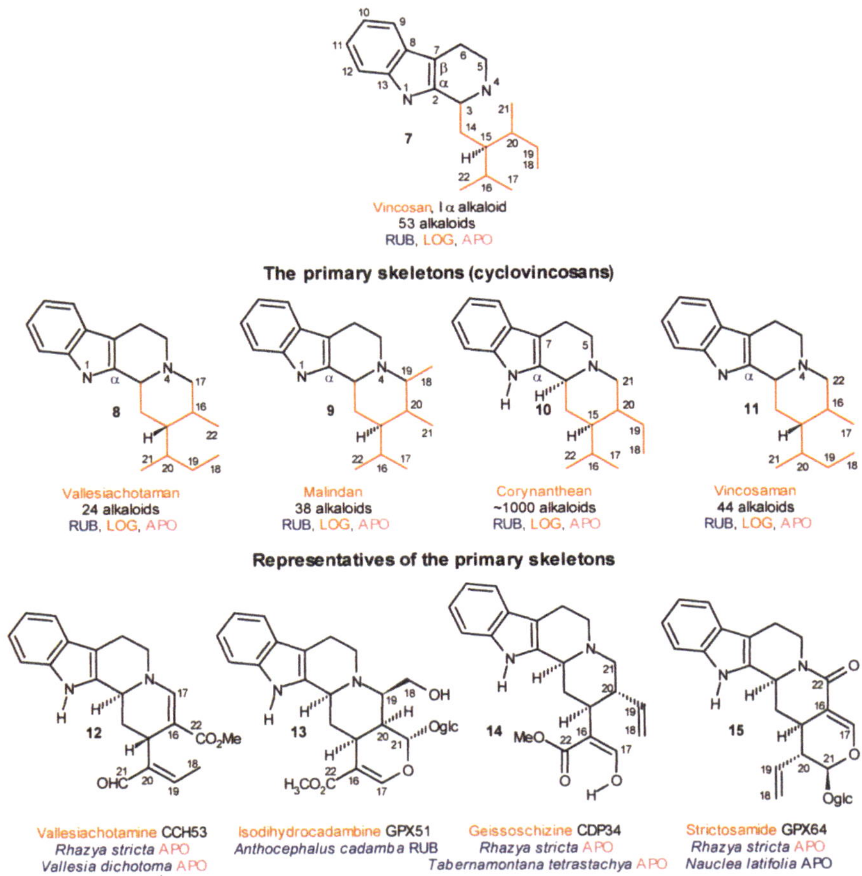

Benoit St-Pierre *et al.*, Figure 3.2 The 'cyclovincosan' skeletons and some representatives. Indole alkaloids types, family of the plant source and number of alkaloids isolated in each type. Isolated compounds are provided with Chapman and Hall (C&H) code numbers. *Rubiaceae* (RUB), *Loganiaceae* (LOG) and *Apocynaceae* (APO). The biogenetic numbering system shown in formula of strictosidine was used. *Reprinted from Szabó (2008); Published under the Creative Commons Attribution License by MDPI AG.*

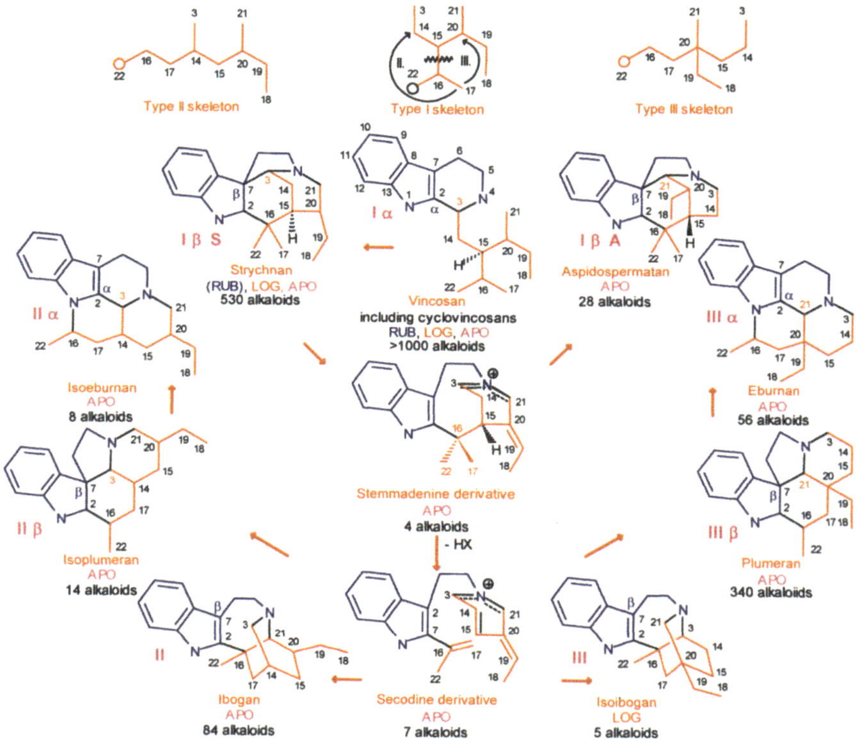

Benoit St-Pierre *et al.*, Figure 3.3 System of monoterpenoid indole alkaloids derived from secologanin. Indole alkaloids types, family of the plant source and number of alkaloids isolated in each type. Type numbers indicate basic skeleton type and subtypes. *Rubiaceae* (RUB), *Loganiaceae* (LOG) and *Apocynaceae* (APO). Arrows (red) indicate the chemotaxonomic connections. The biogenetic numbering system shown in formula of strictosidine was used. *Reprinted from Szabó (2008); Published under the Creative Commons Attribution License by MDPI AG.*

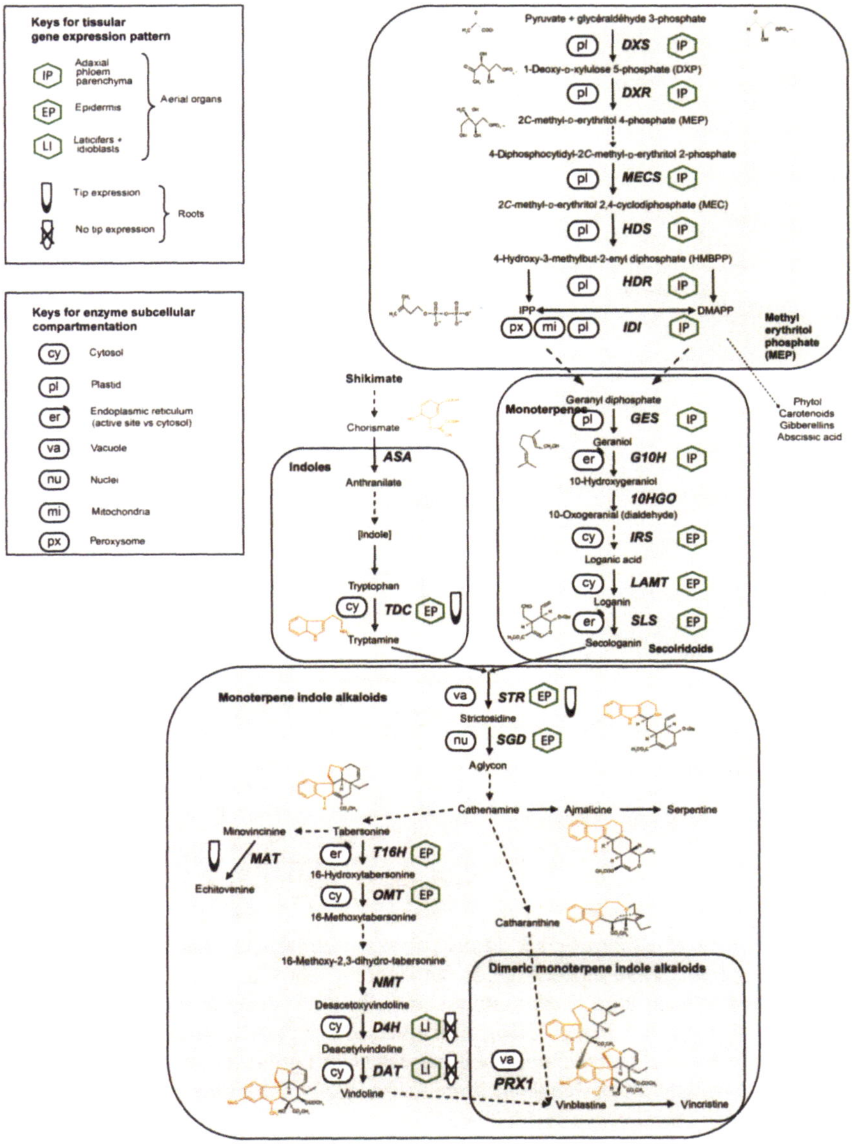

Benoit St-Pierre *et al.*, Figure 3.5 MIA biosynthetic pathway in *C. roseus*, subcellular compartmentation of enzymes and tissular localisation of transcripts. *(Data from Costa et al., 2008; Geu-Flores et al., 2012; Guirimand et al., 2009, 2012; Guirimand, Ginis, et al., 2011; Guirimand, Guihur, et al., 2011; Guirimand, Courdavault, Lanoue, et al., 2010; Levac et al., 2008; Mahroug et al., 2007; Murata et al., 2008; Oudin, Mahroug, et al., 2007; Simkin et al., 2013; Courdavault, unpublished results).*

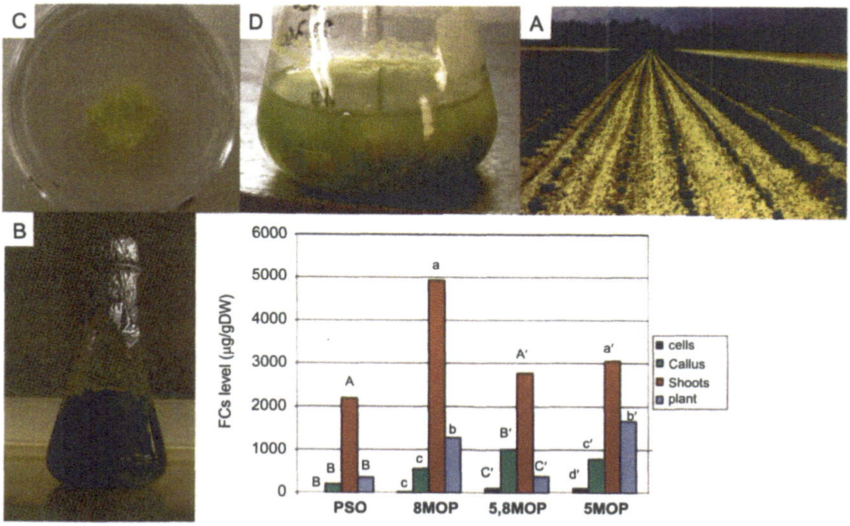

Thi Khieu Oanh Nguyen *et al.*, Figure 8.4 Furanocoumarin levels in *Ruta graveolens* entire plants (A) and *in vitro* cultures (shoots (B), callus (C), suspended cells (D)). PSO, psorelen; 8MOP, methoxy psoralen; 5MOP, 5methoxy psoralen; 5,8MOP, 5,8-dimethoxy psoralen.

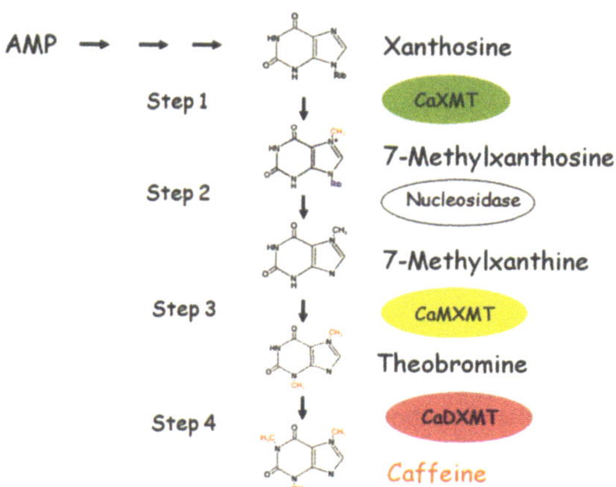

Hiroshi Sano *et al.*, Figure 10.1 Caffeine biosynthesis in coffee plant. Caffeine is successively synthesized from the precursor, xanthosine through three methylation and one ribose removal steps. The first (1), third (3) and fourth (4) steps are N-methylation, and the second step (2) is ribose removal. The enzymes involved are *Coffea arabica* xanthosine methyltransferase (CaXMT) (step 1), 7-methylxanthosine nucleosidase and/or xanthosine methyltransferase (CaXMT) (step 2), 7-methylxanthine methyltransferase (CaMXMT) (step 3) and 3,7-dimethylxanthine (theobromine) methyltransferase (CaDXMT) (step 4).

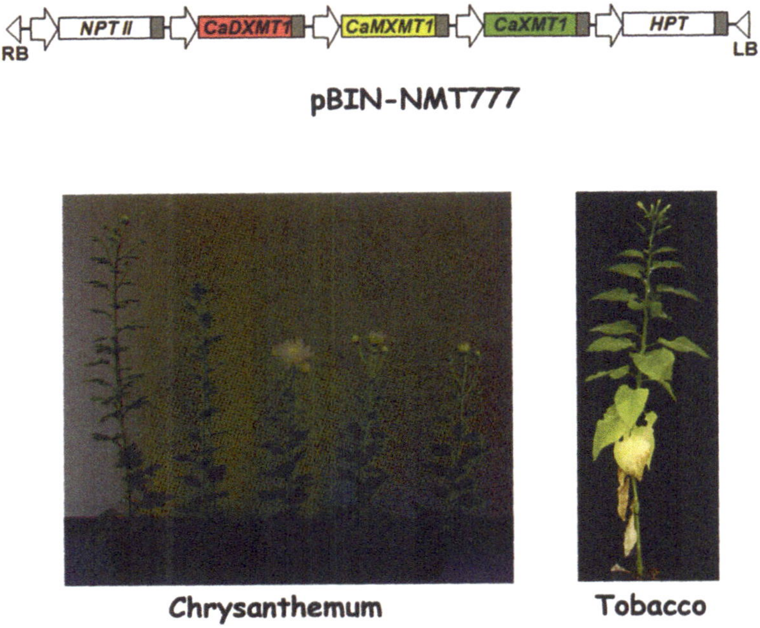

Hiroshi Sano *et al.*, Figure 10.2 Transgenic plants-producing caffeine. A multi-gene transfer vector (pBIN-NMT777) was constructed using *CaXMT1*, *CaMXMT1* and *CaDXMT1*, all were driven by the *35S* (cauliflower mosaic virus *35S* RNA) promoter and *NOS* (nopaline synthase) terminator. The T-DNA region of the plasmid also contained genes for neomycin phosphotransferase (*NPT II*) and hygromycin phosphotransferase (*HPT*) as the antibiotic-resistance marker. The promoter is indicated by open arrows, and the terminator is indicated by shaded squares. The left and right borders are indicated by LB and RB, respectively (upper illustration) (Uefuji et al., 2005). The pBIN-NMT777 was introduced into chrysanthemum (bottom left) (Kim, Lim, Yoda, et al., 2011) and tobacco (bottom right) (Uefuji et al., 2005). Mature plants were apparently normal, but in chrysanthemum, dwarfism and early flowering were occasionally observed.

Hiroshi Sano *et al.*, Figure 10.3 Biotic stress tolerance. *Resistance to insect larvae*: Tobacco cutworm (*S. litura*) larvae at the third instar were subjected to a choice test between three leaf discs from caffeine producing (5 μg/g fresh tissue) (TR) and three leaf discs from wild-type (WT) tobacco plants (Uefuji et al., 2005). *Resistance to aphids*: The third instar cotton aphids were subjected to a choice test between transgenic (3 μg/g fresh tissue) (TR) and wild-type (WT) chrysanthemum whole plants for a week, then leaf detached and photographed (Kim, Lim, Kang, et al., 2011). *Resistance to fungal pathogen*: Detached healthy mature leaves of transgenic (3 μg/g fresh tissue) (TR) and wild-type (WT) chrysanthemum plants were inoculated with spores of grey mould (*B. cinerea*), incubated at 20 °C for 15 days and photographed (Kim, Lim, Yoda, et al., 2011). *Resistance to bacterial pathogen*: Healthy leaves from transgenic (2.3 μg/g fresh tissue) (TR) and wild-type (WT) tobacco plants were inoculated with *P. syringae* pv. *glycinea* and necrotic lesion was observed 24 h later (Kim & Sano, 2008). *Resistance to viral pathogen*: Healthy leaves from transgenic (2.3 μg/g fresh tissue) (TR) and wild-type (WT) tobacco plants were inoculated with tobacco mosaic virus, kept at 30 °C for 2 days and then transferred to 20 °C (temperature shift). Lesions on whole leaf were observed 3 days after temperature shift (Kim & Sano, 2008).

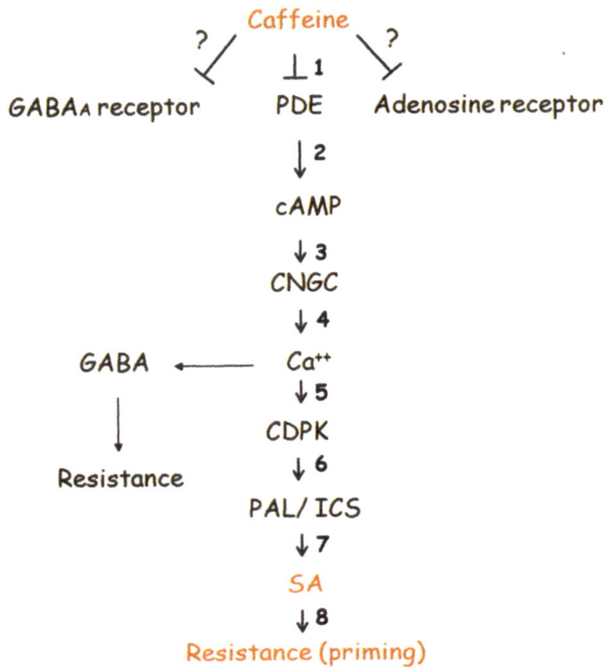

Hiroshi Sano *et al.*, Figure 10.4 Caffeine signal cascade. The molecular cascade of caffeine signal is predicted based on available literatures and experimental data. *Step 1*: Caffeine directly blocks phosphodiesterase (PDE) (Francis et al., 2011). *Step 2*: Degradation of cAMP is inhibited and its level increases (Ma et al., 2009). *Step 3*: Increased cAMP activates cyclic nucleotide-gated channel (CNGC) (Talke et al., 2003). *Step 4*: Activated CNGC increases cytosolic Ca^{2+} level (Kaplan et al., 2007). *Step 5*: Increased Ca^{2+} activates calcium-dependent protein kinase (CDPK) activity (Harmon et al., 2000). *Step 6*: Activated CDPK directly phosphorylates phenylalanine ammonia lyase (PAL) and/or isochorismic synthase (ICS) (Cheng et al., 2001; Coca & San Segundo, 2010). *Step 7*: Activities of phosphorylated PAL and ICS increase, resulting in acceleration of salicylic acid (SA) production (Dempsey et al., 2011). *Step 8*: Increased level of SA primes defence response (Conrath, Pieterse, & Mauch-Mani, 2002). The $GABA_A$ receptors and adenosine receptors have not been identified in plants. However, increased Ca^{2+} induces GABA accumulation, which enhances defence reaction (Bown & Shelp, 1997).

Kolibri 360 GmbH
Ferdinand-Jühlke-Straße 7
99095 Erfurt, Deutschland
produktsicherheit@kolibri360.de